Close Relationships

Close Relationships

HAROLD H. KELLEY, ELLEN BERSCHEID,
ANDREW CHRISTENSEN, JOHN H. HARVEY,
TED L. HUSTON, GEORGE LEVINGER,
EVIE McCLINTOCK, LETITIA ANNE PEPLAU,
and DONALD R. PETERSON

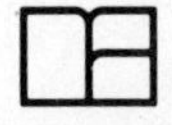

W. H. FREEMAN AND COMPANY
New York

Project Editor: Judith Wilson

Copy Editor: Mary Anne Stewart

Designer: Eric Jungerman

Production Coordinator: Linda Jupiter

Illustration Coordinator: Richard Quiñones

Artists: Catherine Brandel, Victor Royer

Compositor: Jonathan Peck, Typographer

Printer and Binder: The Maple-Vail Book Manufacturing Group

LIBRARY OF CONGRESS CATALOGING IN PUBLICATION DATA

Main entry under title:

Close relationships.

Bibliography: p.
Includes index.
1. Interpersonal relations. 2. Intimacy (Psychology)
I. Kelley, Harold H.
HM132.C535 1983 158′.3 82-25128
ISBN 0-7167-1442-6
ISBN 0-7167-1443-4 (pbk.)

Printed in the United States of America

2 3 4 5 6 7 8 9 0 MP 1 0 8 9 8 7 6 5 4

Contents

Preface

In August 1978, the nine of us met in Los Angeles for three weeks to plan a book on close interpersonal relationships. We shared a dissatisfaction with the status of close-relationship research and a sense of its potential form and importance. Our goals were to assess the current state of the field and to determine directions for future work. Given our collective familiarity with the relevant literature, we expected that during that three-week period we could develop a set of chapter topics and outlines that would make it possible to write such a book during the following year.

From our very first meeting, though, our discussions moved in unanticipated directions. Rather than proceeding directly to topics and outlines, we found ourselves constructing unordered and seemingly unorderable lists—of phenomena, of concepts, of popular and scientific questions, and of methodological issues. We soon became convinced that our joint enterprise would advance only if we could "get back to basics." That is, we felt a need to get beneath the confusing, multilayered, tangled details and down to the essential phenomena of close interaction.

In retrospect, some of the sources of our early confusion and dissatisfaction are clear. Because the phenomena of close relationships have been the subject of common thought and discussion for millennia, their scientific study is permeated with popular ideas and pseudoscientific conceptions. No close-relationship researcher approaches a problem without a deeply ingrained set of ideas and attitudes, and only with great difficulty can these preconceptions be put in their proper place. Perhaps the greatest source of difficulty is that these common conceptions inextricably interweave description and explanation. As a consequence, scientific analysis must cut through a thicket of popular vocabulary, experiential "facts," affective associations, and lay causal theories before its work of definition and measurement can begin. In the absence of such a ground-clearing effort, relationship researchers have not agreed on their definition of concepts, have not established precise operational definitions, and have often failed to understand the broad context of their separate limited labors. The result is that research has been disappointingly slow in cumulative growth.

Those are among the reasons why this book took three and a half years in preparation, why our work group found it necessary to meet again on two occasions (in August 1979 at the University of Massachusetts and in June 1980 at the University of California, Los Angeles), and why the book has its present form. As our difficulties became apparent, we decided we could not merely summarize the existing research literature. It was necessary first to be explicit about the core phenomena (for example, the criteria for a "close" relationship) and about the broad causal framework within which those phenomena occur. This meant writing, discussing, and rewriting our basic orientation, a statement of which, in its last incarnation, constitutes Chapter 2 of this book.

Our joint authorship of Chapter 2 reflects its evolution through repeated discussions and correspondence. We must emphasize that Chapter 2 does not constitute a theory of close relationships. It merely spells out, in what are intended to be descriptive, nontheoretical terms, what investigators of close relationships will inevitably study and, in general, how such phenomena must be studied. In keeping with the orientation of Chapter 2, the rest of the chapters emphasize *how* to think about various topics rather than *what* to think about them. Thus, each chapter primarily analyzes how to approach a particular problem area and is only secondarily a summary of current knowledge about that area.

The preparation of the later chapters proceeded in parallel with the development of the framework in Chapter 2. As a consequence, they use that framework in varying ways and with differing emphases. The primary responsibility for each chapter lay with the listed author(s), but every chapter went through several drafts—each time receiving commentary from most of us and extensive advice from at least two group members. To reflect this collaborative process, we have listed all nine of us as authors of the entire book, while at the same time attributing each chapter to the person(s) most responsible for it.

Every book must find its own audience, but we might mention whom we had in mind during our work. The book is addressed primarily to our professional colleagues—those who think about, investigate, and work with close relationships. Secondarily, it is intended for those students who, in advanced undergraduate classes or graduate seminars, are preparing themselves for research or casework on close relationships. Being psychologists, we have approached close relationships from a psychological perspective. Certainly, one of our aims has been to heighten awareness of the close relationship as an important specialty within psychology—one that cuts across such areas as clinical, developmental, and social psychology. We hope to inspire greater intellectual interchange among these areas as they deal with different types and features of close relationships. Equally important, we believe, is the relevance of our framework for the work of other behavioral and social scientists and practitioners. The framework is meant to capture all levels of analysis of the close dyad, including the social and cultural as well as the personal and interpersonal. It should, therefore, be useful to close-relationship researchers and workers in sociology, family life, social work, family medicine, and related fields.

Our meetings and the preparation of this book were supported by two small grants from the National Science Foundation, BNS 77-27165 and BNS 80-01173. The book would not have come into existence without the support of the Foundation and the facilitation of the grants by Kelly G. Shaver and Robert A. Baron. Certain burdens of our work fell on some shoulders more than on others. Kelley was primarily responsible for drafting and redrafting Chapter 2 as it developed from our many discussions. The tasks of preparing grant proposals, coordinating meetings and correspondence, and final assembling and editing the manuscript were carried out by the UCLA members of the group, Christensen, Kelley, and Peplau. Other persons and grants that played a supportive role for particular chapters are gratefully acknowledged there. We wish to give special thanks to Mary Anne Stewart for her contributions to the consistency and clarity of the entire manuscript through her intelligent and careful editing.

February 1983 HHK, EB, AC, JHH, TLH, GL, EMcC, LAP, DRP

Close Relationships

CHAPTER 1

The Emerging Science of Relationships

ELLEN BERSCHEID and LETITIA ANNE PEPLAU

Relationships with others lie at the very core of human existence. Humans are conceived within relationships, born into relationships, and live their lives within relationships with others. Each individual's dependence on other people—for the realization of life itself, for survival during one of the longest gestation periods in the animal kingdom, for food and shelter and aid and comfort throughout the life cycle—is a fundamental fact of the human condition.

THE CENTRAL ROLE OF CLOSE RELATIONSHIPS IN HUMAN LIFE

Most people are acutely aware that their relationships play a crucial role in shaping the character of their lives. Klinger (1977) found that almost all respondents to the question "What is it that makes your life meaningful?" said that friends were important, most mentioned parents or siblings or relationships with opposite-sex partners or with their own children, and most also mentioned the importance of "feeling loved and wanted." In contrast, less than half said that occupational success or religious faith was an important source of meaning to them.

It is not surprising, then, that people also believe that their personal happiness is integrally bound to the state of their intimate relationships. In a national survey, A. Campbell, Converse, and Rodgers (1976) found that most people consider it very important to have "a happy marriage," "a good family life," and "good friends." Less importance was given to work, housing, religious faith, and financial security. Other studies of what people believe is crucial to their well-being and happiness find similar results. For example, Freedman (1978) concluded from two large-scale surveys of factors associated with happiness:

> There is no simple recipe for producing happiness, but all of the research indicates that for almost everyone one necessary ingredient is some kind of satisfying, intimate relationship. Sex is not far behind in importance, and marriage, that venerable institution that is to some extent a combination of the two, is still, despite all the changes in our attitudes, a crucial factor in many people's happiness. People who are lucky enough to be happy in love, sex, and marriage are more likely to be happy with life in general than any other people. Those who are unhappy in this aspect of their lives are the least likely to have found general happiness. (p. 48)

That close relationships are indeed vital to well-being has been increasingly corroborated in recent years by research on factors associated with mental and physical health and longevity. For example, from their review of available data, Bloom, Asher, and White (1978) concluded that there is "an unequivocal association between marital disruption and physical and emotional disorder" (p. 886). Divorced adults are at severely greater risk for mental and physical illness, automobile accidents, alcoholism, and suicide. The mortality rate of divorced white American men under 65 as opposed to their married counterparts, to take just one comparison, is double for strokes and lung cancer, 10 times as high for tuberculosis, 7 times as high for cirrhosis of the liver, and double for stomach cancer, according to the American Council of Life Insurance (1978). Premature death from heart disease, too, is significantly more frequent among the "loneliness-prone"—the divorced, widowed, and single, both old and young (Lynch, 1977). In fact, people who lacked social and community ties were found to be twice as likely to die from any cause during a 9-year period as were people who had such relationships (Berkman & Syme, 1979). Studies directly assessing feelings of loneliness and social isolation further document the harmful consequences of deficient social relationships (see review by Peplau & Perlman, 1982).

People's personal relationships have implications that extend far beyond those directly experienced by the individuals themselves. Our lives are shaped not only by our own relationships, but also by those of other people. For example, the social and economic costs of divorce affect the entire society. The consequences of premarital sex and teenage pregnancy influence not only the adolescent parents and their children, but also social welfare programs and the community at large. In international relations, personal diplomacy and friendship between world leaders can change the course of history. The effectiveness of military combat units and of sports teams is influenced by the strength of group solidarity. On the job, personal relations among workers influence morale and productivity. All human society has a stake in the nature of people's close relationships. We all benefit from the existence of successful relationships and share, at least indirectly, the costs of relationship deficiencies.

Family Relationships

The family has a special place in thinking about close relationships. Family relationships are central to human existence, health, and happiness—a fact that is almost universally recognized. Over 90 percent of all Americans marry at some time in their life, and most people spend most of their adult life in a husband–wife relationship. In the family are found the very prototypes of the close relationship—the relationship between parent and child and the relationship between husband and wife. The family is also an important unit of social structure, a point widely acknowledged by American political leaders. As President Carter observed in convening the White House Conference on Families (1980), the family is "the foundation of American society and its most important institution," or, as President Reagan (1981) declared, "Work and family are at the center of our lives; the foundation of our dignity as a free people" (p. 4).

It is not surprising that family relations have often been the focus of public discussions of close relationships, or that scientific investigations of close relationships have so often examined the family. In recent years, interest in family relations has been spurred by dramatic changes in the character of the American family. These changes have given added impetus to public concern about the health of the family and provide a general sociohistorical context for the examination of close relationships in this book.

The traditional portrait of the American family (Skolnick, 1978) depicts the family as living comfortably in a single-family house in suburbia provided by the ambitious husband who sprints off each morning to his job, leaving behind his contented wife who prides herself on being an excellent homemaker and mother. Both husband and wife are confident that their relationship, founded on love and mutual understanding, will last until death does them part, and each works hard to create a happy home life for themselves and their two children, basically good kids who honor, love, and obey their parents as they struggle with the normal pains of growing up.

Although it is commonly assumed that this family pattern is fundamental to the American way of life, the so-called "traditional" American family is actually a relatively recent social invention that emerged in the 19th century (Ryan, 1979). In earlier times, the family was first and foremost an economic unit in which husband, wife, and children engaged in productive labor. Affection was less likely to be the basis of marriage (Degler, 1980), and the relationships among family members were considerably more formal, less companionate, and less child-centered than today (Gadlin, 1977). In the 19th century, however, the site of economic production shifted from the household, in which all family members contributed, to a physically separate workplace for each spouse. Increased specialization in the roles of husband and wife was the result, with the husband taking over the primary economic role as provider (Bernard, 1981b), thereby reducing his participation in childrearing and family life. As the wife's relative economic contribution to the family decreased, greater emphasis was given to her specialized skills in childcare and homemaking (Bernard, 1981a). In sum, "in the nineteenth century, popular culture for the first time deemed 'work' a male prerogative and in turn glorified the woman for her domestic and maternal functions" (Ryan, 1979, p. xvi).

The "traditional" American family composed of a breadwinner husband and a homemaker wife appears to have reached its cultural zenith in the 1950s (Skolnick, 1978), when it was frequently portrayed in the mass media in such television shows as *Ozzie and Harriet* and *Father Knows Best.* These depictions of family life exalted the virtues of family loyalty, love, and "togetherness." They also portrayed, often amusingly, relationship problems that, in retrospect, have an appealing simplicity. Parents worried about comforting their teenage daughter who failed to be elected homecoming queen or about telling their son the "facts of life"—not about their daughter's pregnancy or their son's arrest for selling drugs. Family crises involved dad's bringing the boss home for dinner or mom's upcoming speech to the garden club—not his alcoholism or her extramarital affair and certainly not their battle for custody of the children.

Idealized images of the American family represent the standard against which contemporary family relationships are often judged—and come up short. Family relationships in the United States today are straining to accommodate a divorce rate that has increased 700 percent since the turn of the century, with most of this increase occurring in the past two decades (Glick & Norton, 1977; Levinger & Moles, 1979). In 1978 alone, there were roughly a million divorces in the nation (Spanier, 1981), and the United States currently leads the world in the rate of divorce (United Nations Demographic Yearbook, 1982). Although this rate shows signs of leveling off, it has been estimated that about 40 percent of recent marriages will end in divorce (Glick & Norton, 1977). Thus, the dissolution of the husband–wife relationship, once a social rarity, is now commonplace.

Other far-reaching changes in the relations between husband and wife and between parent and child are reflected in the fact that American women have entered the paid labor force in ever-increasing proportions (Almquist, 1977;

Bane, 1976). Until the middle of this century, most women, if they worked at all for wages, did so before marriage or after the loss of a husband. In 1940, only 15 percent of married women worked for pay. By 1960, however, that figure had more than doubled, and, by 1975, 44 percent of married women were in the work force. Today, the figure is over 50 percent. This change has occurred not only among mothers with grown children but also among mothers of small children. As a result, only 16 percent of American families today fit the traditional image of a family consisting of a husband as sole wage earner, a wife as homemaker, and their children (H. S. Ross & Sawhill, 1975).

Alternative Relationships

Still other major changes in the social form and presumed substance of many close relationships have occurred. Several of these are reflected in dramatically increased sexual freedom. In the 1950s, for example, the sexual double standard was firmly entrenched. Whereas men were accorded greater sexual latitude both before and after marriage, "nice girls" maintained chastity before and fidelity afterwards. Thus, Americans were shocked to learn from Kinsey, Pomeroy, Martin, and Gebhard's (1953) landmark study that half of the married women surveyed had experienced premarital coitus (although often only with their future husband) and that over 25 percent of wives had had an extramarital affair by age 40. The public had been less surprised or distressed to learn in an earlier volume (Kinsey et al., 1948) that most men had had premarital coitus and half had had an extramarital affair.

In the 30 years since Kinsey et al.'s research, sexual attitudes and behavior have become increasingly permissive. The majority of young people today believe that premarital sex is acceptable for both women and men, especially in a "love" relationship (DeLora, Warren, & Ellison, 1981; M. Hunt, 1974). The actual incidence of premarital coitus has increased, most notably since 1965 (Hopkins, 1977). In 1980, about 75 percent of women and 90 percent of men were estimated to have had sexual intercourse prior to marriage (Reiss, 1980). For both sexes, age at first intercourse has substantially decreased, while the average number of premarital partners has increased.

Cohabitation between men and women has also increased. Between 1970 and 1979, the number of unmarried couples sharing living quarters more than doubled (Glick & Spanier, 1981), with a 41 percent increase in just the brief period from 1977 to 1979 (Spanier, 1981). Roughly half of all cohabiting couples are never-married adults, most of whom will eventually marry either their current partner or someone else, and 30 percent are divorced persons who will ultimately remarry. But, as the rise in cohabitation suggests, more and more persons are experimenting with alternative life styles—"alternative," that is, to traditional marriage (see Yankelovich, 1981).

Proponents of "swinging," "open marriage," child-free marriage, perpetual singledom, homosexual relationships, and so on have campaigned actively for increased social acceptance, with at least some success.

These statistics provide only the barest skeleton of some of the changes that have taken place in family relationships in recent years. Behind the statistics are millions of individuals who are currently experiencing fundamental changes in the ways they relate to other people. The rapidity of these changes suggests that basic aspects of American social life are in transition, and the end is not yet in sight. Americans are of two minds about the meaning of social changes in the family. Some, perhaps those who most value stability and security, interpret change in an imagery of decline, destruction, and loss. Characteristic are newspaper headlines lamenting that "census shows families are a dwindling species." Others, however, perhaps those who thrive on novelty and independence, view social change in terms of growth and positive movement toward a better life. There is probably some truth to both positions. Social changes often occur as solutions to old problems, but in turn create new problems of their own. Thus, divorce is a solution to individuals' desire to escape from an unsatisfactory marriage, but divorce raises new dilemmas about the stress of relationship dissolution and of life in single-parent families. Women's paid employment lessens the financial worries of many families and permits greater self-expression for women who find full-time homemaking stifling, but working wives create new problems of childcare and the realignment of responsibilities and power in the family. The diversity of family types creates greater options for individuals who feel dissatisfied in traditional marriage, but creates a greater need for public tolerance of diversity.

Current variations in family patterns and the increase of alternative relationship forms make it useful for researchers to expand their focus from "the family" to a more general examination of close relationships. The context of social change adds impetus to the quest for knowledge about relationships and lends a sense of urgency to the enterprise. The diversity in close relationships today also facilitates investigations of relationship dynamics. As Lewin observed (cited in Deutsch, 1954), the best time to study a phenomenon is when it is in the process of change, for it is when an entity is moving and changing that its dynamics reveal themselves most clearly.

This book reflects our belief, shared by others in the behavioral and social sciences, that many of the answers to questions now insistently raised about close relationships ultimately lie in the development of a science of relationships. Such a science will incorporate a body of knowledge about human relationships that can account for the varying forms that relationships take and will identify the forces that shape and are shaped by personal relationships.

A SCIENCE OF RELATIONSHIPS

The desire to understand close relationships is probably as old as humankind. "It seems likely that people have been listening to each other's family problems and responding with commiseration and advice as long as there have been families" (Broderick & Schrader, 1981, p. 5). Poets, philosophers, and religious leaders have long commented on human relations and offered prescriptions for interpersonal conduct. What is relatively new is the effort to study close relationships scientifically, to replace casual observation and intuition with systematic data collection and theory building.

One barrier to the development of a science of relationships has been a long-standing social taboo against systematic investigations of close relationships. For example, as sociologists E. W. Burgess and Wallin (1953) observed, prior to the First World War, most people refused to answer questions about their marital relationships, considering the topic too intimate, personal, and sacred to be discussed. Even as recently as the 1950s,

> love and marriage were regarded as belonging to the field of romance, not of science. The theory of romantic love held full sway, the predominant view was that in some mysterious, mystic and even providential way a person was attracted to his or her pre-destinate . . . that young people fell in love, married, and lived happily ever afterwards, as the result of some mystic attraction. Even when marriages turned out unhappily, the disillusioned explained their failures as being due to their having mistaken infatuation for love. Or else they placed the blame on bad luck or fate. (E. W. Burgess & Wallin, 1953, p. 11)

Today, however, there is much greater public interest in and support for scientific studies of relationships. Indeed, the popularity of books and articles on love, sex, marriage, and parenting; the continuing appeal of workshops on intimacy; and the increasing use of couples counseling all demonstrate that Americans are eager for factual information about close relationships.

The Interdisciplinary Origins of a Science of Relationships

A science of close relationships will be enriched by research and theory from many disciplines. Because relationships are shaped by their social environment, the work of anthropologists, historians, and other social scientists is important in describing and explaining variations in relationships across time and space. Because human relationships are influenced by each individual's biological capacities and predispositions, the work of biologists, ethologists, and other life scientists is also essential. Current work on the biological bases of attachment and dominance and on the evolutionary functions of altruism and parental involvement in the care of the young illustrate research that may provide new insights into human relations. At present, however, the groundwork for a science of human relationships has been most fully developed in the fields of sociology, marriage and family therapy, and psychology.

Sociology

Sociology has traditionally devoted much attention to social relationships. For example, Durkheim's (1897/1951) classic study of suicide presented the first evidence that marriage and social integration provide the individual immunity against suicide. In 1909, Cooley (1909/1962) drew attention to the "incomparable" influence primary groups have on people's lives. Sociologists have continued to examine social relationships in two major ways, the first focusing on the processes of social interaction and the second examining the social institution of marriage and the family.

With respect to social interaction, at least four major sociological perspectives can be distinguished. Structural functionalism (e.g., Merton, 1968; Parsons & Bales, 1955) examines how various components of society, such as the family, the educational system, and the economy, are interrelated and how each helps maintain the larger social system. Social exchange theory (e.g., Blau, 1964; Homans, 1961) analyzes social interaction in terms of such concepts as rewards, costs, and investments. Symbolic interactionism (e.g., Blumer, 1969; Mead, 1934; Stryker, 1980) emphasizes how people define social situations, give meaning to their own actions and to those of others, and create and negotiate roles in social interaction. Finally, a conflict perspective (e.g., Coser, 1954; Simmel, 1955) assumes that conflicts of interest occur not only between groups but also within them and focuses on such questions as who benefits from existing social patterns.

In addition to these efforts, sociologists have been especially interested in the institution of marriage and the family. Family research has grown prodigiously in the past 20 years (see Berardo, 1980), as illustrated by the fact that the National Council on Family Relations recently established a computerized data bank on family resources containing nearly 35,000 citations. Today, family research investigates a broad range of topics, including linkages between the family and the larger society, as seen in the effects of social class, urbanism, and industrialization on family life, and studies of sex roles, family violence, and nontraditional family forms, to name just a few current endeavors by sociologists to understand close relationships.

Marriage and family therapy

The field of marriage and family therapy has also made important contributions to our knowledge of close relationships. The early roots of relationship-oriented therapy can be found in the social work movement, the family life education movement, the work of pioneering "sexologists," and the development of social psychiatry (see historical reviews by Broderick & Schrader, 1981; Guerin, 1976; Kaslow, 1980; Olson, 1971). By the 1920s, physicians, lawyers, educators, and other professionals began to engage in marriage counseling as an adjunct to their regular practice. In the 1950s, professionals, many trained in analytic theory, began to formulate specific therapeutic approaches, including "conjoint marital therapy" (Jackson, 1959) involving both spouses and "family group therapy" involving the entire family. In the last 25 years, therapeutic work with couples and families has undergone extraordinary growth (Gurman & Kniskern, 1981). Today there is increasing agreement in the field that the family is best conceived as a social system, that relationship variables are critical to understanding the family, and that the observation of actual family interaction is often essential to successful intervention.

Work in marital and family therapy has generated new concepts, such as psychological symbiosis, pseudo-mutuality, and the double bind. Advances in theory development, perhaps most notably in family systems theory (e.g., Holman & Burr, 1980; Kantor & Lehr, 1975) have also been made. Therapy-oriented research has expanded in recent years and has become considerably more rigorous. For example, Patterson, Weiss, and others at the University of Oregon (e.g., Patterson, Reid, Jones, & Conger, 1975, and R. L. Weiss, Hops, & Patterson, 1973) have combined a family systems approach with a behavioral learning approach to study problems of aggression in children and of distress in married couples. Raush, Barry, Hertel, and Swain (1974) have examined patterns of interaction among couples in conflict situations. More recently, Gottman (1979) has used techniques of sequential analysis to investigate marital interaction. The burgeoning discipline of marriage and family therapy, which combines the efforts of sociologists, psychologists, psychiatrists, anthropologists, and others, constitutes one vital contributor to the developing science of relationships.

Psychology

Psychology is a third major contributor to the development of a science of relationships. Much of its contribution is currently indirect and lies in its effort to identify and understand the nature of the human animal. Such knowledge is fundamental to understanding human relationships for, as ethologist Hinde (1979) observes, one animal's responses to another and, therefore, many of the regularities in their interaction over time, are heavily determined by the mental and physical features, processes, and capabilities of the animal. In turn, of course, since many human characteristics are determined by the nature of social relationships, the knowledge contributed by a science of relationships is ultimately critical to the full development of psychology as well as many other of the behavioral and biological sciences.

Apart from this basic contribution to an understanding of human relationships, several subdisciplines of psychology have focused directly on relational phenomena. The contributions of clinical and counseling psychologists to marital and family therapy have been mentioned. Developmental psychology is increasingly examining the role that relationships play in human growth and development (e.g., Rubin & Hartup, in press). Special attention has always been given to the child–parent relationship and its effects on the child (e.g., Bowlby, 1973), but recent efforts to understand socialization have recognized that the direction of influence in such relationships is not unilateral. Thus, psychologists have sought to understand the mutual influence and interaction patterns between the child and its caretakers (e.g., Bell, 1968). In addition, the contribution of other early relationships, especially peer relations, to children's social and emotional growth and development is currently the focus of much investigative effort (e.g., Asher & Gottman, 1981; Hartup, 1983). As this focus implies, there is also an emerging interest among developmental psychologists in examining relationships per se, even apart from identifying their specific effects on individual development.

Social psychology, often regarded as the study of social influence and encompassing such traditional lines of inquiry as social power, attitude change, and social cognition, also engages questions of social relationships. Like other areas of psychology, however, social psychology often has taken the "individualistic" view of social phenomena (see Steiner, 1974). This view tends, first, to focus on the causes and consequences of a single person's responses to social stimuli at a single point in time and, second, to attribute the causes of individual responses to factors within the person rather than to factors in the social and physical environment. At least two major exceptions to this individualistic emphasis in social psychology exist, however. The first can be traced to the work of Lewin (1948), who viewed behavior as importantly influenced not only by the characteristics of the individual but also by the interaction between the individual and the environment. This "field" approach to human behavior, in which a change in any one part of the field reverberates to change other portions of the field, can be seen in Thibaut and Kelley's (1959) influential analysis of behavior in social relationships. A second exception has been a continuing interest by social psychologists in interpersonal attraction (see reviews by Berscheid, in press-a; Huston & Levinger, 1978). Recently, this line of inquiry has expanded to emphasize the processes of relationship formation, maintenance, and dissolution, and to investigate such relationship phenomena as self-disclosure, equity, power, and conflict (e.g., R. L. Burgess & Huston, 1979; Duck & Gilmour, 1981; Levinger & Raush, 1977).

THE CONCEPTUAL ANALYSIS OF RELATIONSHIPS

The sheer variety of sources that contribute to an understanding of close relationships has created obstacles to the development of a science of relationships. No two disciplines approach the analysis of relationships in precisely the same way or focus on precisely the same things. Each discipline differs importantly in the theoretical framework it brings to a relationship issue. Each framework operates as a lens through which a topic is viewed. As a consequence, even the same phenomena are frequently perceived in very different ways within different disciplines. Disciplinary differences in terminology, theoretical orientations, and levels of analysis can produce a situation in which "the conceptual jungle chokes the unwary" (Hinde, 1979, p. 6). It is to this central problem, the conceptual analysis of relationships, that our efforts in this book have principally been addressed.

The bedrock on which any science rests is observation and description of the phenomena of interest. Since descriptive analysis necessarily precedes causal analysis, we have emphasized its importance to an ultimate understanding of close relationships in each of the chapters to follow. Description provides the basis for a comprehensive causal analysis of relationship phenomena.

Descriptive Analysis

Description requires a descriptive language, commonly understood and tied to observables, that can be used to represent symbolically the phenomena of interest. A central problem that has impeded the development of a science of relationships is not the absence of a descriptive language, but rather that there are too many descriptive languages for relationships. Each of us in our daily lives uses an extensive "common sense" language to discuss relationship phenomena. Each discipline that treats the subject has also developed its own language of relationships that is heavily influenced by the terminology, concepts, constructs, and theories traditional to that discipline and to its particular focus on relational phenomena. Even subdisciplines within disciplines can vary considerably in the words used to describe a single relationship phenomenon.

When all of these languages meet under one interdisciplinary roof, it becomes painfully apparent that there is no one commonly understood descriptive language of relationships. Further, even the most skilled and dedicated efforts to translate between languages are all too often doomed to failure. First, the words in one language frequently have no clearly specified referents. Second, even when referents are specified, they frequently are not tied to observables, but rather only to other concepts and abstractions that themselves often have unspecified or nonempirical referents.

This problem is readily apprehended when one simply considers the variety of definitions and meanings currently given the word *relationship* and its qualifier *close.* Such words as *love, trust, commitment, caring, stability, attachment, one-ness, meaningful,* and *significant,* along with a host of others, flicker in and out of the numerous conceptions of what a "close relationship" is. The words used to explain the phrase *close relationship* often carry clouds of ambiguity, and so people are not infrequently driven to concrete single-case illustrations or to highly abstract analogies and metaphors to try to communicate what they mean by the term, often with little success. When investigators cannot agree on an issue so fundamental as when two people are in a "relationship" with one another, or on the basis for classifying a relationship as "close" versus "not close," then the development of a systematic body of knowledge about close human relationships becomes problematic indeed.

To circumvent these problems, we collectively agreed that our first task must be to identify the concepts that appear to be necessary and fundamental to the description of relationships, regardless of the disciplinary perspective, special interests, and other particulars of a specific investigator. Our approach to this task was to identify the basic data of relationships—the nature of the events that necessarily must be described and subsequently causally analyzed if relationships are to be understood. This approach quickly led to a consideration of the basic meaning of the word *relationship.*

When the myriad conceptions and usages of the term *relationship* are collected and carefully compared, it becomes apparent that the term essentially refers to the fact that two people are in a relationship with one another if they have impact on each other, if they are "interdependent" in the sense that a change in one person causes a change in the other and vice versa. Thus, as we discuss in Chapter 2, "Analyzing Close Relationships," the study of relationships is concerned with the interdependence between two people—with describing the quantity and quality of that interdependence over time and with identifying the causal factors that both affect and are affected by that interdependence.

It follows, then, that the basic data of relationships, the facts that must be recorded, described, and ultimately understood, concern the ways in which two people affect each other. These data must (1) identify the activities (e.g., the thoughts, feelings, actions) of each person that affect and are affected by the activities (thoughts, feelings, actions, and so on) of the other and (2) specify the nature of the effects of each person's activities on those of the other. In order to do the latter, that is, to identify the causal connections between the two persons' activities, it is necessary that observations be made of the pair over a considerable time period. Only in this way can a determination be made of which activities of each person are consistently affected by those of the other. Thus, the description must provide details of the temporal sequence of the two persons' activities. In brief, the descriptive analysis of relationships focuses on describing the number, nature, and temporal patterning of the interconnected activities that form the substance of social relationships.

Descriptive analysis necessarily precedes the classification of relationships into types, for example, as "close" or "not close." Classification presumes the observation and description of a series of interconnections and the assessment of certain of their properties. Precisely what properties of the interconnected pattern an observer will regard as important depends on the observer and the aims of the investigation. Chapter 2 outlines a number of properties that we believe many will regard as important.

For the purposes of this volume, the basis for classifying a relationship as "close" assumed special importance. Such a classification must be made on the basis of certain properties of the interaction pattern. We believe that a relationship may be profitably described as "close" if the amount of mutual impact two people have on each other is great or, in other words, if there is high interdependence. A high degree of interdependence between two people is revealed in four properties of their interconnected activities: (1) the individuals have *frequent* impact on each other, (2) the degree of impact per each occurrence is *strong,* (3) the impact involves *diverse* kinds of activities for each person, and (4) all of these properties characterize the interconnected activity series for a relatively long *duration* of time.

Whether or not the reader agrees with this classification scheme and the rationale presented in the next chapter that supports it, it should be understood that this is the referent for *close* used throughout this book. *Close,* as we use the term, is virtually synonymous with *influential;* people in close relationships have a great deal of impact on each other. Whether the impact is for good or ill for the individuals involved is a separate issue from classifying the relationship as close or not. So, too, is the question of whether the two people subjectively feel close or verbally report that they are close. There is little doubt that through interaction the individuals involved develop beliefs about their relationship (e.g., whether it is "close" or "superficial" or "happy" or "destructive") and about the partner (e.g., whether he or she is "sincere" or "loving"). The degree of correspondence between the participants' beliefs about the relationship and an investigator's description of the properties of actual relationship activities encompasses a large set of interesting questions. However, Chapter 2 emphasizes that, to be useful, relationship descriptors must ultimately be tied to properties of the interconnected activity pattern that can be recorded and agreed on by impartial investigators.

Causal Analysis

When a relationship is observed for a long time, there ordinarily will be detected certain regularities in the patterning of interconnected activities and, on certain occasions, major changes in these patterns will also be observed. It is these regularities and changes in interaction patterns that a science of relationships must ultimately predict and explain. Chapter 2 addresses this issue of causal analysis. As we elaborate, a causal analysis of interaction regularities requires the inference of relatively stable "causal conditions" that act on the relationship to produce and maintain these regularities. Similarly, an explanation of change in an interaction pattern requires the identification of changes that have taken place in previous stable causal conditions. For example, a descriptive analysis of one couple might reveal that, while one partner virtually always accedes to the requests of the other, the other rarely does so. A number of causal conditions may be tentatively invoked to account for this regularity, ranging from one person's greater "power," stemming from an ability to affect the other's economic outcomes, to "social norms" that prescribe deference on the part of the compliant person, or to such personality dispositions as "nurturance" or "autonomy." Each of these tentative explanations carries the implication that if the presumed causal condition changes (e.g., one person loses his or her source of income), then the interaction pattern will also change.

The sheer number and variety of causal conditions that may be invoked to explain specific regularities in interaction in close relationships stretches toward the infinite. Causal conditions may, however, be classified into several major groups. As discussed in Chapter 2, there are *personal* causes, e.g., relatively enduring characteristics of the individuals, such as their personality traits or abilities. When an interaction regularity reflects a particular combination of the dispositions of the two persons (e.g., a "nurturant" individual paired with a "succorant" other), or when the activity pattern is the product of their interaction (e.g., a particular mutual understanding or shared expectation that the two have evolved), we refer to *relational* causes. Finally, *environmental* causes refer either to features of the social environment or of the physical environment within which the relationship is embedded. Causal explanations that invoke such factors as "societal norms" or the "restrictive living quarters" shared by participants in a relationship are examples.

Thus, a relationship's causal context is formed by the dispositions of each person, by factors emerging from their combination or interaction, and by features of the social and physical environment. The interplay between two or more causal conditions, both within and across types, is, of course, frequently presumed to be responsible for many of the regularities of relationship patterns. It is doubtful, in fact, that any interaction pattern can be fully explained without reference to factors in all of the types of causal conditions and their interplay with one another.

The framework we present in Chapter 2 is intended to provide a general basis for the description and causal analysis of human relationships. In major part, all of the chapters that follow are concrete illustrations, elaborations, and extensions of that core chapter to various aspects of close relationships. The basic framework of Chapter 2 transcends, we believe, any one specific theory, any particular disciplinary perspective, and any single relationship phenomenon. Choices about theory, disciplinary approach, and topic are at the discretion of the individual investigator and must be made before any specific investigation can proceed. In contrast, we believe that the basic framework itself is not so discretionary and is largely dictated by the common principles and assumptions that underly all scientific endeavors. A careful reading of Chapter 2 is, thus, critical for understanding all of the topical discussions in the chapters that follow.

TOPICS IN CLOSE RELATIONSHIPS

Certain relationship phenomena have traditionally captured more attention than others, and, similarly, certain causal factors have been thought to account for more of the regularities in interaction patterns than others. In the remaining chapters of this book, we discuss some of the substantive and methodological issues that are currently of special interest to those attempting to understand close relationships. Although the range of topics we cover is broad, it is not at all encyclopedic. The necessity of limiting discussion to some issues and to specific facets of those issues at the expense of others was keenly felt. For this reason, each of the chapters should be considered not as a comprehensive review of the topic but rather as largely illustrative of the kind of conceptual analysis that each topic seems to require and of the kinds of approaches that currently seem promising.

Interaction

In Chapter 3, "Interaction," McClintock examines the ways in which social scientists have typically observed, recorded, and analyzed overt behavioral events to provide descriptions of the changes and regularities in activity patterns over time. This chapter also considers the covert cognitive activities that take place when people interact and that, together with affective and overt behavioral activities, comprise the dynamic process of interaction. In addition to demonstrating the importance of interactional analysis for any systematic understanding of relationships, this chapter also discusses some of the causal factors that may account for the recurring activity patterns that typically characterize close relationships.

Emotion

In Chapter 4, "Emotion," Berscheid discusses some of the problems and possibilities involved in understanding emotional phenomena as they occur—and sometimes inexplicably fail to occur—in close relationships. Close relationships are, of course, generally recognized to be the setting for the most dramatic and intense of human emotions. Unfortunately, the popular association between emotional experience and close relationships is most often manifested in the tendency to define a "close" relationship as one characterized by strong positive emotional experiences. We argue, however, that both the intensity and the positivity of an individual's emotional experiences in a relationship are inadequate and misleading indices of the closeness of the relationship. Using our conception of closeness and what is known about the dynamics of emotion, Chapter 4 introduces the concept of "emotional investment," this being the potential, rather than the actuality, of experiencing intense positive or negative emotion in a relationship. The chapter also discusses a number of other factors involved in the prediction and understanding of emotional phenomena within relationships.

Power

The concept of power is frequently invoked to explain a wide variety of relationship interaction patterns, yet this fundamental concept has been used in diverse and contradictory ways. In Chapter 5, "Power," Huston draws on the conceptual framework of Chapter 2 to clarify how previous researchers have used the concept and to provide a more comprehensive analysis of power. Huston examines patterns of influence in the momentary give and take of interaction and considers the causal conditions that enable one person to exercise power or intentional influence over another. The chapter concludes with a discussion of husband and wife decision making in marriage and of influence processes in parent–child interaction.

Roles and Gender

Close relationships are characterized by relatively consistent and comprehensive patterns of activity or roles. The distinctive roles of husband and wife are illustrative of such patterns. In Chapter 6, "Roles and Gender," Peplau uses the conceptual framework from Chapter 2 to describe the general nature of roles in close relationships and to outline the types of causal condition that influence role patterns. This conceptualization of roles is contrasted with previous perpectives on roles. To illustrate the description and causal analysis of roles, the chapter provides a detailed discussion of gender-based roles in dating and marriage and considers several explanations for gender-based role specialization.

Love and Commitment

Love and *commitment,* the focus of Chapter 7, are surely among the most frequently used words in discussions of close relationships. Each carries, however, a wide variety of meanings, both in popular parlance and among theorists and investigators. In this chapter, Kelley draws a distinction between love and commitment, showing the partial overlap between the two concepts that has led to confusion in their usage. Love and commitment are each analyzed in terms of the observable phenomena believed to be their characterisic manifestations, the current causes believed to be responsible for these observed phenomena, and various ideas about their origins and developmental course. The goal of this chapter is to illustrate how such complex phenomena as love and commitment can be dissected in terms of the basic conceptual framework presented in Chapter 2 in order to reduce their current ambiguities and to direct further theoretical and empirical effort.

Development and Change

Few issues are as challenging as those posed by an examination of the temporal development of close relationships. In Chapter 8, "Development and Change," Levinger proposes a temporal sequence for analyzing relationship development, beginning with acquaintance and ending with deterioration and termination of the relationship. The chapter considers the factors that may propel a relationship from a mere acquaintanceship to the high degree of interdependence characteristic of a close relationship. The fact that there are multiple influences on the developmental course of relationships is emphasized, along with the fact that there are multiple paths to increasing and decreasing interdependence. As part of this discussion, the chapter reexamines certain classic issues in relationship formation, such as filter models of mate selection and the importance of personality complementarity, and also discusses the impact of such events as parenthood and serious illness on relationship development.

Conflict

In Chapter 9, "Conflict," Peterson discusses some of the conditions that influence the initiation of conflictual interactions, as well as the conditions that affect the avoidance versus engagement of conflict and its escalation or resolution. Five possible outcomes of conflict, from separation of the partners to structural improvement of the relationship, are discussed. The function of conflict in the development of relationships is also considered, and ways in which patterns of progressive alienation or relationship growth may become established are illustrated. Finally, ways in which unnecessary conflicts can be reduced and other conflicts may be put to constructive use are suggested.

Intervention

Conflict, especially in marriage, often drives partners to seek outside intervention to improve and preserve their relationship. Intervention presupposes that a descriptive analysis of the relationship has been made, that the causal conditions responsible for current interaction patterns have been identified, and that these conditions can be effectively changed or modified through available intervention procedures. In Chapter 10, "Intervention," Christensen presents a conceptual analysis of dysfunction in close relationships. He also examines the treatment of relationship dysfunction, addressing issues of the client–therapist relationship, the assessment of close relationships, and intervention strategies used to alter those relationships. Current approaches to the treatment of distressed relationships are selectively reviewed, including those that focus on intervention with the individual, with both partners, and with the social environment. The chapter concludes with a brief review of the empirical literature and a discussion of the implications of our conceptual framework for clinical work with distressed relationships.

Research Methods

Methodological issues are raised throughout this volume. Such issues are unavoidable and central to an understanding of close relationships, since the adequacy of our methods determines the extent of our knowledge. In Chapter 11, "Research Methods," Harvey, Christensen, and McClintock discuss the major methods currently available for descriptive and causal analysis of relationships. Several observational and participant-report strategies for describing close relationships are discussed; the strengths and weaknesses of these approaches are examined using the conceptual framework presented in Chapter 2. Turning to causal analysis, correlational and experimental designs and several variations of these classic strategies are discussed and illustrated. The chapter uses the framework from Chapter 2 to illustrate that major questions about close relationships can be examined with existing methodologies. The chapter concludes by noting the reciprocal ties between methodological and substantive research.

A Science of Relationships

It can be argued that the development of a science of relationships is important not only in its own right, but because it is essential to progress in related sciences such as psychology and sociology. In Chapter 12, "Epilogue: An Essential Science," Kelley makes a case for this view. The framework of Chapter 2 and several research examples are used to show that close relationships must be taken into account if the dynamics of psychological and social change are fully to be understood. Furthermore, this understanding of the influence of close relationships, whether on the individual or on the society, requires that internal relationship dynamics be determined through longitudinal investigations of interaction processes.

The epilogue highlights again, as do many of the earlier chapters, the great need for painstaking descriptive analysis. At the same time, it is clear that the idea of dealing with time-series data, with the emergent properties of relationships, and with the reality that in ongoing relationships each variable is both independent and dependent—that many variables act and interact to affect each other and, then, ultimately themselves—is an alien idea to many of us. Further, in our roles as editors and peer reviewers, many of us traditionally have derogated the value of simple description and have placed a high premium on causal analysis, so high a premium, in fact, that causal analysis is sometimes encouraged even when it is premature. Along with new methodologies, new technologies, and new theories, perhaps some of us will also need a new attitude—about what we can determine and about how fast and how precisely we can determine it.

The quest to solve the mysteries of close relationships is a formidable task. This book may sensitize readers, even those who have long grappled with relationship issues, to the enormity of the effort of developing a science of relationships. But, against the difficulties that surround it, there lies the guarantee that the work is worthy of the effort it demands. No attempt to understand human behavior, in the individual case or in the collective, will be wholly successful until we understand the close relationships that form the foundation and theme of the human condition.

It is our hope that the reader of this volume will acquire not only an appreciation for the complex outlines of a science of close relationships, but also a sense that the task before us is "do-able"—not soon, and certainly not by any one discipline, but ultimately through the concerted efforts of many investigators. The emergence of a science of relationships represents a new frontier—perhaps even the last major frontier—in the study of humankind. The uncertainties and frustrations of exploration in this domain are surely matched by the excitement, the challenge, and the satisfaction of discoveries about familiar, yet little understood, phenomena.

CHAPTER 2

Analyzing Close Relationships

HAROLD H. KELLEY, ELLEN BERSCHEID,
ANDREW CHRISTENSEN, JOHN H. HARVEY,
TED L. HUSTON, GEORGE LEVINGER, EVIE McCLINTOCK,
LETITIA ANNE PEPLAU, and DONALD R. PETERSON

Any layperson or scientist comes to interpersonal relationships with a large set of preexisting ideas, concepts, labels, implicit and explicit theories, beliefs about causes of important phenomena, and expectations about consequences of various states or events. The complexity of existing scientific information bearing on interpersonal relationships can be seen by examining a recent propositional inventory of research on the family (Goode, Hopkins, & McClure, 1971). Some 3,000 propositions are listed that bear on the interpersonal relations found within and around the family (dating, husband–wife, parent–child, siblings, and so on). Each of these propositions is of the form "X leads to Y," or "X is associated with Y." For example, "the less the husband and wife depend on each other, the greater are their chances of splitting up"; "maternal possessiveness is negatively correlated with education of the mother"; "status differentials in the type of labor will frequently be a source of conflict between siblings." The 3,000 two-variable propositions in this format could be generated by a combination of as few as 78 variables (X's and Y's) or by as many as 6,000 variables (if each X and Y in each proposition were unique to it). A rough estimate suggests that the actual total of X and Y variables in Goode et al.'s inventory approximates 700. Thus, in this one encyclopedic source (whose 2,000-item bibliography is hardly complete), the student of interpersonal relationships is confronted with some 700 variables (terms, concepts, factors) and their possible interrelations. A dozen examples will suggest their variety: favoritism, aggression, ordinal position, schizophrenia, discipline, economic security, father absence, privacy, affinal relations, achievement, marital adjustment, and social change.

Faced with such a profusion of variables and propositions in the literature on relationships, we have found it necessary to go back to basics. We have adopted the approach of a visitor from outer space, attempting to specify what it is that such an alien, not having the extensive experience with human interpersonal relationships that all earthlings have, would see and hear, and what such an alien would do in the way of analysis, interpretation, and inference in order to make sense of these data.

Using this approach, we try to characterize the essentials of what the scientist does in the study of interpersonal relations. We especially focus on demarcating as clearly as possible the line between *description and data,* on the one hand, and *interpretation and theory,* on the other. This demarcation is both essential and difficult in any science, but it is particularly troublesome in interpersonal relations. As lifelong participants in these relations and as a daily audience for the extensive lore about them, earthbound scientists have their heads full of labels, theories, and so on, in which data and concepts are inextricably intertangled. Our common experience and common ideas inevitably afford materials for scientific insights and hypotheses. Some of our a priori ideas are undoubtedly useful leads to the truth, but others are wrong and take us down blind alleys. When we might choose to rely on them for economy's sake (they reduce the necessity for preliminary observation, piloting, and pretesting), we have few means of distinguishing the more useful from the less useful preconceptions. In fact, we cannot choose *not* to be influenced by them; they inevitably affect our work.

We present, then, a general approach to the study of interpersonal relationships. This is *not* a "theory" of interpersonal relations, but rather an outline of what one sees and hears, and thereby has available as data about interpersonal relations, and what one does with those data in the way of inference, interpretation, and theory building and testing. We have found it impossible to write about *what* we look at, and the terms in which we analyze and interpret our data, without at the same time implying *how* we do it—that is, the general methodology of research. In Chapter 11, "Research Methods," we discuss specific methodological issues in research on close relationships; but here, as we outline the logical nature of the data pertaining to these relations and the logic of their interpretation, we characterize the basic operations involved in such research.

STUDYING RELATIONSHIPS

The approach to be developed in this chapter can be illustrated by following an imaginary visitor from outer space who happens to become interested in the inner spaces of close relationships. The visitor, Dyas, is a descendant of Aphrodite, goddess of love, and Hermes, god of science.

Why would Dyas identify relationships, as between pairs of people, as something to study? Relationships do not have the clear identity and boundedness of physical objects (persons, rocks, flowers, animals). However, they *are* observable. Dyas would see two persons often being physically close, moving together, orienting toward each other, touching and talking to each other, and so on. Once such pairing had been detected and several or many instances of such pairings had been observed, Dyas could readily form the concept of "relationship," become interested in these entities as objects of study, and bring them under investigation in much the same ways as Dyas' scientific colleagues do for other entities.

Description

What might Dyas' research on relationships consist of? As in all science, it would certainly begin with *description.* Dyas' descriptive efforts would undoubtedly center on gaining information about the phenomena characterizing and associated with relationships. These would be the phenomena that caught Dyas' eye in the first place—the phenomena of interaction. In this book, we consider the elementary phenomena of interaction to be interpersonal patterns of events. An *event* is any change in a person, for example, in actions, speech, facial expressions, that an investigator of relationships may consider important. When the events for two persons are seen to occur in an interpersonal pattern, with each person's events being associated in some patterned way with the other's events, an observer has evidence of interaction between the two. Examples of such patterning include the look of one person toward the other and the resulting mutual eye contact, one talking and the other listening, one touching and the latter moving closer or pulling away, the exchange of tender or angry feelings, and the joint moving of furniture. Dyas might choose to study patterns of small (brief, simple) events or patterns of large (long, complex) ones. At the small extreme, Dyas might examine words, head movements, smiles, and so on. At the large extreme, Dyas might examine instances of being together; a task one undertakes and completes on behalf of the other; a joint activity, such as tennis or sexual intercourse; an explanation one gives the other; and so on. Dyas' initial choice of size of "unit" would be quite arbitrary, but later it might be guided by an analysis of recurrent patterns of the events Dyas initially chose to identify. For example, as the statistical structure of language became evident, Dyas might move from the events constituted by phonemes or words to those constituted by sentences or remarks. Similarly, Dyas' description of the activity associated with tennis might move from a stroke-by-stroke description to a description of games or sets, perhaps with special notation of more significant elements within these larger units. In these decisions and analyses, Dyas would be faced with the problems discussed in Chapter 3 ("Interaction") and might discover some of the solutions described there.

After listing the events and patterns for a given pair, Dyas would very likely try to summarize the list. Dyas would aggregate the data in some manner, as by counting frequencies of particular events and patterns of events, and calculating percentages, averages, and so on. The aggregation process would then permit Dyas to describe the relationship in terms of its properties. A *property* is any summary description of interaction that an investigator may choose to devise. This description may summarize the frequency, rate, duration, and so on of various events or of various patterns of events. Examples include the average length of time spent together each day, the frequency of use of "we," the percentage of time devoted to instrumental tasks versus socioemotional activities, the amount of eye contact and touching, or the patterning of behavior during sexual intercourse. Then, once a number of relationships had been characterized as to their properties, Dyas would be able to compare them, classify them into types, study the interrelations among different properties, and so on. At this point, Dyas might distinguish some relationships from others, and, if we looked at Dyas' criteria for doing so, we would call some relationships "close" and some not.

So far we have been rather vague about the sources of our visitor's information about the events in relationships under study. Dyas can, of course, observe the pair directly or have a third party make observations. Those observations can include the full range of events illustrated above, not only the overt actions of the participants, but their verbalizations as well. Dyas would soon realize that the latter include reports of the participants' observations of events and patterns of events in their own interaction, as when they express inner events (feelings, thoughts), recall past patterns (some of which Dyas may not have observed), or remark upon current ones. Among these reports would be some that resemble the products of Dyas' aggregation operation, these being statements that reflect the participants' own impressions of regularities and trends in the relationship ("You always . . ."; "We never . . ."; "You've been more . . . lately"; "We're spending too much time . . ."; "Overall, I'm quite pleased with the way we have . . ."; and so on). Finally, our visitor would soon learn that many participants in relationships are willing to make these reports directly and even to answer many of Dyas' questions about the properties of the relationship. Participants may not be able to provide assessments of all the properties Dyas might have identified (e.g., percent of time with eye contact or with intimate touch), but they will usually make valiant attempts to answer Dyas' questions, they will often have available the words with which to express their global assessments, and these will often be tantalizingly similar in meaning to the properties Dyas has observed (e.g., "I'd say we're more intimate than most couples I know"). Dyas will be faced with a set of complex issues about participants' reports about their relationships—about their meaning, correspondence with third persons' observations, and the scientific use to be made of them.

Causal Analysis

Dyas' study of relationships may end with description in terms of properties. However, if Dyas shares the curiosity of earthling scientists and their interest in understanding and forecasting, Dyas is likely to move on to *causal analysis.* The operations involved in causal analysis are quite complex, so our characterization must be even more sketchy than that for the descriptive operations. Causation can be identified at many different levels, but for simplicity we will emphasize only two: (1) *causal connections* between different events within interaction and (2) *causal links* between various conditions outside the interaction and the properties of the interaction. In both cases, the main impetus to causal analysis is provided by evidence of *covariation.* In causal connections, certain events regularly occur with others; in causal links, certain properties vary with changes in external conditions. Causation is never observed directly; it is always inferred. When evidence of covariation can be supplemented with certain other necessary evidence (temporal precedence of the causal factor, observations that preclude alternative causal explanations), Dyas can infer the existence of a cause-and-effect relation.

In observations of the patterns of events within interaction, Dyas might find that one person's harsh words usually precede the onset of the other's weeping, or that episodes of tender interaction usually precede sexual activity. Those observed patterns would permit the visitor to infer in each case that the first event plays some causal role in the occurrence of the second. We will use the term *causal connection* to refer to causal relations between different events in the relationship. A prominent feature of a "relationship" is that events associated with one person are causally connected to those associated with the other person. Indeed, this is a *necessary feature* of "relationship" as we define it. We (or Dyas) would probably never consider a "relationship" as an entity for analysis unless the pattern of events observed at the outset led us to suspect that causal connections existed between two persons, the two thereby constituting part of a dynamic whole. Later in this chapter, we will analyze closely the nature of these connections and the properties of the relationship that, in the aggregate, they comprise.

Dyas may also make a causal analysis of the relations between properties of the interaction and various conditions outside the interaction. For example, Dyas may observe that the use of "we" decreases each time the mother of one of the participants comes to stay in their home; or that the husband's influence on buying decisions declines when the husband is not gainfully employed. From regular changes in properties and associated prior changes in the mother's presence or the husband's employment, Dyas may infer that the latter play a causal role in relation to the former. We will use the term *causal condition* to refer to factors, such as "mother's presence" or "husband's employment," that are postulated or shown to produce changes in the properties of interaction. In contrast to events, which have brief causal effects *within* the interaction, causal conditions are more stable causal factors, produce longer-lasting changes, and can be viewed as impinging on the interaction from *outside* it. Causal conditions may affect the properties of interaction, but they can also be understood as affecting the event-to-event causal connections (which are inferred from certain event patterns, as described above) and, in some cases, as directly affecting other causal conditions. These different possibilities will be explained later. Here, it is important to note that we use the term *causal link* for all the various causal relations a causal condition may have, thus distinguishing those relations from the *causal connections* between events.

Causal conditions are brought under scrutiny when there are changes in relationship properties, as in the examples above, but they also operate during periods of stability. Thus, one person's harsh words may always be followed by the other's weeping throughout the course of Dyas' observations. This regularity and the inferred causal connections between the two types of events is potentially explainable in terms of a causal condition that gives rise to the recurrent pattern of events (e.g., the second person's emotional vulnerability to the first person's anger) and that happens to remain constant throughout the particular period of observation.

The identification of operative causal conditions will pose many problems for Dyas, just as it does for us. The criterion of covariant precedence is rarely as clear as our examples suggest. Multiple causation is the rule rather than the exception. A contributing cause may long antedate the observed change, requiring a second cause to trigger the actual change. These problems can be somewhat attenuated if Dyas has control over causal conditions and can introduce or remove them at will. However, the limits of Dyas' probable control are obvious. With changes occurring at various levels (in events, in properties) and in great numbers, Dyas may be confused about which ones to attempt to explain. In this matter, as in the descriptive stage, Dyas can get advice from the participants. They have their own ideas about what is worth trying to explain. They already have their own explanations for many things and may not consider these phenomena deserving of scientific analysis. For the causal questions Dyas chooses to put to them, they will often have ready answers—theories about what factors are responsible for common changes, explanations for shifts in their own and others' behavior, and so on. Dyas, like ourselves, can get much advice from the participants about the causal analysis, but, just as we do, Dyas will face many difficult questions about what use to make of it.

At the end of these efforts, Dyas will know much about interpersonal relationships: their common properties, their varieties, their internal dynamics, and the factors that produce major changes in them. Dyas will be able to some degree to predict their course and, in certain respects, to modify conditions to change them. In short, Dyas will to some degree achieve the scientist's goals of understanding, prediction, and control.

Elementary Concepts

This thought exercise with our imaginary visitor has served to introduce the elementary concepts that we regard as the minimal essentials in the study of interpersonal relationships. We will develop these concepts further in the remainder of this chapter and will use them throughout this book.

1. *Events.* Interpersonal patterns of events constitute the basic data of interaction. Events consist of any change in a person that is considered important by a particular investigator. Thus, events include such phenomena as actions, reactions, emotions, and thoughts. Events constitute the elements in the dynamics of interaction because they are changes that are causally connected with other changes. The changes in one person are caused by other changes, in that person, in the partner, in the environment, and so on. The changes in one person also cause further changes, in that person, in the partner, in the environment, and so on. All the information we obtain about a relationship, whether it depends on observation or report, ultimately refers to these events.
2. *Properties.* By the operations of aggregation, numerous observations or reports of single events or event-patterns are assembled and summarized to provide descriptions of cumulative properties of relationships. These properties describe such features of the relationship as its emotional tone, the frequency and intensity of interaction, the extent to which the two persons think about each other, and their relative influence on the course of their interaction.
3. *Causal connections.* By the operations of causal analysis, in which the temporal patterning of events within the interaction is observed and/or controlled, we infer certain events to be causally connected with other events. The unity of a pair relationship, its existence as an "entity," derives from the fact that many events associated with each person are causally connected to events associated with the other person.
4. *Causal conditions.* The operations of causal analysis also enable us to infer the existence of certain more or less stable and enduring causal conditions. These conditions are identified through observing that certain attributes of the persons (e.g., employment, abilities, attitudes) or of their social or physical environments (e.g., in-laws, number of friends, housing facilities) are causally linked to properties of the interaction.

Causal conditions are responsible both for the stability of the relationship (insofar as the conditions are stable) and for changes in the relationship (insofar as the conditions eventually change).

5. *Causal links.* By the operations of causal analysis, certain properties of the interaction are inferred to be caused by certain causal conditions. And, as we will see below, the interaction is often inferred to cause changes in certain causal conditions. The causal relations so inferred are described as causal "links" in contradistinction to interevent causal "connections."

We now explain these concepts further, and show how they provide a useful perspective for understanding and organizing current research and theory about close relationships. We will not confine ourselves to any particular theory or to a particular level of analysis. Our only commitment is to a description of relationships in terms of the interaction between their members and an explanation of that interaction in terms of causal conditions that, relative to the flux of events in the interaction, are stable. Given this orientation we seek a general view of the close relationship that will enable current theory and data to be placed in relation to one another and that will highlight the gaps in the current work—the unasked questions, the needed ideas, and the unsolved methodological problems.

Our analysis in this chapter will focus on the close *pair* relationship. This focus is dictated by practical considerations. Although very complicated, the conceptual analysis of the dyad is manageable. Added persons greatly increase the complexities and diminish precision. The importance of relationships of more than two persons goes without saying. Here we will take account of such larger collections by considering how other persons and groups impinge upon the pair and affect their interaction and how, in turn, the pair selects and shapes its social environment.

DESCRIPTION OF RELATIONSHIPS

The Basic Data of Dyadic Relationships

The relationship between two persons, P and O, can be described in many different ways and conceptualized in many different terms. However, *all the various descriptions and conceptualizations will explicitly or implicitly refer to two chains of events, one for P and another for O, that are causally interconnected.* Thus, the basic data of a dyadic relationship can be described schematically, as in Figure 2.1, by two chains of events that are located along a time line and

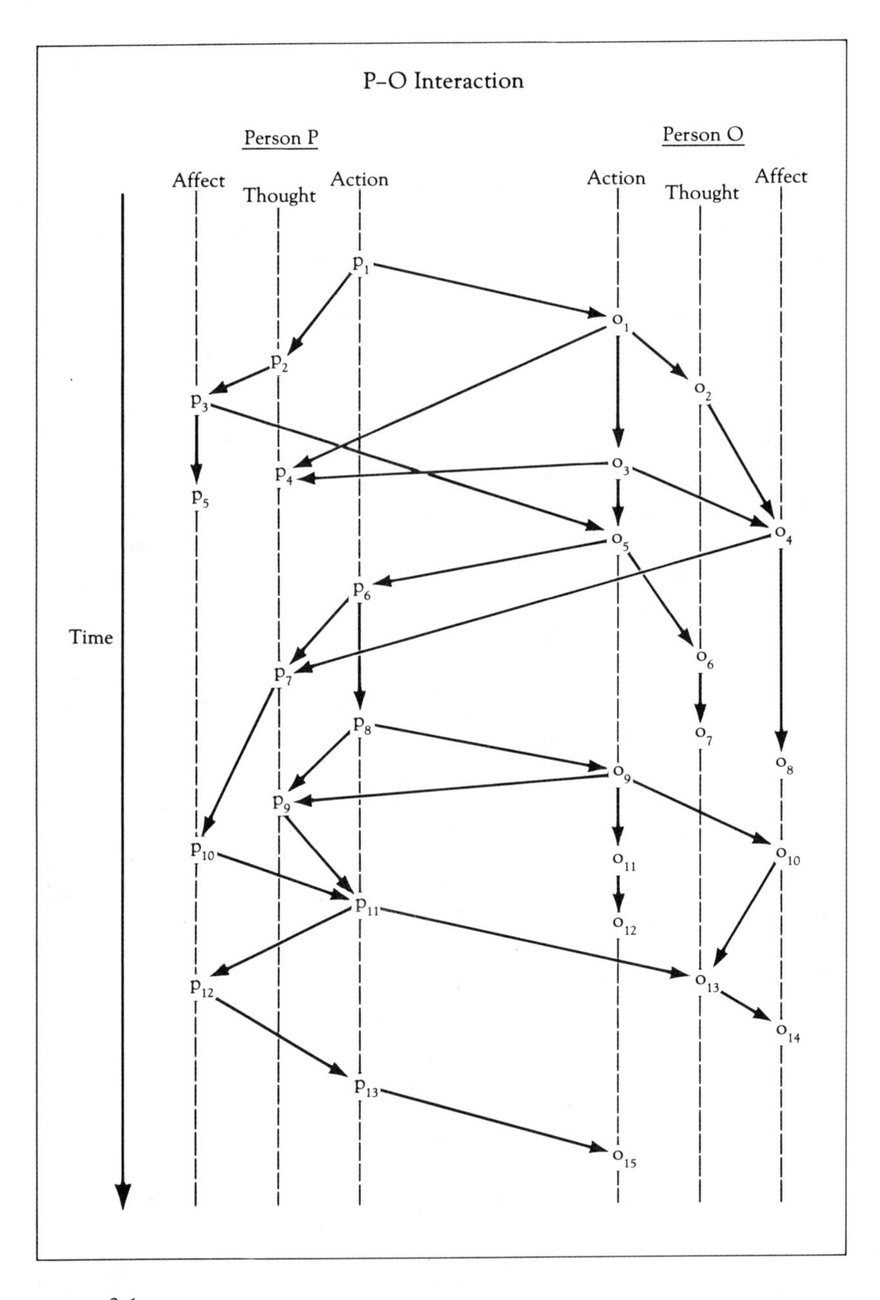

FIGURE 2.1
The basic data of a dyadic relationship. Each person has a chain of events, each chain including affect, thought, and action. The events are causally connected within each chain (shown by arrows from one p to another or from one o to another) and the two chains are causally interconnected (shown by arrows from a p to an o or from an o to a p). The interchain connections constitute the essential feature of interpersonal relationships.

are related by arrows indicating causal connections. The events in person P's chain are indicated by p_1, p_2, and so on, and the events in person O's chain by o_1, o_2, and so on. Our diagram shows that (1) each person's chain of events consists of multiple strands (several things go on simultaneously for each person, such as acting, thinking, and feeling); (2) events are causally connected within each person's chain; and (3) events are causally connected between the two persons' chains, this last being the basic feature of interpersonal relationships.

Figure 2.1 suggests, though perhaps inadequately, the complexity of P–O interaction. A simpler example, cast in terms of specific events, is shown in Figure 2.2. (Any particular illustration of our general description necessarily has special properties and, therefore, may be somewhat misleading. However, we take this risk in the interests of concreteness and intelligibility.) In Figure 2.2, we see one possible description of an interaction between a young woman (P) and a young man (O). The interaction begins with the first event in the young woman's chain—a verbal act consisting of a compliment to O about his appearance. Her event p_1 causes event o_1, his visible autonomic response—blushing. Meanwhile, P thinks about her remark and wonders whether it was appropriate. Shortly thereafter, P perceives that O has blushed and interprets it to mean that he likes what she said. Simultaneously, O becomes aware of his own emotional response to the compliment. When P subsequently smiles (following her interpretation of his blush), the smile, along with his awareness of his own reaction, causes him to think that P noticed his reaction, and this in turn produces an increased blushing as well as a smile. P notices the smile and feels good. And so forth.

This example shows some of the kinds of events that might be used to describe an interaction. We use "event" as a neutral and general term to refer to any change in P and O that may be regarded as important by a particular investigator or theorist. Thus, as in our illustration, an "event" may be a voluntary action, a conditioned response, an affective reaction, a perception, or a thought.

The causal connections *within* each chain reflect how earlier events in a person produce or affect later events in that person. In Figure 2.2, P's compliment at p_1 leads to her thought at p_2. O's realization at o_3 that P noticed his reaction leads to further blushing (o_4) and a smile (o_5). Each person's chain of events usually has some degree of temporal organization and structure produced by such intrachain causal connections. Thus, needing something and knowing it can be obtained lead to appropriate action. The well-learned skills involved in driving a car or playing tennis are reflected in causal chaining that produces organized sequences of motor activity. The regular phonemic structure of the words and the regular grammatical structure of the sentences produced by an experienced speaker similarly reflect intraperson chaining.

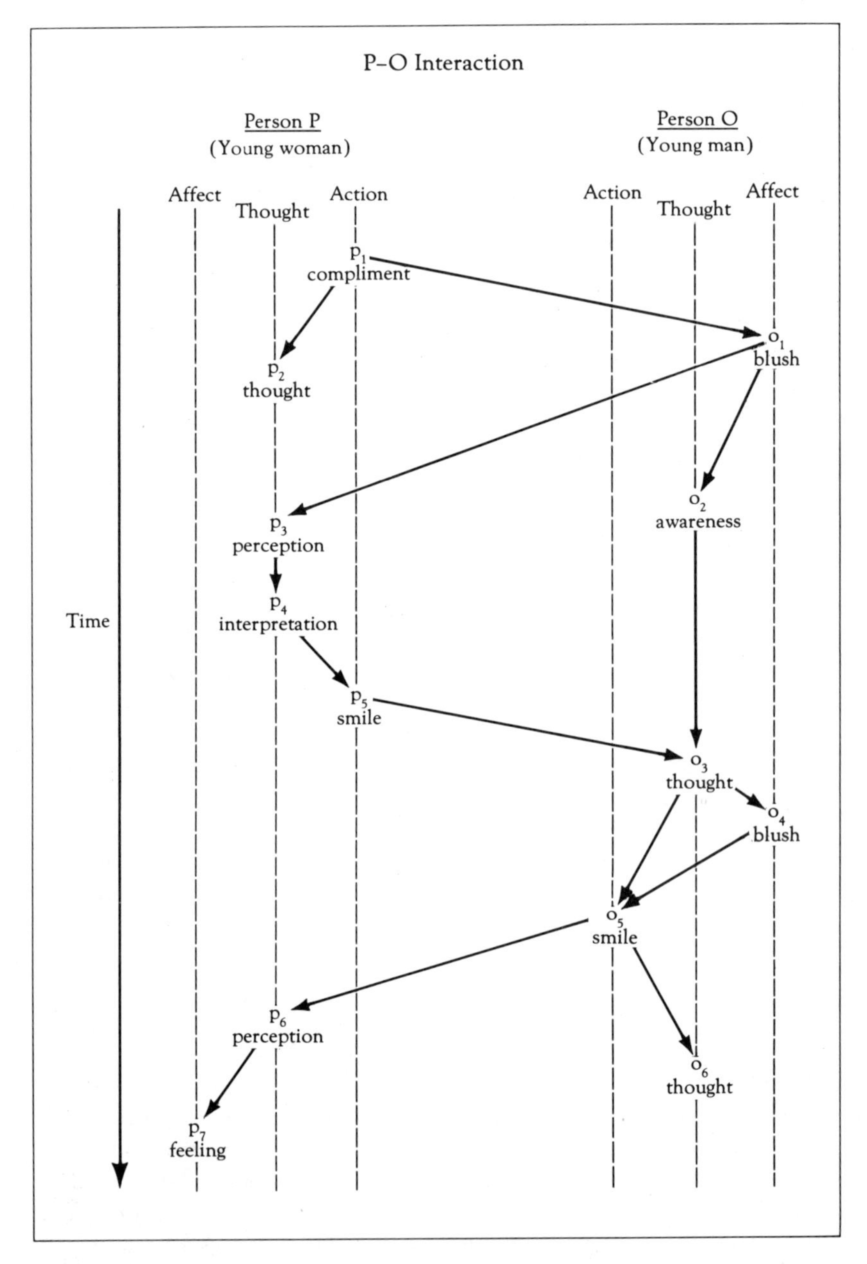

FIGURE 2.2
Brief portion of interaction between a young woman and man. The events in each person's chain of affect, thought, and action tend to produce further effects within that chain (shown by the p-to-p and o-to-o causal connections) and effects in the other person's chain (shown by the p-to-o and o-to-p causal connections).

Most important for our understanding of relationships are the causal connections *between* the two persons' chains. We can speak of there being a "relationship" between P and O only when we detect that their two chains of events are, to some degree and in some manner, causally interconnected. In general terms, this means that the events in P's chain play some causal role in relation to the events in O's chain *and* that events in O's chain also play some causal role in relation to the events in P's chain. Interchain causal connections are implied by the term "interaction," which broadly characterizes interpersonal process. As a defining characteristic of interpersonal relationships, the interchain causal connectedness is well summarized by the term "interdependence," which refers to causal connections in both directions between P's and O's chains.

Prominent instances of causal interconnections are communication and interpersonal perception. For example, Figure 2.2 includes instances of both verbal and nonverbal communication. Other instances would be physical effects, as when a mother gives candy to a child or moves the child from one place to another. In all cases, the connections between the two chains are overt (via sight, sound, touch, and so on) and therefore, in principle, are accessible to a sentient observer of the interaction. However, because important effects are often covert and because unimportant visible effects often occur in both chains, it is usually difficult for an observer to provide an accurate and detailed description of the interconnecting process.

The sample of interaction described in Figure 2.2 is only a fragment of an episode involving this young woman and young man. There may have been earlier such episodes, and there may be further episodes in the future. At the level of detail used in Figure 2.2, even rather simple relationships may involve thousands or tens of thousands of events and causal connections over a short time span. Few investigations have dealt with interaction at this level of specificity. It is customary to deal selectively with events (as by limiting one's description to verbal events) or to refer to events in more molar terms (e.g., in terms of an exchange of positive evaluations or in terms of getting acquainted), or both. One continuing issue in research concerns the units of analysis—their type and size—that are most appropriate for describing interaction. It must also be noted that investigators often obtain aggregate summaries of interactions, such as the total amount of eye contact or the total number of positive comments exchanged between two persons over some period of time.

The Properties of Interdependence

Given the foregoing view of relationships, as summarized schematically in Figure 2.1, the task of analyzing and describing a relationship becomes that of assessing and characterizing its interdependence. All investigations of dyadic interpersonal relationships deal with *data* that derive in some way from the two causally interconnected chains. All theories and hypotheses about such relationships involve *conceptual terms* that refer in some way to the interdependence between the two chains. Therefore, through a logical analysis of the properties of the interconnections, it is possible to outline the types of data and concepts that may be involved in the study of relationships. This analysis also permits us to specify what we mean by a close relationship.

The reader will understand that the two chains of events and their interconnections lend themselves to a great variety of descriptions, analyses, and hypotheses. However, the following eight categories seem to constitute the most important properties with respect to which interdependence can be analyzed:

Kinds of events

Analyses of the two chains may differ greatly in the kinds of events that they identify. To give a few examples, events may be conceptualized in terms of actions, affects, and thoughts; kinds of resources provided, as in Foa and Foa's (1974) distinctions among goods, services, money, information, status, and love; the nature of the contribution to a discussion, as in Bales' (1950) interaction process analysis, with its distinctions between asking for information, giving orientation, showing tension, and so on; or signals that govern turn-taking in conversation, as in Duncan and Fiske's (1977) distinctions between back-channel signals, speaker continuation signals, and within-turn signals. French and Raven (1959) implicitly provide a taxonomy of "influence" events and hypotheses about their differential consequences (see Chapter 5, "Power"). Among the many analyses, there are great variations in the size and complexity of the events that are identified, ranging from the molecular extreme (e.g., the shift in direction of gaze) to the molar (e.g., the strategy of influence). Closely associated with distinctions among kinds of events are distinctions among kinds of causal connections between the two chains of events. Distinctions can be made between such phenomena as verbal communication, influence by visual cues (as in nonverbal communications and imitation of a model's behavior), touching and stroking, and physical force.

Patterns of interconnections

Causal interconnections can also be distinguished as to pattern. A few of the many logical possibilities are shown in Figure 2.1. An event in one person's chain (for example, P's) may be caused by one or several other events in the same chain (e.g., p_3 versus p_{11}); by one or several events in the other person's chain (e.g., p_6 versus p_5); or by various combinations of events in both persons' chains (e.g., p_9). An event in one person's chain may have no further effect (e.g., p_5); further effects only within that chain (e.g., p_7); further effects only in the other person's chain (e.g., p_{13}); or further effects in both persons' chains (e.g., p_{11}). The further effects within a given chain may be simple (e.g., p_{11} leads to o_{13}, which leads only to o_{14}) or complex (e.g., p_8 leads to o_9, which generates two separate intrachain sequences of further events).

The two chains and their interconnections are constituted of numerous such patterns. A particular relationship may be characterized by a preponderance of certain patterns. For example, in some pairs, P's effect on O may usually be simple (e.g., P's effects on o_1, o_9, and o_{15} in Figure 2.1), but, in other pairs, P's effect on O may usually depend on concurrent events in O's chain (as illustrated by o_{13}). This distinction corresponds to what Thibaut and Kelley (1959) describe as "fate control" versus "behavior control." Similarly, in some pairs, the effects of interchain connections may be few and simple (e.g., o_{15}), but, in other cases, they may ramify within chains and reverberate between chains (e.g., P's effects on o_1 or o_9). The reader will readily imagine an interchange characterized by cross-connections that have limited and isolated effects (e.g., a casual, routinized conversation about the weather) as compared with one in which each cross-connection tends to have extensive remote effects (e.g., a "significant" exchange of self-disclosures, accompanied by much thought and monitoring of self and partner).

Strength of interconnections

Interconnections vary in intensity or strength. Various aspects of strength may be distinguished: The change produced in O may be great, involving single responses of large amplitude, numerous responses, or long chains of responses. Strength is also indicated when the change is produced with short latency or with high dependability. It is also possible to define properties related to the efficiency with which P produces changes in O, these taking account not only of the magnitude of change(s) in O but of the associated changes in P (e.g., units of change in O per unit of change on P's part; rewards P gives O relative to the costs P incurs or to the rewards P foregoes).

Aggregating all the interconnections that characterize a relationship, we may distinguish relationships in which the interconnections are generally strong from those in which they are generally weak. This distinction would correspond to such concepts as cohesiveness (Cartwright and Zander, 1968) and degree of interdependence (Thibaut and Kelley, 1959). At an aggregate level, it is also possible to compare the strength of the intrachain connections with the strength of the interchain connections. Thus, the events in one person's chain may be mainly determined by intrachain causal connections, whereas those in another person's chain may be mainly determined by interchain connections. The first person might be characterized as relatively independent of the partner and the second person as relatively dependent. Kelley and Thibaut (1978) suggest a related concept, the degree of dependence, defined in terms of the proportion of the variance in a person's outcomes that is controlled by the partner and the proportion that is controlled by the two acting jointly.

Frequency of interconnections

Over any given time span, the number of interconnections between the two chains may be few, many, or intermediate in magnitude. At an aggregate level, we may distinguish relationships in terms of *rate,* that is, the number of interconnections per unit of time. Those with low rates would be of several different types—for example, two people who interact only intermittently, two who have few ways of affecting each other, or two persons who interact at a leisurely pace.

Diversity of interconnections

Dyads may be distinguished in terms of the number of different kinds of events that are interconnected. The two persons may affect each other in a number of diverse ways (sexual activity, recreational activities, joint work, intellectual discussions, and so on) or in only one or two different ways. The distinction here is between broad, richly textured interaction and single-theme, unidimensional interaction. Hinde (1979) refers to the former as "multiplex" and the latter as "uniplex."

Interchain facilitation versus interference

A particular portion of P's chain can often be characterized as "movement toward a goal" or, in the terms of Chapter 4 ("Emotion"), as an "organized action sequence." In this common case, the causal connections coming from O to P's chain may *facilitate* the directed movement or sequential organization or may *interfere* with it. Or, of course, the O-to-P connections may have no effect on this portion of P's chain. Facilitation versus interference refers to the relation between interchain causal connections and the organization or sequencing of intrachain connections. In facilitation, interchain causal connections promote the organization of intrachain connections; in interference, the former hinder or disrupt the latter.

Interchain facilitation and interference can be conceptualized in a number of ways; different theoretical perspectives will point to different sorts of interchain effects. Some examples of interference include (1) O's action changes P's state or location so that P can no longer as easily reach P's goals; (2) O's behavior does not "mesh" with P's in the sense that it is not directed in accordance with P's ongoing goals (Hinde, 1979); (3) O's action disrupts the usual internal organization of P's chain of events (e.g., incompatibility of moods; Thibaut and Kelley's, 1959, conception of interference versus facilitation); (4) O says things that confuse P, unsettling P's beliefs, creating attributional uncertainty, maligning P's self-image, or inducing cognitive dissonance; (5) O's activities interrupt some ongoing plan or activity of P's, thereby causing emotion.

It should be emphasized that facilitation or interference need not be symmetrical between the two persons. The interchain effects may be facilitative for P but interfering for O. For example, P's helping O to change a flat tire on her car may be facilitative of P's self-image as a protective and competent male but may be interfering with O's desire to be self-sufficient in mechanical matters. It should also be noted that mutual facilitation does not always promote positive interaction. Sometimes facilitative interchain connections lead to positive effects for P and O, as when the partners wind up their tennis match feeling happy and relaxed. But sometimes facilitative interchain events lead to negative interaction, as when a political discussion becomes an escalating argument in which each person's expression of views stimulates the partner in reeling off well-learned counterarguments. A "conflict-adapted" couple may display great interchain facilitation by the effective manner in which the fighting tactics of each support the similar tactics of the other.

Symmetry–asymmetry of interconnections

In regard to any of the preceding properties, the interconnections from P to O may be similar to those from O to P (symmetry) or different (asymmetry). In one type of asymmetry, the kinds of events that are affected by the partner are different for the two persons. For example, Blau (1955) describes a P who gives technical advice to O and, in return, receives approval and deference from O. There may also be differences between the two directions in the strength, frequency, and diversity of the connections. The qualitative differences between P and O in the kinds of effects each has on the partner find their parallel in differences between the two in the type of social influence they exercise (French and Raven, 1959) and in their enacted roles. Quantitative differences between the P-to-O and O-to-P connections in their strength and frequency relate to differences between P and O in degree of dependence, amount of influence, and so on.

Duration of interaction and relationship

The duration of any particular interaction episode and the duration of the relationship as a whole can be measured by the length of time during which various indices (e.g., frequency or strength of interconnections) are above some threshold level. Thus, a relationship may be said to "begin" when the two persons first affect each other to some specified degree and to "end" when they no longer do so.

A number of conceptual and operational problems arise in connection with determining the duration of a relationship. One concerns the definition of "relationship" in cases of extreme asymmetry. If the causal connections mainly go from P to O, we may not wish to consider there to be a P–O relationship at all. This would be the case for an O who admires a P from a distance, closely following P's activities and career and being affected by P's actions without P having any knowledge of the effects. Following our earlier statement that there exists a relationship between P and O only when their two chains of events are *inter*connected, the duration of the relationship will depend on when the causal connections in both directions surpass some criterial levels of strength and frequency.

A second problem has to do with distinguishing between (1) temporary breaks or "time-out" in the interaction and (2) disruptions of the relationship that are followed by its renewal. In few, if any, relationships is there continuous interaction. Often there are long periods of noninteraction, due to vacations, work responsibilities, hospitalization, and so on. The problem, then, is to distinguish the temporary discontinuities from the "permanent" ones that happen to be followed by the relationship's beginning anew. This distinction will very likely require evidence about the temporal course of external causal conditions that affect the propinquity of the pair and about the shared understandings and expectations that provide the basis for psychological continuity despite physical separation.

Properties of Interdependence: Subsequent Discussion

The preceding eight categories of properties delineate what appear to be the major features of interdependence. We present this list as a useful itemization of distinguishable properties, without any illusion that ours will be the final word on the matter. In the subsequent chapters of this book, these properties are considered at greater length as they become important to various topics in close relationships.

Chapter 3, "Interaction," analyzes the interaction process at the level of events and organized sequences of events. It deals with both the objective description of overt interaction and its subjective interpretation by participants and observers. The chapter also illustrates the major factors responsible for different interaction patterns.

Chapter 4, "Emotion," focuses on the particular interaction sequences that, through interchain interference, generate emotional experiences. Importantly, this chapter analyzes the features of interdependence that create the potential for the development of emotional exchanges.

Chapter 5, "Power," analyzes the specific portions of interaction that constitute the intentional use of interchain connections by one person to influence the other. This chapter also examines the overall features of relationships, described in terms of dominance, in which there is asymmetry between the P-to-O and the O-to-P connections over a broad diversity of events.

Chapter 6, "Roles and Gender," deals with the recurrent intrachain sequences that are important for the life of the relationship. The chapter gives particular attention to the asymmetries in heterosexual relationships that are related to gender-linked roles and to the causal factors underlying these asymmetries.

The phenomena of love and commitment, outlined in Chapter 7, "Love and Commitment," raise broad questions about the strength, frequency, diversity, and duration of the bonds between P and O. Of particular interest are the causal conditions that differentially affect the stability versus the strength of the interchain connections.

In the present chapter, we will briefly indicate how the eight properties may be used to distinguish types of relationship and, more specifically, to define what we mean by a "close" relationship. We will also briefly suggest how stages in the development of relationships can be distinguished in terms of their characteristic properties of interaction. This topic is considered at length in Chapter 8, "Development and Change," which describes the different features of interaction typical of beginnings, middles, and endings of close relationships and the changing causal conditions involved in their developmental progression.

Chapter 9, "Conflict," and Chapter 10, "Intervention," return to the phenomenon, initially considered in Chapter 4, of interaction sequences characterized by interference. Chapter 9 emphasizes conflictual interactions, their successive stages, and their various consequences. Chapter 10 examines interaction patterns associated with relationship dysfunction and describes several treatment approaches to altering these patterns.

Chapter 11, "Research Methods," delves further into the methodological problems encountered in describing the properties of relationships and identifying their causal antecedents. In making a case for the value of a science of relationships for psychological and social science, Chapter 12, the epilogue, illustrates how the careful study of interaction processes is necessary if the role of these relationships in relation to individuals and society is to be fully understood.

Types and Stages of Relationships

Relationships can be distinguished and classified in terms of the properties outlined above. For example, from Wish, Deutsch, and Kaplan's (1976) data, we might infer that the relationships of close friends and of husbands and wives are characterized by high strength, frequency, and diversity; symmetry; and high mutual facilitation. These relationships are to be contrasted with the relationships of business rivals and personal enemies, which have medium strength (probably along with low frequency and diversity);

symmetry; and high mutual interference. The relationships between parent and child and between master and servant are both characterized by high asymmetry, but the former are usually stronger, more frequent, and more diverse, and involve different kinds of events (more socioemotional content and less task-oriented activity).

Stages in the course and development of a relationship can also be distinguished and classified in terms of the above properties. The "career" of a P–O relationship can be described in terms of the succession of different "types of relationship" through which P and O move from the beginning to the termination of their relationship. When a relationship changes markedly in any property, it is reasonable to say that it has moved to a new stage or level.

In line with this view, Wish et al. (1976) obtained descriptions in the same terms of different relationships and of different "stages" of the same relationship. "Stages" were childhood relations versus current adult ones. The typical person's relationship with her or his mother (or teacher/professor) shifts from asymmetry to midway beween asymmetry and symmetry ("unequal" versus "equal" in the terms used by Wish et al.). Relationships with siblings move from midway between interfering (competitive) and facilitative (cooperative) toward the facilitative pole.

To suggest that relationships and stages of relationships be described in the same terms, using the eight types of property, is not to imply that moving from one relationship to another involves the same dynamics as moving from one stage to another. However, the use of common terms probably has considerable heuristic value through enabling direct comparisons of (1) between-relationship variations and (2) within-relationship changes.

Defining "Close" Relationships

Our focus is on a particular class of relationship referred to as "close" relationships. By a close relationship, we mean one of considerable duration (measured in months or years rather than hours or days) in which the causal interconnections between P's and O's chains are strong, frequent, and diverse. That is to say, *the close relationship is one of strong, frequent, and diverse interdependence that lasts over a considerable period of time.* Examples of such relationships are friendships, serious love affairs, marriages, and parent–child relations.

All the relationships we regard as "close" will, by definition, be characterized by the four properties of strength, frequency, diversity, and duration. It must be remembered, however, that they may be distinguished in terms of other properties of interdependence as well, such as facilitation or interference. They also may go through stages defined by, say, shifts in degree and kind of asymmetry, while all the time remaining "close."

We have given much thought to our choice of the term *close* to characterize the relationships of special interest here. On the one hand, it has some connotations of intimacy and positive emotion that are not entirely appropriate to the full range of relationships we wish to include. Relationships need not involve the exchange of intimate information or produce regular intense positive feelings in order to be tightly interconnected in the ways we would regard as defining closeness. For example, close co-workers may never share intimate details of their personal lives; spouses may feel great hostility for each other but continue to have strong effects on each other. On the other hand, in two of its other connotations, "closely connected" and "physically close," *close* seems to be exactly the right term for our meaning. We are interested exclusively in relationships in which the lives of the two persons (as represented by the two chains of events described earlier) are closely intertwined. The two are tightly bound together by virtue of many strong causal connections between them. Physical closeness figures prominently in contributing to this close causal connectedness, not as a necessary condition but as a factor that greatly promotes extensive interconnections.

Level of positive affect is often proposed as a criterion for the "closeness" of a relationship. Close relationships are commonly believed to be characterized by strong positive emotion and high affective involvement. As noted above, close relationships as we have defined them do not necessarily involve positive feelings. Moreover, as we will see in Chapter 4, "Emotion," close relationships have high *potential* for affect, but at any point in time may not manifest much affect. This point relates to the property of facilitation versus interference. Being characterized by many strong interchain connections, a close relationship always has potential for interruption of one or both persons' intrachain organization of behavior. In that limited sense, our view is that close relationships are characterized by high affective "involvement." However, many such relationships go for long periods of time without the occurrence of serious interruptions and, hence, with little actual affect. This matter is considered in some detail in Chapter 4.

We recognize the possibility of defining closeness in terms other than those stated above. Huston and Burgess (1979) summarize many of the features that have been suggested to characterize close or intimate relationships. Beyond the kinds of properties specified in our definition (frequency, duration, diversity, intensity), their list refers to a number of factors that, in the next section, we will identify as "causal conditions." These factors include shared norms (about communication, responsibilities); attitudes (liking, love, trust); beliefs about the relationship (its uniqueness, importance); and relations with other persons. The first three of these refer to characteristics of P and O that, on the one hand, presumably emerge from P–O interaction and, on the other, play a role in structuring it. For example, love is sometimes conceptualized as an attitude of P toward O that accounts for the occurrence of particular patterns of P and O events, but it may also be an attitude that results from P and O's interaction with one another. The same point can be made regarding the fourth factor, P's and O's relations with others, as an indicator of the closeness of the P–O relationship. P and O's exclusive association with each other usually identifies them as a close pair. However, we see their relations with others not as a defining property of closeness but as a causal condition that both affects and is affected by the closeness.

While others have defined closeness in terms of psychological (e.g., attitudinal) or extradyadic (e.g., social) causal conditions, we believe there are good reasons to begin with the details of interaction, including both its observable and subjective features. Relationships having the properties of closeness as we have defined them will typically be characterized by certain attitudes, understandings, and social conditions. It will be these factors that promote, enable, and require the frequent, intense, and varied interchain connections and patterns of subjective reactions. On the other hand, independent of the initial reasons for their existence, relationships that are close, as we have defined it, will tend to develop or enhance the causal conditions constituted by particular attitudes, understandings, social connections, and so on. We are less certain of the exact nature of these causal conditions than of the properties of frequency, intensity, diversity, and duration. Indeed, it seems obvious that the causal conditions (e.g., attributes of P and O, relationships with others) are likely to be quite varied for a number of different relationships (e.g., young lovers, swinging couples, traditional spouses, roommates, co-workers), all of which we would wish to term "close." Thus, we elect to anchor our definition of closeness in the interconnections between P and O events rather than in any particular configuration of attitudes, understandings, and so on, on the part of P or O or of their social environment.

The External Causal Connections of the Relationship

Figures 2.1 and 2.2 show how the events in each person's chain cause other events in that chain and events in the other person's chain. The events in each chain are also partly controlled by events external to the two persons and their interaction, that is, by events in the social and physical environments. These events are many in number and heterogeneous in their nature and effects. They include such diverse events as noises that produce startled responses in one or both persons or that interfere with one's hearing what the other has to say; provocation to anger or sexual arousal provided by other persons; instigations to thoughts about one's own inadequacy provided by others' possessions or skillful behavior; and safe and secluded conditions that facilitate P and O's tête-à-tête. These events might be described in "stimulus" or "objective" terms, but, as the above examples illustrate, it is usually more convenient to describe them in terms that refer to their effect on P or O. This description in terms of their effect also tends to limit our attention to the most relevant subset of the many events, namely those that have consequences for P's and O's chains and interconnections.

It is also apparent to an observer of interaction (and to the participants themselves) that events in P's and O's chains have effects external to those chains, that is, in their physical or social environments. P and O make noise, break furniture, turn lights on and off, compliment or criticize other persons, pet dogs, cuddle children, and so on. Once again, these activities are more conveniently described in terms of their external effects than in terms of the specific p or o event that has the effect.

When we take account of these external events, we must draw a more complete diagram of the P–O interaction than that represented by Figures 2.1 and 2.2. Figure 2.3 provides an example of the causal connections between the P–O interaction and its social and physical environment. The symbols e_{soc} and e_{phys} are used, respectively, to refer to events in the social

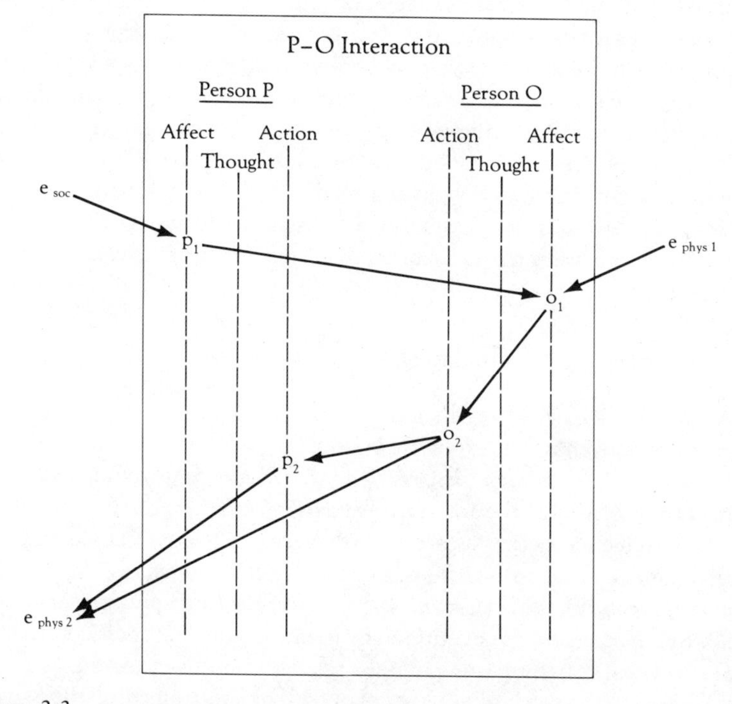

FIGURE 2.3
Illustration of external causal connections of the P–O interaction. An event in the social environment (e_{soc}) causes an affective event in P's chain (p_1). The latter event, together with an event in the physical environment ($e_{phys\,1}$), produces affect (o_1) in O that leads O to action, o_2. O's action stimulates P also to act (p_2), and their actions jointly serve to cause a change in the physical environment ($e_{phys\,2}$).

and physical environments. The example shows that e_{soc} (perhaps the remark of a third person in the hallway) causes some event in P's chain (a nervous reaction) that, along with $e_{phys\,1}$ (perhaps a scene from a television show O is watching), causes an event in O's chain (a sharp experience of anxiety). This event (o_1) leads O to do something (o_2) that both stimulates P to action and, jointly with p_2, affects the physical environment, $e_{phys\,2}$ (they join together in closing and locking the doors and windows).

As these examples show, external events and their causal connections vary in the same way as do internal ones—in type, pattern, facilitation, strength, and asymmetry. This similarity is obvious in the case of the connections between P's chain and that of a third person, Q. The P–Q relationship can be described in the same terms as the P–O relationship. The same is true for the connections between P's chain and any distinguishable portion of the physical environment.

More generally, the relationship between P (or O) and any part of the social or physical environment can be described in terms of properties similar to those used to describe the P–O relationship. The relation between P and E may be symmetric or asymmetric in terms of frequency and strength of causal connections. If asymmetric, we may observe that P exercises a good deal of control over E or that E tends to control P. Both social and physical environments may be facilitative or interfering in the way they impinge on P's chain of events. It should also be noted that the P–O relationship is sometimes described and evaluated relative to other relations, for example, P–Q and O–R relations. Thus, it is possible to speak of P's relative dependence on persons O and Q or the relative diversity of the interconnections.

The Importance of Event-Level Analysis

We can now see why the P and O chains of events and their causal interconnections must be, explicitly or implicitly, the focus of any analysis of the close relationship. The first point is that these chains and their interconnections constitute the interface between P and O. All the mutual and unilateral influence occurs as a result of events in the two chains that are causally connected. The events and their connections comprise the reality of the relationship, for without them there is no relationship. The quality and type of the relationship is constituted by them. Events and their connections are no less real for the participants in relationships than for our scientific analysis. As a self-conscious observer of oneself in a relationship, one is aware of actions, thoughts, and feelings in one's own chain. As an observer of the partner, one is aware of the actions and other overt responses in the partner's chain. As an observer of the interplay itself, one is aware of the interconnections—that one acts and the partner reacts, that the partner initiates and that one resists or follows, and that what each says affects the other and, often, results in visible response. As scientific observers, we record these events and interconnections and aggregate them to provide descriptions of a relationship's properties. As informal observers of our own or others' relationships, we form and report summary impressions of them (e.g., of closeness, equality, conflict) that in some way reflect these events and connections. In short, all the descriptions we obtain of relationships are based in one way or another on information about interaction as defined at the event level.

The second point is suggested by Figure 2.3. Various factors, such as attributes of P and O, other people, and their physical environment, affect the relationship *only* as they affect the events in or connected with the two chains and the interchain connections. Similarly, the relationship affects other factors (the participants, other people, its environment) *only* by events in or connected with the chain that (1) are in some manner affected by interchain connections and (2) have effects on the other factors. We now turn to a consideration of these latter factors—the causal conditions that, on the one hand, affect and shape the relationship and, on the other hand, are affected and shaped by it.

ANALYSIS OF CAUSAL CONDITIONS

In the preceding section we used general terms to describe one striking aspect of dyadic relationships, namely the interaction—the "give and take"—between two persons. We used the term *event* to refer to the various discrete occurrences that we, both as participants in and observers of relationships, recognize to happen in them. We indicated the causal connections that exist among those events, these connections also being a salient part of our knowledge of interaction—that "one thing leads to another," that each person affects, stimulates, and influences the other.

In the present section we pursue the implications of a second salient aspect of interaction, namely, its *regularities*. The observer of any sizeable portion of the interaction between two persons soon detects regularities and recurrent patterns in the "give and take." The participants themselves are aware of many of these regularities, particularly those that involve following explicit interpersonal plans and schedules. To both participants and observers, regularity is especially salient (1) when it changes, that is, when there is a shift from one level or type of uniformity to another, or (2) when the regularity in a particular relationship or type of relationship contrasts with that observed in other relationships or types.

An account of these regularities requires that we posit certain relatively stable causal factors that act on a relationship. An account of changes in observed regularities requires that we look for changes in such causal factors. An account of differences between relationships in their respective regularities requires that we identify differences in the causal factors that impinge on them. These causal factors are to be distinguished from the causes referred to as "events" and, indeed, they play an important part in determining the events and their interconnections. We will refer to these more stable (though occasionally changing) causal factors as *causal conditions.*

The remainder of this chapter is devoted to an analysis of causal conditions. It will serve to identify the broad causal context within which the dyadic relationship exists. An understanding of this context is necessary if we are to answer questions about the origins of relationships, the differences among them, the dynamic interplay between them and their environments, and the changes and trends in relationships that occur during their course.

The Inference and Identification of Causal Conditions

Each dyad is characterized by recurrent events that often distinguish it from other dyads. Thus, a given pair may be characterized by one person's frequent remarks on certain topics, certain joint leisure-time activities, one person's work on certain tasks, and one or both members' actions directed toward certain third persons. Also notable in any relationship are the recurrent connections between the two persons' event chains, that is, regularities in what leads to what. These include one person's frequently having a certain effect on the other (e.g., remarks that make the other feel guilty) and sequences of P-to-O and O-to-P connections (e.g., P regularly criticizes O's appearance, and O regularly responds with weeping, which never fails to make P angry).

The properties discussed in the preceding sections describe some aspects of these regularities. An interaction pattern may be said to be strong, asymmetrical, or facilitative only if most of its interchain connections are strong, asymmetrical, or facilitative. Of course, more specific descriptions are possible. We can describe in detail a particular pattern of asymmetry by observing that P regularly cooks the meals while O washes the dishes.

While it is possible to stay at the descriptive level and simply note the regularities in events and event-to-event sequences, most conceptualizations of relationships (and indeed of all behavior) assume that regularity implies the existence of *causal conditions.* These causal conditions will be defined more completely below, but here we note that they are relatively stable attributes or states that are presumed to determine what events and event-to-event connections are likely to occur. The events and connections may be any of the kinds shown in Figure 2.3. The variety of causal conditions can be illustrated by (1) P's habits, (2) emotional support for P that exists in the social environment, and (3) sources of noise in P's physical environment. P's habits would be reflected in P's regular response to certain e or o events; emotional support in the social environment, by regular e_{soc} events that produce feelings of security and confidence in P; and noise, by regular e_{phys} events that create disruptions of certain otherwise dependable interchain connections (e.g., failures of communication). In these three examples, a condition of P, of the social environment, and of the physical environment account, respectively, for the three regularities in event-to-event connections.

The dangers in naming and explaining

Observation of regular patterns of interaction both leads to *descriptions* of the regularities and motivates *inferences* about the causal conditions that account for them. Some causal conditions are easily distinguished from the regularities for which they are presumed to account, as is the case when some external factor, such as age, noise, or outside friends, is found to have some particular effect on the interaction pattern. However, description and explanation are often easily confused. It is to avoid this confusion that we have adopted the term *properties* for description and *causal conditions* for the inferred causal factors. The two are to be carefully distinguished. The danger is in going directly and simply from a single observed property of a relationship to an inference of a causal condition. The risks here are subtle. Consider a case in which we observe the property of asymmetry in a particular respect, e.g., that P gives instructions and O follows them more often than the other way around. The first type of risk occurs when, as often happens, the property is given a seemingly innocent label, such as "P's dominance." A label of this sort is tempting because it is familiar, easy to remember and explain to others, and it seems to describe adequately what has been regularly observed. The problem is that most labels of this sort have causal connotations. In their use, we slide unwittingly from naming to explaining. In doing so, we create for ourselves all the possible difficulties entailed in explicitly using the label as an explanation, but with little likelihood that we (or our readers) will be aware of them.

A second type of risk occurs when, as is also common, we conclude that the observed pattern of asymmetry between P and O is caused by P's dominance. That is to say, we explicitly infer "P's dominance" to be the causal condition underlying the observed property. That is perfectly appropriate as a *hypothesis* about the true causal condition, but, in the absence of further information bearing on the inference, it must be treated *only* as a hypothesis. Causal conditions are not merely explanatory constructs; they are simplifying and organizing conceptual tools, serving to impose order on complex sets of observations. To invoke a causal concept, such as "P's dominance," is to imply many different regularities that form a certain pattern. In this particular case, the term implies a causal property of P that will be manifest not only in the particular observed asymmetry in relation to O, but also in other facets of their relationships. To invoke "P's dominance" (in any more than a hypothetical manner) in the absence of further knowledge about P and O or about other relationships is to engage in circular reasoning ("P influences O a great deal because P is dominant, and P is dominant because P influences O a great deal") and, worse, to risk incorrect identification of the true causal condition(s) underlying the observed regularity. For example, if we know only the one fact, there is no basis for ruling out alternative exlanations, such as that the asymmetry in influence reflects a norm governing the P–O relationship or a special vulnerability of O (rather than power of P).

The general point is that the inference of causal conditions and the identification of the true ones is a very complex and often tedious process. To a great extent, it is what science is all about. We find these difficulties easy to accept when we imagine studying some esoteric subject (such as molecular paleontology), but the complexities and problems of causal analysis are easy to forget when we deal with the all too familiar domain of interpersonal relations. Labels and explanations readily leap to mind for almost everything we observe. Naming and explaining are blurred together because the common terms used to describe the interpersonal phenomena evoke vivid causal metaphors. Only by sustained consciousness of the risks in this process and close self-criticism at every step of the route from description to causal inference can we break out of the limitations of lay language and conceptions and establish an objective science of interpersonal relations.

Evidence regarding change

An important data pattern, other than observed regularity, that suggests an inference about causal conditions is evidence about *change* in event-to-event regularities. A simple example is provided by the evidence we would use as a basis for inferring the existence of an acquired habit as a causal condition located in P, as when P learns to respond aggressively to O's passivity. Not only is P observed to respond dependably with a certain subset of p's to certain stimuli provided by O (the two subsets being determined by response-and-stimulus generalization), but, at an earlier time, prior to conditioning or reinforcement, P did not do so. Furthermore, during acquisition of this response, there was to be observed a certain recurrent sequence of events (e.g., the application of the unconditioned stimulus or reinforcement by O) that could plausibly be interpreted as the cause for the development of P's regular responding. Evidence of the weakening of an acquired habit would come from evidence of a decline in the regularity of a particular type of o-to-p pattern, and especially so if this were accompanied by evidence of appropriate causal conditions for "extinction," such as nonreinforcement by O. The point here is that the imputation of a causal condition is often based on a complex pattern of recurrence and regularity, with shifts in this pattern following variation in factors (reinforcement and extinction conditions) that may be inferred to instate or terminate the focal causal condition.

An analogous example is provided by a "pair norm," such as an agreement between P and O that controls their performance of household chores. The data pattern pointing to a norm's existence as a causal condition of the relationship includes the following: (1) There is observed to be uniformity between the two persons in their behavior, including their comments about what one "ought" to do and their application of sanctions for conforming and nonconforming behavior. (2) This uniformity contrasts (a) with the event patterns in the same pair of people at some earlier time and (b) with that in other pairs who are similar in other respects. (3) The uniformity in (1) can be accounted for by other causal conditions existing for the particular pair (e.g., proximity, intercommunication, and attraction) that may be inferred to promote norm development.

The paragraphs above emphasize how causal conditions are inferred from regularities in the events and sequences relating to the dyad. It is important to point out that causal conditions are also inferred on other grounds. When there is known to be a change in some feature of the pair's environment (e.g., change in employment) or in some attribute of one or both members (disability, aging), it is often reasonable for the investigator to believe that some causal conditions have changed. The reasonableness of this belief is based on the investigator's prior knowledge about the particular dyad or similar ones. Ideally, it is possible for the investigator to document the belief by obtaining evidence about changes in regularities relating to the dyad and by showing those changes to be plausibly explained by the alleged condition changes. Another strategy for the study of causal conditions involves comparing samples of dyads for which certain conditions are believed to be different but which are highly similar in other respects. Again, in the ideal case, it is possible to document that the only systematic difference between comparison samples is in regularities related to the alleged causal condition differences.

Sometimes, the investigator is able, directly or indirectly, to control the changes in conditions for a given dyad. This control enables the nature and extent of the changes to be more precisely defined and the time of the changes to be known in advance and accurately located on the time scale. The latter makes possible before-and-after assessments of the regularities believed to be controlled by the conditions. Sometimes investigators can directly manipulate conditions, but more often they will work with the dyad in enabling and directing them to modify their own conditions. Intervention in relationships constitutes attempts of this sort to promote the two persons' modification of their shared physical and social environent, the interpersonal sequential habits of each one, their interpersonal routines, and their relationship's norms.

Proximal versus distal causes

Jessor and Jessor (1973) suggest that the environments that determine interaction should be ordered along a proximal–distal dimension. At the proximal end are such factors as other persons' expectations or evaluations of the relationship and at the distal extreme, such factors as climate, social structure, and culture. This distinction is important in the causal analysis of close relationships. Consider an example: through careful observations, an investigator may infer that a husband's unemployment creates conflict within a marriage. That is, unemployment constitutes a causal condition that affects the level of conflict between husband and wife. It does not detract from that level of explanation to ask about the more proximal conditions, themselves determined at least in part by unemployment, that produce the manifestations of conflict. There are numerous possibilities. For instance, the resources that the husband controls may be reduced, with the consequence that his wife no longer defers to his wishes and conflicts of interest more often result in open disputes. In this case, the proximal causal condition is the husband's resources. Alternatively, unemployment may cause the husband to become depressed, and that personal causal condition leads him to be more sensitive to criticism and more likely to respond aggressively to his wife's remarks, which are no different than they have always been. Or, as a third example, unemployment may simply throw the two together for a large portion of the day, with the consequence that they have greater difficulty in coordinating their various activities. Thus, a particular distal condition may affect quite different proximal conditions, and these, in turn, are likely to differ in their specific impact on the relationship. Our understanding of the causal dynamics becomes complete only when the operative proximal condition or conditions are identified.

The causal analysis is also incomplete if an investigator focuses on proximal causes to the exclusion of more distal causes. For example, a therapist may observe that a couple is having sexual difficulties based on the wife's extreme anxiety about sexual intercourse. Rather than ending the causal analysis with the proximal cause of "anxiety," it might prove useful to examine less immediate causal factors. For example, the wife's anxiety might be traced to fear of pregnancy based on a set of factors, such as health problems preventing her from using reliable contraceptives, the unavailability of abortion facilities in her community, her husband's recent loss of his job, and the existence of four other children in the family. Alternatively, it might be discovered that the woman's anxiety is unique to her sexual encounters with this particular partner and result from her lack of trust in him or his habitual clumsiness in love making, or both. The point to be made by such illustrations is that both proximal and distal causes need to be taken into account in order to develop a complete understanding of close relationships.

Contemporaneous versus historical explanation

The proximal–distal distinction has a parallel in the difference between contemporaneous and historical explanation. The explanation of current interaction in terms of the past experiences of the two persons, separately or together, is one valid mode of causal understanding. However, a more complete understanding requires identifying the contemporaneously existing residues of the experiences (the present attitudes, motives, shared understandings) and determining their effects on the interaction (Lewin, 1943). The latter requires identifying the events and interevent connections to which they give rise, whether events in P's or O's chains or environmental e_{soc} or e_{phys} events.

The Concept of Causal Condition

It is appropriate now to explain in more detail what we mean by "causal condition." The term *condition* was selected to refer to a kind of causal factor that is distinguishable from the class of "events." We use *condition* to refer to a broad class of such causal factors, without commitment to any particular kind, such as trait, state, propensity, or disposition, within that class. As the earlier examples of habit, social support, and noise illustrate, a term is needed that applies equally well to P, O, their social environment, and their physical environment.

Almost all aspects of the dictionary definition of *condition* are appropriate for the present usage. Condition refers to a particular state or form of being, including a particular state in regard to circumstances, position, or social rank and a particular form of being or nature. A condition is something that must exist if something else is to be or to take place, an affecting influence, something that limits or modifies the existence or character of something else.

From this definition, the reader will understand the purposes that causal conditions serve in the understanding of interpersonal relationships. On the one hand, they are relatively stable causal factors that exist over relatively long time periods (relative, in both cases, to the brief causal elements we call "events"). Causal conditions affect or influence the occurrence of events and sequences of events, and, because of their relative stability, causal conditions account for the *recurrence* of events and sequences. On the other hand, when they change, causal conditions produce noticeable shifts in the properties of the relationship. Insofar as a particular causal condition influences a number of events and sequences and does so over a period of time, its change has ramifying effects, both over a variety of occasions and over a broad class of events and event-to-event connections.

The stability of causal conditions is only a matter of degree. They are stable relative to the short time span of the type of causal entity we have labeled "event." Some conditions exist over a period of years, others over periods measured in days, and others over periods measured in minutes. There is a continuum of length of existence (duration) along which various causes may be located. For example, emotion may be brief (an acute experience of fear), or several hours long, or a chronic state of anxiety (susceptibility to feelings of fear). By the term *event,* we capture the briefest of these phenomena—the kind that occurs in the course of interaction, for example, as responses to other brief stimuli and as stimuli themselves to subsequent brief responses. The term *condition* is particularly applicable to the long-term states that are responsible, during their existence, for the occurrence of certain brief events and event-to-event sequences. Thus, a personal condition of susceptibility to anxiety is evident in the fact that certain cues provided by the partner or environment regularly result in brief experiences of fear.

The preceding examples illustrate how causal conditions affect events and sequences. However, causal conditions are also often affected by these events and sequences. P's habits of responding to O are acquired by virtue of the occurrence of certain sequences of events in their interaction. The support provided to O by the social environment is promoted by what O does and how these actions affect other persons. The noise in P and O's physical environment may sometimes be of their own making, a product of certain events in their chains that cause e_{phys} events of "noise," as when they buy a noisy washing machine.

Because they may be affected by events and sequences as well as affect them, causal conditions account for observed long time delays in the causal connections between earlier and later events and sequences. For example, O does something to P that has an immediate effect (e.g., O's insulting remark followed by P's anger). Several days later, P reacts to some minor action on O's part in a way that reflects the earlier sequence (e.g., P become inappropriately angry at some innocent comment by O). The assumption that the initial o-to-p sequence resulted in some change in a causal condition in P (e.g., P's memory, attitude, belief) provides the necessary causal account of the delayed effect.

Kinds of Causal Condition

Some causal conditions can be located *in* the environment or *in* one or the other person. Other causal conditions can be more accurately characterized as existing in the relation *between* environments and persons or in the relation *between* the two persons.

Determining the "location" of a cause can be a difficult matter. At one level, all causes are relational. If P causes an effect in O, a comprehensive analysis must examine both the "potency" of P and the "vulnerability" of O. However, in much of our thinking, we give differential attention to the qualities of either P or O. Consider three examples of an arrow wound resulting in the death of a man. If an arrow wound killed a healthy man, we would focus more attention on the arrow (its sharpness) than on the qualities of the man (that he, like all human beings, shares a susceptibility to sharp objects). If, however, an arrow wound killed a hemophiliac man, a causal analysis might focus more attention on the qualities of the man (hemophilia) that make him and other people with that attribute uniquely susceptible to all wounds. Finally, if an arrow wound killed Achilles, a causal analysis might focus equal attention on the well-shot arrow and the unique susceptibility of Achilles. In each example, the death required both an arrow and a vulnerable man. But the focus of our causal analysis varied, contingent on our knowledge about the response of all people to certain events, about the differential response of people with particular qualities to certain classes of events, and about what constitutes a unique response.

We did not take the examples above from interpersonal interaction for obvious reasons. Such clear-cut data patterns about common and unique responses are not so easily identified in interpersonal interactions. Most causal conditions account for only part of the total interactional variance, and complex combinations of multiple causal conditions are necessary for a complete account of interaction. Furthermore, causal conditions both influence and are influenced by other causal conditions. This mutual influence, in which the causal links go in both directions, further complicates any attempts to locate the causal conditions that account for event-to-event regularities.

In the sections below, we discuss and illustrate causal conditions that are often located *in* environments or persons and causal conditions that are often located in relations *between* environments and persons. The determination of location is somewhat arbitrary. The reader should remember that all causal effects are ultimately relational, based on the relation between "potency" in one location and "vulnerability" in another. However, one can conceptually analyze each set of attributes separately.

Environmental and personal conditions

Causal conditions of the physical environment (identified by the symbol E_{phys}) generate regularities in the e_{phys} events that impinge on the relationship. For example, weather conditions provide recurrent rain or thunder, and working conditions provide recurrent noise, poor lighting, and regular availability of certain tools. The conditions of the social environment (identified as E_{soc}) produce regularities in the e_{soc} events that affect P's and O's chains of events. For example, the social conditions at P's working place determine that P is regularly exposed to directive instructions from a supervisor and distractions from co-workers. The economic conditions of the pair are responsible for both the weekly paychecks and the monthly bills. The availability of alternative partners as a condition of P's social environment may be responsible for P's receiving regular personal compliments, eye contact with smiles, and invitations to private interaction. "Support" as a condition of the social environment is responsible for such regular events as provision of advice, help with tasks, and offers of loans of money. "Social norms" constitute causal conditions that result in the regular presence of behavioral models who exhibit uniformity and provide consistent exhortation to conforming behavior.

For all practical purposes, environmental conditions are also responsible for the p and o events that are closely connected to the e events in the e-to-p or e-to-o sequences. In Figure 2.4, the environmental events $e_{phys\,1}$ and $e_{soc\,1}$ might, in certain circumstances, be considered responsible, respectively, for p_1 and o_2. Thus, a p event that is dependably (consistently for each person), generally (for all persons), and uniquely caused by a certain e event may be considered part of the consequences of the environmental condition (E_{phys} or E_{soc}) responsible for the e event. In these cases, the description of the environmental condition often includes references to the p events. For example, we might say that the environment of the workers in a certain manufacturing plant is "stressful," this term being a characterization of their working conditions. The term refers to the fact that the e_{phys} events that regularly affect the workers in that plant (e.g., noise, fumes, pace of work) dependably produce symptoms of stress (e.g., heightened blood pressure).

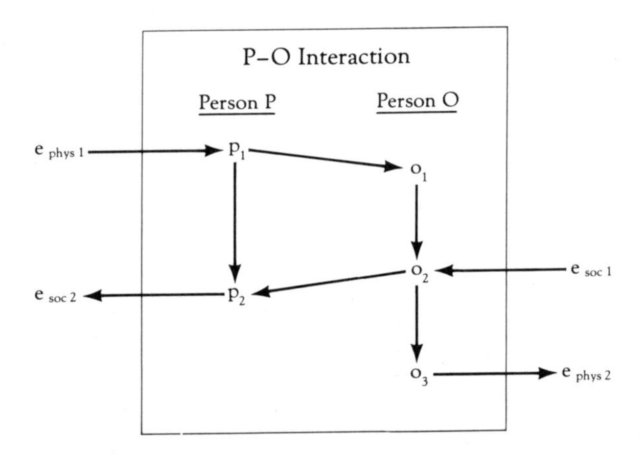

FIGURE 2.4
Physical and social environmental events that affect or are affected by the P–O interaction. All arrows represent causal connections.

Similarly, we often describe conditions of the social environment in terms of dependable p effects, as when we speak of a "supportive" or "distracting" social environment. However, we must not overlook the necessity ultimately to identify the specific e events that any particular environmental condition produces. Just as the public health researcher must identify specific "stressors," our causal analysis will be incomplete—and wholly "psychological"—unless we identify the e event in even the highly dependable e-to-p sequence.

Environmental conditions are also responsible for certain of the events that occur as a result of p or o events, for example, the p_2-to-$e_{soc\,2}$ and o_3-to-$e_{phys\,2}$ sequences of Figure 2.4. It is the environmental condition of task ease or task difficulty that causes all persons to do well or poorly in physical or social tasks. In general, it is some condition of the environment that causes some particular e to occur consistently (for each person), generally (for all persons), and indiscriminately (for a large class of p events). Thus, a social environment characterized by ethnic prejudice will be responsible for the fact that for many persons and for many p's, the resulting consistent social environment event will be rejection and hostility.

Causal conditions associated with the person (referred to as P or O conditions) are responsible for the events and sequences of events that regularly appear in the P or O chain. P's habits of conversational interaction determine how P responds, both verbally and nonverbally, to the other person's questions and comments. O's hearing impairment reduces O's appreciation of musical performances and limits O's participation in conversations under noisy circumstances. P's state of anxiety may be responsible for recurrent fear responses both to events in the physical and social environment and to events internal to the person (e.g., thoughts of phobic objects). As the last example illustrates, some personal conditions are responsible for regularities in the p-to-p sequences in a person's chain. These include such causal conditions as "thought habits" (resulting in regularity in the way one thought leads to another); "writing style" (generating the regular organization of written verbal output); and "motor skills" (producing sequential patterns of motor behavior, as in typing or ice-skating). Similarly, labeling and attribution tendencies cause regularities in e-to-p-to-p sequences, as when a stimulus leads to a certain percept, which leads to a certain verbal label or explanatory response.

Just as environmental conditions are responsible, in a practical sense, for the p events dependably caused by certain e events, so personal conditions may be considered responsible for the e events dependably linked to relevant p events. The personal conditions of skill and strength generate recurrent p-to-e_{phys} sequences that are characteristic of P. Conditions associated with appearance (beauty, disfigurement, obesity) may produce certain p-to-e_{soc} sequences, the e_{soc} occurring for most observers in the social environment, consistently so (for each observer), and uniquely so (not for other p events). Analogous to the cases in which environmental conditions are identified in terms of their dependably associated psychological consequences, certain personal conditions are identified in terms of their dependable effects on the environment. The conditions of being "lovable," "attractive," or "repulsive" refer to kinds of e_{soc} events that are dependably linked to the specific p events caused by the conditions. As before, a complete analysis of the causal links requires identifying the relevant p events in these recurrent p-to-e_{soc} sequences.

Psychological traits, such as aggressiveness, dominance, and introversion, have classically been regarded as P causal conditions. They were assumed to govern a person's behavior in a wide variety of settings and relationships. The more recent "situationist" view of traits reflects the growing evidence that the regularities are more situation-specific than had earlier been assumed (Bowers, 1973). Issues of this sort can be decided only by the details of research results, and, unfortunately, the facts are not always simple. For example, recent work by D. J. Bem and Allen (1974) suggests that some people possess traits to a greater degree than do others, inasmuch as they show greater cross-situational consistency in their behavioral tendencies. The evidence showing the situational specificity of behavior relating to such traits as honesty and shyness directs our attention to causal conditions defined by the relation between E and P. We next consider the general class of such conditions.

Relational conditions

We emphasized above that all causal effects are ultimately relational. Yet, it has been convenient, and appropriate under certain specified conditions, to identify certain conditions as "environmental" and others as "personal." However, there are certain conditions that can be understood to exist only in the *relation between* environment and person or in the *relation between* two persons. Whereas the symbols E_{phys}, E_{soc}, P, and O have been used for simple causal conditions, which can reasonably be said to exist *in* the environment or the person, we will use the symbols $E \times P$ and $P \times O$ for these relational or "joint" causal conditions. Relational conditions are constituted by pairings of E and P (or of P and O) and produce effects that are not predictable from either factor alone.

Consider, for example, the $P \times O$ conditions of propinquity, speaking the same language, attitude similarity, and personality complementarity. Each of these conditions is responsible for recurrent p-to-o and o-to-p sequences: Propinquity is a basic condition that governs the number and types of such sequences that can occur; common language is responsible for sequences entailing successful communication; attitude similarity, for sequences in which opinion expression leads to expression of agreement; and personality complementarity, for sequences in which one person acts in a manner that fulfills the other's needs. In none of these cases can the causal condition be located in either person. Each condition is defined by the conjunction of attributes of the two persons—their respective spatial locations, language skills, attitudes, and personality dispositions. Similar relational conditions can be identified for combinations of E and P, as when we consider the fit between P's training and the requirements of P's job, or between P's social skills and the expectations of P's social group.

Some relational conditions exist for a specific E and a particular P and generate sequences that are more or less unique to that pair. For example, the condition of possessing a special allergy causes a recurrent e-to-p sequence that is consistent for the person but rare for other persons. There is some particular potency of the environment (the allergen) that exists only with respect to the particular vulnerability of the person, and, similarly, the vulnerability of the person exists only in relation to this particular potency of the environment. A parallel to the special environment–person interaction caused by a rare sensibility is that caused by a special talent. Consider the idiot savant who is able to rapidly multiply mentally pairs of seven-digit numbers: This ability exists only in relation to a specific kind of task, and the task has this special tractability only in relation to this unique ability. Analogous relational conditions specific to a particular P and O are responsible for ways in which they uniquely influence and respond to each other. It is to these P × O conditions that we refer when we say that P and O "strike it off unusually well" or "have a special chemistry for each other" or when outsiders don't understand what P and O "see in each other." P × O conditions are also seen in the unique ways in which P and O manage to aggravate each other.

Some relational conditions, as in most of the examples given above, are present at the outset of the relationship. These relational conditions are based on preexisting properties of P and O and on the way those properties dovetail or fit together. Other relational conditions are *emergent,* arising from the interaction between P and O. Thus, a pair may develop special interpersonal habits, one or both persons learning specific behavior that is uniquely elicited by the other person's events. These habits will cause recurrent chains of p-to-o and o-to-p sequences, as in greeting, love-making, conversation, and fighting routines. For example, P and O may exhibit a regular pattern in which P criticizes O, O cries, and P feels guilty. This reflects an emergent P × O condition (perhaps P's ambivalent attitude toward O) if the pattern is unique to the P–O relationship, for example, if P feels badly about O's response in a way that differs from P's feelings about others persons' similar reactions to P's criticism.

Norms, agreements, and shared understandings are sometimes emergent P × O conditions, existing between P and O and having no existence independent of the relationship. Such conditions generate recurrent sequences, for example, of giving and receiving, leading and following, and coordinating activities. The overt behavior in these sequences is usually accompanied by special (intrachain) private events, such as P's expectations of O's behavior, P's awareness of O's similar expectations of P's behavior, and P's knowledge that O knows of P's expectations and of P's awareness of O's expectations. Shared perception of attitude similarity, another emergent $P \times O$ condition, also entails not only knowledge of the similarity but awareness of the partner's similar knowledge. These expectations and perceptions make very salient to P and O that the understanding or similarity exists as a condition "between" them. Thus, these expectations and perceptions become important aspects of the "subjective" meaning of the close relationship.

Changes in Causal Conditions

We have analyzed the causal processes of interpersonal relationships at two levels—at the level of the fleeting causal phenomena described as *events* and interevent connections, and at the level of relatively stable causal phenomena described as *causal conditions.* Figure 2.5 shows in schematic form the relations between and within each of these levels. The interevent *causal connections* shown at the lower level are to be contrasted with the *causal links* that connect the two levels and that provide direct causal relations between various causal conditions. (In this and subsequent similar diagrams, for simplicity we omit representation of the specific connections between environmental events and events in the two persons' chains. The causal links going to and from the interaction are, of course, joined to it by specific connections, such as e-to-o or p-to-e connections.) We may use the diagram in Figure 2.5 to analyze how changes in the various causal conditions may occur. This topic relates to later chapters (Chapter 8, "Development and Change," and Chapter 10, "Intervention"), so the discussion here will be brief.

It is obvious that many of the conditions affecting the dyad change for reasons that have nothing to do with the dyad itself. The environmental conditions, and even many of the personal conditions, are themselves embedded in causal systems outside the dyad and are subject to change as those systems change. Wars, social movements, economic recessions, drought, and exhaustion of natural resources are but a few of the broad conditions that affect E_{phys} and E_{soc} and, often, P and O themselves. One aspect of this broad picture is that changes in some of the conditions affecting the dyad often produce changes in other relevant conditions. A change in economic conditions that affects the resources P brings to the relationship may also cause changes in the social environment of P and O, as when friends move away to seek new jobs or when a member of the extended family loses a job and becomes dependent on P and O. In the upper portion of Figure 2.5,

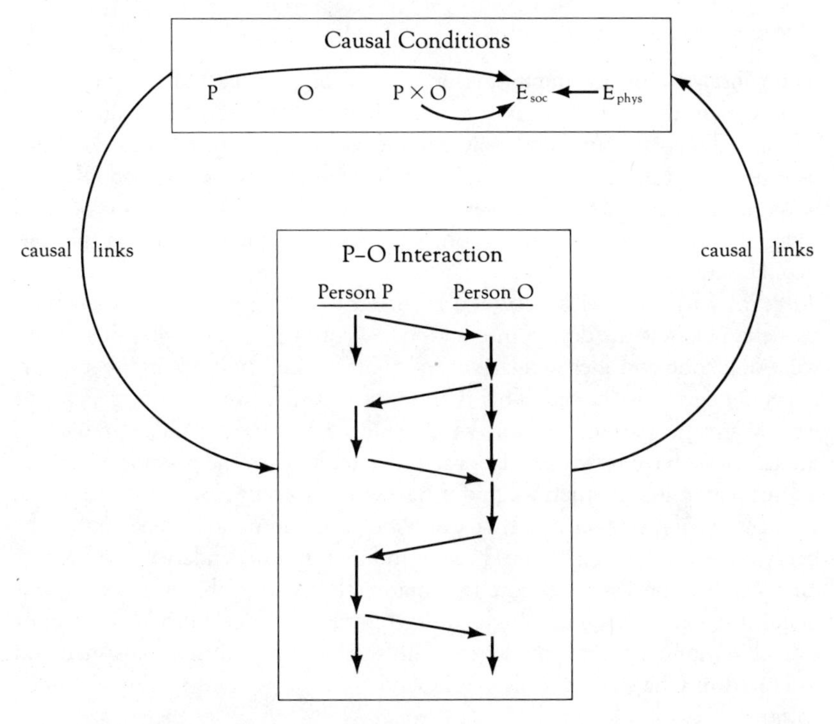

FIGURE 2.5
The causal context of dyadic interaction. The arrows within the interaction represent interevent causal connections within and between P's and O's chains. Causal conditions affect the interaction (through the downward causal links), and the interaction affects the causal conditions (through the upward causal links). Causal conditions are also often linked directly to each other, as in the upper part of the diagram.

we have illustrated such intercondition causal links by arrows that connect one causal condition with another.

However, some of the causal conditions are affected by the dyad. The members of a dyad take actions that modify its physical and social environments. They move their residence, construct and modify their living quarters, drop old friends and make new ones, change their membership in social groups, make regular deposits in savings accounts, and so on. Through the interaction within the dyad and with the external environments, the P and O conditions become modified, as when the persons acquire new individual skills, habits of thought, and needs. And through interaction, the pair can modify the P × O conditions, as in learning new interpersonal routines, agreeing to follow different rules as to division of labor, and adopting new schedules of joint and individual activity.

Direct and indirect interdependence

The causal conditions impinging on a dyad may be affected by their interaction, as in the examples above, but also by one or both persons' individual actions. Thus, the husband may gamble away the joint savings with disastrous consequences for the couple, or the wife may maintain good relations with neighbors, who then make possible the pair's occasional use of a ski lodge. In their respective effects on such causal conditions, the two persons may be said to be *indirectly* interdependent. This can be contrasted with the *direct* interdependence constituted by the causal connections between their two chains of events during interaction. According to our earlier definition, we would not consider a relationship characterized only by indirect interdependence to be "close." This is the way in which we are interdependent with many people with whom we never interact and of whose individual identities we have no knowledge, as, for example, with the prior occupants of a forest campsite at which we find it necessary to stop. Observations that the world is getting smaller and that everyone is becoming interdependent with everyone else refer to increases in such indirect interdependence. For present purposes, it is necessary to note that many relationships that are close by our definition (i.e., that have high direct interdependence) also involve considerable indirect interdependence. This distinction is further explained and discussed in Chapter 6 ("Roles and Gender") as it pertains to gender-linked roles.

Causal loops

Inasmuch as the dyad is both affected by its causal conditions and able to modify them, there exist causal loops in which certain conditions affect the internal process in ways that then affect the initial or other conditions.

Causal loops may be illustrated by examining the contribution of two initial P × O conditions (propinquity and objective attitude similarity) to the formation of a dyad and indirectly, through their effects on its process, to the development of further conditions that promote its continuation. Let us imagine that two people are thrown together for a brief period of time, for example, while traveling, at work, in a classroom, or at a party. They begin to interact, extend the period of being together, and arrange to get together later on. At a gross level, we see in this example the causal loop shown in Figure 2.6. Environmental conditions change the P × O condition (propinquity), which then, through the downward causal link, affects certain aspects of their interaction. In general, being in physical proximity serves to make possible certain causal interconnections between the two persons. These causal interconnections have effects on the relationship (via processes *within* it, as described below) such that, through the upward causal link, the interaction acts on the condition to maintain it or even to change it further in the direction of its initial shift. Having been moved together by some

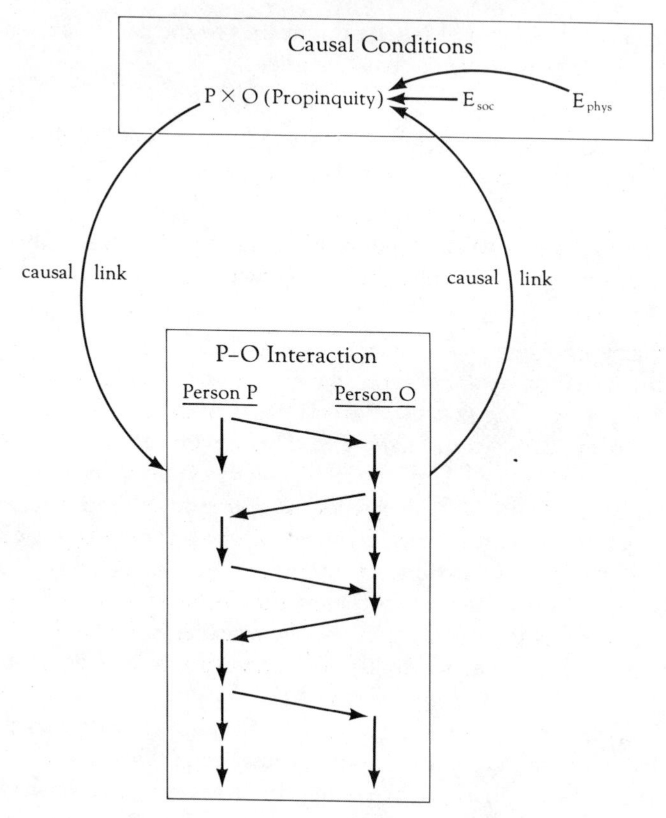

FIGURE 2.6
Illustration of a causal loop. The P × O causal condition of propinquity, initially caused by the social or physical environment, has effects on the P–O interaction (through the downward causal link) which, in turn, acts (through the upward causal link) to maintain or increase the degree of propinquity.

external event, the pair continues to move closer together and to maintain and regularize the proximity.

This example illustrates positive feedback. An initial change produces further changes in the same direction. We can easily imagine a contrasting scenario in which the opposite occurs. Upon being thrown together, the two find each other to be disagreeable and take action to move apart again. This sequence constitutes a causal loop with negative feedback. The dyad acts to restore the original condition of separateness.

A full understanding of a causal loop, such as that pictured in Figure 2.6, requires determining the dynamic processes *within* the intraction that, so to speak, *close the loop.* This determination is an important aspect of the proximal–distal problem of causal analysis described earlier. In the present case, any account of the internal process that relates to the pair's increasing or maintaining their own proximity must identify:

1. Events in P's and O's strings that move them together (or that are incompatible with movement apart, keeping them together once they become so)
2. Causal connections between P and O that lead to such events and that are made possible by their initial being together

The exact type of events and connections specified for a given problem depend on the investigator's *theory* about the events and causal connections within the interaction. One possible elaboration of these processes for our propinquity problem is shown in Figure 2.7. This elaboration is developed along the lines of Altman and Taylor's (1973) theory of social penetration. Alternative elaborations could be developed according to other theoretical orientations, such as a strict reinforcement view, in which approach behavior is reinforced during the interaction, or an "imprinting" model, in which events in the initial interaction serve as cues to elicit behavioral patterns that each person acquired early in life. However, the example in Figure 2.7 will serve for our present purpose, which is to illustrate the role of *interaction dynamics* in the causal loop shown in Figure 2.6.

In Figure 2.7, the two interconnected chains of events are shown in the center box. For convenience, the input from causal conditions is shown at the left, and the output to them, on the right. However, as our arrows show, the conditions are causally linked with both persons' chains of events. Two causal conditions, propinquity and objective attitude similarity, are shown to exist initially on the input side. These factors exist throughout this interaction and make possible the interconnections shown. (It goes without saying that the events and causal connections listed in the interaction process in Figure 2.7 are but a truncated version of what would ordinarily be involved in an interchange of this sort.)

As Figure 2.7 shows, while being together, when some unspecified external or internal factor stimulates P to express an opinion on some matter, O hears it. O finds that it agrees with O's own views, then feels good and expresses agreement. P notices O's positive affect, notes that O shares and supports P's opinions, and, in turn, feels good about it. (Perhaps the "feeling good" reactions reflect P and O conditions relating to their respective needs for opinion validation.) P's reactions are noticed by O, who infers that P is aware of the agreement and is pleased by it. This set of p and o events provides a first impetus to the development of a new condition, namely a shared perception that the two hold similar attitudes. Meanwhile, the prior events in P's chain cause P symbolically to "approach" by making smiling eye

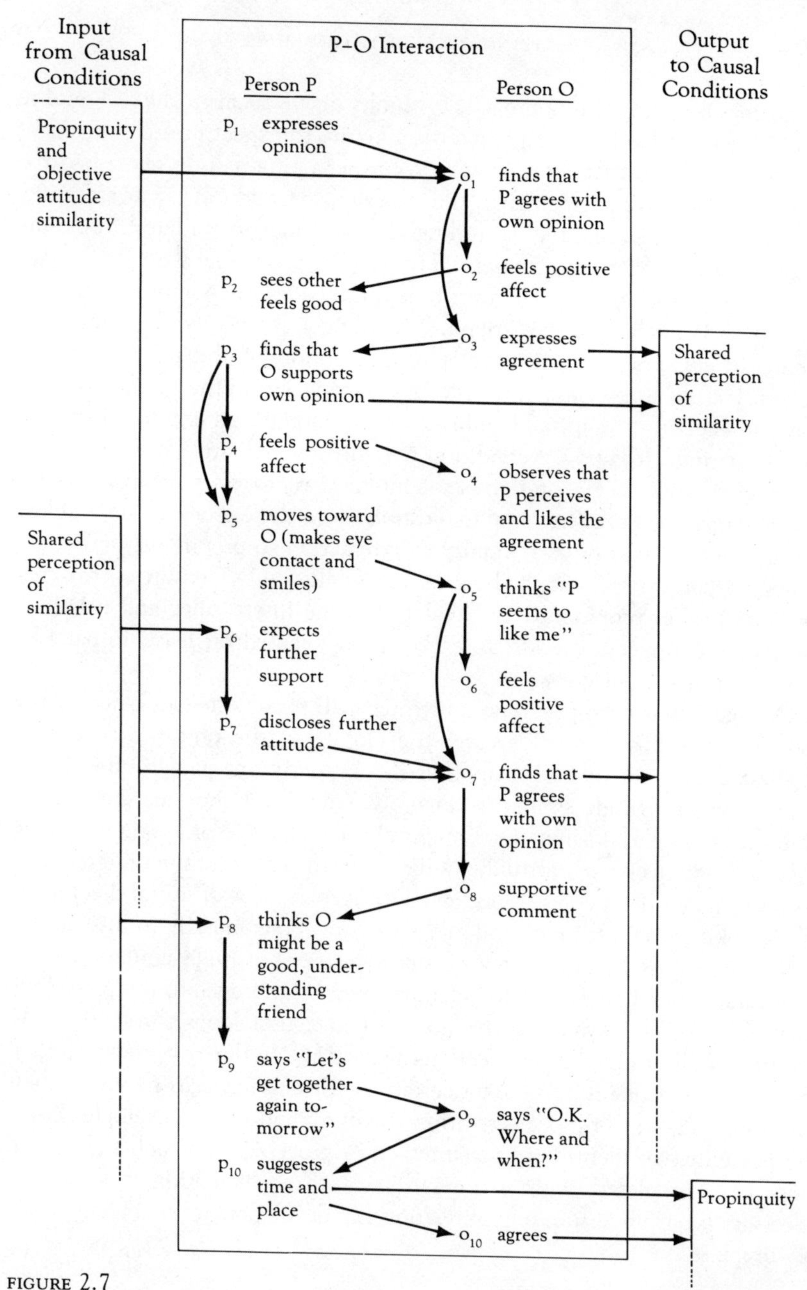

FIGURE 2.7

Illustration of an interaction process that closes the causal loop by which initial causal conditions (propinquity and objective attitude similarity) are sustained or strengthened and lead to a new causal condition (shared perception of similarity). The interconnected chains of events for the two persons are shown in the center. Input from the causal conditions to the interaction is shown by arrows entering from the left, and output from the interaction to the causal conditions by arrows leaving to the right. Interaction closes the causal loop by including direct or indirect connections between those events affected by the input and those events responsible for the output.

contact. Observing this approach, O thinks that P seems to like O, and so on. When P is led by a perception of similarity to expect further agreement and support and, therefore, to state a further opinion (one related to the domain of perceived agreement), the statement strengthens O's awareness of their attitudinal similarity. Accordingly, O responds to P's latest disclosure with a warm, supportive comment. This response, together with P's increasing certainty that the two hold attitudes and values in common, leads P to think that O might be a good friend. Accordingly, P suggests a later meeting. Similarly encouraged by a developing perception of the relationship, O agrees. It is, of course, the latter effects occurring within the interaction that have an effect on the initial condition of propinquity, serving to reinstate it at a later time. It is now a condition partially under P and O's control.

This hypothetical elaboration of the interaction processes occurring during this first meeting between P and O illustrates how these processes might close the causal loop linking propinquity to interaction to propinquity. The two necessary conditions described earlier are fulfilled: (1) Certain events that promote further propinquity finally occur in the interaction, and (2) these events are connected causally to earlier events caused (at least, in part) by the initial propinquity.

Our particular example shows only one such possible means whereby the loop may be closed. In this case, the closure of the propinquity loop is facilitated by loops involving another P × O condition, namely, the shared perception of attitude similarity. Figure 2.7 does not show it, but, in the course of this relationship, there might also be a closure of a loop involving the condition of *objective* attitude similarity. In the course of their interaction, P and O may influence one another's attitudes in areas of initial difference. Thus, their initial attitude similarity may contribute causally to interaction that is then causally linked to a further increase of attitude similarity.

Several general points are to be emphasized by reference to the preceding example: (1) The dyad may be involved in causal loops through which certain conditions having an effect on the dyad are, in turn, affected by it; (2) the loops may make it possible for certain conditions *indirectly* to affect other conditions, by way of processes within the dyad; and (3) the example shows the possibility of identifying the interaction processes by which the causal loops are completed. Full investigation of these mediating processes is necessary if our understanding of relationship development and change is to be complete.

Research examples

Existing research on interpersonal relationships provides a few examples of the type of investigation implied here. Snyder and his colleagues report two studies, each dealing with a P condition (one person's preinteraction attitudes or beliefs about the other) and the positive feedback causal loops in which it is involved. In the first, by M. Snyder, Tanke, and Berscheid (1977), male subjects interacted, via an intercommunication system, with a woman who, they were led to believe, was physically attractive or unattractive. As compared with those in the "unattractive condition," the men who believed the partner to be attractive talked with her in a way that independent raters judged to be more sociable, bold, and attractive. Apparently as a consequence, the woman responded in a way that was (again, according to independent raters) more sociable, poised, and socially adept. In short, the man was led by his expectations of the partner to act toward her in a way that confirmed those expectations. The entire causal loop is not documented; the investigators present no evidence on whether the man's expectations about the partner were changed by the interaction. However, it is likely in such cases that the man's initial positive expectations would create a positive feedback loop with the confirming evidence strengthening his positive orientation toward the partner and that, in turn, heightening her own positive response to him.

In contrast, M. Snyder and Swann (1978) demonstrate some of the elements of a positive feedback loop involving initial *negative* attitudes. Person P's expectations of being treated hostilely by an opponent in a competitive game led P to act in a more hostile manner toward the opponent than otherwise. P's manner led the opponent also to behave aggressively, resulting in P's strong belief that the opponent was a hostile person. Thus, the hostility between P and O set in motion by P's initial beliefs may easily escalate, with each person becoming convinced of the other's negative attitude.

In contrast to these positive-feedback-loop scenarios, in which the initial direction of interaction becomes accentuated by its effects on causal conditions, one can also find examples of a negative feedback loop. Thus, one person's initial hostile feelings toward the other may lead to events that cause the person to withdraw from interaction and thereby permit the hostility to subside. Negative feedback loops of this sort probably occur as reponses to the equitableness of allocation of costs and rewards between persons in stable relationships. As evidence from Walster, Walster, and Traupmann (1978) suggests, in most heterosexual couples, there is a rough balance between the two persons in what they get out of the relationship relative to what they put into it. Theories about the equity "process" (e.g., Walster, Walster, and Berscheid, 1978) describe the dynamics of an equity-restoring process that is set in motion by perceived departures from the balance point. Both the overbenefited and underbenefited person feel discomfort if perceived inequity is too great and act so as to restore it.

Causal conditions and stages

As long as the causal conditions remain unchanged, the relationship is likely to be stable—on a "plateau" or in a "groove." However, when one or more of the important conditions change, the relationship will tend to move to a new "stage." That is, it will change in many respects, taking on new properties (e.g., of strength, diversity, asymmetry), and exhibiting them for some considerable period of time. For example, in the early period of a love affair, two persons may come to trust each other, reach an understanding about one another's feelings, and make commitments; this change in P × O conditions will have a stable and ramifying effect on their relationship. They will have moved to a new stage of their relationship, in terms of degree and extent of interdependence, reflected in changes in many of the ways their two chains of events are interconnected. Similarly, if P undergoes a change in a personal causal condition, as in somehow becoming physically disabled, the P–O relationship will enter a new stage as, to marked degrees and for some considerable time period, the properties of their interdependence are changed. Similarly, a relationship may enter a new stage if, through change in environmental conditions (employment and location of residence), there are general and semipermanent changes in the interdependence.

The Causal Context of Dyadic Interaction

As Figure 2.5 and our various examples suggest, the internal causal structure of the dyad and its context of causal conditions are very complex. No investigator studies the entire framework in all its complexity. Nor does any existing theory attempt to analyze it in a manner that is both comprehensive in scope and detailed in level. However, all investigators, all hypotheses, and all theories relating to dyadic interaction refer in one way or another to this broad framework or to its components.

To appreciate the variety of types of questions that may be asked within this framework, let us consider specific issues relating to the formation and development of relationships. Many hypotheses and studies of interpersonal attraction deal only with the "downward" causal linkage in Figure 2.5. For example, Byrne (1971) and his colleagues have investigated how the P × O condition of attitude similarity affects the initial attraction between P and O. This type of research has been criticized for dealing only with "first impressions" and not considering interaction between P and O (i.e., the causal processes within the dyad).

A contrast is provided by studies that investigate relationship development over a long time span. As described in Chapter 8, "Development and Change," many of the hypotheses about long-term development and change consider how shifts in causal conditions may be responsible for modifications in interaction properties. This kind of question is posed, for example, by theories that relate age and occupational changes to the internal dynamics of the dyad (e.g., increase in the wife's power as related to increasing age of the couple, Kelley, 1981). Other hypotheses concern the direct causal links between various causal conditions, as when changes in available friends (changes in E_{soc} conditions) are thought to produce shifts in the needs that P brings to the P–O relationship (a personal condition).

Studies of the "upward" causal linkage in Figure 2.5 are those that focus on how, in the course of its development, the dyad changes its various causal conditions. Examples are provided by investigations of the changes in love (a P or P × O condition) as a person interacts with, gets new information about, and thinks about the loved one (Bentler and Huba, 1979; Tesser and Paulhus, 1976); of the internalization of their fathers' values by sons who experience rewarding interaction with them (Payne and Mussen, 1956); and of reduction in outside opposite-sex contacts as a heterosexual pair moves to commitment (Leik and Leik, 1977).

Newcomb's study (1961) of the process of getting acquainted is an admirable example of research that analyzes the causal loops that link the dyad to its causal conditions. Newcomb finds that initial perceived similarity leads to attraction and interaction. However, the initial perceptions are often incorrect. With further interaction, the P × O condition of perceived similarity changes. Attitudes become perceived with increasing accuracy so that, finally, perceived similarity corresponds closely to actual similarity and interaction is most frequent between persons who initially held objectively similar attitudes. Newcomb did not obtain much direct evidence about the interaction processes within the dyad. However, from his evidence on the increasing convergence between objective and perceived similarity, we can infer that the crucial events and interchain connections involved the disclosure of important attitudes but little change in these attitudes.

CONCLUSION AND ISSUES

This chapter has attempted to show that any study of the dyadic relationship and any theory about it will deal, explicitly or implicitly, with some portion of the conceptual framework we have outlined. This framework includes *events* in the interaction, *properties* of the interaction, *interchain causal connections* (internal causal dynamics), *causal conditions,* and various *causal links* among the conditions and between them and the events. We present our framework as one within which investigators can locate their particular problems. We believe it is important for the investigators to have this broad framework in mind in order to identify wisely the boundaries of their special interests and to remain aware of the ways in which those interests may border on the work of other researchers. It is important to know the larger context of particular projects and, especially, the "neighborhood" of related studies and ideas in which they are located.

Our analysis highlights conceptual and methodological issues that are common in research on interpersonal relationships. The first issue is the existence of *causal loops.* Even though much work on relationships will continue to be linear in its causal analysis ("A causes B"), much can be gained by considering each linear link as a possible portion of a larger feedback process ("A causes B, which causes A, etc."). In this way, otherwise discrete parts of knowledge become interrelated. For example, the finding that "similarity leads to attraction" becomes seen as related to research showing that "attraction leads to similarity." Most important, the identification of causal loops will increase understanding of the dynamics of relationship change and resistance to change, inasmuch as these loops characterize the ways in which interaction is both a product of its causal conditions and a possible mechanism for the modification of those conditions.

Second, we would emphasize the importance of detailed investigation of *causal processes within the dyadic interaction.* It is possible to measure input to and output from the dyad and to fill in the intervening processes by speculation. There has been a tendency to treat the dyad as a "black box," with much theorizing about its contents but little effort to determine them. Unfortunately, this tendency leaves us with many gaps and flaws in our understanding. For reasons of inaccessibility and the complexities mentioned earlier, the analysis of internal processes is difficult. However, it should not be avoided.

Related to the black-box problem is a third issue raised by our framework, concerning *distal versus proximal analysis* of causation. One form of the problem is how to translate causal conditions into their effects within the relationship, where they become the proximal causes for events in the interaction. For example, our analysis suggests the importance of maintaining a sharp distinction between propinquity as a distal causal condition and the proximal internal events that it affects (see Figure 2.7). As we have emphasized, causal conditions affect the events and causal connections (both interchain and intrachain) within the relationship. Causal conditions should not then be equated with those internal events or connections. Attitude similarity as a causal condition affects the causal link between disclosure and agreement, and it is by those internal events that attitude similarity has its effects on the relationship. Similarly, when the respective genders of P and O are observed to affect their interaction, we must not use gender as a direct explanation for the effect. Rather, we must ask what specific internal events and connections are modified by the condition of gender.

Finally, a fourth issue highlighted by our analysis concerns the relative importance to the course of the relationship of the "downward" and "upward" *causal links* in Figure 2.5. The dyad is located in a set of causal loops that interlink it with the "external" causal context constituted by P, O, P × O, E_{phys}, and E_{soc}. In part, the dyad is a creature of the external factors that condition and shape its internal processes. However, insofar as it acts to select and modify the conditions, the dyad is also partly a creator of its own causal environment. An important continuing issue for every dyad (as, indeed, for every individual) is the question of the degree to which it is to be master of its conditions rather than a victim of them. How manageable are the causal conditions and how much can they be modified by the dyad itself? This question takes on special importance in relation to the P × O conditions, such as norms, interaction habits, and understandings, which to an important degree are products of the interaction itself. These conditions can play an important part in controlling and eliminating conflict between P and O, and it is in this feature of the relationship that the question of modifiability of causal conditions becomes most significant. In almost every interpersonal conflict, the central causal question is whether the process is accounted for by immalleable conditions (e.g., incompatible backgrounds, stable personality traits, impossible economic circumstances) or by malleable conditions (e.g., poor communication conditions, inadequate interpersonal skills, changeable occupational roles). This important type of causal question can be answered only when the causal loops of the dyad and the effect of internal interaction processes on external conditions are well understood.

CHAPTER 3

Interaction

EVIE McCLINTOCK

Tillie Olsen (1956) in the novella *Tell Me a Riddle* poignantly sketches the relationship between David and Eva:*

> For forty-seven years they had been married. How deep back the stubborn, gnarled roots of the quarrel reached, no one could say—but only now, when tending to the needs of others no longer shackled them together, the roots swelled up visible, split the earth between them, and the tearing shook even to the children, long since grown.

Having retired, David wants to sell the house and join the Haven, a cooperative for the aged. Eva won't consider it.

> "What do we need all this for?" he would ask loudly, for her hearing aid was turned down and the vacuum was shrilling. "Five rooms" (pushing the sofa so she could get into the corner) "furniture" (smoothing down the rug) "floors and surfaces to make work. Tell me, why do we need it?" And he was glad he could ask in a scream.

---

*The following excerpts are from the book *Tell Me a Riddle* by Tillie Olsen.

During the preparation of this chapter the author was supported by grant OCD 90-C-620 from the Administration of Children, Youth, and Families and grants MH31882 and MH33137 from the National Institute of Mental Health.

"Because I'm use't."

"Because you're use't. This is a reason, Mrs. Word Miser? Used to can get unused!"

"Enough unused I have to get used to already. . . . Not enough words?" turning off the vacuum a moment to hear herself answer. "Because soon enough we'll need only a little closet, no windows, no furniture, nothing to make work, but for worms. Because now I want room. . . ."

He continues his effort to influence her.

Over the dishes, coaxingly: "For once in your life, to be free, to have everything done for you, like a queen."

"I never liked queens."

"No dishes, no garbage, no towel to sop, no worry what to buy, what to eat."

"And what else would I do with my empty hands? Better to eat at my own table when I want, and to cook and eat how I want."

After years of having to manage on very limited resources, she has found peace in solitude, and

She would not exchange her solitude for anything. *Never again to be forced to move to the rhythms of others.*

He is violating her peace with his constant campaigning, talking, reasoning, trying to manipulate, to influence.

And it came to where every happening lashed up in a quarrel.

"I will sell the house anyway," he flung at her one night. "I'm putting it up for sale. There will be a way to make you sign."

Then Eva gets sick, takes to her bed. She has cancer. David has to give up his dream, borrow money, and try to give her a few restful months before she dies. He takes her to visit her children and then to California. As she becomes weaker, he takes care of her day and night. She becomes delirious and all her old pains gush out:

"Paul, Sammy, don't fight."

"Hanna, have I ten hands?"

"How can I give it, Clara, how can I give it if I don't have?"

He talks back to her:

"You lie," he said sturdily, "there was joy too."

Then suddenly he sees her as she used to be, and takes her "in his arms, dear, personal, fleshed. . . ." In the morning their granddaughter finds him asleep holding his wife's hand in his. He stays with her till the end, witnessing her agony.

Tillie Olsen's poignant novella illustrates concisely two tasks that are at the heart of any systematic understanding of close relationships, whether literary or scientific: (1) the *description* of the interactional regularities that characterize a relationship at any point in time and (2) the explanation of these regularities by positing *causal conditions* and *processes* that give rise to, maintain, or change them. To convey the quality of the relationship, the author narrates critical *events:* words, acts, thoughts, and feelings of the two spouses. These events are tightly linked to each other. In the narrative, one can easily infer *causal connections* from David's feelings to his words, from his words to Eva's words, from his words to her feelings, and so on. The conflict that characterizes their relationship is conveyed through their quarrels, which consist of recurrent interlinked *event sequences.*

In addition, the author invokes the *causal conditions* that she sees as underlying and shaping the observable pattern of conflictual interaction. Some of these causal factors are contemporaneous, for example, the conflict of interest (when David wants one thing and Eva wants another) that gives rise to recurrent quarrels. Other causes are historical in origin, being the residues of frustrations and deprivations experienced in the long course of the relationship and the lives of the spouses. And when, by a twist of fate, a new condition arises that changes the causal context of the relationship, it alters the interaction between the spouses dramatically. Eva's terminal illness eliminates some of the bases for conflict and imposes new constraints upon their behavior. Her disability makes their interaction increasingly more asymmetrical. He becomes the caretaker and she the dependent. The existing patterns of conflict are replaced by patterns of conciliation and support.

The goal of the present chapter is to consider issues related to the description and explanation of interaction processes in relationships. Dyadic interaction is viewed as a multievent process that encompasses overt observable behaviors and covert subjective interpretations of these overt behaviors. It is assumed that moment-to-moment interaction between partners is an essential constituent of their relationship and, futhermore, that interactions show similar features and functions across a variety of relationships: marriages, friendships, parent–child relationships. Guided by the conceptual framework presented in Chapter 2, we will first examine those relationship *events* that are commonly referred to as *interaction,* and then consider the causal contexts impinging on these events. The first section of this chapter will focus on the overt events of interaction. We will examine some of the features of overt interactions, the ways they have been studied, and their patterning and continuity over the life of relationships. The second section will address the interpretative processes through which interactants make sense of the overt interaction. The final section will consider some of the causal conditions and processes that have been assumed to contribute to the organization and recurrence of interactional patterns in relationships.

THE OVERT EVENTS OF INTERACTION

Analysts of interaction in relationships have rarely examined the organization of its overt occurrences for its own sake. Rather, they have analyzed interaction in order to address substantive or theoretical issues. For example, in a search for the interactional antecedents of child attachment, Ainsworth and Bell (1969) examined how mother–infant interactions at one point in time affect interactions at a later point. Gottman (1979) analyzed patterns of overt interaction to differentiate between distressed and nondistressed marital relationships. Patterson (1979) investigated the dynamics of dysfunctional interactional patterns between aggressive children and their mothers in order to change them. Birchler, Weiss, and Vincent (1975) studied marital interaction in order to identify behavioral correlates of marital satisfaction or dissatisfaction. We are now able to look at the phenomena these and other studies have identified and ask: What common features of interaction are evident across a variety of analyses? How have these features been useful in suggesting lawful patterns of interaction in relationships?

Features of Interaction

The characterization of dyadic process in Chapter 2 suggests that interaction can be viewed as a sequence of causally interconnected events within the behavior chains of two people. This conceptualization, presented schematically in Figure 2.1 of Chapter 2, captures three features of interaction that have been stressed by other analyses. First, it identifies the elementary components of dyadic interaction within the event chains of its individual participants. It focuses our attention on what individuals do and say in the course of interaction and raises questions about how we can identify and quantify these events so as to be able to characterize important properties of the interchange. Second, in showing the parallel flow of events in the chains of P and O, it suggests the possibility of describing what both P and O are doing during each interval of time. Thus, combinations of simultaneous and near-simultaneous behaviors of the interactants may be used to describe interaction. Third, it portrays interaction as a sequence of events with temporal organization.

Individual level units

The identification of individual level units within the behavior chains of each interactant has been a common practice in analyses of interaction. Such units have been referred to as "behavior units" (Gottman, 1979), "behavior phases" (Tronick, Als, & Adamson, 1979), "acts," (Bales, 1950), or "actions" (Harré & Secord, 1972). Their occurrence has been identified and characterized in several ways. In an ethological analysis of mother–child behavior,

Blurton Jones (1972) defined individual level units in terms of some manifest quality of the behavior, for example, "hit," "take," or "smile." Bales (1950) categorized interpersonal behavior in groups on the basis of the inferred goals of the actor, for example, "to give advice," "to show solidarity." Patterson (1979) analyzed the interaction of aggressive children and their mothers by classifying behaviors in terms of their consequences for the recipient, for example, "aversive" or "nonaversive." Gottman (1979) coded behaviors in terms of the sender's inferred affect: positive, neutral, or negative.

An illustrative study employing units of individual behavior to describe interaction was conducted by Blurton Jones and Leach (1972):

> The behavior of thirty-five mothers and their 2–4 year old children was observed during separation at the beginning of a playgroup and during greeting at the end. The frequencies of the various items of behavior (smile, approach, touch, wave, show-give, point, leave, play, etc.) were counted up for each individual, and factor analysis showed the following main dimensions of behavior: [child] crying at separation leading to greeting with either rapid approach with arms raised and touching the mother, or no response except looking at the mother and pointing at an object. Ready departure from the mother went with greetings in which play continued or objects were shown or given to the mother.

In this example, individual level units were identified in the behavior chains of the interactants, and their covariation, identified through correlational analysis, was used to describe qualitatively different patterns of separation and greeting.

Simultaneity of dyadic behavior

The simultaneity of individual behaviors in interaction was originally recognized by communication theorists, who noted that interactants are always behaving in ways that are meaningful to each other, even when one of them is silent or apparently inactive while the other is active (Scheflen, 1973; Watzlawick, Beavin, & Jackson, 1967). Analyses of the moment-to-moment unfolding of verbal and nonverbal behaviors of therapists and patients (Scheflen, 1973) and of speakers and listeners (Condon & Ogston, 1971; Duncan & Fiske, 1977; Kendon, 1970) have demonstrated that interaction involves the *coordination* of actions of the two participants. Thus, in conversations, while P is active talking, O can be observed to emit nonverbal behaviors that signal reactions to what is being said and readiness to respond.

An example of the temporal unfolding of interactions between 3-month-old infants and their mothers, described in terms of the successive combinations of their simultaneously occurring behaviors, is provided by Tronick et al. (1979). They observed "dyadic phases" through which the interaction proceeds. Through careful analysis of videotaped face-to-face interactions filmed using a split-screen technique, they found the following succession of phases to be typical:

> Initiation occurs when the mother's face brightens and she baby talks to a sober baby, or a baby vocalizes and smiles to the mother who has paused too long. Mutual orientation takes place with neutral or bright faces, with the caregiver talking or the infant making isolated sounds. Greeting occurs with mutual smiles and potentially much or little hand and body movement, but with both partners looking at one another. Play dialogue occurs when the mother talks in a burst-pause pattern and the infant vocalizes during the pauses. Disengagement may occur when one of the partners looks away from the other while that partner is still oriented. (pp. 352–353)

Temporal structure

Sequential patterns can be detected when the interaction between two people has been recorded so that it preserves the detailed temporal relationships between events. From recurrent orderings of events across the chains of interactants, the presence of interchain connections may be inferred, and organizations of events at different levels of complexity can be observed.

Several alternative conceptions of interchain connections have been proposed. Raush (1965) noted that P's control over O's behavior is demonstrated when O's actions vary according to P's prior action. In an investigation of the determinants of friendly and unfriendly acts of children in a variety of settings, Raush found that the immediately preceding act accounted for 30 percent of the variance in the subsequent action. In information theory terms, interchain connections are inferred when knowledge of one actor's behavior reduces uncertainty about the subsequent behavior of the other (Gottman, 1979; Wilson, 1975). A number of statistical techniques, including Markov-chain analyses, information measures, and lag-sequential analyses, have been employed to detect the interconnections between the actors' behaviors in interaction streams. Detailed decriptions of these analytic techniques can be found in Gottman and Notarius (1978), Gottman and Bakeman (1979), Bakeman and Dabbs (1976), Castellan (1979), and Sackett (1979). Thomas and Martin (1976) and Thomas and Malone (1979) argued that conceptualizations of interaction should consider not only how O is affected by P's prior behaviors (*interchain causality*), but also how O is affected by O's own prior behaviors (what we have called *intrachain causality*).

Although covariation between events in P's chain and subsequent events in O's implies interchain causality, it is a necessary but not sufficient condition for inferring such causality. Behavioral covariation does not preclude the possibility that the interconnections between events are caused by factors outside the interaction. For example, Jones and Gerard (1967) noted that the interaction between two individuals may be regulated by a script that both partners follow, in the manner of actors performing in a play. The resulting interaction may reveal strong covariation between the two actors' behaviors, yet one actor's behavior does not actually "cause" the other's response. Aware of this problem, Patterson (1979) proposed that the causal status of any particular behavior needs to be tested by experimentally altering its frequency of occurrence and then observing the consequences for the other's behavior.

The orderly temporal succession of events across the chains of interactants has allowed investigators to identify units at different levels of complexity. The hierarchical arrangements exhibited by interactional behaviors, that is, the ways in which lower-level units are arranged into middle-level units, which, in turn, are integrated into still larger units, are either explicitly or implicitly considered by most analysts (Burke, 1974; Gottman, 1979; Harré & Secord, 1972; Patterson, 1979; Peterson, 1979; Scheflen, 1973). Scheflen (1973) noted that in order to study higher-order integrations of behavior, such as recurrent interaction patterns, analysts need first to identify the constituent events so as to be able to reconstruct the relation between them. Depending on how the analyst looks at it, any given event can be considered to be either an integration of smaller acts or a unified entity that occurs as a constituent of a larger configuration. For example, Gottman (1979) identified the "thought unit" as the minimal unit in the conversations between spouses. A number of thought units that are similar in content make up a "behavior unit," which is the building block for more intricate sequences. As we have seen, in sequential analyses, an event in one actor's chain is linked to a temporally contiguous event in the other's. The resulting two-unit sequence has been called an "*interaction unit*" (Patterson & Moore, 1979), a "*couplet*" (Goffman, 1976; Peterson, 1979), and a "*dialogue unit*" (Williams Moore & McClintock, 1979). In the course of an interaction, several dyadic interaction units can become combined into *interaction sequences.*

The manner in which interactants' behaviors are organized sequentially has been extensively examined by Gottman (1979) and his colleagues in an investigation of the ways in which clinic couples seeking counseling and nonclinic couples interacted when solving problems. The husbands and wives who participated in this study were asked to come to a mutually satisfactory resolution of a troublesome problem in their relationship. Their interaction was videotaped and coded in terms of the overt content of each behavior unit (e.g., agreement, disagreement, communication talk, mind-reading, expressing feelings about a problem) and in terms of the quality of affect (positive, neutral, or negative) with which each of the content codes was delivered (as inferred from the nonverbal behavior of the speaker).

Sequential analyses of the data suggested that the problem-solving interaction could be divided into an agenda-building phase, an arguing phase, and a negotiation phase. Each phase was characterized by different behaviors and was conducted by clinic and nonclinic couples in a different manner. For example, during the agenda-building phase, nonclinic couples were more likely than clinic couples to generate validation sequences, in which one spouse's expression of feelings about a problem was followed by the other's agreement, validating what was said before. On the other hand, nonclinic couples were more likely to produce cross-complaining sequences, in which one spouse's expression of feelings about a problem was followed by the other's counterexpression of feelings about the same problem. During the negotiation phase, nonclinic couples were more likely than clinic couples to engage in contract sequences, in which the problem-solving effort of one was accepted by the other. Clinic couples were more likely to enter counterproposal cycles, in which one spouse's proposals for a solution were followed by the other's counterproposal.

Gottman's work illustrates some of the merits of sequential analysis of interaction, namely its ability to reveal what people actually do in their relationships. The descriptions of patterns of dyadic behavior are useful both for understanding the dynamics of relationships and for developing ways to change certain dysfunctional patterns.

Approaches to the Analysis of Interaction

Having discussed some general features of overt interaction, we will now examine some of the approaches used to study it. First, we will review methodological strategies used to obtain samples of interactive behavior. Then we will consider analytic strategies used to understand interaction in relationships.

Methodological strategies

The analysis of ongoing interaction in close relationships involves a sequence of operations that is common across a wide range of studies. The process starts by defining and recording the events that make up the flow of interaction and then proceeds by coding and aggregating the recorded information in order to characterize the interaction. *Recording* is the registering of the raw occurrences of ongoing interaction. It can be carried out by the participants themselves, by observers, or by mechanical recording devices. The resulting records can contain atemporal counts of events, that is, occurrences recorded without reference to their temporal ordering, or sequences of events recorded in such a way as to preserve the temporal relationships between events. An example of the former would involve recording all instances of supportive verbal or nonverbal actions in the course of interaction regardless of when they happened. An example of the latter would be a narrative description of all the events in the order in which they occurred.

The second operation, *coding,* involves the classification of events in the flow of interaction, using interpretative categories provided for that purpose by the investigator. Coding is closely related to the recording process. In some cases, the participants or observers are trained by the investigator to recognize and record the occurrence of events on the spot as the interaction proceeds. In other instances, narrative or mechanical records are coded by the investigator after the record has been obtained. In all cases, the coding phase transforms raw information into data upon which statistical aggregation procedures can be performed.

Typically, behavior codes describe the content of events, for example, whether these are aversive or nonaversive or reflect agreement or disagreement. An alternative to transforming behavioral occurrences into behavior codes is to rate the degree to which the events of the interaction reflect some underlying quality, for example, positivity or negativity of affect. In other instances, observers are trained to observe segments of interaction and then report their impression of the quality of this interaction on a number of rating scales provided by the investigator. The latter methods reflect attempts to capture stable properties of interaction by asking observers to perform mentally some sort of aggregative or summarizing operation and then to report their conclusions. The relative merits of behavior codes versus ratings have been insightfully discussed by Cairns and Green (1979) and will not be discussed here.

The final phase of interaction analysis involves *aggregating* the data, that is, counting and summarizing the information contained in the coded records. Depending on the goals of the study and the preferences of the investigator, data can be aggregated in ways that preserve their temporal properties or not. Atemporal codes can be tallied and summed. Frequencies and probabilities of occurrence of different codes can be compared, rates of particular codes for selected time intervals can be computed, and the correlations between codes can be examined. Temporally ordered codes lend themselves to a greater variety of aggregations. They can be aggregated without reference to their temporal ordering, or they can be tabulated as antecedent–consequent pairs to indicate the frequency of transitions from one code to the next (Castellan, 1979). In addition, they can be treated as time series, and the relationships of the codes within and across the series can be examined (Gottman, 1979). To illustrate these methodological distinctions, we will briefly review studies representative of four different strategies.

1. ATEMPORAL EVENTS RECORDED BY PARTICIPANTS. An example of events recorded by participants without reference to their temporal ordering is found in a study conducted by Wills, Weiss, and Patterson (1974). One of the goals of the study was to identify the behavioral determinants of global ratings of marital satisfaction. Toward this end, the investigators asked spouses (1) to

count their partner's pleasurable and displeasurable "affectional" behaviors (for example, touching, kissing) during the time both were at home together; (2) to record their partner's pleasurable and displeasurable "instrumental" behaviors (that is, behaviors that involved execution of practical tasks) using the Spouse Observation Checklist; and (3) to rate their own satisfaction with portions of the day during which significant interaction with the spouse had occurred. Spouses were trained to recognize and record the occurrence of pleasurable and displeasurable affectional behaviors the moment they occurred by using a portable event recorder. Categories for coding instrumental behaviors on a daily basis were also provided by the investigators. Spouses recorded and evaluated aspects of their interaction for 14 consecutive days and were found to be reasonably reliable recorders.

The investigators subsequently compiled daily summary scores of pleasurable and displeasurable instrumental behaviors and rates of pleasurable and displeasurable affectional behaviors per hour. These scores showed significant correlations with ratings of the subjective quality of interaction during that day. Displeasurable behaviors accounted for 65 percent and the pleasurable behaviors for 25 percent of the explained variance of ratings of satisfaction. In addition to searching for interactional antecedents of global subjective feelings, this study examined the issue of reciprocity in pleasurable and displeasurable behaviors between spouses. The authors correlated husbands' and wives' scores over the 14 days of the study. While wives' and husbands' pleasurable and displeasurable instrumental behaviors and pleasurable affectional behaviors seemed to be uncorrelated, displeasurable affectional behaviors of spouses had an average correlation of .59. These results suggested to the authors the operation of a reciprocity mechanism in the case of displeasurable affectional behaviors but not in the case of pleasurable ones. Questions about how this pattern of reciprocity is initiated and maintained over time in marital interaction cannot be answered by these data since the critical events in the spouses' chains were recorded independently and atemporally.

2. INTERACTION STREAMS RECORDED BY PARTICIPANTS. In a study of satisfied, average, and dissatisfied couples, Peterson (1979) analyzed accounts of significant interaction episodes that occurred in the couples' lives as recorded in narrative form by each of the spouses (for example, "We went to bed. I hugged my wife. She said she did not want to cuddle. I turned over and went to sleep.") Interaction records were subsequently coded by the investigators. Each recorded interaction was broken down into the "major moves" or "acts" that made up the sequence. Each act was then interpreted, that is, its interpersonal meaning or message was inferred. These messages were subsequently coded so as to account for both the covert experience and the overt behavior of the spouses in terms of the *affect*, the *construal* (the attributions

made by the actor about self, other, and situation), and the *expectation* for the other's subsequent behavior that they expressed. Finally, long sequences of acts were broken down into two-unit sequences called interaction cycles or statement–reply couplets. The affect–construal–expectation codes for each component act of these couplets were then listed and the frequencies of similar combinations of codes were tabulated for disturbed, average, and satisfied couples. Several patterns were identified: Cycles of mutual enjoyment ("Let's enjoy ourselves"—"Yes, let's") were most commonly reported by satisfied couples. Cycles of support ("I'm down"—"I'm on your side") were reported most frequently by average couples. Aggression–injury cycles ("I hate you"—"I'm hurt and wish to avoid you") were common among disturbed couples.

Peterson's research strategy is interesting for several reasons. First, it elicits records of marital interaction with minimal intrusion upon the participants' daily lives and in a form that does not require spouses to perform recording operations while engaged in their daily activities. Spouses record episodes on forms whenever they feel like doing it. Second, it makes explicit the various interpretative steps that are involved in coding ongoing interaction and the types of inferences investigators often make about the function and consequences of overt behavior. Finally, it appears to provide the investigator with accounts that reflect significant patterns of interaction from the life of relationships and that correspond to some underlying quality of the relationship, in this case, the extent of satisfaction with the marriage.

3. OBSERVER OR MECHANICALLY RECORDED EVENTS. The advent of audio and video recording technology has rendered the recording of interaction in vivo by observers less popular. In the sixties, however, many studies of parent–child and marital interactions employed observers as recorders of interactions. In such instances, observers were trained to recognize and assign occurrences of particular behaviors to categories provided by the investigator. Levinger (1969) employed Bales' coding scheme in a study of family interaction designed to compare the strengths and weaknesses of two data collection techniques, interview versus observation. In that study, 31 families consisting of father, mother, and at least one son were observed and interviewed. In the observation portion of the session, each family engaged in a series of tasks presented to them by the interviewer, who remained in the room while the families tried to decide about such matters as how to spend $10,000 and where to go on holiday. The observer coded each distinguishable act into 1 of the 12 categories provided by the scheme (e.g., shows tension, solidarity, antagonism). In aggregating the data, the frequency of occurrence of each code was counted, and sums of selected codes were combined to describe task and socioemotional properties of family interaction.

Since observer presence can have reactive effects on the interaction in close relationships, investigators often record the interaction on audio- or videotapes and subsequently code it. Birchler et al. (1975) videotaped the interaction of distressed and nondistressed couples in a laboratory situation in which spouses were asked to discuss anything they wanted for the first 4 minutes and to resolve differences of opinion for the next 10 minutes. The videotapes were then coded to provide measures of positive behaviors (e.g., agreement, smiling) and negative behaviors (e.g., criticism, disagreement). The rates of these behaviors per minute were used to compare distressed to nondistressed couples. Birchler et al. (1975) concluded that

> distressed marital dyads emitted negatives at a mean rate almost 1½ times that for non-distressed dyads. In addition, relative to non-distressed marriages, distressed couples exchanged significantly less positive behavior in the problem solving setting. . . . (p. 353)

The conclusions of Birchler and his coworkers support previously described findings of Wills et al. (1974), who reported greater reciprocity of negative behavior among distressed couples.

4. OBSERVER OR MECHANICALLY RECORDED INTERACTION STREAMS. Ongoing interaction can be recorded by observers in narrative form (e.g., see Dyck, 1963; Raush, 1965) or as sequences of codes (Bakeman & Brown, 1977). It can also be recorded on audio- or videotapes. These mechanical records are subsequently coded by observers trained for that purpose. When coding ongoing interaction, either at the time when it occurs or from mechanical records, observers first have to segment its flow into units, which are then assigned a label or code on the basis of the specified categories. Coding interaction events involves a process whose features will be discussed in more detail later in this chapter in the section on the interpretation of interaction. In general, investigators assume that observers can be trained to identify boundaries between events at the level of analysis defined by the investigator and that they can classify each of these events into one of a limited number of mutually exclusive categories that are exhaustive of the significant events that may occur in the particular interaction. Interaction streams are thus transformed into sequences of codes. Subsequent to coding, investigators aggregate their data in ways that will allow them to detect the sequential organization or patterning of events.

Analytic strategies

The logic underlying efforts to detect sequences in ongoing interaction can be illustrated by the following example: Imagine that we have recorded a segment of interaction between P and O and that we have coded each actor's behaviors as they occur in temporal sequence. This operation has resulted in a sequence of codes, Pa Ob Pb Oc Pa Ob Pb Oc Pa Oc Pa Ob, in which each code represents an actor (P or O) engaging in an action (a, b, or c). It is possible to aggregate these data in two ways: (1) We can tally the frequency of Pa, Pb, . . ., Oc and report the probability that each will occur in the course of interaction; or (2) we can tally the transitions from one code to the next, preserving the information about their sequential ordering. The result of this operation will be a contingency table pairing all antecedent to all consequent codes. The cross-tabulation or transition table will present in its marginal distributions the unconditional frequency and probability of occurrence of events and, within the cells, their conditional frequencies and probabilities of occurrence. A transition table of the events in our example will show us that certain codes repeatedly follow others. In fact, some of these temporal associations may be so strong that, by knowing the antecedent event, we can predict with considerable certainty what event will follow; for example, code Oc in our example is always followed by Pa. In such cases, we can say that there is a pattern in the data, that one event covaries with another. The presence of such a pattern can be ascertained by contrasting the probability that an event will occur anywhere in the sequence (simple or unconditional probability) and the probability that it will follow another specified event (conditional probability). The difference between the unconditional and conditional probability of an event reflects the degree of temporal sequential association between that event and the one that precedes it (Gottman, 1979).

In addition to the detection of recurrent two-event sequences, it is possible to identify longer sequences in the stream of interaction. For example, Sackett (1979) developed a system called lag-sequential analysis, which was described by Bakeman and Dabbs (1976) as follows:

> The analysis begins by designating one behavior the "criterion behavior" (this procedure can be repeated as many times as there are behaviors, so that each behavior can serve as the criterion). Then for every other behavior a "probability profile" is constructed; each profile shows the lagged conditional probabilities that the behavior will follow the criterion immediately (lag 1), follow as the second behavior (lag 2), follow as the third behavior (lag 3), and so forth. Peaks in the profile indicate sequential positions following the criterion at which a given behavior is more likely to occur, while valleys indicate positions at which it is less likely to occur. If a behavior is sequentially independent of the criterion, then its conditional probabilities at various lags should be about the same as its simple, or unconditional, probability. Several profiles can be examined together to determine which behavior is most likely at each lag after a particular criterion behavior. This information in turn would suggest probable sequences even though actual sequences are not observed for more than two behaviors at once. (p. 339)*

*Reprinted by permission of the American Psychological Association.

In his investigation of the structure of marital interaction, Gottman (1979) has made extensive use of Sackett's lag-sequential analysis. One of the features of marital interaction that has interested Gottman is that of reciprocity, a phenomenon that, as we have seen, has attracted great interest in research on marriage relationships (Birchler et al., 1975; Peterson, 1979; Wills, et al., 1974). Gottman observes that two different definitions and operationalizations of reciprocity are encountered in analyses of interactions. Overall, reciprocity is a pattern of interaction in which similar behaviors, inputs, resources, and so on are in some way matched. Some investigators measure reciprocity in terms of matching rates of similar behaviors between spouses by correlating the rates of positive or negative behavior of each husband and his wife. Other investigators view reciprocity more in terms of contingency between two actors' behaviors, in which the behavior of one alters the probability of a subsequent similar behavior of the other. Gottman advocates the second definition of reciprocity and analyzes records of marital interactions for patterns of reciprocity of positive and negative affect.

Beginning with the expectation that distressed couples will evidence greater reciprocity of negative affect than nondistressed couples, Gottman examines the "affect" with which the interacting spouses deliver their message. As we noted earlier, in Gottman's studies, each behavior unit is coded in terms of both its content and its "affect" (expressed through tone of voice and other nonverbal cues). Gottman then examines sequences of affect codes, using Sackett's sequential analysis technique.

Figure 3.1 presents some of Gottman's findings in graphic form. The graphs depict the probabilities of husband and wife behaviors following a behavior of the husband delivered with negative affect. The solid line is a profile of the probabilities of behavior delivered with negative affect by the husband at each of the lags. The dotted line represents the probability profile of negative affect expressed by the wife at each lag. The examination of the probability profiles of nonclinic and clinic couples makes evident the tendency of all spouses to reciprocate negative affect and the tendency of clinic couples to do so more than nonclinic ones. From these and other data, Gottman (1979) concludes that

> taken together, the analyses on the reciprocity of affect suggest that the reciprocity of negative affect is more characteristic of clinic than of non-clinic couples and that the reciprocity of negative affect is a better discriminator between groups than reciprocity of positive affect. (p. 111)

An alternative approach to analyzing interaction streams that uses dyadic rather than individual units is provided by Bakeman and Brown (1977). In this study, 45 infants and their mothers were observed on the third day of the infant's life while mothers were bottle-feeding their infants in a hospital room. Two observers watched each pair for two half-hour sessions, one

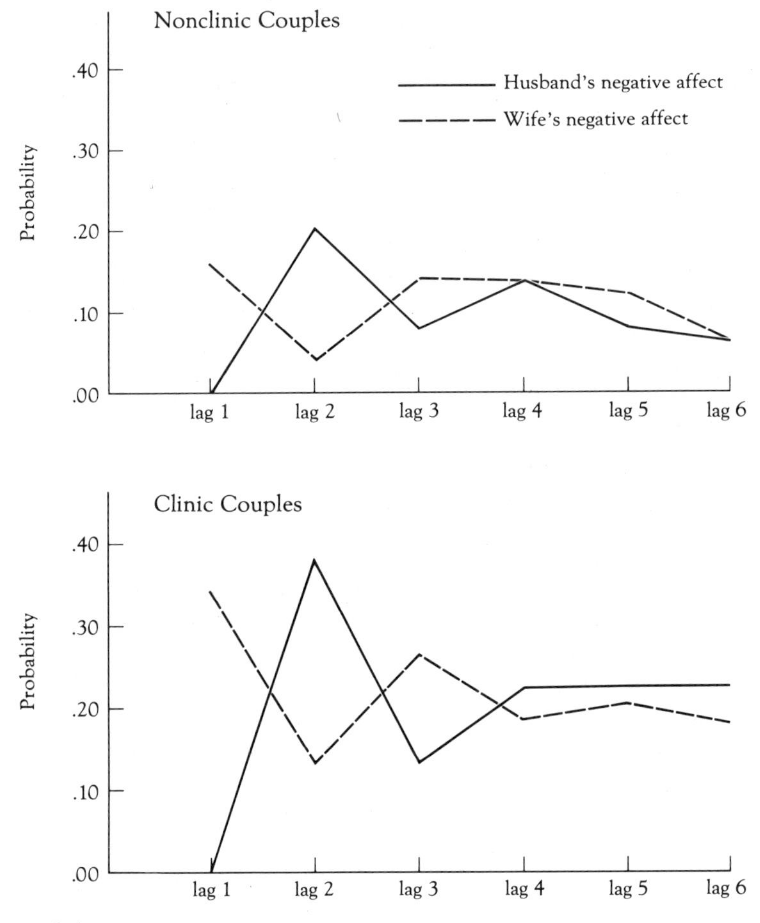

FIGURE 3.1
Conditional probability profiles of husband and wife behaviors delivered with negative affect at each of six lags after a husband behavior delivered with negative affect. (Adapted from Gottman, 1979.)

observing the mother's behavior and the other, the child's. They recorded behaviors continuously using a portable digital recording device. Each observation record was then divided into 5-second intervals, and the behaviors of mother and child during each of these intervals were coded as (1) *active* when the mother or child engaged in communicative acts, i.e., behavior that was responsive to or likely to elicit a response from the other, or as (2) *inactive* when the infant or mother was quiet. Since maternal and child behaviors were recorded separately, each time interval could be categorized into one of four mutually exclusive and exhaustive states: (1) the coacting state, when both mother and child were active; (2) the mother-alone state, when the mother was active but the child inactive; (3) the infant-alone state, when the infant was active and the mother inactive; and (4) the quiescent state, when both were inactive. These states and transitions from one to the next are summarized in the diagram presented in Figure 3.2.

This "state transition" diagram summarizes some interesting aspects of the mother–infant interaction in the feeding situation. The probabilities in the circles show the proportion of time intervals the pairs were categorized as being in each of the four states. Clearly the quiescent state and the mother-active state are the most frequent ones. The arrows represent the transitions from one state to a temporally contiguous one. The probabilities on the arrows are the conditional probabilities that the state at the beginning of the arrow will be followed by the state at the end of the arrow. Overall, the diagram indicates that the mother–infant interaction progressed smoothly. Abrupt transitions, for example, from the quiescent state to the coacting state, were very infrequent and were, therefore, not represented in the diagram. The diagram also indicates that, at this early time in the mother–infant relationship, the mother controls the dyadic system. Not only was the

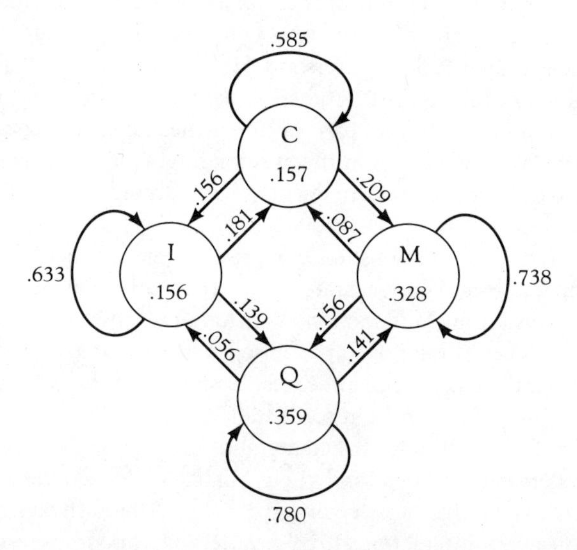

FIGURE 3.2
Mother–child interaction in a feeding situation. Transition diagram for four dyadic states: quiescent (Q), mother active (M), infant active (I), and coactive (C), with simple probabilities shown in the circles and transition probabilities shown by the arrows. (From "Behavioral Dialogues: An Approach to the Assessment of Mother–Infant Interaction" by R. Bakeman and J. V. Brown, Child Development, *1977, 45, 195–201. Copyright © 1977 by the Society for Research in Child Development, Inc. Reprinted by permission.)*

M (mother-active) state more frequent, but the probability that the quiescent state would be broken by the mother was higher than the probability that it would be broken by the infant. On the other hand, the coacting state was more likely to be broken by the infant, leaving the mother acting alone.

Interaction over Time

The enduringness of close relationships suggests that interactions between partners occur repeatedly over long time periods, ranging from months to years. However, people in relationships do not interact all the time. They come together, interact, part to engage in other activities, and then meet again. Thus, although interaction is a continuous stream or flow of interlinked occurrences whenever it takes place, over the life of any relationship it occurs only *intermittently*.

The chunks of interaction that occur between intervals of noninteraction represent *episodes*. The term *episode* as defined by Webster's dictionary is "an occurrence or connected series of occurrences and developments which may be viewed as distinct and apart although part of a larger and more comprehensive series." Episodes can be thought of as time slices of the relationship delimited by the onset and the termination of causal interconnections between the behavior chains of the two members. The present definition of episode is nearly identical to Peterson's (1979) conception of interaction sequence, which he sees as starting when an act by one participant is followed by contingent action on the part of the other, continuing as long as reciprocal contingency prevails in the ensuing interaction, and ending when reciprocal contingency ceases, frequently with physical separation of the participants.

The view that close relationships comprise a large number of interaction episodes between P and O, separated from each other by time periods of alternative activities and interactions with others, provides a natural sampling unit for studies of relationships. Episodes sharing similar properties can be sampled from the daily life of close relationships, and the interactions that take place within them can be recorded and analyzed. For example, Ainsworth and Bell (1969) and Bakeman and Brown (1977) observed the interactions between mothers and their infants in the course of episodes during which the mother was feeding the baby. Alternatively, researchers may want to sample interactions from a variety of episodes to establish the stability of interaction patterns across episodes. Along these lines, Christensen (1979) unobtrusively recorded interactions in the home by activating a tape recorder randomly within predetermined time periods, for example, in the morning around breakfast time and in the evening around dinner time. Finally, episodes of interaction have been elicited in laboratory situations in which investigators give couples or parents and children certain tasks to perform in order to observe styles of mutual influence and conflict resolution (Gottman, 1979; Raush, Barry, Hertel, & Swain, 1974).

Episodes of interaction have an *internal temporal structure* that has been compared to that of chess games. Gottman (1979) noted that, like chess games, episodes of interaction contain three phases—opening, middle, and closing—which, in problem-solving interaction, correspond to agenda-building, arguing, and negotiation phases, respectively. In his research, Gottman (1979) found each of these phases to be characterized by distinct sequences of interaction. E. D. Miller, Hintz, and Couch (1975) studied the opening phase of interaction episodes between dyads of established friends and dyads of strangers. They observed that, during the opening phase, all dyads acted so as to establish mutually compatible goals and roles. Friends proceeded through this phase much faster than strangers, acting as if they were invoking agreements and agendas established through past interactions. Endings have also received some research attention. Albert and Kessler (1978) found that, in the closing phase of interaction episodes, interactants engage in behaviors that involve summing up, disengagement, and reassuring each other of the continuity of the relationship in the future. The beginning, middle, and ending of an episode, despite their distinctive sequences and events, are organized into a coherent whole. Episodes can be regarded as higher order organizations of their component *phases*. Phases, in turn, integrate one or more *interaction sequences* made up of one or more *dyadic units*, which are pairings of *individual behaviors*. Thus, episodes are molar units of interaction in close relationships.

Each episode as a whole and each of its components separately often have a *thematic coherence* that is easily recognized by observers and participants. The thematic coherence of episodes is in many ways analogous to that of stories. Stories, like episodes, are made up of sequences of interrelated events. Like stories that have a topic or theme that integrates their various parts, episodes also have themes that integrate their various phases, interaction sequences, and so on. Inasmuch as interaction has not yet been extensively analyzed at the level of episodes, we may borrow some ideas from the analysis of written texts (stories) to understand the coherence of behaviors evident in interaction episodes. Kintsch and van Dijk (1978) proposed that the coherence of a total story is preserved by macrostructural rules that impose the constraint that each part of a text "must be connected relative to what is intuitively called a topic." The topic acts as a global constraint and establishes a meaningful whole by linking all the parts. A similar process can be assumed to obtain during the enactment and/or interpretation of interaction episodes, pulling together diverse behaviors and sequences into a coherent whole. There is presently little evidence about the types of theme that contribute to the coherence of interaction episodes. Watson and Potter (1962) identified a number of specific themes that appeared in brief episodes between strangers at a cocktail party. The work of Schank and Abelson (1977), Wish, Deutch, and Kaplan (1976), and Forgas (1976) suggests that most overt topics of episodes may be variants of a relatively small number of underlying interpersonal themes, such as positivity–negativity, intimacy–distance, and dominance–submission.

Episodes of interaction are often characterized by *continuity over time.* Although interaction in close relationships is intermittent, it is often experienced by the members as an uninterrupted continuum. Several temporally distant episodes may be linked by the same theme and develop as if no interruption had occurred. For example, a quarrel between two spouses may start in the morning at the breakfast table, be interrupted when they go to work, and be resumed when they return home in the evening. The continuity between episodes is reflected in the cross-episode stability and repetition of particular patterns, such as conflict, symmetry–asymmetry, and so on. Such concepts as roles are often used to describe these molar patterns of behavior (see Chapter 6, "Roles and Gender").

THE INTERPRETATION OF INTERACTION

While an examination of the overt events of interaction provides useful information for understanding close relationships, any analysis of interaction will be incomplete if it fails to consider the participants' interpretations and their cognitive processes. Interpretations of one's own and others' behavior form an integral part of the interaction process. Similarly, investigators of close relationships must consider the processes by which observers of interaction interpret what they witness.

Making sense of interaction events involves two often-simultaneous processes: (1) the recognition and classification of overt behavioral cues as representative of some class of events and (2) the assignment of underlying stable goals, feelings, and dispositions to the actor who emits such cues. Social-psychological analyses of social perception and cognition have mostly focused on the latter process, through which perceivers aggregate events and attribute stable characteristics to others (Heider, 1958; Kelley, 1979). The processes through which the behavioral cues occurring in the chains of interactants are classified and given meaning have not yet been fully explained. Consequently, the question How do we know what others are doing while they are acting? has yet to be fully answered (Newtson, 1976).

The absence of a general theory explaining how people's actions are understood does not reflect a lack of interest or work on the topic. Philosophers, psychologists, sociologists, linguists, and many others have written much about how people make sense of the world around them. However, the complexity of the phenomena and the lack of communication between disciplines have precluded the development of a generally accepted approach. In the present discussion, we will undertake to describe and explain some of the intepretative processes that take place in the course of interaction by using concepts from diverse disciplines.

Partitioning the Flow of Ongoing Interaction

We have already noted that interactions occur continuously over time. How do P and O or observers to their interaction perceive what is taking place? Existing work suggests that (1) perceivers view interaction as a sequence of discrete events and not as a continuous flow; (2) the perceived discrete events are attributed meaning by being recognized as instances of general classes of behavior; and (3) the same events are used to make further attributions about the stable characteristics of actors and relationships.

The partitioning of the continuous flow of interpersonal behavior into a sequence of discrete events has been viewed as a consequence of the limitations of the human perceptual apparatus. It is generally believed that the way we process information constrains us to segment perceptual inputs into units, which we then categorize into a relatively small number of classes. This appears to be true for the perception of sounds, tastes, sizes, colors, temperatures (G. A. Miller, 1956), words (Cole & Jakimik, 1978; Pendse, 1978), written texts (Kintsch & van Dijk, 1978), as well as ongoing behaviors (Dickman, 1963; Newtson, 1976; Newtson, Engquist, & Bois, 1977).

Dickman (1963) attempted to determine whether the stream of individual behavior is seen as a continuum or as a sequence of discrete units, and, if the latter, whether different people perceive reasonably similar units. He first showed subjects a brief film and then presented the same events described verbally on a sequence of cards. Each card represented a small segment of the film. Subjects were asked to indicate where they perceived boundaries of units of action. He found that there were boundaries of molar events that are perceived by all subjects. However, some subjects perceived these molar units as being subdivided into smaller subunits. For example, one person would see the sequence P lifts a cup to her lips, takes a sip, and replaces the cup on the table as one action, "drinking a sip of coffee," bounded by the first and last events. Another person would perceive the same outside boundaries of the action but would further subdivide it into three component acts: lifting, sipping, lowering.

More recently, Newtson (1973, 1976; Newtson & Engquist, 1976; Newtson et al., 1977), using films rather than verbal descriptions of action, demonstrated that perceivers are able to segment sequences of filmed motor behaviors into units with considerable consensus and that the boundaries specified by the perceivers systematically correspond to significant variations in the configuration of stimulus features. Newtson's research supports the view that actions in the behavior stream are experienced by perceivers as a series of discrete units. The perceiver's action boundaries, labeled "breakpoints," were found to have distinctive properties not shared by other parts of the behavior stream. Breakpoints, according to Newtson, are points at which one action is differentiated from another. They are selected because a "meaningful change" has occurred in the stream of behavior relative to the immediately preceding breakpoint, and they suggest to the perceiver a coherent interpretation of the action; that is, they contain cues that are critical for the recognition of the action.

In trying to understand dyadic interaction, perceivers (observers as well as participants) probably need to monitor and coordinate information derived from the behavior chains of both interactants. Our knowledge of how this monitoring and coordination is accomplished is currently very sparse. E. McClintock and Baron (1979) proposed that, in decoding interactive behaviors, observers and participants use *social interaction grammars*—sets of rules that focus the perceiver's attention on particular features of the interactive situation and define what changes in the situation should constitute boundaries of interaction units. Investigators of early mother–child interactions (Bruner, 1975; Tronick, Als, Adamson, Wise, & Brazelton, 1978) suggested that infants acquire social interaction grammars that allow them to mesh their behavior with that of their mother and so construct coherent interaction sequences. Similar social interaction grammars can be assumed to operate in the decoding of other interactive behavior.

Such grammars can be thought of as sets of hierarchically organized rules that direct the observer's attention to particular features of the interactional situation: (1) changes in the actor–recipient configuration; (2) changes in the actor's behavior; and (3) features of the actor's behavior. The actor–recipient configuration is defined by the perceived origin and directionality of overt behavior. In other words, the person in whom the behavior originates is seen as the actor, and the one toward whom the behavior is directed is seen as the recipient (Bales, 1950). Changes in the actor–recipient configuration define boundaries of units commonly perceived as "turns" (Duncan & Fiske, 1977) or "floor switches" (Gottman, 1979). Within a stable actor–recipient configuration or turn, the observer's attention is directed to changes in the actor's behavior. Changes in the actor's behavioral cues define breakpoints of units perceived as individual actions (Newtson et al., 1977). In addition, social interaction grammars provide observers with dictionaries of behavioral categories that can be used to label or categorize the observed actions. Thus, these grammars can be conceived of as cognitive guidelines aiding perceivers in partitioning the flow of ongoing interaction and labeling its component parts.

In everyday interactions, actors and perceivers use implicit rules of which they are not conscious, and the partitioning process is fast and automatic. Consequently, it has been difficult to study the rules empirically. However, examples of explicit social interaction grammars can be found in the coding manuals developed to guide observers in coding interaction for research purposes (Bales, 1950; Dyck, 1963; Gottman, 1979).

Recognition and Labeling

The process through which the perceiver attributes meaning to the discrete behaviors that make up the flow of interaction is *recognition.* P watches O moving her arm in a certain way and leans towards her to be caressed, or he moves away to avoid a blow. In each instance, P recognizes configurations of O's behavior as members of very distinct classes of events, a caress versus a blow. Currently, our knowledge of how perceivers recognize constantly changing stimulus configurations is primarily derived from analyses of comprehension of spoken language or written text. Since the requirements placed upon the perceiver of human speech, that is, the need to understand events while they are happening, are quite similar to those placed upon the perceiver of interaction, we may possibly extrapolate from theories of language or text comprehension (Schank & Abelson, 1977; Schlesinger, 1977) to the understanding of interpersonal events.

Recognition is generally believed to be an inferential process that associates some perceptual input to a general concept, class, or category (Cherry, 1978). Recognition is not regarded as a passive process during which the perceiver matches inputs to cognitive templates, like fitting a key to a lock. Rather, it is viewed as an active process in which the perceiver develops hypotheses about what is to be observed and uses the stimuli received as evidence to support or refute these hypotheses or beliefs (Cherry, 1978; Neisser, 1976).

Perceivers formulate hypotheses about what kinds of events are happening and what events are likely to follow on the basis of their general knowledge of close relationships and their particular knowledge of a specific relationship or person. This body of factual knowledge emerges gradually as the perceiver collects and stores information about the social world. It is organized into cognitive structures that have been referred to as "functional schemata" (Norman & Rumelhart, 1975); "frames" (Minsky, 1975); "causal schemata" (Kelley, 1971); and "implicit theories of relationships" (Rands & Levinger, 1979). J. D. Becker (1973) and Foa and Foa (1975) propose that cognitive structures hierarchically interrelate and organize different aspects of the perceiver's body of knowledge. For example, the work of Triandis, Vassiliou, and Nassiakou (1968) supports the notion that observations of social behavior result in a body of associations between pairs of actors (e.g., husband–wife) and clusters of actions (e.g., to help, respect). These associations between pairings of people and clusters of actions form a basis from which perceivers generate expectations or hypotheses that they use in the interpretation of interpersonal behaviors.

Schank and Abelson (1977) examined the way in which cognitive structures are employed for understanding events and proposed that specific knowledge about a particular situation is organized into a "script." Scripts describe appropriate sequences of events in particular contexts; for example, there are "restaurant" scripts and "post office" scripts. Scripts involve repeated and stylized everyday situations that elicit stereotyped sequences of action. To interpret less stereotyped situations, perceivers first have to infer the goals of the actors, that is, the objects, states, or ends that the actors strive to attain in the particular situation. Then perceivers construct plans, that is, sequences of actions that might allow the actors to realize their goals. The background information for inferring actors' goals in specific situations is provided by themes. Themes, according to Schank and Abelson, are bundles of predictions about how one person is likely to act towards another in various situations. Schank and Abelson found three general categories of themes: role, interpersonal, and life themes. Each is seen as representing a particular type of predisposition of an actor, which, when identified, helps the perceiver to make sense of the actor's behavior by providing a context for the behavior.

Interpersonal themes are important in interpreting interactions in close relationships. For example, if we know that John loves Mary, we will be able to predict and understand his behavior in a wide variety of situations. Schank and Abelson further suggested that there are three dimensions underlying most interpersonal themes: positive–negative; intimate–distant; and dominant–submissive. When any of these theme dimensions is invoked, goals and plans are inferred to aid the perceiver in decoding ongoing interaction. Similar dimensional schemes underlying people's perceptions of interpersonal relationships were identified by Wish et al. (1976), Triandis (1972), and Triandis et al. (1968). The stability of these dimensions across a number of additional studies led Wish et al. (1976) to suggest that they may represent a way in which knowledge about relationships is structured cognitively by most people.

To recapitulate, events in the flow of interaction do not automatically convey meanings. They are given meaning by the perceivers (Cherry, 1978; Darley & Fazio, 1980). Perceivers use the knowledge they possess about close relationships in general (e.g., mothers love their children); about the situation in which interactants find themselves (e.g., in restaurants people order food and eat it); and about the specific history of the relationship (e.g., Mary and John do not get along with each other) to form expectations or hypotheses about what is to occur. On the basis of these hypotheses, the first events in an interaction episode are interpreted. From this first interpretation of events, a scenario may be constructed by the observer, incorporating scripts and plans that give rise to expectations about what is to happen next (Schank & Abelson, 1977; Schlesinger, 1977).

The molar units of dyadic interaction described earlier in this chapter can be thought of as hierarchically organized contexts of meaning for interpreting interactional behavior. The theme of an episode can provide the context for understanding the interaction sequences that occur within it. If, for example, the theme of an episode of interaction is a quarrel, then a sequence of struggling between the interactants will be perceived as an exchange of blows; alternatively, if the theme of the episode is one of friendly teasing, struggling will be perceived as "horsing around." Furthermore, the themes of particular sequences within episodes can provide the context for understanding component interaction units, and finally, each interactant's behavior is interpreted in the context of the other's preceding act. When two actors are exchanging criticism, a smile by either one of them is perceived by the other as sarcasm, but, in a more neutral interchange, the same smile might be interpreted as expressing interest or amusement.

Attribution Processes

The interpretation of events in the chains of the interactants involves more than simple recognition. Perceivers, whether outside observers or participants viewing their own and their partner's behavior, engage in additional inferential steps. They make sense of observed events by attributing them to underlying plans and intentions of the individual (personal causes) or to the causal influences of prior events in the interaction stream (relational causes). Inferences about underlying plans and intentions may lead, in turn, to the attribution to the actor of stable dispositional characteristics (Kelley, 1979; Kelley & Thibaut, 1978). Such attribution has implications for subsequent interpretations and behavior. Most important is the fact that attributed traits and dispositions create expectations of disposition-related behavior in the future. These expectations are not restricted solely to similar behaviors. Darley and Fazio (1980) note that dispositional attributions are linked to the perceiver's implicit theory of personality.

> Following ordinary attribution processes, a perceiver may infer a trait or disposition of the target person. In the perceiver's naive theory of personality, this trait will be connected to other traits, and so the perceiver expects the target to exhibit behaviors dispositionally related to those inferred traits as well. (p. 870).

In addition to attributions about stable personality traits of individual members, interpretations of events often lead to inferences about stable properties of the relationship. Watzlawick et al. (1967) and Bateson and Jackson (1964) were among the first to observe that one of the most significant ways in which interactants and observers make sense of their interaction is by *punctuating* its continuous flow, that is, pairing events in a way that introduces the impression of cause–effect interconnections. Punctuating an interaction sequence requires establishing an arbitrary starting point, after which the sequence is unitized and organized in such a way that particular causal orderings of events become obvious to the perceiver. The resulting pairings of events depend upon where in the sequence the starting point was established. A given sequence of P–O interactions can be viewed as initiations by P and responses by O if the starting point for the sequence is set at P's action. Or, if the starting point is established by O's action, the sequence may be perceived as an interruption by P of O's ongoing activity and O's response to this interruption. The recurrent observation of particular pairings of events leads to inferences about relationship properties. In the above example, the first punctuation of the interaction stream will lead the perceiver to conclude that the relationship is asymmetrical, with P frequently initiating and O responding. Alternatively, the same relationship may be seen as conflictual if P is seen as frequently interrupting O's plans.

The fact that interactants punctuate their interactions is particularly obvious in situations of conflict, when disagreements concerning how sequences are punctuated are expressed openly (Bernal & Baker, 1979). Watzlawick et al. (1967) illustrate this with the following example:

> Suppose a couple have a marital problem to which he contributes passive withdrawal, while . . . [she contributes] nagging criticism. In explaining their frustrations, the husband will state that withdrawal is his only defense against her nagging, while she will label his explanation a gross and willful distortion of what "really" happens in their marriage: namely, that she is critical of him *because of* his passivity. Stripped of all ephemeral and fortuitous elements, their fights consist in a monotonous exchange of the messages "I withdraw because you nag" and "I nag because you withdraw." (p. 58)

Differences in the punctuation of interaction do not arise only in conflict situations. On many other occasions, spouses are found to process the events of the interaction stream in a different manner and attribute responsibility in different ways. Komarovsky's (1967) interviews with blue-collar couples revealed that husbands' and wives' views of what caused and alleviated bad moods in the spouse tended to be quite similar with regard to the husbands' moods, but quite disparate in the case of the wives' moods. Husbands tended to underrate their contributions to their wives' bad moods and to exaggerate their helpfulness in restoring good moods. Such egocentric biases in attribution seem to be quite prevalent, although their cognitive bases are not fully understood (M. Ross & Sicoly, 1979).

The actors in interaction often have different information as the bases for their judgments, and, as a consequence, they reach different conclusions about the cause–effect connections between events. Consider an interaction sequence in which the wife is upset and cries, the husband puts his arms around her and kisses her, and she stops her emotional display. He will probably punctuate this sequence in such a way that his intervention had a causal role in changing her mood. On her part, the wife may operate with a different set of informational inputs. As compared with observers, actors tend to be more aware of their private feelings (M. Ross & Sicoly, 1979) and to attach less importance to the constraints placed upon them by social or situational factors (Darley & Fazio, 1980). Consequently, the wife might react to his gesture of affection and concern by saying to herself, "Look how nice he is to me, I'd better pull myself together and stop crying." She then stops crying and smiles at him. Her perception and recollection of the sequence will include her own effort and contribution to her mood change, elements absent from her husband's perception. Therefore she is less likely to perceive him as the main cause of her mood change and more likely to be aware of her own contributions.

If punctuational vagaries can lead members to erroneous characterizations of their relationship, scientific analyses of interaction can also result in erroneous conclusions by mispunctuating interaction sequences. Indeed, for years, the prevailing zeitgeist in developmental psychology led investigators to systematically mispunctuate interactions between parents and children and focus only on the effects of the former upon the latter (Parke, 1979a). Current investigators view interaction as a reciprocal process and, consequently, they tend to examine their data using alternative punctuational schemes. They establish instigation–response pairings between P's and O's behaviors by using P's behavior as the instigation and O's behavior as the response, and then they reverse the order and examine O's behavior as the stimulus and P's behavior as the response (Gottman, 1979; Patterson, 1979). By this means, they are able to document both the reciprocal linkages between P's and O's behaviors and any asymmetries in their interaction.

Misunderstandings

As partners in interaction, P and O often possess distinct dictionaries of categories that they use to classify the interactive events that they enact or perceive. Foa and Foa (1975) have observed that misunderstandings between P and O may arise when behavioral occurrences are mapped into different categories by P and O (mismapping) or when the cognitive categories that each interactant possesses are not matched (mismatching). Mismapping occurs very often in cross-cultural encounters in which a behavior that is given a particular meaning in one culture has a different meaning in the other. It also occurs in marital relationships in which, because of socialization of personality differences, the same action is mapped into different categories by the two spouses, in spite of the fact that the spouses possess the same categories. Mismatching can occur when two people differentiate unequally within or between categories. For example, in the course of their study of behavioral determinants of marital satisfaction, Wills et al. (1974) instructed the husbands to increase their output of pleasurable, affectional behaviors. Most wives reported a significant increase in pleasurable, affectional behaviors but also an increase in pleasurable instrumental behaviors. The authors attribute this finding to the husbands' tendency not to distinguish between affectional and instrumental behaviors. They reported an anecdote that illustrates this failure:

> The experimenter contacted one husband at the beginning of Day 14 to express his concern that the subject's wife had not reported any increase in his pleasurable affectional behavior on the preceding day. The husband replied that he certainly had complied with the instruction [to increase pleasurable affectional behavior] by *washing his wife's car* (an instrumental behavior). (p. 810)

Obviously, misunderstandings and misattributions can have serious repercussions for the development and maintenance of close relationships. They represent failures in communication that, in the short run, might give rise to corrective interaction sequences. These interaction sequences tend to interrupt the overall theme of an episode inasmuch as they include behaviors aimed at eliminating discrepancies in interpretation, for example, explanations, apologies, or retractions (Jefferson, 1972). Alternatively, misunderstandings may generate conflictual sequence, such as those described in Chapter 9, "Conflict." In addition, misattributions may alter the course of a relationship without necessarily instigating overt conflicts.

Construction of Interpretative Contexts

Our preceding discussion has focused on the cognitive processes used by interactants and observers to interpret events that occur during ongoing interactions. Members of relationships, however, are not solely decoders or interpreters of their interactions. They often engage actively in the creation of contexts that will allow them to interpret each other's behavior. A good illustration of such processes is found in Pollner and Wikler's (1979) analysis of the interactions of family members with a severely mentally retarded child whom the family believed to be of normal intelligence. Their analysis illustrates some of the contextual strategies the family members used to interpret the child's behavior as consistent with their belief.

One strategy, that of "framing," involves one actor prestructuring the situation in a way that maximizes the likelihood that the other's action will appear to have a certain meaning. In some cases, framing is explicitly performed by one of the actors, as in this example provided by Pollner and

Wikler (1979) from their observations of the interaction between Mary, the mentally retarded child, and her sister.

> Mary is sitting on the table; sister is sitting on a chair nearby. Sister pulls another chair up just behind Mary: (saying here, you wanna sit down, but not on the table; you sit down). Sister points to chair; Mary stands with hand on chair, closed in by the table and the chair. (p. 182, translated from the German)

In this "frame," the sister allows Mary very few degrees of freedom in reacting to the sister's prompt. Framing is usually not so explicit, but all cases share a common characteristic: An actor contrives or selects a situation that constrains the other's alternatives so that the resulting behavior can be used as evidence of underlying skills, feelings, and dispositions consistent with the actor's initial expectations.

Another contextual strategy is "post-scripting." This again involves P's imposition of a physical and verbal context for O's ongoing behavior that allows it to be interpreted as consistent with P's a priori beliefs about O. In one form of post-scripting, described by Pollner and Wikler (1979) as "commanding the already done," P requests O to engage in an activity that O has already initiated. When done quickly and subtly, the inversion in temporal sequence of the events is hardly noticeable to either the interactants or outside observers. This process may encourage the impression of congruence between P and O or may provide a basis for the inference that P exercises control over O when, in fact, neither of these is true.

The interface between overt and covert events in the course of interaction is an active, two-directional one. On the one hand, overt events in the chains of the interactants' behaviors influence covert interpretations. On the other hand, interpretations at any point in time condition overt events at subsequent times and thus alter the course of ongoing interaction. Furthermore, members of close relationships actively participate in staging interpretative contexts that make their own and the other's behavior meaningful in terms of their a priori views of each other and of their relationship.

THE CAUSAL CONTEXT OF INTERACTION IN CLOSE RELATIONSHIPS

Interaction in relationships is characterized by order and organization. This order and coherence not only exists at each point in time but extends over time. Across interaction episodes, the interaction of P and O tends to be organized in stable ways and to evidence consistent sequences and patterns. The recurrence of interaction patterns, such as patterns of symmetry–asymmetry, conflict or coordination, implies the presence of relatively enduring causal factors that exercise constraints on P and O across time and limit the randomness or variability of their behaviors. In Chapter 2, these factors were labeled *causal conditions.* Causal conditions consist of relatively enduring features of persons, of the relationship itself, and of the physical and social environments of a relationship.

Personal Conditions

In examining the traits, dispositions, needs, habits, abilities, and so on that P and O bring to their interaction, we will first discuss some of the genetic propensities that make interaction possible. We will then consider some of the rules or norms that P and O internalize in the course of their socialization and that allow them to organize their interaction in an orderly sequential manner. Furthermore, we will consider other stable characteristics of P and O, such as age, gender, education, and social and cognitive skills, that have been found to influence interaction. Finally, we will discuss certain more transient factors that operate at the time of interaction. These factors constrain the course interaction takes but are constantly modified by the ongoing events.

Genetic bases of social responsiveness

There seem to be certain built-in dispositions that make people responsive to social stimuli. These dispositions promote the establishment of causal connections between the behavior chains of P and O and make interaction possible. Because innate behaviors can be strengthened or changed through learning, investigators interested in studying built-in propensities have tried to isolate their effects in two ways: (1) by examining signaling and response behaviors that are common across cultures and (2) by identifying in the early interactions of infants and their caretakers those predispositions that underlie social responsiveness before they are affected by learning.

The cross-cultural approach is illustrated by the research of Eibl-Eibesfeldt (1972), who addressed the question of whether there is a signaling code common to all cultures. After observing and filming in several cultures the facial expressions of people meeting each other, he identified a common set of nonverbal cues, which he labeled the "eyebrow flash." This pattern involves opening the eyes wide and slightly raising the eyebrows. It occurs during greeting and seems to signal readiness for contact. Another apparently built-in mechanism that promotes interaction is the orienting reflex (Diebold, 1968), which causes individuals to turn their faces and attend to a source of sound. This reflex contributes to the alignment of facial-visual and vocal-auditory channels and renders interactants accessible to each other for communication.

The ontogeny of social responsiveness has been studied in the context of infant–mother attachment. Bowlby (1969) reviewed research that indicates that neonates show strong preference for patterns as opposed to plain colors and have selective sensitivity to soft sounds and human voices. These preferences make the human infant particularly sensitive to the human face, and a clear preference for faces over other objects becomes established by the fourth week of life. There is evidence of differential responsiveness to people versus things as early as a few weeks of age (Bruner, 1975).

We will not attempt to provide a complete listing of innate propensities that may contribute to social interaction. They function primarily to focus actors' attention upon each other and thus make interaction possible. During infancy these tendencies facilitate parent–child interaction, which in turn renders the infant more susceptible to cues and rewards from the social environment and more responsive to future social interactions. Their absence, as, for example, in the case of infants with congenital blindness or deafness, can contribute to serious misalignments in parent–child interaction and lead to severe impairments of the children's subsequent responsiveness to social and nonsocial stimuli (Fraiberg, 1979).

Internalized rules of social responsiveness

It is generally believed that, through early interactions with caretakers, infants acquire further skills and expectations necessary for generating and maintaining interaction with others. Bruner (1975) notes that, from early interactions with parents, the child learns the *management of joint attention,* a requisite for joint action. Bruner's observations of mother–infant interactions showed that mothers continuously teach their infants both to focus their attention upon the mother or other objects and to alternate their behaviors with those of their mother's. Alternations of "acting-waiting for response-acting" were also observed in play interactions between toddlers in the absence of any adults (H. S. Ross & Goldman, 1977). Thus, humans appear to learn at an early age the *rules of turn taking* that regulate their interactions (Duncan & Fiske, 1977). These rules foster interchain linkages by providing guidelines for coordination of the actions of an initiator and a recipient.

For the initiator, rules of turn taking prescribe the following behaviors: The initiator's action should be addressed or directed to a recipient, and that person should be given an opportunity to respond at the completion of the initiator's action (Lakoff, 1972; Phillips, 1976). For the recipient, the rules prescribe that the recipient should attend to the initiator's behavior and should respond when given an opportunity. These rules reduce the likelihood of interactional symmetries, such as monopolizing the conversation, that interfere with the generation of longer interactive sequences.

Another set of rules regulates the thematic coherence of interaction. Rules of semantic entailment require that each act in a sequence should be meaningfully connected with the one that precedes it (Kintsch & van Dijk, 1978). Infant–caretaker games and interactions evidence this type of alignment of the content of successive actions. Expectations of meaningful sequential linkage of behaviors seem to be acquired quite early. For example, Tronick et al. (1978) have documented reactions of distress and withdrawal from interaction in infants as young as 3 months when their mothers did not respond in a familiar and meaningful way to their greetings and other attempts to elicit responses. As children grow older, the semantics of interaction with others become increasingly complex. As the child's understanding of the world becomes more sophisticated, misalignments of meaningful sequences are often perceived as humorous rather than distressing.

Stable characteristics of P and O

P and O bring to their relationship personal characteristics that increase or decrease the likelihood of particular patterns of interaction. Age, gender, personality traits, and skills have been linked to the types of interaction patterns that emerge in close relationships. The personal attributes of interactants constrain interaction in two ways: (1) by directly affecting P's or O's behaviors and (2) by affecting P's or O's expectations about or reactions to each other.

Examples of the effect of age can be found in the parent–child literature. The child's age is clearly associated with changes in motoric, cognitive, language, and social skills that affect parental reactions to the child and condition the parent–child interaction (Bell, 1968). Age has also been associated with the types of interaction patterns that children evolve in their peer relationships. Gottman and Parkhurst (1980) analyzed tape-recorded conversations of children with friends and with strangers. They reported a number of differences between younger children (2–4 years) and older children (5–6 years). With both friends and strangers, older children's conversations were characterized by more activity and talk and less fantasy play. In both cases, younger children were much less likely to leave their messages unclarified. With their friends, younger children were more likely than older children to compare their preferences and experiences in ways that established common grounds. Also, younger children were worse at resolving squabbles with strangers than were older children.

The influence of gender on patterns of interaction has been repeatedly documented. Komarovsky's (1967) case study of blue-collar marriages is illustrative. She found, for example, that, although in the majority of couples the level of disclosure between spouses tended to be similar, in one-fifth of the couples studied, women disclosed more than their spouses, while in only one-tenth of the marriages did the husband disclose more. Blue-collar husbands with less than high school education seemed particularly to have problems disclosing concerns and feelings. Komarovsky attributes the reticence of the less educated husbands to a "trained incapacity to share."

> The ideal of masculinity into which they were socialized inhibits expressiveness both directly, with its emphasis on reserve, and indirectly by identifying personal interchange with the feminine role. Childhood and adolescence, spent in an environment in which feelings were not named, discussed or explained, strengthened these inhibitions. In adulthood they extend beyond culturally demanded reticence. The inhibitions are now experienced not only as "I shouldn't," but as "I cannot." (p. 156)

Komarovsky's respondents reported similar differences in conflict resolution strategies. Husbands reported to be more likely to withdraw in the face of conflict, while wives were more likely to argue, "holler," or cry.

Physical attributes of P and O, such as attractiveness (M. Snyder, Tanke, & Berscheid, 1977), or impairments, such as blindness, can provide opportunities and impose constraints upon interaction. Observations of interactions between blind infants and their mothers suggest the pervasive influence that the infant's state has upon the interaction. According to Fraiberg (1979), parents of blind infants are much more likely than parents of sighted infants to engage in extensive gross tactile and kinesthetic stimulation with their babies.

Emotional and thought disturbances that often accompany different forms of mental illness also influence the way in which P and O behave or react to their partner. As will be discussed below, during interaction, P and O have to calibrate their overt behavior so that, on the one hand, it is aligned with their personal goals and self-definitions, and, on the other, it conveys to their partner information about goals, expectations, and views of the self and the relationship. The process of mapping all these divergent meanings onto overt behavior is one of encoding. The skill with which the actor encodes behavior plays an important part in the development of the interaction and in the patterns that it may take. Some research suggests, for example, that encoding deficits arising from underlying thought disturbances may contribute to the pathological interaction patterns between schizophrenic children and their parents (Liem, 1974).

Cognitive and affective systems

P and O enter the interactional arena with built-in propensities to respond to each other's cues, armed with a set of rules that prescribe a regulatory structure for their interchanges, and possessing a set of stable personal characteristics that affect their give and take. However, the presence of all these personal conditions is not sufficient to explain the orderly moment-to-moment unfolding of the interaction between P and O. A set of more variable and alterable causal factors can be assumed to operate at the time of the interaction and to control its flow during any particular episode. These factors are generated by stable cognitive and affective systems that control behavior and interaction.

Intrapersonal processes controlling behavior have been discussed by G. A. Miller, Gallanter, and Pribram (1960) and Powers (1973). Similar notions of behavior control systems have been used by Bowlby (1969) to explain attachment behavior and by Heise (1979) in his theory of interpersonal behavior. The "control systems" approach assumes the presence of cognitive or affective guidelines that act as criteria for evaluating ongoing behavior. This approach also postulates the presence of a testing mechanism that monitors overt behavior and compares it to the underlying criterion. If the behavior matches the criterion, then the actor proceeds to execute the next stage of the guideline. If there is an incongruity, then new behaviors are elicited and enacted until the incongruity disappears. We will briefly review some of the cognitive and affective systems that generate the variable guidelines that have been proposed to control interaction. As we will point out in our ensuing discussion, there is a circular causal loop between these guidelines and the behavior that they control: On the one hand, they regulate the course of interaction, and, on the other, they are also altered by it.

P and O engage in interaction with particular goals as to the behavioral alternatives, outcomes, or states that they desire to attain in the course of any particular interaction episode. These are *situation-specific goals* generated by the interactants' underlying needs, motives, and values; by the exigencies of the situation; by external pressures; and so on. Goals give rise to *contingency plans* that provide each actor with a blueprint for action—scenarios of those behavioral sequences that will foster the attainment of goals (Schank & Abelson, 1977). These plans can be assumed to be part of the cognitive structures or schemata that, as we discussed in the context of the interpretative processes, are employed by the interactants for interpreting the events and sequences that they observe in the course of interaction (Schank & Abelson, 1977). Sometimes P and O develop plans that are unique to the particular context of their interaction. On other occasions, they employ stereotyped and routinized programs (Scheflen, 1973) or scripts (Schank & Abelson, 1977).

For a particular interaction, each actor's plan of action prescribes a set of appropriate behaviors for that actor and a set of complementary behaviors for the other that enable the actor's goals to be attained. Thus, if P wants to play with O, she might offer him a toy or invite him to a game, but she is not likely to hit or insult him. She will also expect him to accept or reject her invitation but not to attack her.

Inasmuch as each actor's goals and associated plans prescribe both a course of action for that actor and a course of action for the partner, the attainment of the actor's goals require the partner's cooperation and coordination. Thus, when P and O interact, they each need to convey to the other their expectations and to constrain the other's behavioral alternatives so that their personal goals will be attained. The process of controlling or constraining the other's behavior has been described in different but convergent ways by symbolic interactionists and social psychologists.

Symbolic interactionists Weinstein and Deutschberger (1963) use the term *altercasting* to describe the control tactic through which one actor attempts to constrain the other's behavioral alternatives. They see it as involving P casting O in a particular identity by emitting behaviors that constrain O to take a particular role vis-à-vis P.

Social psychologists Jones and Gerard (1967) suggest two processes through which control can be exercised in interaction: cue control and outcome control. When exercising cue control, P emits behavioral cues that will elicit in O responses or previously learned behavior sequences facilitative of those of P. P and O exercise cue control over each other in many and subtle ways, for example, by dressing or acting in particular ways to attract the other's attention or to elicit sexual interest. Outcome control, on the other hand, involves the manipulation of rewards and punishments in order to constrain the other's behavior (Thibaut & Kelley, 1959).

The influence efforts that P and O attempt in the course of their interaction are also conditioned by their perceptions and expectations of each other's preferences and stable dispositions. Darley and Fazio (1980) propose that P develops expectancies about O, either because of past observations of O or because of the type of person O is seen to be. They further suggest that these expectancies are more likely to be about stable intentions and dispositions that generate general classes of behaviors rather than about specific behaviors. In the course of interaction, P and O will align their behavior with these expectancies of each other as well as with their personal goals and plans.

In interactive situations, the adequacy of a behavior in attaining P's desired goals is dependent upon O's response, and O, of course, operates under the same constraints. O has goals and plans and expectancies; O interprets P's behavior in the light of this cognitive schemata and then responds to P. O's response to P's behavior can have a number of different consequences that may affect the course of their subsequent interaction. If O's response facilitates P's expectations and plans, it will allow P to proceed to the next stage of P's plan. If it interferes, it might generate on P's part additional efforts to influence O, or it might produce changes in P's contingency plans or even cause P to alter situational goals. Thus, in ongoing interaction, there is a circular causal loop between overt interpersonal behaviors and the underlying cognitive systems that control them. The actor's goals, plans, and expectancies condition interactive behavior. The behavior and its consequences, in turn, can affect and alter the output of underlying cognitive systems.

Although the line between cognition and affect is often blurred, some actions by P and O have clear hedonic consequences for each other. Some of their actions may be inherently rewarding or aversive to the other, while other actions may facilitate or interfere with the other's plans. The rewards accrued and the costs incurred by each interactant in the course of interaction may be referred to as their "outcomes" from the interaction. Caldwell's (1979) analysis of negotiation processes in sexual encounters suggests some of the ways in which these outcomes may be aggregated by the interactants and exercise constraints upon their future interaction. Caldwell hypothesized that couples may keep track of their outcomes in a manner analogous to a bookkeeper's accounts. In interviews with a diverse sample of couples, she found that actors acted as if they were tallying the positive and negative outcomes of their interaction and as if they kept a balance sheet that revealed the relative indebtedness of each to the other. This indebtedness, Caldwell conjectures, constitutes the basis for determining what each actor can expect to receive or deliver and how each should go about giving or getting resources.

Caldwell further suggests that the running balance of P's and O's outcomes at any time will be judged on the basis of an existing rule about how outcomes should be distributed in the relationship. If the running balance departs greatly from the criterion suggested by the rule, then actions may be taken to redress the balance of outcomes. For example, when P and O find themselves in a conflict-of-interest situation in which their individual goals do not correspond (e.g., P wants to have sex but O does not), if P is indebted to O for past favors in similar situations, P might be constrained to assent to O's plan in order to redress the balance of their outcomes.

Heise's (1979) affect–control theory of interactive behavior also highlights the affective component of the systems that control behavior. He proposes that actors enter interactive situations with a repertoire of identities (e.g., man, banker, father, son). Associated wtih these identities are sets of fundamental sentiments, which, in Heise's scheme, are represented by Osgood, May, and Miron's (1975) three dimensions of affect: potency, activity, and evaluation. In the course of interaction, interpersonal events elicit transient feelings that may be compatible or incompatible with the identity-related fundamental sentiments. Heise assumes that actors will always choose from their repertoire of behaviors those that allow transient feelings to remain aligned with fundamental sentiments. If events disturb or deflect an individual's transient feelings away from fundamental values, a course of action will be chosen by that actor to restore these feelings to their original baseline. Alternatively, an actor may manage affective deflections by "reinterpreting" interactional events or by redefining the identities of the actors involved in the interaction.

The following example illustrates Heise's theory of how underlying stable or fundamental attitudes in conjunction with temporary feelings aroused in the course of interaction may determine the unfolding of interactional sequences. Two college students, a man and a woman, meet at a party after a football game. They both see themselves and each other as witty and attractive. They are introduced, and she makes a friendly comment, which he counters with a witty reply. Both laugh and experience feelings in line with their fundamental attitudes toward each other and themselves. If they continue to find each other's comments interesting and funny, their interaction will continue along this vein. If, however, her next comment is thought by him to be "tacky," his transient feelings about her will diverge from their initial baseline. Consequently, he might redefine his original impression of her as witty. On her part, her perception that her comment was not favorably received will deflect her transient feelings about herself away from their fundamental level, and she might see herself as boring and unattractive; or, alternatively, her perception of him might be changed, and she might see him as pedantic and uninteresting. If the latter occurs, she might stop interacting with him; if the former occurs, she might choose from her repertoire of actions one that will elicit a positive reaction and will reinstate her feelings about herself close to their original baseline.

From our preceding discussion emerges the view of interaction as an active and constructive process during which interactants select and calibrate their behaviors according to underlying cognitive-affective guidelines. Critical to this alignment are the interpretative processes discussed in an earlier section of this chapter, through which interactants evaluate their own and their partner's behaviors. The outcomes of these evaluations provide data enabling judgments of the degree of compatibility of ongoing events with underlying guidelines. They also provide information for changing and updating the guidelines. By this view, interaction is not a sequence of stimulus–response pairings or the automatic enactment of internalized scripts. Rather, it is the active creation of chains of causally linked events resulting from the interplay between the interactants' cognition, affect, interpretations, and behavior.

Relational Conditions

Relational causal conditions are those that arise out of combinations of P and O attributes or that emerge from past P–O interactions. An example is provided by Altman and Haythorn's (1967) investigation of the effects of isolation on pairs of men. The investigators created pairs of men compatible and incompatible in their preexisting social needs. Altman and Taylor (1973) describe some of the results of this study as follows:

> Different forms of incompatibility were associated with different modes of environmental use. Men incompatible on need dominance [where both men were high on desire to control and dominate others] became very territorial over time in their use of the environment and were also highly interactive. These were volatile groups who had great difficulty in the situation [two of the three groups who could not complete the isolation period came from these conditions] and men were literally "at each other's throats." They gradually came to divide up the room and to have their own territories, as reflected in very exclusive use of chairs, beds, and sides of the table. Thus they tended not to use the environmental sector of the other, although they still interacted with one another in a very active, perhaps often competitive way. Compatible dyads, on the other hand, exhibited high territoriality during the early days, but this gradually lessened over time, indicating more and more use of one another's space and objects as days progressed. (pp. 111–112)

Emergent shared understandings, beliefs, and agreements that are generated in the course of interaction influence the frequency and patterning of subsequent events. McCall (1970) has called the emergent norms of expected conduct the "culture" of the relationship. Watzlawick et al. (1967) and Jackson (1977) have labeled them "relationship rules" and view them as determining interaction by prescribing and limiting the members' behaviors and organizing their interchanges into a reasonably stable system. A related concept, discussed by Ferreira (1977), is that of family myths. These are well-systematized beliefs, shared by all the family members, about their mutual roles in the family and the nature of their relationships. The function of these beliefs is to provide people with blueprints for action and to eliminate the need for constant negotiation.

A different type of relational condition is suggested by behavioral analyses of interaction that identify the genesis of "interpersonal habits": recurrent patterns of interaction elicited and maintained by the particular ways P and O reinforce each other. These interpersonal habits are maintained by overlapping schedules of reinforcement embedded in the P–O interaction. Patterson's (1979) research with families of aggressive children reveals some of the dynamic processes that sustain patterns of mutual coercion observed in these families. Consider an interaction sequence that is quite common in the interchange of mothers and young children. The mother is engaged in an activity, the child makes a demand, the mother ignores the child, the child whines, the mother begins to pay attention, and the child stops whining. Children's interruptions of maternal activities can be quite aversive to the mother, and the mother's ignoring the child's request is aversive to the child. Patterson and Whalen (1978) found that maternal ignoring is a stimulus that controls the occurrence of child whining. In an experiment conducted in a home, they were able to manipulate the rate of a mother's ignoring and demonstrate that changes in the rate of ignoring were associated with changes in the rate of the child's whining. Child whining, in turn, is quite aversive for the mother. The mother's response of attending to the child, however, has positive consequences for the child: It removes the aversive stimulus (maternal ignoring) and in doing so increases the likelihood that the child will whine on future occasions. In response, the child stops whining, and, by discontinuing the aversive behavior, the child affords a positive outcome for the mother, thus reinforcing her acquiescence to the whining. Consequently, in this sequence, the mother reinforces the child for whining, and the child reinforces the mother for giving in to whining. In the short run, both interactants maximize their payoffs by stopping aversive stimuli. In the long run, however, they reinforce each other's tendencies to behave in an aversive way. Patterson has called the dynamics that give rise to these mutually coercive patterns "reinforcement traps" and has demonstrated how these patterns can be changed when one of the two interactants is taught to produce different consequences for the other's aversive behavior. Patterson's research illustrates some of the complex interpersonal dynamics that can give rise to conflictual as well as mutually gratifying patterns in relationships.

Another relational condition discussed by Kelley and Thibaut (1978) and Kelley (1979) is the structure of outcome interdependence that characterizes a relationship. Interdependence between members refers to the ways that they each control one another's outcomes, that is, rewards and costs. Patterns of interdependence can be analyzed in terms of several properties that these authors assert are related in systematic ways to various aspects of interaction processes observed in different kinds of dyads. Kelley (1979) notes that

> the varieties of interdependence structure reflect both the different types of problems that persons encounter in their relationships and their means (via mutual influence) for dealing with those problems. Any particular pattern of interdependence has latent within it certain possible courses of interaction—plausible scenarios of action and reaction, communication (requests, complaints, threats, promises), and the associated feelings. These scenarios are not necessarily fully acted out by participants. Their mutually recognized possibility and appropriateness to the situation govern the actual course of its events. (p. 43)

Social Conditions

Close relationships are embedded in a network of other relationships, some preceding and others following the development of any particular relationship. Prior research suggests that the involvement of P and O in other relationships affects the frequency and patterning of their interaction. For example, Bott's (1957) study of English urban families revealed some very interesting connections between the social network of the spouses and the patterns of interaction within their marriage. When both husband and wife were integrated in close-knit kin and friendship networks before marriage, each continued, after their marriage, to be deeply involved in outside activities, each received support and help from their respective networks, and consequently each made fewer demands upon and had lower expectations for support and companionship from their marriage partner. On the other hand, when the spouses' networks were loose and fragmented because of geographic and social mobility, each spouse sought help from the other. The first pattern of network integration generated less frequent interaction between husband and wife, fewer joint recreational activities, and greater role specialization in the division of household labor. The second pattern of network integration resulted in lesser role specialization, greater joint organization in carrying out family tasks, and more joint recreational activities. Bott interpreted her findings as showing that social networks are the causal condition for role specialization. (For other interpretations, see Chapter 6, "Roles and Gender.")

Hoffman and Manis' (1978) cross-sectional study of couples at different stages of family life illustrates the effects of children on marital interaction. Their findings, congruent with those of other studies, suggest that the advent of children contributes to and sustains over long periods of time major changes in the interactions of married couples. Specifically, (1) it decreases the frequency of interaction and joint leisure time for the spouses; (2) it accentuates role specialization, with the woman assuming more of the domestic tasks and the man continuing his role as provider; and (3) it provides the couple with a shared focus and joint satisfactions that bring them together and contribute to their marital satisfaction.

Physical Environmental Conditions

Daily interactions in close relationships take place in a variety of physical environments that influence the content and patterning of dyadic interchanges. The positive effects of environmental arrangements have been investigated in a number of studies exploring the relationship between propinquity and relationship formation. In some instances, propinquity has been observed to facilitate frequent interaction between P and O and to promote the development of relationships. Under different circumstances, physical propinquity has been found to have detrimental consequences upon interaction. When the physical space available to each participant is highly limited, a condition usually called "crowding," proximity increases the probability that interaction will be frequent and that it will elicit mutual behavioral interference (Murray, 1974).

The literature generally suggests that conflictual interaction patterns are more likely to emerge in situations in which the environmental conditions require people to interact and prevent them from escaping and thus avoiding confrontations. An example is provided by P. E. Mitchell's (1971) study of the effects of crowding in high-rise buildings in Hong Kong. He found that residents of upper floors reported feeling much more hostility and experiencing more debilitating mental health symptoms than residents of lower floors, even holding constant the size of the living quarters. He interprets this finding as sugesting that lower-floor dwellers could more easily escape their living conditions because of their easier access to outside areas.

Certain features of the physical environment elicit expectations, needs, or overt behavior sequences in P and O. Ecological psychologists posit that a considerable proportion of individual behavior is controlled by the environment (Barker, 1963). Although there is no empirical evidence bearing upon this issue, it can by hypothesized that many interaction sequences between a closely related P and O are evoked and supported by properties of the environment. For example, patterns of marital role specialization become associated with various environmental props ("her" kitchen, "his" workshop) that continue to generate and sustain sex-typed patterns on a daily basis. Sometimes environmental features elicit incompatible goals or sequences in P and O. We are all familiar with the scenario of a mother trying to pull her child away from the candy machine while the child tries to pull the mother toward it. In their daily life, certain attributes of the environment may interact with stable dispositions of P and O and contribute to mutually interfering behavior sequences. In the previously mentioned Altman and Haythorn (1967) study of the effects of isolation upon dyads, physical space restrictions and social isolation acted in statistical interaction with personality characteristics of the participants to elicit particular interaction patterns. Pairs of persons both high on need dominance showed highly territorial behaviors and conflict early in their interaction, whereas pairs with other personality patterns showed different patterns of interaction.

Many of the environmental constraints that operate on members of close relationships are the product of their own past interaction. In developing their relationships, people often select or construct particular contexts for their interaction. Married couples may purchase and furnish a home that then becomes the primary setting in which they and their friends subsequently meet and interact. Couples also stage their interaction, using environmental props to elicit particular reactions in each other, as when P prepares dinner and serves it by candlelight to put O in a romantic mood. In this way, environmental factors not only function as independent variables causing particular interaction patterns, but, as Altman and Taylor (1973) observed, they are actively used to influence the interaction in the same way that words or actions are used.

The view of members as active contributors to their interaction suggests again that, although interaction is constrained by personal, relational, social, and environmental causal conditions, it also acts to effect changes in these conditions. Members establish the rules and agreements that guide their future interaction, actively shape and maintain particular social and physical environmental conditions, and modify their personal conditions. Hence, dyadic interaction is enmeshed in circular causal loops, through which it influences the conditions that in turn constrain it.

CONCLUDING COMMENTS

Until recently the majority of close relationship researchers have largely ignored interaction and have focused on global aspects of relationships as perceived and evaluated by their members, such as reports of marital satisfaction, perceptions of marital distress, and broad evaluations of power, role specialization, and closeness. In this chapter, we have attempted to document some of the theoretical and empirical gains that result from interactional analysis.

Interactive phenomena are complex, encompassing a variety of forms of overt and covert events. Hence, a number of approaches have been developed to study their various aspects. One class of approach has focused on the organization of properties of overt interaction; another has centered its attention on the interpretative processes through which interactants make sense of overt events. Both approaches have yielded useful and promising results. For example, from analyses of observed interaction streams, we have learned more about the qualities and patterns of overt interaction that distinguish between different types of relationships, for example, happy versus unhappy marriages or coercive versus noncoercive parent–child dyads. From studies of interpretative processes, we are currently discovering how close relationship participants construe and summarize their interactions and how these construals in turn affect the participants' subsequent behavior. To date, very few studies have addressed the intricate connections between overt and covert classes of events within interaction. The interface between overt behaviors and their interpretative antecedents and consequences defines a very important area for future research.

The study of interaction in close relationships is not a coherent and unified field of investigation. It is characterized by a proliferation of concepts and methods whose diversity constitutes both an advantage and a drawback. The existing variety of descriptive and causal analyses of interaction has allowed us to view and understand interactive phenomena from a number of different perspectives—through the eyes of observers and through those of participants, at very molecular levels of measurement and at more molar ones. At the same time, the lack of agreed-on procedures for the measurement of interaction properties often limits the generalizability and comparability of results from different studies. The need for generally accepted definitions and measures of the phenomena of interest is obvious, and the field seems to be slowly moving in this direction.

The development of new methodologies has played an important role in the study of interaction. To a great extent, our current understanding of what takes place in relationships is dependent on the power of the methodological tools we have available to study it. With the advent of less obtrusive recording methods and more sophisticated data analysis techniques, we should be able to obtain and examine heretofore inaccessible samples of interaction, thereby gaining access to a much richer empirical data base. This, in turn, should permit us to make more valid and generalizable descriptions of interaction, enabling us to further refine our theoretical models of dyadic process so as to better understand how close relationships develop, are maintained, and change.

CHAPTER 4

Emotion

ELLEN BERSCHEID

INTRODUCTION

The landscape of close relationships presents so vivid a panorama of human emotion that the very phrase "close relationship" carries the implication of passions spent or anticipated, of feelings of every size, shape, and description, of, at the very least, some experience of *affect*—an antiseptic term, but one that encompasses without prejudice the entire range of quality and intensity of human emotion and feeling, from mild irritation to raging hatred to blinding joy to placid contentment. Indeed, close relationships are not infrequently defined with special reference to their affective nature. Many do not consider a relationship beween two people to be "close" unless there exist strong positive affective ties between the participants.

Despite the popular association between the closeness of a relationship and strong positive affect, our conception of a close relationship (see Chapter 2) makes reference neither to the magnitude nor to the quality of the affect that may or may not be experienced by the persons in the relationship. Rather, *relationship* is simply defined in terms of the causal interdependence that exists between two people, and *closeness* is defined in terms of certain properties of that interdependence.

The author wishes to thank Gene Borgida, Bruce Campbell, Susan Fiske, George Mandler, Robert Sternberg, and Auke Tellegen for their helpful comments.

Despite the considerable intuitive appeal of using the magnitude and quality of affect characteristic of a relationship as an aid in the determination of its closeness, we avoided doing so for several reasons. These center around the grave conceptual difficulties posed by the use of an index of the emotion experienced within a relationship to classify it as close or not close and, relatedly, popular misunderstanding of the nature of emotion itself and of its significance within the relationship, particularly within relationships of any duration. As elaborated below, these reasons combine to make affect, as a classificatory variable of closeness, a seductive but dangerous siren to theorists and researchers attempting to navigate the sea of close relationships.

Problems in the Use of Affect as a Criterion of Relationship Closeness

If the affect two persons experience in their association with one another is to be used as an aid to the classification of their relationship as close or not close, one must identify within both P's and O's chains of events those that are "affective" in nature and that also possess either direct or indirect interchain causal connections (i.e., affective events that occur "within" the relationship). When these are identified, the affective events within a relationship may be evaluated along a number of dimensions.

The popular view of close relationships as characterized by strong and positive affect suggests two potentially important dimensions that may be used to classify a relationship as close, as does the work of many theorists who have proposed conceptual schemes for simplifying and organizing observations of emotional phenomena: (1) *magnitude of affect* (or "affective intensity" or "emotional energy" or "level of activation") (e.g., Block, 1957; Davitz, 1969; J. A. Russell, 1979) and (2) the *hedonic sign of affect* (or its "positivity vs. negativity," "pleasantness vs. unpleasantness," "pleasurableness vs. unpleasurableness") (e.g., Block, 1957; Davitz, 1969; Nowlis & Nowlis, 1956).

The magnitude and sign of the affect two persons experience in their association with one another may be assessed in a number of ways. Following a procedure similar to that used by D. S. Holmes (1970), for example, relationship participants might be asked to keep a log over some period of time of each occurrence of feeling and emotion associated with their partner and also to indicate its intensity and its sign. The first difficult decision, of course, is the choice of time period over which the measurements are to be taken. Should the time period encompass the affect the participants have experienced during the whole of the relationship, or within the past day, the past week, the past month, or the past year? If, arbitrarily, the past year is selected, the relationship participants might then be placed in two-dimensional space, the one dimension representing the extent to which they typically experience positive or negative affect in the relationship and the other representing the extent to which they intensely (and/or frequently) experience affect, regardless of hedonic sign, within the relationship.

Using such a scheme, we would have little trouble classifying persons located in the quadrant defined by positive and intense and/or frequent affect as being in a "close" relationship with their partner. Classification of such relationships as close is problematical neither in terms of the popular view of close relationships nor in terms of the conceptual analysis to be developed in this chapter. Classified on other, nonaffective grounds—specifically, the frequency, diversity, and strength of the causal interconnections characteristic of the relationship—there is little question that such relationships would be considered by most observers, theorists, and investigators to be close. The problem, rather, lies in classifying the relationships of individuals who fall into the other quadrants as "not close."

Emotionally quiescent relationships

There is, first, the problem represented by those individuals who experience little affect of any hedonic sign within their relationship. Although it is not clear what the rationale is to be for deciding that emotionally intense relationships are closer than emotionally tranquil ones, it is obvious that a compelling rationale is in order for eliminating the latter from the domain of close relationships. We might become highly discomfited by a scheme that classified a relationship as not close on the basis of relatively little affect experienced over a long period of interaction if, for example, we were to observe one partner following a sudden and irrevocable separation from the other. If death of the partner was the cause of the separation and if the loss precipitated a long period of extremely intense emotion in the form of grief, we might suspect that the relationship had been very close indeed. On the other hand, if we had classified a relationship as close on the basis of frequently experienced intense affect and observed that the loss of the partner precipitated only a relatively mild and short-lived emotional reaction on the part of the survivor, we would probably suspect that the relationship had not been close—at least not as close as the relationship in our first example.

That a scheme that classified emotionally quiescent relationships as not close might produce such discordant surprises with some regularity is suggested by the emerging evidence on emotional reactions to separation and divorce. It is apparently not unusual for those who believed their relationship to be affectively moribund to experience emotions of unanticipated frequency and strength upon actual separation from their spouse. According to M. Hunt and Hunt (1977), who studied the postseparation experience, "one of the most surprising things to many newcomers [to separation] is that they are subject to enormous mood swings that may continue not just for the first few hours, but often for many days or even weeks" (p. 41). Strong and conflicting emotions toward the spouse are not uncommon, nor is a sudden and inexplicable revivification of sexual feeling for the partner (Kressel & Deutsch, 1977). R. S. Weiss (1975), too, documents the intensity of emotion (sometimes distressful, sometimes euphoric, most often a bewildering mixture of each) often precipitated by the partner's absence—emotions not necessarily matched in frequency and intensity to those previously experienced in the partner's presence. In sum, what little data are available suggest that the magnitude of affect experienced in a relationship on a daily basis imperfectly predicts the magnitude of affect the participants may experience upon dissolution of the relationship.

Magnitude of emotional reaction to separation

If it is true that at least one of the difficulties of using magnitude of affect experienced in the relationship over some unit of time as a criterion of closeness is that we are unable to distinguish relatively affectless relationships that are close (using other criteria) from affectless relationships that are not, it is tempting to simply discard this particular magnitude-of-affect measure in favor of another—the magnitude of affect the participants experience upon a separation from each other that is perceived by them to be irrevocable. The magnitude of emotional reaction to separation is an attractive indicant of the closeness of the relationship for several reasons, including its face validity. More importantly, many social scientists have found the strength of emotional reaction to separation from the partner to be useful in the study of infrahuman social relationships and human parent–child relationships (e.g, Ainsworth, 1969, 1972; Bowlby, 1969, 1973). With respect to the latter, the affective responses a child makes upon separation from the mother (most notably, crying, but also such other symptoms of distress as disturbances of sleeping and eating), as contrasted to its affective reactions to separation from others in its immediate social environment, are often considered to be an important index of the strength of "attachment" of the child for the mother. Attachment theory and research may be of special interest to those interested in adult close relationships if only because the concept of attachment approximates, at least on a metaphorical level, the concept of closeness. In addition, the emotional effects that have been observed to occur upon separation in heterosexual adult relationships bear marked similarity in their gross outlines to those experienced by children upon separation from their attachment figures.

Whatever illumination attachment theory may ultimately shed on adult close relationships, even a cursory look at an "emotional reaction to separation" variable reveals a number of problems with respect to its potential to serve as an index of closeness in adult relationships. The first of these is that, although the supposition that the strength of the attachment bond between two people ought to reveal itself when it is broken by separation, the proper operational definition of attachment, and thus its measurement, is by no means entirely clear even in the relatively restricted case of mother–child relationships. (See L. J. Cohen, 1974, for a discussion of problems in the operational definition of human attachment, and Masters & Wellman, 1974, for a procedural critique of studies of infant attachment.) Second, some relatively drastic revisions in theoretical conceptions of attachment behavior would seem to be necessary for its measurement in adult relationship separations since many of these separations are relatively voluntary. For example, attempts to maintain proximity to the attachment figure are considered central to attachment behavior (L. J. Cohen, 1974), but separations in adult relationships (M. Hunt & Hunt, 1977; R. S. Weiss, 1975) often have been instigated and/or maintained by the person who is also experiencing the strong affective reactions to the separation. In addition, of course, partners' perceptions that the separation is irrevocable may be less likely in the case of adults (except when death is the cause of separation) than it is in the case of infants. Further, and as L. J. Cohen (1974) has noted, attachment behavior becomes more difficult to measure as an individual matures and maintenance of proximity to the attachment figure increasingly becomes an internalized symbolic process.

Third, and in addition to these as yet unresolved problems, the use of measures of magnitude of affect upon separation as indicants of relationship closeness presents at least one very practical problem for researchers of adult relationships. Given the unethicality of experimentally inducing partners' perception of irrevocable separation from each other, or even of threat of separation, we are left with a measure that will allow us to determine the closeness of a relationship only posthumously. It is the strength of the partners' emotional reaction to separation that many researchers, and laypersons as well, seek to understand and to predict *before* the fact of its occurrence. Its understanding and prediction require the identification and investigation of the variables of which it is a function. These, we believe, and as we shall elaborate in this chapter, lie in the properties of causal interdependence characteristic of the relationsip.

Affectively negative relationships

Those individuals who experience much affect within their relationship, but affect that is negative, also pose a problem for a scheme that automatically classifies such relationships as not close. Consider, for example, a man whose diary indicates intense and frequently experienced negative affect toward his wife. His quantitative expression is corroborated not only by his own impassioned qualitative statements to the effect that his life's one sincere desire is to terminate his strife-torn association with his mate (and he would do so immediately were it not for financial and other exigencies), but also by his wife's affect diary and qualitative statements, all of which suggest that she vigorously reciprocates the sentiment. The self-reports of both partners may even receive external corroboration from outside observers who testify that, indeed, the interaction between the two consists primarily of silence broken only by fights, arguments, and needling—events all clearly permeated with negative affect for both partners. And, yet, such relationships may endure for years. In stark contrast to the relative fragility of many more pleasant and pleasurable affective relationships, the stability of the interaction between some of these couples is wondrous to behold.

In addition to their endurance, relationships characterized by negative affect sometimes present another puzzle on the occasion on the death of, or separation from, the partner. Despite the survivors' past avowals that they fervently desired to dissolve the relationship, they may be thrown into an inexplicable (at least to them) depression when the relationship is, in fact, dissolved. One case in point is the man who, according to newspaper accounts of his trial for homicide, had been severely nagged by his wife of some 30 years. He had left her countless times for this very reason but always returned. Finally, as he explained to the judge, during one of her more vociferous diatribes, he shut her up by strangling her to death. He confessed to newspaper reporters later, however, that he now missed her very much.

Thus, many relationships that observers using nonaffective criteria would classify as close may possess a predominantly negative affective theme. Indeed, the family—the social constellation that contains the prototypes of the close relationship—is the setting for many strong negative emotional experiences, as investigators are currently documenting. Family sociologist Straus (e.g., Gelles & Straus, 1979; Straus & Hotaling, 1980) has pronounced the family to be the most physically violent group or institution that a typical person is likely to encounter. Fitz and Gerstenzang (1978) found that most of the anger and hostility men and women experience in their daily lives is directed toward a blood relative, with heterosexual partners (girlfriend, boyfriend, or spouse) running a close second, and that the anger experienced towards relatives and heterosexual partners tends to be more intense than anger experienced in other relationships.

It is clear that strong negative affect experienced more or less regularly, perhaps even exclusively, in a relationship many would consider as close on other grounds, is not unusual. At the least, a classification scheme that excluded such relationships from the domain of close relationships would exclude many family relationships.

Ambivalence of affect in close relationships

Although it is physically violent, the family also provides the most loving and supportive relationships many people ever know. It is perhaps for this reason that such relationships are popularly viewed as the source of only positive emotion and feeling. One consequence of this widely accepted view has been that researchers long overlooked the negative affect and destructive behavior that often occurs in family relationships. Another result is that little is known about ambivalence of affect within relationships (Berscheid, 1982). And yet another consequence is that for persons who subscribe to the prevailing view that close relationships can be virtually defined with reference to positive affect, the occurrence of intense negative feeling within a relationship must inevitably lead to doubts about the closeness of the relationship. Or, as the noted psychoanalyst Reik (1972) pointed out decades ago, the popular belief that close relationships are characerized only by positive affect

> brings us unnecessary unhappiness and causes much misery. . . . We are then very intolerant of feelings of resentment and hatred that sometimes rise in us against beloved persons. We feel that such an emotion has no right to exist beside our strong affection, even for a few minutes. We are afraid that the appearance of so unwelcome a guest seriously endangers our feeling of affection. (pp. 99–100)

It was Reik's thesis (as well as others, e.g., May, 1969) that indifference, or the lack of any emotional involvement with the partner at all, is the true enemy of relationships, and that hate is always the silent (and sometimes not so silent) partner to love.

Unfortunately, the belief that the experience of negative emotion toward another is foreign to the experience of positive emotion is often shared and maintained by many researchers of interpersonal attraction (Berscheid, in press-a), as well as by many researchers of marital satisfaction. Most instruments used to assess the affect one person feels toward another, for example, assume a bipolar unidimensional view of affect that implies that the further respondents are toward the positive end of the scale (e.g., +3, "Love/like very strongly"), the more distant they are from the negative pole (e.g., −3, "hate/dislike very strongly"), and, thus, the less likely they are to experience negative affect within the relationship.

Summary affect statements

Although most traditional scales used to assess the affective quality of a relationship provide no means for the assessment of ambivalence of affect within it, it can be argued that they were not meant to; they simply elicit a global "summary of affect" statement from the respondent. But what such scales require respondents to do is to "sum over" (or use whatever cognitive algebra occur to them as appropriate) a number of emotional experiences in order to arrive at a statement of their general affective experience within the relationship and/or their general affective attitude toward it. That is, it is generally assumed (e.g., Matlin & Stang, 1978) that when asked such questions as "How happy are you?" (with the partner, the relationship, the job, or life in general), the respondents review their experiences and feelings and come up with an index accurately summarizing these. How valid this assumption of accurate summary is, however, or what the true referents of such summary statements actually are, is a mystery. Most traditional affect assessment scales have received little or no consideration of the cognitive gymnastics their respondents must typically perform in order to review and summarize extraordinarily individualistic and complex sets of data.

At the least, however, it seems apparent that persons who experienced joy associated with their spouse on Saturday, anger on Sunday, love on Monday, and hatred today have a different problem in answering a summary affect question than do persons who have experienced anger toward their mate every day for the past several months. Further, persons who scan a number of strong positive and negative affective experiences and arrive at a "neutral" assessment of the affective quality of their relationship probably have a very different relationship than persons who arrive at the same neutral point because they have had no strong emotional experiences within the relationship. And there are yet other grounds for speculating that the exploration of the relationship between emotional events and affect summary statements will reveal that the relationship between the two is not straightforward. R. L. Weiss, Hops, and Patterson (1973), for example, report that spouses' daily records of the number of behaviors their mates performed that were pleasurable or displeasurable to them were significantly related to their Locke–Wallace marital satisfaction score, but the amount of variance in marital satisfaction these discrete events accounted for was less than impressive.

Future exploration of the correlation between emotional event reports and abstract affective statements may reveal individual differences in the tendency for persons' affect statements to correspond to their actual emotional event experiences, as well as differences in how well each variable will predict to other variables of interest for each individual. In addition, the manner in which persons proceed from the event level of analysis to more abstract levels may be interesting and informative in itself. For example, a number of investigators (e.g., Barnett & Nietzel, 1979; Wills, Weiss, & Patterson, 1974) report that displeasurable daily events associated with the spouse account for far greater variance in global satisfaction-with-spouse ratings than do pleasurable events, also suggesting that people's affective mathematics may be more complicated than frequently supposed.

Finally, it should be noted that not only does an overreliance on summary statements to assess affect within a relationship mask many questions of potential interest to close relationship investigators, but it seems doubtful that answers to such questions with regard to enduring close relationships will provide enough variance to account for many phenomena of interest within those relationships. Most studies of happiness (e.g., Gurin, Veroff, & Feld, 1960) consistently find that most persons report their marriages (as well as their jobs and their lives overall) to be happy, with less than 5 percent stating that they are dissatisfied or unhappy. The referents for these responses are not clear, but it seems highly unlikely that an affective event–level analysis of these relationships would produce so homogeneous a picture.

Conclusions and Overview of Chapter

The preceding arguments and evidence suggest that neither the hedonic sign nor the intensity and/or frequency of emotion experienced within a relationship serve as a satisfactory classificatory variable of the closeness of that relationship. However, these affective dimensions have considerable interest as dependent variables, both in themselves and as concomitants and integral components of almost all close relationship phenomena of current interest. The initiation of an adult heterosexual relationship, for example, is often associated with the emotions of joy, romantic love, and ambivalence; the development and maintenance of a close relationship is not infrequently characterized by anger, hatred, and contempt, as well as satisfaction, contentment, and happiness; and the dissolution of a close relationship is often the occasion for grief, melancholy, and depression.

The position taken here is that knowledge of the various properties of interdependence characteristic of the relationship, combined with an understanding of what is known of the antecedent conditions for emotional experience, may further our understanding of the great range, in both quantity and quality, of human emotion as it occurs in the close relationship context. We assume, in fact, that the affective phenomena that occur in a relationship are a direct function of, and therefore predictable from, the various properties of interdependence that characterize the relationship.

Therefore, in this chapter we shall

1. Briefly outline what is known about the antecedents and consequences of human emotional behavior;
2. Attempt to relate the causal antecedents of emotion to the properties of interdependence characteristic of close relationships, illustrating how relationships we have classified as "close" (on the basis that the causal interconnections between the participants' chains of events are strong, frequent, and diverse) satisfy the necessary, although not sufficient, conditions for the experience of intense emotion within the relationship and satisfy the necessary and sufficient conditions for intense affect upon irrevocable separation;
3. Illustrate how individuals who are particiants in a relationship characterized by strong, frequent, and diverse causal interconnections are "emotionally invested" in the relationship, whether or not they are aware of it and whether or not they frequently experience affect—positive or negative, intense or mild—within the relationship;

4. And, finally, outline how knowledge of the specific properties of interdependence characteristic of a relationship and of the antecedents and consequences of human emotional behavior may promote understanding and prediction of some certain emotional phenomena as they may (or may not) occur within the relationship.

THE PSYCHOLOGICAL STUDY OF HUMAN EMOTION

There is no problem in human behavior whose unraveling does not to some extent, often to a large extent, depend upon a psychological understanding of the dynamics underlying emotional phenomena. As a consequence, the problem of human emotion ranks as one of the oldest and most central in psychology. Unfortunately, it has also proved to be one of the most difficult. The problem of understanding human emotion has been so difficult, in fact, that psychologists have periodically thrown up their hands in frustration and disgust at its intractability (Arnold, 1960; Hebb, 1949; P. T. Young, 1927). Fortunately for all students of human behavior, however, and especially for students of close relationships, who cannot avoid confronting questions of emotion, the psychology of emotion has experienced a theoretical and experimental renaissance within psychology over the past two decades (Berscheid, in press-b)

Features of Theories of Emotion

Contemporary theories of emotion contain four common elements: (1) They take an evolutionary perspective; (2) they postulate a physiological component to emotional experience; (3) they recognize the vital role of cognition in emotion; and (4) they take a "continuous-loop" view of the dynamic interaction of the physiological and cognitive components of emotional experience. Each of these elements has its source in historical milestones in the psychological odyssey toward an understanding of emotion.

The evolutionary perspective

Emotion was among the first psychological phenomena to be removed from the arena of philosophical speculation. Its formal introduction as a matter for scientific inquiry was made by Darwin, whose interest in the origin and evolution of Homo sapiens led him directly to an interest in the role of emotion in furthering human welfare and survival. In his classic treatise *Expression of the Emotions in Man and Animals,* Darwin (1899) reported a series of observations begun by himself in 1838, as well as observations made by others. In addition to stimulating an interest in the expression of emotion that continues to the present day (e.g., Izard, 1977), Darwin's work left subsequent investigators two important legacies. The first was an insistence that no theory of emotion would be likely to be valid if it did not illuminate what purpose emotion serves for the survival of the species. Today, virtually all theorists of emotion agree that the experience and expression of emotion has served, and probably continues to serve, an important function in the survival of humans. From the evolutionary perspective, then, emotion is not an "irrational" and frivolous component of human behavior. To the contrary, its rationality and dignity derive directly from its service in the life and death struggle of each individual to survive and to survive as comfortably and as happily as possible (e.g., Plutchik, 1980).

Darwin's second legacy was his recognition that there was some sort of association between strong "nervous excitation" and the expression of emotion (if not the experience of emotion as well). Darwin (1899) concluded his report, in fact, by saying that the subject of emotion "deserves still further attention, especially from any able physiologist" (p. 366).

The physiological component

What the association between excitation and emotion was, Darwin did not say. That was left for the American psychologist James, who was dismayed by the disarray and confusion already characteristic of the study of emotion, and particularly by the tendency to study each of the emotions individually. James emphasized the importance of developing a general theory of emotion that would account for all the emotions and then proceeded to offer his own. He proposed (1884) that our perception of an "exciting fact" caused bodily changes to occur and that our subsequent perception of the internal bodily changes taking place in us constitutes the emotion. To experience an emotion is thus, according to James, to sense a particular pattern of internal physiological changes that we then recognize as "fear" or "anger" or "love" or as one of the other emotions. Those organic sensations occurring in the viscera (the organs in the cavities of the body) were thought by James to be most important to emotional experience. One year later, Lange (1885/1922) advanced a similar theory, the principal difference being that Lange believed that the sensations of change in the circulatory system (e.g., blood pressure), rather than the viscera, were most important in emotional experience.

The influential James–Lange theory stimulated much research that, unfortunately for those who believed their quest for a general theory of emotion was over, culminated in a devastating critique by Cannon (1927; 1929). On the basis of the many research findings that had become available, Cannon argued that the richness and variety of the subjective experience of emotion was not at all matched by a comparable richness and variety in patterns of bodily visceral changes. The viscera had been revealed to be relatively insensitive structures, for one thing, and, for another, the same bodily changes appeared to occur in very different emotional states and, worse, they also sometimes occurred in nonemotional states as well. Thus, the subjective experience of emotion could not be accounted for by the sensing of internal physiological changes. Cannon's attack on the James–Lange theory proved fatal to it, and the problem of emotion underwent another eclipse in psychology.

Despite the inadequacy of his general theory of emotion, James made a vital contribution to our understanding of emotion. Today, most contemporary psychological theories of emotion assume that peripheral physiological processes are necessary to emotion, that a necessary (although not sufficient) condition for emotion to occur must be a physiologically aroused organism (e.g., Strongman 1978). It is now also generally agreed by most that the pattern of physiological arousal is relatively undifferentiated from one emotion to another and is not responsible, therefore, for the great variety of subjective emotional experiences.

The cognitive component

While James' gift to an understanding of emotion was to draw attention to the role of peripheral physiological arousal in the experience of emotion, there remained the problem of the discrepancy between the fact of relatively undifferentiated physiological arousal and the fact of the richness and variety of subjective emotional experience. This was the problem that stymied researchers and suppressed their interest in emotion until Schachter (1959; 1964) and others (e.g., J. Hunt, Cole, & Reis, 1958) proposed that cognitive factors are major determinants of emotional state.

Schachter assumed that an emotional state must be some function of physiological arousal. He further proposed that people interpret and identify their aroused physiological states in terms of the meaning of the situations that precipitated the arousal. Or, "cognitions arising from the immediate situation as interpreted by past experience provide the framework within which one understands and labels his feelings" (Schachter, 1964, pp. 50–51). Schachter and his colleagues (e.g., Schachter & Singer, 1962) performed a series of brilliant experiments in which arousal was manipulated (through the injection of a drug whose effects mimic arousal) independently of the surrounding situation and the interpretive cues it provided. These investigators found that the nature of the emotional state subsequently experienced (e.g., anger versus euphoria) by those in an aroused state was a function of the surrounding situation, and that persons placed in the same situation but who were not physiologically aroused tended not to experience any emotion at all. These experiments supported Schachter's hypothesis that for people to subjectively experience an emotion, two events must occur: First, they must perceive that they are physiologically aroused. Second, they must make a cognitive evaluation of that arousal and arrive at an emotional "label" that explains or describes what they are experiencing. They are most likely to evaluate the arousal by examining the situation they believe precipitated it.

Schachter maintained, therefore, that there is no internal physiologically distinctive core to each of the various emotions. The richness and variety of emotional experience is provided primarily by the variety of the external situations in which arousal is experienced and in the "emotional meaning" of those situations to us. Most importantly, Schachter's experiments suggested that neither physiological arousal alone, nor the application of an emotional cognitive label by itself, is a sufficent condition for the subjective experience of emotion. Both appear to be necessary.

Schachter's fundamental view of emotion was extremely influential, and, today, virtually all contemporary psychological theories of emotion posit an interaction between physiological and cognitive determinants of emotional state. (See Candland, Fell, Keen, Leshner, Plutchik, & Tarpy, 1977, and Plutchik, 1980, for a variety of current views.)

The continuous feedback loop among components of emotional experience

It was the nature of the interaction between the physiological and cognitive components of emotional experience, however, that plagued and discouraged emotion researchers for years. The "sequence problem," as it came to be known (Bindra, 1970; Plutchik, 1962), was another of James' legacies, and an unfortunate one. Given that both cognition (or as James put it, "the perception of the exciting fact," such as a gun pointed at our head) and physiological arousal (or, "our feeling of the same changes as they occur") are necessary to emotional experience, in what temporal order do these components occur? James proposed that the cognitive appraisal of the situation occurred first, the physiological reaction (as well as overt molar actions) occurred second, and, finally, the feeling of emotion occurred. Other theorists proposed differently, and the sequence problem dominated emotion research for decades.

A pithy estimate of the value of the debate is offered by Candland (1977), who points out that

> there is no special advantage in assuming that there is a temporal sequence among the three processes, and much waste has come from believing that the appropriate way to untangle our confusions about emotion is to sort out the temporal sequence. We have created a monstrous problem by assuming that explanation requires uncovering a temporal sequence. . . . (p. 66)

The "monster" currently has been reduced to negligible size and ferocity by the simple, sensible, and expedient proposition of a "continuous loop" of feedback among the elements of emotional experience:

Neither cognitive nor physiological factors are antecedents. Because the combined cognitive and physiological components of the emotional experience can be activated quickly, they are viewed as feeding back upon and modifying both continuing emotional responses. This feedback or continuous-loop system affects continuing and future responses in three ways: (1) emotional experiences, through their cognitive and physiological qualities, can feed back and modify the perception of eliciting stimuli and thus the emotional reactions; (2) emotional experiences can modify the cognitive appraisal of emotional stimuli, and thereby the continuing and future emotional experiences will be different; and (3) emotional experiences can modify the physiological reactions elicited by continuing stimulation and thereby modify the continuing emotional experience. (Candland, 1977, pp. 66–67)

The Place of "Feelings" in Emotion Theory and Research

With only a few exceptions (e.g., Pribram, 1967), most contemporary emotion theorists and researchers have adopted the view espoused by Darwin and James that "feelings" are simply "little emotions." A few theorists have given attention, albeit passing, to the question of the manner in which an emotion may be "little" and thus properly termed a feeling. Most of these views seem to parallel Jung's (1968): Emotions are seen as intense states accompanied by bodily change, whereas feelings are simply good or bad values assigned to things. Emotions thus are accompanied by physiological arousal. Arousal is absent in feelings, but a cognitive evaluation of the goodness or badness of the stimulus (usually in terms of the individual's welfare) remains. This distinction—that *emotions* necessarily involve the sensation of physiological change, whereas *feelings* are thought (or remembered) but unfelt experiences—is tacitly accepted by most contemporary theorists and investigators (e.g., Candland, 1977).

Even though emotion and feeling lie on the same continuum of physiological arousal, and it is, therefore, sometimes difficult to distinguish between an emotion and a feeling, it is worthwhile to preserve the distinction between them. The distinction between emotion and feeling is particularly important for the study of affect in close relationships for a number of reasons, only one of which can be mentioned here—that is, that communication and understanding are made unnecessarily difficult in discussions of affect within close relationships both by the lack of clear referents of the terms *emotion* and *feeling* and by the great variety of referents for these terms even when they are made explicit. Thus, one person assumes he is currently experiencing the "emotion" of love (and answers a questionnaire to that effect) even though he has not experienced physiological arousal within the relationship for the past several years. Similarly, another person may describe in the most pale, colorless, and purely "cognitive" way the strong "emotions" she believes she is suffering or enjoying in her relationship, and this self-report (and the person's terminology) is accepted at face value. Conversely, a person whose knees shake and hands tremble in the presence of his partner may report that he has some mild "feelings" toward her.

Throughout the remainder of this chapter, therefore, the term *emotion* will refer exclusively to "hot" emotion, or to a state accompanied by physiological arousal and by the perception of that arousal. The word *feeling* will refer to states in which: (1) a cognitive "emotional" appraisal is unaccompanied by actual physiological arousal (or by only very weak arousal) and unaccompanied by the perception of arousal as well; (2) there is actual physiological arousal as well as the perception of that arousal, but these are unaccompanied by a cognitive "emotional" appraisal; and (3) a cognitive "emotional" appraisal is accompanied by the perception of physiological arousal but not by actual arousal (see Berscheid, in press-b), for further discussion of these points).

Thus, the position taken here is that any cognitive evaluation alone (e.g., on any good–bad dimension, such as love–hatred, like–dislike, approach–avoidance), no matter how strong, is not an "emotion" if it is unaccompanied by physiological arousal. The "emotion" of love, for example, is conceived to be different from a "feeling" of love or from an "attitude" of love. The latter may involve extremely positive evaluations of the partner on various personality rating scales, approach tendencies toward the partner, doing favors and conferring other benefits upon the partner, and the representation of such cognitive evaluations and behavioral tendencies on "affect" scales. Similarly, the "feeling" or "attitude" of hatred, which may result in the most heinous and destructive acts toward another, is viewed not to be the same as the "emotion" of hatred, which also may result in similar negative acts.

A Contemporary Theory of Emotion

In *Mind and Emotion,* Mandler (1975) outlines a theoretical framework for viewing human emotion that may be particularly useful for unraveling the tangled skein of emotional phenomena as they occur—and sometimes surprisingly fail to occur—in close relationships. His theory incorporates all of the basic features of most contemporary approaches to emotion outlined in the previous section and illustrates in detail how these may interact to produce the experience of emotion.

The role of autonomic nervous system arousal

Peripheral physiological arousal, specifically that of the autonomic nervous system (ANS), plays a central role in Mandler's theory. The ANS is concerned with those bodily functions that are generally termed "visceral"—the muscles of the heart, the smooth muscles of the intestines, the blood vessels, the stomach, and the genitourinary tract—and with such glands as the salivary glands. These functions are largely involuntary, or "autonomous" of the individual's control. Further, the ANS appears to act in a more or less total fashion, and its output appears to be fairly undifferentiated from one discharge to the next. (See Tarpy, 1977, for a discussion of the nervous system and emotion.)

The ANS is composed of two subsystems: the sympathetic nervous system and the parasympathetic nervous system. The specific action of the sympathetic nervous system is likely to result in an increase in heart rate, blood pressure, and blood flow; glucose conversion to increase blood sugar level; dilation of pupil size; and inhibition of digestion, among other effects. The physiological symptoms so often associated with emotion—the pounding heart and dry mouth, the perspiring hands and clammy skin, the butterflies in the stomach, and the flushed face and tremulous hands—are all products of ANS discharge, particularly of the sympathetic nervous system. The action of the parasympathetic nervous system is opposite to that of the sympathetic. Acting with the sympathetic nervous system to maintain the individual's homeostatic balance, the parasympathetic nervous system has as its principal function the conservation and restoration of the individual's energy following energy expenditure, and it produces such effects as decreased heart rate and increased salivation. When people perceive they are in a state of physiological arousal, it is the action of the sympathetic nervous system component of the ANS, rather than of the parasympathetic component, that is likely to be occurring.

Autonomic nervous system arousal has long been considered to be of physiological adaptive significance to humans. Cannon (1929) proposed that the responses of the sympathetic nervous system are part of a bodily "emergency reaction." The notion of "general adaptation syndrome" responses that promote the individual's "fight or flight" from harmful stimuli was later developed by his student Selye (e.g., 1974). Selye and many other physiologists believe that the body reacts adaptively with a single general syndrome of physiological response to stress (or change in its internal or external environment), whether that stress be a severe loss of blood, an infectious disease, a snarling dog, a new job, or a threatening spouse. This set of physiological responses appears to prepare the individual to cope bodily with the stress. Increase in blood sugar level, for example, gives the individual an extra boost of energy to fuel the flight or the fight against the stressor.

Mandler argues that ANS arousal has psychological adaptive significance as well as physiological adaptive significance. Specifically, he hypothesizes that our perception that we are experiencing ANS arousal warns us, if we are not already aware of it, that there is some stimulus in our external environment to which we must pay attention in order to initiate appropriate action if it should prove that our well-being is at stake. Autonomic nervous system arousal is thus conceived to be a psychological alarm bell, a signal system that tells us—often loudly and insistently: Something important is going on out there that may affect your well-being! Pay attention! Something needs to be done! The perception that we are experiencing ANS arousal, then, is believed to prompt us to direct our attention to the stimulus situation in which we find ourselves in order to determine the meaning of the situation, especially its implications for our welfare, and what actions, if any, we should take to protect our well-being.

Thus, Mandler theorizes that the ANS acts as a "secondary support system" for our analysis of the meaning of a stimulus situation. The ANS is only a "backup" warning system that something has happened that may have implications for our welfare, because the ANS is relatively slow; it may not respond until 1 to 2 seconds after the stimulus event has been perceived. During the brief period between our perception of the external stimulus event and our perception of the internal event of ANS discharge, there is opportunity for us to have determined the meaning of the external event and to have already selected an appropriate action in response to it. As a consequence, our hands need not yet be sweating for us to move out of the way of a coiled brown object that suddenly appears in our path. We may feel our heart pound only after we have swerved our car out of the way of an oncoming truck and shiver only after we have caught ourself from falling. However, if we have not already determined the meaning of the external stimulus event and taken action in those few seconds, our perception of the secondary internal event of ANS discharge provides a backup signal that alerts us that the situation demands that we turn our attention to it to try (or continue to try) to determine its meaning. In short, the psychological adaptive significance of ANS arousal is to ensure that we will attend to, and cognitively evaluate, stimuli for which appropriate action in the interest of personal well-being and survival might need to be taken. At the same time, ANS discharge may ensure physiologically that, should our meaning analysis of the situation lead us to fight or take flight, our bodies will be prepared to do so.

There is much evidence that ANS discharge is associated with attentional processes (Berlyne, 1960, 1974) and that these, in turn, often prompt cognitive activity (see Berscheid, in press-b, for further discussion of this point). Attentional processes, in fact, are conceived by most modern cognitive theorists to be integral to an understanding of how our brain processes and stores information. Mandler and others (e.g., Neisser, 1967) believe, for example, that the permanent storage of information in memory requires an act of attention, and Mandler further argues that the concept of "focal attention" and the concept of "consciousness" appear to refer to much the same processes. The objects and events to which we pay our direct attention, for example, are the stuff of consciousness, or the things we are "aware" of. Consciousness seems to have the adaptive function of permitting us to react reflectively instead of automatically in transactions with the environment, for one of its functions appears to bring two or more mental contents into direct juxtaposition, where they may be deliberately compared and analyzed. Consciousness thus appears to play an important role in the appraisal of situations and in the choice of potential action alternatives.

Emotion and "mind"

Autonomic nervous system discharge, believed to be essential to the experience of emotion, is thus also conceived to be integrally related to the functioning of the mind. The word *mind* simply refers to the cognitive processes and mental structures that are hypothesized to intervene between the sensory input the individual receives (any and all perceptible events, whether they take place inside or outside the skin of the person) and observable outputs (e.g., verbal reports, measurable physiological arousal, actions). Mandler terms the organized system of mental structures that is believed to process the sensory input the individual receives the *cognitive–interpretive system.* This system of mental structures allows us to determine the "meaning" of any stimulus input.

The cognitive–interpretive system and its operations, then, is a theoretical fiction. It cannot be directly observed. Its nature must be inferred from observations of the relationship between the stimulus input to the system and its output in the form of observable behavior (see Anderson, 1980, for an overview of the assumptions and methods of cognitive psychology). Thus, the mind is not available to inspection by introspection, just as Freud's "unconscious" was inaccessible to introspective examination. Introspection can only yield the conscious results of the cognitive–interpretive system's previous work on the stimuli processed by it—not the actual working on, or the processing of, the sensory data. Since consciousness/focal attention in humans is severely limited, the relatively small window of consciousness displays the results of the workings of the mind much as a typewriter displays the results of a computer's work on the input that it has been given. Consciousness, in other words, is not the same as "mind," nor is it synonymous with "thought."

One important implication of this view for close relationship researchers is that an individual's self-report of events necessarily can draw only upon what appears in consciousness. Unfortunately, as Mandler (1975) points out, "events and objects in consciousness can never be available to the observer without having been restructured, reinterpreted, and appropriately modified by structures (unconscious cognitive structures of which we are not aware) that are specific to the individual doing the reporting" (p. 51), and, for this reason, "people's reports about their experiences, their behavior, and their actions are very frequently, and may always be, fictions or theories about those events" (p. 52). The special methodological problems this presents to investigators of relationships are discussed in Chapter 11, "Research Methods."

Meaning analysis

One of the functions of the cognitive–interpretive system is to organize past experience, thereby permitting current stimulus events to be quickly integrated with past history. Any stimulus input to the cognitive–interpretive system is thus assumed to be subjected to an analysis of its relationship to the individual's existing cognitive structures in order for its meaning to that individual to be determined.

Mandler assumes that the extent of the meaning analysis to which a stimulus input will be subjected depends on: (1) the complexity of the individual's mental structure (e.g., the more complex an individual's mental cognitive structure is, the more implications a particular stimulus input will have for a variety of past experiences and thus the more complex the meaning analysis will be and the "richer" the meaning of the stimulus to the individual) and (2) the state of the individual at the time of input, particularly attention and task requirements (e.g., the degree and depth of the meaning analysis will covary with the degree and intensity of attention, so that the more a stimulus has commanded our attention, the more quickly and thoroughly we will attempt to discover its meaning).

Mandler also hypothesizes that the process of meaning analysis has a "stop-rule" that goes into effect whenever the requirements of the task are apparently fulfilled. For example, once we have determined that the large black animal charging toward us is a bull and not a cow, we are unlikely to try to further determine whether it is an Angus or a Hereford or whether its conformation is such that it will win a prize at the county fair. The requirements of this meaning analysis task were fulfilled when our analysis of the relationship between the stimulus input (large snorting black animal with horns moving toward us fast) and existing cognitive structures (e.g., determined by our knowledge of certain properties of large animals with horns who sometimes move fast, such as their ability to outrace a human who does not have a big head start) helped us understand the "meaning" of this stimulus event. Our determination of the meaning of the event allows us to select an appropriate action alternative (e.g., climb a nearby tree) that is now more in our best interests than the action we had been engaging in (e.g., picking daisies). Thus when our meaning analysis of the stimulus input produced "bull," it also may have clearly indicated an action alternative. The "stop-rule" for that meaning analysis is likely to go into effect at this point, and we will take the action indicated. (This, of course, does not imply that later, at our leisure, we won't perhaps go back and mull the series of events over and further flesh out their meaning.)

Thus, it is assumed that actions we have executed in the past (or actions we have read or heard about) are represented in the cognitive structures. These cognitive representations of actions make it possible for us to cognitively manipulate actions without actually performing them; that is, once a representation of action has been developed, these cognitive stuctures can be used for elaborate choice and decision processes (e.g., whether there is time to make it to the fence or whether climbing the tree is the better alternative) prior to actually engaging in the action. Current actions, then, are conceived to be activated by particular action structures, or "action systems," within the total organization of cognitive structures.

Organized action sequences and plans

Some of these action systems will be *organized action sequences* of behavior, which are of special importance for predicting emotion. An organized action sequence is a series of actions that are emitted as a whole or as a single unit. When the first action in the sequence is made, the others in the series tend to follow. Klinger (1971) describes the development of such sequences:

> When organisms attempt new sequences of motor responses, their performances are at first labored and error-prone, requiring close attention and concentration. With repetition, the sequences become smooth and efficient, requiring less conscious attention and less sustained effort. As the sequence becomes overlearned, it can be performed virtually unconsciously, and the organism may find it awkward to initiate the sequence without completing it, as when the owner of a standard-transmission car tries to drive with automatic transmission and finds his left foot still searching out the clutch during traffic stops, even after several trials with the clutchless vehicle. Until his foot comes crashing down upon the empty floor, the driver is unlikely to have been aware of his action and is unlikely to have experienced it as effortful. At such a point of overlearning, the response sequence has attained the properties of a unitary response and may be considered "integrated" [or "organized," in Mandler's terms]. (p. 164)

What is especially important about an organized action sequence, then, is that, once it has started, and no matter how long or complex it is, it often has an inevitability of completion. Furthermore, the activation of the underlying action structure on a particular occasion is independent of what its consequences or goals turn out to be. While a set of consequences (i.e., reinforcers) undoubtedly established the behavior in the first place, on this later occasion it is the controlling stimulus for the initial behavior in the sequence that starts the specific action sequence, and, once initiated, the entire sequence of behavior tends to "run off" regardless of its appropriateness or consequences on this occasion.

Much of the interaction that takes place in close relationships is undoubtedly made up of organized action sequences. They can be clearly seen, for example, in spouses' interactions in marriages of any duration. Some long-term relationships, in fact, may consist almost entirely of many organized response sequences on the part of each spouse. A problem often arises when partners wish to change their behavior, perhaps because they have come to perceive it as undesirable or inappropriate. Because of the "inevitability of completion" feature of organized action sequences, people may experience great difficulty in modifying their behavior once the initial controlling stimulus has appeared and the sequence has begun.

In addition to their tendency toward completion once started, another important feature of organized behavior sequences that makes them difficult to change is that the individual does not give full conscious attention to the behaviors as they are being executed. The performance of each bit of behavior in organized behavior sequences tends to be "thoughtless," "automatic," or relatively "unconscious." Further, of course, people may not be even aware that a number of their behaviors *are* organized into a sequence—that even though the behaviors may occur over considerable time and space, they are emitted as a unit, and, therefore, changing the middle or end responses is extremely difficult once the sequence has been triggered.

An important antecedent condition for the development of organized behavior sequences is the frequency of use of the sequence, with organization developing gradually. The continuing exercise or execution of a particular action system tends to generate a more tightly organized and invariant structure. Thus, to discover an individual's (or a couple's) organized action sequences, the frequency with which a series of actions is performed is a good clue, as are the invariability of the nature of the actions performed and the smoothness of their performance. Another useful clue is a decrease in the actor's conscious attention to the behavioral acts during their performance and, relatedly, the actor's superior reaction time to extraneous stimuli (e.g., Keele, 1968) and ability to conduct a second activity simultaneously with the series of acts (e.g., Schmidt, 1968). The spouse who can eat his breakfast grapefruit, read the morning paper, and simultaneously "converse" with his wife is probably exhibiting a highly organized action sequence.

Many organized action sequences are parts of other organized action sequences that are parts of higher-order *plans.* The tightly organized behavior sequence of driving to work in the morning, for example, may be a part of the much higher-order plan to earn a living. The smoothly run-off behavior sequence of making toast and coffee may be part of the organized action sequence of preparing breakfast for one's spouse that may be part of a higher-order plan to preserve one's marriage. Plans undoubtedly have a hierarchical structure, some being more encompassing of other plans and behavior sequences than others. Like organized behavior sequences that have been performed in the past, plans also probably vary in their degree of organization and detail (e.g., the vague plan to buy a house "someday" versus a well-thought-out step-by-step series of actions that will culminate in buying a house in 3 years). Higher-order plans, then, can be viewed as response sequences initiated and in some stage of completion.

Plans involve "goals" or objects and/or events that an individual not only values but also is committed, at least in some degree, to striving for (Klinger, 1977). As noted in Chapter 1, Klinger (1977) found that, when people were asked whether their lives were "meaningful," most interpreted the meaning of "meaningfulness" to be akin to "purpose" or "having aims" or "plans" and, when asked to tell precisely what made their lives meaningful, close to 90 percent mentioned one or another kind of close personal relationship as something that contributed meaning to their lives. For many, it was the only thing mentioned. Given the prominent role close relationships play in most people's plans and purposes, and given the relationship between emotion and organized action sequences and plans (as shall be discussed in the next section), it is not surprising that close relationships provide the setting for a range and intensity of emotion unmatched by any other context.

Interruption: An antecedent condition for emotional experience

Mandler argues that the interruption of some ongoing activity or plan, particularly the interruption of highly organized behavior sequences in the process of execution, is a sufficient, and possibly necessary, condition for the occurrence of ANS arousal. Interruption is thus a sufficient and possibly necessary condition for the experience of emotion. When an individual experiences emotion, an interruption of some ongoing activity (either an organized behavior sequence or a higher-order plan) has probably occurred.

Support for Mandler's interruption hypothesis can be derived from a number of sources. From the evolutionary point of view, interruption is an important event to any organism. Interruption signals that important changes have occurred in the environment. These environmental changes often lead to altered circumstances of living, adapting, and surviving. The fact that arousal occurs whenever current activity is interrupted has important survival value for, as previously noted, it has preparatory flight and fight functions that will help the organism adapt to the changed environment; it may also serve as a signal system that leads to increased attention and information seeking.

Support for the interruption hypothesis is also provided by the relationship between ANS arousal and attention and the known characteristics of stimuli that elicit attention. If Mandler is correct, these should be stimuli that tend to be interruptive of the individual's ongoing activities—which, in fact, appears to be the case. Such attention researchers as Berlyne (1960, 1974), for example, have observed that stimuli that possess such characteristics as novelty, surprise, complexity, and incongruity win the perceiver's attention, other things being equal, over other stimuli, and they also appear to be associated with higher levels of "activation" or arousal. In addition to stimulus novelty, the relevance of the stimulus to the individual's ongoing activities and plans is also known to be a factor in attentional processes; attentional mechanisms appear to select personally relevant materials and events. The factors that define such personal relevance are generally unknown, but under conditions of high novelty, the degree to which another person is perceived by the individual to have the power to affect the individual's comfort and well-being is directly related to the degree to which that person can dominate the individual's focal attention/consciousness (cf. Berscheid & Graziano, 1979; Berscheid, Graziano, Monson, & Dermer, 1976).

With respect to emotion as it occurs in social relationships, the interruption hypothesis also receives support in the anecdotal and clinical literature addressed to such specific strong emotions as romantic love. For example, that interruptive obstacles seem to act to heighten passion rather than to decrease it has been frequently noted, the most well known literary instance probably being the saga of Romeo and Juliet. Further, the results of the few studies that have even tangentially examined the effects of obstacles on intensity of romantic love (e.g, Driscoll, Davis, & Lipitz, 1972; Walster, Walster, Piliavin, & Schmidt, 1973) seem to support Bertrand Russell's view: "I think it may be laid down that when a man has no difficulty in obtaining a woman, his feeling toward her does not take the form of romantic love" (cited in Kirch, 1960, p. 11). Of course, he could as well have said that a woman's feelings for a man are likely to be similarly affected by obstacles.

As Mandler conceives of them, the subjective experience of emotion and the occurrence of so-called emotional behavior occur between the time a highly organized response sequence or plan is interrupted and the time at which either the interrupting stimulus is removed or the individual has found an available substitute response that allows the sequence to be completed. During this period, initiated by the onset of an interrupting stimulus: (1) ANS discharge occurs; (2) attention is likely to be focused on the interrupting stimulus or on a search for it; and (3) the interrupting stimulus is likely to undergo a meaning analysis with the aim of selecting an action that allows completion of the interrupted sequence or plan. During the period in which the meaning analysis is occurring, ANS arousal should be increasing. When an action permitting completion is found, ANS arousal should disappear and, with it, the subjective experience of emotion.

In addition to subjective emotional experience, another consequence of interruption is assumed to be a tendency to try to complete the organized response sequence as long as the situation remains essentially unchanged, perhaps by more forcefully performing the actions previous to the blockage; that is, the organism will keep trying to complete the action sequence, frequently with extra vigor. For example, when the key that has always unlocked the front door in the past now unexpectedly fails to open it, we most likely will persist for some time in twisting the key harder and harder, finally kicking and pounding on the door and perhaps simultaneously showing such "emotional behaviors" as crying and cursing. Or, if the cookie that usually brings a child out of the doldrums fails, two cookies may be tried, and, if that doesn't work, perhaps a dish of strawberry shortcake will. Increased activity and the more vigorous repetition of the old response that is now failing to have the expected effect have long been observed to occur upon our discovering obstacles on our well-worn path to a goal (Klinger, 1977). Or, as Mandler (1975) puts it, "organized sequences tend to follow the dictum: 'If at first you don't succeed, try again' " (p. 156). During the period of redoubled effort, however, the individual should be experiencing emotion as a consequence of the interruption and the failure to complete the sequence. If and when completion occurs, ANS arousal should be "turned off" and emotional behavior should cease.

Another possible consequence of interruption is response substitution, or the completion of the sequence or plan by other organized sequences or plans that are more or less specific to the one that has been interrupted. For example, if upon driving to work we find that our usual route is blocked, an alternative route we have used in the past may be quickly substituted to complete the sequence. Or, if one's husband is called out of town in the middle of a gardening project requiring four hands, a friendly neighbor who has helped before might be enlisted before the plants die. Some organized sequences also have interchangeable parts (e.g., writing several letters to friends), and the sequencing by which the parts are executed is unimportant (if one letter is interrupted because of failure to immediately find the address, another may be substituted). Mandler assumes, then, that whenever we are prevented from completing any one organized sequence, we will tend to try to complete the sequence by substituting another, even minimally relevant, organized sequence.

Finally, Mandler suggests some general principles that probably govern the relationship between interruption and degree of arousal (and thus the probable intensity of the emotional experience):

1. Degree of ANS arousal upon interruption should be related to the degree of organization (i.e., the tightness and invariance of structure) of the behavior sequences and plans that are interrupted.
2. Degree of arousal should vary with the discrepancy between the interrupting event and the interrupted structure.

3. Substitute behaviors available at the time of the interruption should decrease arousal effects.
4. Degree of arousal probably depends on the place of the organized behavior sequence in the hierarchy of organized behavior sequences and higher-order plans (e.g., if the organized behavior sequence of "driving oneself to work" is part of a much higher organized behavior sequence of "earning a living," interruption of the low-level plan of driving to work may not produce much in the way of arousal because the executive plan in existence at the time is a much "higher" one and because, since alternatives are available, there is nothing in the interruption of the low-level plan that necessarily interrupts the higher one).
5. The more highly organized behavior sequences and plans should also be the ones that are resumed if the situation permits, and, thus, although the degree of arousal may be high if they are interrupted, the arousal should be short-lived if the interruption is noncontinuing.

Summary

The two principal components of Mandler's theory of emotion are ANS arousal and the cognitive–interpretive system. ANS arousal is believed to be a necessary condition for emotional experience. One function of ANS arousal is to act as a secondary signal system for initiating a meaning analysis of a stimulus situation. In Mandler's view, the prime preprogrammed, "wired into the species" releaser of ANS arousal is interruption of an ongoing activity, particularly organized behavior sequences and higher-order plans. Interruption is thus a sufficient, and possibly necessary, condition for emotional experience. ANS arousal has the effect of constricting focal attention/consciousness to the stimulus situation to discover the stimulus that is the source of the arousal and/or its nature and meaning. ANS arousal thus initiates a meaning analysis of the stimulus situation if one has not already been initiated. The meaning analysis is directed toward determining whether action should be taken and, if so, identifying the nature of an action that will protect our interests. During the meaning analysis, and before an action alternative is determined (if one is available), ANS arousal should be increasing. Thus, the interruption of an organized behavior sequence or higher-order plan produces a state of arousal that, in the absence of substitute responses to complete the sequence or plan, then develops into one or another emotional experience depending on the occasion of the interruption. Interruption, then, may lead to the experiences of fear, anger, joy, euphoria, love, or another emotion, depending on factors other than the interruption itself. The resumption of an organized response sequence or plan is one "stop-rule" for ANS arousal and a meaning analysis of the stimulus event, and thus it is also a "stop-rule" for the emotional experience.

EMOTION IN RELATIONSHIPS

The term *relationship*, as discussed in Chapter 2, refers to two chains of events, one for P and one for O, that are on the same time dimension and that are causally interconnected. For two people to be in a relationship with one another, then, some of the events in P's chain of events must be causally connected to some of the events in O's chain, and, similarly, some of the events in O's chain must be causally connected to some of the events in P's chain. The phrase "emotion in the relationship" refers, therefore, to an event in either P's or O's chain of events: (1) whose immediate causal antecedent(s) lies in an event(s) in the other person's event chain and (2) that may be termed an "emotional event."

Emotional Events and Their Assessment

The identification of an event as "emotional" may be made in at least three ways, each of which defines a specific level of analysis and requires the availability of a certain type of behavioral data: (1) physiological behavior; (2) overt molar behavior; and (3) verbal self-report of inner experience. Each of these may be conceived as three separate event chains running along a single time dimension for a single individual. Some combination of these types of data, rather than each taken alone, is most useful for determining whether an emotional event has occurred. Further, given the status in contemporary emotion theory of physiological arousal as a necessary condition for the experience of emotion, the physiological level of analysis assumes special importance in the classification of events as emotional or nonemotional.

Confidence in the classification of an event should be highest when data are collected at all three levels of analysis and the evidence from each is congruent with the others. For example, if P shows a sharp rise in all indicants of ANS arousal immediately following O's announcement that O's mother will be making an extended visit (or following O's touching P, or asking P a question, and so on), then the sequencing in time of these two events (O's action and P's rise in ANS arousal) suggests that a causal relationship exists between an event in O's chain and a particular kind of event in P's chain that we know satisfies a necessary condition for P to experience emotion in the relationship. (In lieu of physiological monitoring devices, persons in relationships, like good clinicians, often attempt to detect the readily visible symptoms of ANS arousal, such as sweaty palms, flushed face, and so on.) If, in addition to showing a sharp rise in arousal upon O's announcement, P were to kick the cat, yell at the kids, slam the door, or otherwise engage in overt behaviors most observers would term "emotional," then the data collected at the molar behavioral level of analysis also would suggest that O's action was causally related to an emotional event in P's chain of events. And, finally, we might be especially confident in identifying the event as an "emotional" one if we were to overhear P telling a friend how "angry" P was because of the visit. Data from all three levels of analysis, in this case, would corroborate the categorization of the event as "emotional."

More difficult is the instance in which, by all sensitive assessment devices, P may have clearly shown a sharp rise in ANS arousal, but there is no evidence of emotion whatsoever at any other level of analysis. P, for example, may reveal a perspiring forehead, flushed face, and, generally, all signs of imminent apoplexy upon O's announcement that O's mother will be making an extended visit, but may not behave overtly in any other way a reasonable outside observer would categorize as "emotional." Further, when asked about being "angry" or "upset," P may show genuine surprise that the question even has been asked.

Difficult, too, is the case in which P has shown a rise in arousal, has kicked the cat, yelled at the kids, and refused to eat supper because of a sudden attack of nausea, but, when asked by O if the impending visit has angered and upset P, P replies, "Angry? Who, me?" Under these circumstances, the investigator (not to mention P's spouse) has to make some decisions. The first of these will probably be whether P's self-report is truthful. P may be very aware of being angry, but, just as it is often useful and instrumental to pretend emotion when none exists, it is also sometimes useful to deny emotion when it does.

It is entirely possible, however, that P's self-report that no emotion was experienced was an honest one, with no motive or intent to deceive. Physiological arousal is a necessary but not sufficient condition for emotion. In addition to arousal, P must perceive arousal (e.g., notice the heart pounding) and interpret that internal event as an emotional one. To do so, P must locate the external cause of the internal event (e.g., my mother-in-law is going to visit for 2 months), decipher the personal meaning of that stimulus event (e.g., just when I was counting on some peace and quiet around here, I'll have to move out of the bedroom, chauffer her around, and so on), and apply an emotional cognitive label to the event (e.g., I am "angry").

Thus, P's self-report that O's announcement did not cause anger may be an honest one for a number of reasons. First, P may not have noticed the physiological arousal. There are undoubtedly wide individual differences in the extent to which people are attuned to their internal physiological events because of psychological, sociological, and constitutional reasons. Second, even if P noticed the arousal, P may have attributed it (perhaps correctly) to the 10 cups of coffee drunk at work and thus may have given a nonemotional interpretation to the arousal. Third, P may have noticed the arousal and ascribed an emotional label to it, but made a different attribution as to the source of the arousal and thus arrived at an emotional interpretation different from the one the investigator and O suspect is the appropriate one (e.g., I was not "angry" about the visit; I felt "fear" because, while you were talking,

I remembered that I forgot to make a deposit today to cover all those checks I wrote). (See London & Nisbett, 1974, for a discussion of these and other points.)

Not infrequently, investigators and observers use the rule that the final authority on whether P has had an emotional experience is P, and so self-report data often assume a special status in the categorization of events as emotional ones. On the other hand, some investigators and therapists might decide on the basis of the physiological and overt behavioral data that P is not "in touch" with his or her emotions, and might work to help P become more attuned to internal events, to learn to draw causal connections between external events and internal events, and to practice labeling of these events so as to become able to express such emotions. The point to be made here is simply that self-report of emotion is generally a valuable but also problematical kind of data. Certainly, self-report is not the royal road to the classification of events as "emotional" or "nonemotional" that some consider it to be. In fact, collecting data only at one level may lead to unreliable and invalid judgments of events as "emotional."

Discrepancies in classification of a single event as "emotional" or "nonemotional" among investigators (and also among participants in close relationships) thus may be the result of focusing upon different event chains. Even when all three types of data are collected, discrepancies in classification may result from the use of different classification rules with respect to discrepancies between events in the different chains. Both sources of discrepancy may result from adopting different theories of emotion and/or from the differing availability of measuring instruments at different levels of analysis. The propensity for investigators to use different bases for classification, when combined with the failure to specify the basis upon which an event is classified as emotional or nonemotional, can be expected to create the same problems in close relationship research as it has in emotion research (e.g., see Strongman, 1978). It is necessary that close relationship researchers be aware of the possible reasons for discrepancies and that explicit decision rules be developed to evaluate such discrepancies and to arrive at a judgment that an individual has experienced emotion. In the discussion that follows, when an event is termed "emotional" or "nonemotional," it is assumed that evidence at all three levels of analysis is clearly in accord with the classification (unless otherwise noted).

The Antecedents of Emotion in Relationships

Since the substance of a relationship lies in the causal connections that exist between P's and O's chains of events, and since physiological arousal is a necessary condition for the experience of emotion, it is clear that for emotion to be experienced "within" a relationship, some of the causal connections from O's event chain to P's event chain must cause physiological arousal to occur for P (and/or some of the events in P's chain must cause physiological arousal for O). If we assume that interruption is a sufficient and possibly necessary condition for the occurrence of physiological arousal, then it is also clear that a causal connection from O's chain to P's must "interrupt" something to generate physiological arousal for P. And, as outlined previously, what it probably must interrupt is a highly organized behavior sequence and/or a higher-order plan. Or, in the terminology introduced in Chapter 2, what it must interrupt is an *intrachain sequence,* that is, a sequence of events within P's chain of events in which each event is causally connected to the next event. Further, since this intrachain sequence must be interrupted by an event in O's chain, it is clear that for P to experience physiological arousal as a result of an interchain causal connection (and thus for emotion to occur "within" the relationship), some of the events in O's chain must be causally connected to an intrachain sequence within P's event chain.

Interfering interchain connections to intrachain sequences

The preceding argument can be summarized by reference to Figures 4.1 and 4.2, which show, respectively, the necessary and the sufficient conditions for arousal (and thus emotion). In both figures, the $p_1 \rightarrow p_2 \rightarrow p_3 \rightarrow p_4$

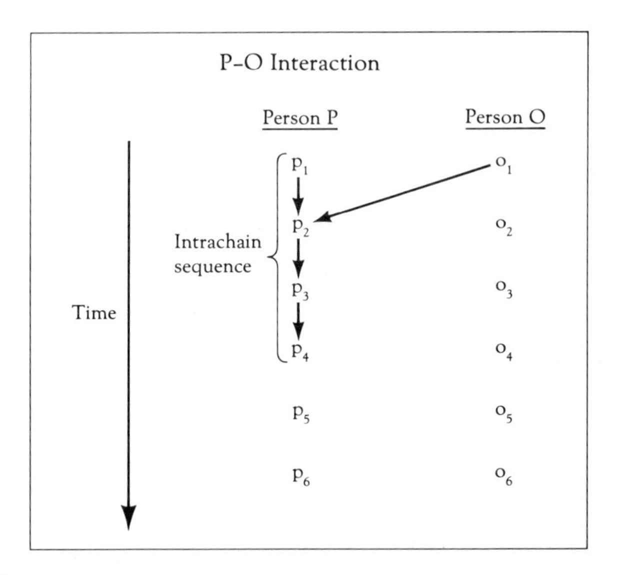

FIGURE 4.1
Illustration of an interchain causal connection from O to an intrachain sequence in P's event chain, satisfying a necessary but not sufficient condition for P to experience emotion within the relationship. The event o_1 affects an event in P's intrachain sequence but does not interfere with the completion of the sequence.

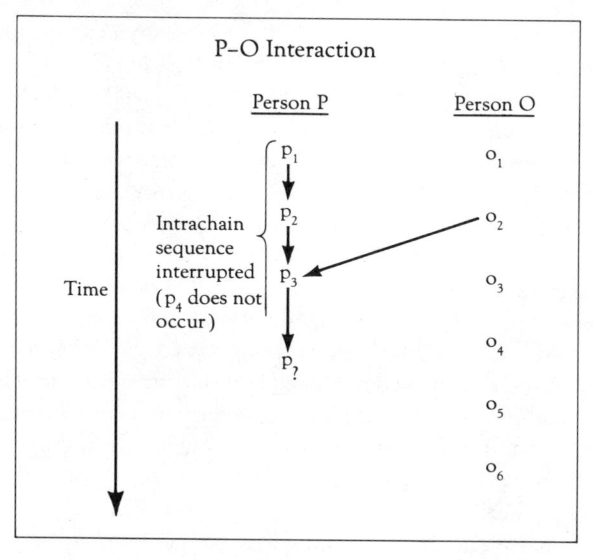

FIGURE 4.2
Illustration of an interfering interchain causal connection interrupting the completion of an intrachain sequence, thus satisfying a sufficient condition for P to experience emotion within the relationship. Event $p_?$ is likely to be some form of emotional behavior.

intrachain sequence is one that has been observed on prior occasions to occur dependably. In Figure 4.1, an event o_1 in O's chain has an effect on an event p_2, which is part of P's intrachain sequence. An interchain causal connection of this sort, from O's chain to one of P's intrachain sequences, is a necessary condition for P to experience emotion. However, it is not a sufficient condition. It is not sufficient to cause emotion because the o_1-to-p_2 connection does not always interfere with P's intrachain sequence. In Figure 4.1, for example, it can be seen that, while o_1 (along with p_1) influences p_2, both p_3 and p_4 occur as usual. For emotion to occur, the interchain connection must interrupt or interfere with the occurrence of the remaining events in P's intrachain sequence, as is illustrated in Figure 4.2. That is, the events that would have occurred in P's event chain (had O *not* performed o_2 or had o_2 *not* been causally connected to p_3) did *not* occur (or they occurred more weakly, in a distorted form, or otherwise showed significant variance in strength or form from what would have been expected on the basis of P's previous performance of this intrachain sequence). Figure 4.2 shows the o_2-to-p_3 connection as preventing the occurrence of p_4, the last element in P's sequence. The subsequent events in P's chain, indicated by $p_?$, are likely to include some form of emotional behvior.

For example, the $p_1 \rightarrow p_2 \rightarrow p_3 \rightarrow p_4$ intrachain sequence may represent P's regaling his poker group with his favorite story (told and retold and finely polished over the years) of the time the family was on a camping trip and he was awakened in the night by a big black grizzly bear licking his face and what he bravely did about it. In the midst of the tale, P's wife may have wandered into the room to replenish the beer, and o_2 may represent her interjection, "That *wasn't* a bear, George! Unless grizzly bears have long wagging tails, and you know the Anderson's black lab was always getting loose!" P's intrachain sequence constituting the remainder of the story undoubtedly will not run off as it has so many times before.

Or, to take another example, P may be a skilled skier who, in the interest of "togetherness" with her new husband, who recently became enamoured of the sport, takes him to her favorite ski resort. For P, who has expertly skied the trails for years, the entire day is filled with many tightly intrachained and (usually) smoothly executed behavior sequences that are begun only to be interrupted—from the time her husband has trouble getting on the ski lift, to his discovering that her favorite run is too steep, to their mutual trudge to a lesser slope, to her finally flying down that trail only to look back and discover her groom clinging to a tree with one hand and to a loose ski with the other. The stage has been set for P to experience emotion "within" the relationship.

It should be emphasized, in the context of the above example, that O's behavior is not in *itself* interfering and interruptive; it is cause for emotion only in relation to the fact that the events in P's chain to which it is causally interconnected are highly organized and intrachained. If they were not, if P were also a novice skier, there would be little for O to interfere with. P would be in search of the next action to execute as much as O so obviously is. It can be seen, then, that the same behavior on the part of O may be interruptive (and sufficient cause for emotion) in one relationship and noninterruptive in another relationship.

In the case of the skiing newlyweds, the emotion that results for P from O's interference probably will be negatively labeled (e.g., mild anger or "irritation"). However, it need not always be the case that interrupted intrachain event sequences result in negative emotion. P's arousal upon being interrupted by the sight of her husband wrapped around a tree may be interpreted and experienced as "fear" that he's broken a leg, a wave of "love" at his endearing helplessness, and so on. Or, for another example, O may break into P's highly organized toilette sequence, which includes absentmindedly reaching for the electric shaver while sitting in the bathtub (which, this morning, didn't drain properly) by shouting, "Don't touch that!" P may be delighted to have been interrupted. We shall later discuss some of the determinants of the hedonic sign of emotion, but here it should simply be observed that, while one can predict that an interrupted sequence probably satisfies a sufficient condition for emotion to occur, the positivity or negativity of the emotion (if it does occur) is not determined by the fact of the interruption.

Facilitative interchain connections to intrachain sequences

Interchain causal connections to intrachain event sequences are not always interruptive and interfering. Indeed, they may facilitate and augment the performance of the next event in the sequence. Every time George has told the bear story in the past, his wife may have interjected, "Yes, and it was such a *huge* bear!" at the appropriate point in the tale, serving as the stimulus for George to speak of the depth of his terror. Many domestic routines in daily marital life are instances in which the individual's intrachain sequence could not be performed well, if at all, were it not for the occurrence and appropriate timing of the partner's responses that help stimulate and make possible the next response in the individual's intrachain sequence.

Meshed intrachain sequences

Our discussion to this point has focused on the case in which P is engaging in an intrachain sequence and O has made responses that either interfere with or facilitate P's intrachain responses. In many cases, however, *both* P and O are simultaneously engaging in highly organized intrachain sequences. Furthermore, events in each person's chain may facilitate the performance of the other's sequence. In this case (illustrated in Figure 4.3), the two intrachain sequences may be described as *meshed.* This state is to be contrasted with that of *unmeshed* sequences, in which there are no causal connections between the

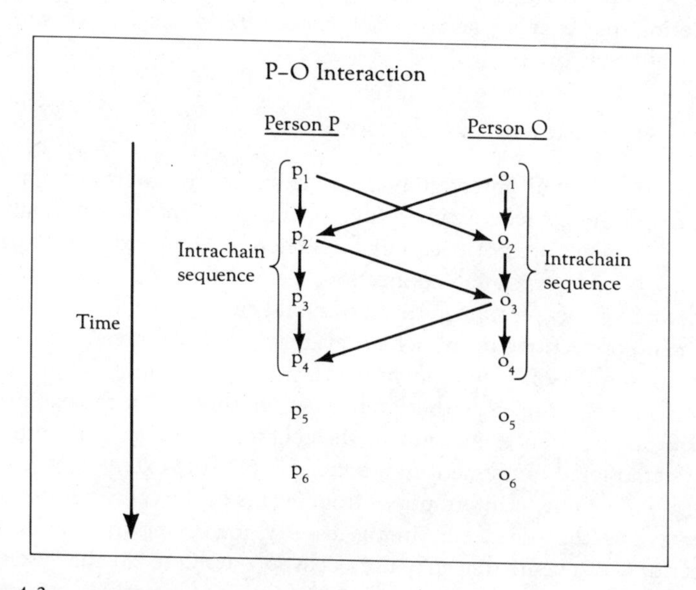

FIGURE 4.3
Illustration of meshed intrachain sequences. The P-to-O causal interconnections facilitate O's intrachain sequence and vice versa.

two intrachain sequences simultaneously occurring (when they are "disengaged," so to speak), and with *nonmeshed* sequences, in which the causal connections between the two intrachain sequences interfere with the enactment of one or both. Meshed intrachain sequences are commonly seen in such mundane tasks as getting the kids ready for school in the morning and paying bills and in more complex interactions, such as social entertaining and sexual intercourse. Meshing makes it possible for many such interconnected and highly organized intrachain event sequences to be performed without the slightest interruptive hitch.

Meshed event sequences show all the characteristics of other highly organized behavior sequences: (1) Given the initial stimulus, the full chain of responses, on both P's and O's side, tends to run off, thereby revealing their character as a single unit rather than as discrete responses; (2) little variation in the nature of the sequence or the form and timing of the responses is observed; and (3) the behaviors are performed smoothly, rapidly, and without much conscious thought, sometimes allowing other activities to be pursued simultaneously. These characteristics, as well as their repetitive occurrence in P's and O's event chains, permit their identification as meshed event sequences. Meshed sequences can also be identified as having existed when one person, through sickness or absence or for other reasons, fails to perform her or his part and the partner's sequences become disorganized. On these occasions, the failure of one person to provide the appropriate responses at the appropriate time in sequence is interruptive of the other person's completing the sequence and probably constitutes a sufficient condition for emotional behavior.

Emotional Investment in Relationships

It can be seen that interchain causal connections to intrachain event sequences satisfy a necessary condition for the experience of emotion in the relationship. They are not necessary for a relationship to exist, as there may be many interchain causal connections between P and O but none to intrachain sequences—though this seems unlikely. The extent to which such interchain connections to intrachain events do exist, however, defines the extent of each person's emotional investment in the relationship. The term *emotional investment* refers to the extent to which there exists the potential for P and O to experience emotion in the relationship. One may think of a person's emotional investment in a relationship as the extent to which that person is vulnerable to interruptions from events in the other person's chain (either events that do occur, in the case of interfering interchain causal connections, or events that may not occur in the future, in the case of the facilitative connections involved in currently meshed intrachain sequences).

The degree to which a person is emotionally invested in a relationship does not necessarily correspond to the frequency and intensity of emotion the person may actually experience in the relationship on a daily basis. A person's emotional investment in a relationship is defined by *both* the facilitative and the interfering interchain connections, while the degree of emotion actually experienced in the relationship is associated with only the interfering, or interruptive, interchain causal connections.

The partners' emotional investment in the relationship may be symmetrical or asymmetrical even though the overall frequency, strength, and diversity of interchain causal connections may be the same for both. That is, P may have many more interchain connections to his or her intrachain sequences than O does; thus, O's "interruptibility power" and O's power to precipitate emotion in P are greater than P's power to affect O emotionally even though their relative power to affect each other's nonemotional behavior remains equal.

Relationships characterized by meshed intrachain sequences

If the relationship is characterized by a great number of meshed intrachain sequences, there should be little occasion for the experience of emotion within the relationship on a daily basis. Such a relationship should appear to be, to both observers and participants, emotionally "tranquil" and "serene" (or, depending on the observer's values, "routine" and "humdrum"). The genesis and development of such relationships is an empirical question, but two obvious hypotheses present themselves.

First, the participants, by virtue of their similarity of attitude, values, background, interests, skills, and so on, are able to develop well-coordinated, compatible, or (in our terms) meshed interactions with relative rapidity and ease. In marital interaction, for example, two persons who come together with similar understandings of the purposes and goals of their marriage and the roles each partner should play, who share similar educational backgrounds, intellect, and skills, as well as similar attitudes toward the persons, objects, and events that impinge upon their lives, ought to be able to mesh their interactions with little disruption of each individual's organized behavior sequences and higher-order plans. This hypothesis is, of course, no more than what marriage and family texts and pastoral counselors have been saying for years—"similarity" and "friendship" and "liking" and "compatibility" are the best bases for achieving a satisfying long-term interaction with another person. The only point to be made here, and one we will return to later, is that relationships characterized by meshed sequences are not likely to be emotion-producing interactions, at least so long as the sequences stay meshed.

A second hypothesis is based on the observation that most persons do not meet and marry their clone, a person of similar background, interests, and perceptions of the world. In such cases, one can expect initially a good deal of interruption and thus emotion. Over time, however, the partners may "hammer out" their interactions so that they become meshed with one another. One might predict, then, that relationships that have endured for many years may be more likely than relationships of lesser longevity to be characterized by facilitative interconnections.

The analysis developed here does not imply that all emotionally tranquil relationships are necessarily characterized by facilitative interconnections. It *does* imply, however, that a relationship may be emotionally tranquil for two very different reasons: (1) Little or no emotion may be experienced in the relationship because all interchain causal connections are facilitative; or (2) little or no emotion may be experienced in the relationship because there are *no* interchain causal interconnections to intrachain sequences, either facilitative or interfering. In the former case, the relationship is very much emotionally "alive"; under its tranquil surface lies enormous latent potential for the participants to experience the most intense emotion a person can ever experience in a relationship with another human being. But under the tranquil surface of the latter relationship lies nothing; the relationship is emotionally infertile, without prospect of producing emotional experiences of any sort.

Ironically, a relationship may become emotionally sterile in the interests of emotional tranquility within the relationship and in the participants' well-meaning desire for harmony with each other. As the participants discover that they interrupt each other in particular interactions, they may sever, bit by bit, their interchain causal connections to those intrachain sequences. She now skis alone and he plays tennis with others; she disciplines the children without his intervention; and she stays out of the finances, which have become his exclusive domain. And finally peace and harmony prevail; two people live their individual lives parallel to one another in time and space.

It is for this reason that conflict within a relationship is truly a "crisis" in the full meaning of the word and deserves the attention relationship investigators have devoted to it (see Chapters 9, "Conflict" and 10, "Intervention"). Derived from the Greek word meaning "opportunity," crisis is defined as a "turning point," a stage in a sequence of events at which the trend of all future events, especially for better or worse, is determined. The conflict occasioned by a nonmeshed interaction may be resolved in two ways: (1) The participants may work to mesh their interaction, thereby preserving the fact of interconnection; or (2) they may sever it. In both cases, harmony can be restored to the relationship. But the former mode of conflict resolution retains the emotional investment of the participants, while the latter means of conflict resolution weakens the emotional fabric of the relationship. When two people have trouble agreeing on an issue—where to spend their vacation, for example—well-meaning friends and counselors sometimes say, "That's easy; you go to the mountains and he can go to the seashore and everyone can be happy." It's not that simple. If conflict is always resolved through severing the interchain causal connections that are the source of the interruption and disturbance, the relationship may be slowly emotionally "gutted" over time. Both persons may lose their emotional investment in the relationship, and, ultimately, dissolution of the relationship may be a matter of indifference to both, although, of course, such relationships may endure indefinitely in peace and harmony. Participants may not be aware of the implications for the relationship of severing interconnections, and, if they were, they might not always wish to purchase their emotional tranquility at that expense. On the other hand, of course, some interconnections may have to be severed to preserve the remainder.

With respect to relationships characterized by meshed sequences, another point might be made. To some extent, perhaps even to a great extent, participants may be unaware of this property of their relationship. Our emotional system, it will be recalled, operates in the service of our survival and well-being. It is for this reason that events (and people) who interrupt us not only are the source of our physiological arousal, and thus our subjective experience of emotion, but they can successfully compete for our attention and can command and dominate our thoughts. Since the emotional system largely appears to be a "trouble-shooting" system, where there is no trouble, where there are only meshed facilitative interactions, there also should be relatively little attention to, and conscious awareness of, the interactions. Over time, and as an interaction becomes increasingly integrated and organized, it should recede from awareness and become "automatic" and unthinking.

One may hypothesize, therefore, that, while participants in relationships characterized by nonmeshed intrachain sequences may be acutely conscious of the relationship overall and of the partner as well, those in well meshed relationships may not be so aware. They may, in fact, be seduced by the emotional tranquility of the relationship to discount their emotional investment in it, only to discover, when the relationship is dissolved, how much the relationship meant to them, as evidenced by their severe emotional upheaval upon the relationship's termination. Meshed intrachain sequences, then, are often the hidden ticking emotional bombs in a relationship. They explode when they are severed, and the individual experiences, as a consequence, great interruption in the performance of those intrachain behavior sequences and plans that were interconnected.

It is now not difficult to see why emotional reaction to dissolution of a relationship (or to irrevocable separation) may be a good index of the "closeness" of the relationship. Upon separation, all of the hidden meshed interconnections are unmasked to wreak their emotional vengeance upon the individual. Myriad facilitative interconnections are interrupted, and, in the case of death of the other, there is no possibility of completion ever again.

Each interruption causes the absent partner and the former relationship to painfully surge into the surviving partner's conscious awareness, and so it is not hard to understand the observation that the full meaning of a relationship wtih another is often realized only after the relationship has dissolved. The old plaint that "I love you more, now that you're gone" is not only understandable in terms of what is known about emotion, but, in the case of relationships characterized by facilitative interconnections, it is predictable. Thus, the nature of our emotional apparatus is such that over time we may lose our awareness and appreciation of the depth of our emotional ties to another. But, aware or not, the emotional price will be paid upon dissolution of the relationship.

Just how high the price will be depends not only on the extent to which the relationship was characterized by meshed sequences but also on how rapidly (and if) the partner can find a substitute means of completing those sequences. M. Hunt and Hunt (1977), for example, observed that those men and women who had established a relationship with a third person prior to their separation and/or divorce seemed to escape to some extent the emotional turmoil that appeared to be typical of the postdivorce period. Presumably, these people had already made substitutions, and so the fact of separation occasioned little interruption (and undoubtedly was facilitative of sequences with the third person). For this reason alone, knowledge of an individual's social network and relationship alternatives is crucial to predicting the severity of emotional reaction to dissolution of a relationship (or to any nonmeshed interaction). When few or no substitutions are possible, then the emotion precipitated should be severe and of some duration. In this regard, it might be noted that, as a relationship endures over time, not only may the number of facilitative interconnections be increasing, but the opportunity for substitutions to be made may be decreasing or increasing unequally for the partners, which should affect the severity of each partner's emotional reaction to the dissolution of the relationship.

The fact that facilitative interconnections recede from conscious awareness also illuminates the fact that people sometimes experience bewilderment at the ferocity and ambivalence of their emotional reactions to the dissolution of a relationship they actively worked to terminate and fervently desired to end. Such a relationship is frequently characterized by much conflict, and it is its interfering causal interconnections that are likely to command attention and awareness, while the facilitative interconnections may have so receded into unconsciousness that the individual has forgotten they exist. We may hypothesize, then, that in ongoing relationships, people tend to be more acutely aware of what is wrong with the relationship than of what is right. Some tangential support for this hypothesis is provided by a study reported by R. L. Weiss, Hops, and Patterson (1973), which found that "displeasurable" events accounted for far greater variance of spouse ratings of the pleasantness of their interaction with the spouse than did "pleasurable" events. The investigators comment that this finding "suggests that in married couples, tracking aversiveness may be better established than tracking positivity" (p. 315). The analysis here suggests that this tendency is not limited to married couples and to marital interaction. When the relationship is severed, however, all the interconnections go and people may, for the first time in a long time, appreciate what was good about the relationship as well as what was bad. It is little wonder, then, that voluntary relationship dissolution is not infrequently characterized by vacillation with respect to the question of "to terminate or not to terminate" (R. S. Weiss, 1975), by ambivalence toward the relationship and toward the other, and by intense emotion even in cases in which, before the actual separation, there was only a resolute and single-minded determination to rid oneself of the source of one's unhappiness.

One final point might be made. The fact that potential trouble commands our attention and awareness is also one reason why the amount of attention the partners pay to each other is not a good index of the extent of the emotional ties that exist within the relationship or their attraction to each other (Berscheid & Graziano, 1979)—despite the folklore popularity of the notion that attention to and attraction to another are directly related. The complaint "You never pay attention to me anymore; [therefore] you don't love me anymore" expresses the commonly perceived relationship between these two variables. While this might be true in the beginning of a relationship, when novelty and unfamiliarity are still producing their documented effects on selective attention and conscious awareness, it should not be true as these factors decrease in the relationship and as interactions become better meshed. To be "taken for granted" may be an unpleasant by-product of well-meshed relationships, but the lack of attention a person receives from the partner, like the absence of emotion that probably accompanies the inattention, masks the power the partner may have to become a potent source of emotional experience for the other—and to become once more the focal point of attention.

Emotional investment in close relationships

A close relationship has been characterized as one in which the properties of interdependence are such that the causal interconnections between the participants are frequent, strong, and diverse. It can now be argued that, while close relationships should not be defined in terms of either the magnitude or the hedonic sign of the affect that may be typically experienced within the relationship by the participants, such relationships *are* generally characterized by emotional investment. That is, there is considerable potential for emotional experience within a close relationship, whether that potential has been actualized and whether the participants themselves are aware of it. The proposition that close relationships are emotionally invested relationships rests on the assumption that as the number, strength, and diversity of the interchain causal connections in a relationship increase, the probability that a significant number of these will be connected to intrachain event sequences also increases.

To summarize, close relationships are, virtually by definition, characterized by emotional investment of the participants in the relationship. The emotional investment of each partner is defined by the number and strength and diversity of the interchain connections to that individual's intrachain event sequences and represents the potential degree of emotion the individual may experience within the relationship. The emotional investment of each individual in the relationship may be unrelated to the frequency, intensity, and quality of the emotion the individual typically experiences in the relationship or to the frequency, intensity, and quality of emotion the individual has *ever* experienced in the relationship. Further, the participants may or may not be aware of the degree of their emotional investment in the relationship or the emotional liability they carry in the relationship. Thus, the absence of frequent or intense emotion within a relationship is a poor index of either the closeness of the relationship or of the degree of emotional investment of the participants. Relationships may be emotionally quiescent because: (1) there are no interchain connections to intrachain event sequences (and thus no emotional investment in the relationship); or (2) there are many interchain connections to intrachained events and the relationship is highly emotionally invested, but the interconnections are facilitative. In the latter case, the full degree of the partners' emotional investment in the relationship is most likely to become evident upon dissolution of the relationship.

SPECIAL PROBLEMS ASSOCIATED WITH EMOTION IN RELATIONSHIPS

There are a number of special problems associated with the prediction of emotion in relationships. These include predicting the hedonic sign of an emotion, the course of emotion over time, and the effect of emotion experienced outside the relationship on the relationship. Some of these represent perennial problems in emotion theory and research but assume special importance for close relationship researchers, while others are unique to understanding emotion as it occurs within a relationship. The first problem discussed below is an example of the former and represents one of the thornier issues to confront emotion theorists.

The Hedonic Sign of the Subjectively Experienced Emotion

The hedonic sign, positive or negative, of the emotion that is experienced within a relationship has critical implications for the survival of the relationship, as well as for the behavior of the participants on almost all dimensions of interest. Further, it is the positive emotions that so many seek within the realm of a close relationship and that sometimes prove to be exasperatingly elusive, while, in contrast, the negative emotions often appear to be all too frequent, intense, and lasting.

The analysis of emotion outlined in the previous section suggests that to predict the occurrence of a particular kind of emotion for a participant in a relationship, one needs not only to identify that person's highly organized behavior sequences and higher-order plans and their hierarchical organization but also to know how a specific interruptive event will be interpreted by the person and whether substitute behaviors are readily available. Despite the degree of specificity of knowledge such predictions may require in the individual case, some emotion theorists and researchers have attempted to discover the general principles that may underlie whether arousal is likely to result in a pleasant or unpleasant subjective emotional experience.

Traditionally, many have taken a behaviorist point of view, suggesting that all emotions can be understood in terms of the consequences of punishment, positive reinforcers, or negative reinforcers. Unfortunately, the behaviorist analysis has been regarded as unsatisfactory for a number of reasons, among them the fact that experimental evidence (if not our own daily experience) suggests that the application of reinforcers does not always produce emotion. For example, the 20th-anniversary gift to our spouse of the 20th spoon for her spoon collection (a "positive reinforcer") may produce little or no emotion at all of any hedonic sign (although it may well produce quiet satisfaction). Most certainly, it will not produce as much emotional delight as it did on the first occasion or even on the second. Similarly, after the repeated delivery of an aversive stimulus, the application of that stimulus may also produce very little emotion. For example, in a conflict-ridden relationship, one spouse's yelling at the other may not evoke an emotional reaction. Mandler (1975) argues that positive or negative stimuli produce very little emotion when they are not interrupting—when their onset (or offset) is expected, for example. And, when they are not interrupting, they should not be the occasion for the subjective experience of emotion; for a stimulus to be the occasion for emotion, it must interrupt something, and an interruption is almost by definition a surprise—it is not expected, and so it frequently is not easily and quickly circumvented.

Mandler gives only cursory attention to predicting the probable hedonic sign of an emotion. He does observe, however, that negative emotion is a frequent response to interruption of any kind; that is, knowing nothing else about the situation in which the interruption occurs and its meaning to the individual, one might predict that the emotion that will be experienced will be of a negative hedonic sign. Since the interruption is unexpected, the individual frequently has few other situationally relevant responses available to perform (or at least few such responses high in the response repertoire at the moment of interruption). It is this lack of control over the interrupting event in the environment—this "helplessness" the individual feels in the face of the interruption—that is presumed to be responsible for the high frequency of negative emotion that frequently occurs in response to interruption. Further, since interrupting stimuli tend to be stimuli over which the individual has no control, at least momentarily, and since the emotion that is experienced is in the service of the individual's regaining control over the stimulus and regaining mastery of the environment so that the individual may act to serve his or her own best interests, it seems reasonable that many of the "hotter" emotions we experience are negative.

Not all of the strong emotions we experience are negative, however, and it is the positive emotions—such as "elation," "joy," "love," and "delight"—that present a special problem for emotion theorists. They are a problem principally because they do not follow directly from a general "emergency" theory of emotion as the negative emotions do. Nevertheless, as Leeper (1948) and others have observed, the positive emotions must be included within any theory of emotion that has any pretensions of comprehensiveness and general vaidity. As a consequence, it is how emotion theorists manage to shoehorn the fact of positive emotion into their framework that frequently separates many theories—when, that is, positive emotion is treated at all, for, as Candland (1977) notes, it is mainly the negative emotions that have received attention in emotion theory and research.

Most theorists (e.g., Arnold, 1960; R. S. Lazarus, 1968) agree that the hedonic sign of emotion depends on the consequences of the individual's meaning analysis of the stimulus event and whether it is viewed as "benign" or a "threat" to the individual's welfare. If it is true that to produce an emotion, a stimulus must be interrupting, then to produce a positive emotion, the stimulus not only must be interrupting, but it must be seen by the individual to be "benign." Herein, however, lies what we may propose to be *the fundamental paradox of positive emotion:* If a stimulus is interrupting of an individual's organized action sequences and/or higher-order plans (each of which presumably has been developed in the service of the individual's welfare and survival), in most instances the stimulus should *not* be seen as benign but rather as a potential threat to the individual's well-being.

The "control" hypothesis

Mandler suggests that one class of events sometimes associated with positive emotion consists of those that, although they are interrupting, are perceived by us to be events we have some control over—specifically control over their onset and offset. Mandler points out that sometimes we have control over whether to allow ourselves to be exposed to such interrupting, surprising, or novel stimuli as magicians, rollercoaster rides, and "dangerous" situations, in which some elements are unpredictable and thus expose us to risk. In such situations, not only do we control the onset of the stimulus event, but often it can be terminated when we please. For example, should the magician become too frightening, the individual may leave the theater. When the individual has no control over the onset or the offset of the interrupting stimulus, however, Mandler suggests that the emotional label applied will be a negative one.

The "control" hypothesis would seem to make good sense in terms of satisfying the two conditions we have posited here as necessary for the occurrence of positive emotion; the event is *interrupting,* but the control over its onset and offset allows it to be seen as relatively *benign.* Furthermore, one direct implication of the control hypothesis is that positive emotions are rarely experienced as intensely as negative emotions. In negative emotions, the degree of arousal should be higher since, during the time the individual is searching for a way to "turn off" the interrupting stimulus, ANS arousal should be increasing. In the controlled exposure situation, by contrast, the individual already knows what to do to terminate the stimulus; the "alarm bell" of arousal serves little function, and it is questionable that it will ring so loudly and insistently.

Both implications of the control hypothesis, that positive emotions arise in situations in which people feel they have control and are often less intense than negative emotions, receive some support from Davitz' (1969) extensive study of the "language of emotion." He found that, in describing their experiences of various emotions, people tend to use items suggesting personal competence and "enhancement" to describe the positive emotions, such as love and joy–for example, "I have a sense of sureness"; "a sense of more confidence in myself"; "a feeling that I can do anything"; "I feel powerful, stronger, bigger." In contrast, the items used to describe such strong negative emotions as anger and fear suggest "inadequacy"—for example, "a sense of being totally unable to cope with the situation"; "there's a sense of not knowing where to go, what to do"; "I feel vulnerable and totally helpless"; "a sense that I have no control over the situation." (It should be noted that, according to Davitz' analyses, such negative emotions are to be differentiated from depression, apathy, and grief, characterized by low arousal. See Berscheid, in press-b, for a discussion of the relevance of this point to the present discussion.) Strong negative emotions, then, appear to be associated with a loss of control over the situation and the interrupting stimuli, while positive emotions appear to occur in circumstances in which one perceives that one's sense of control and mastery over one's situation and future has actually increased.

Davitz also analyzed the degree of activation or physiological arousal implied by the physical sensations used to describe various emotions. The results suggest that the degree of activation in the case of positive emotions is somewhat less than it is in the case of such negative emotions as fear and anger. The activation items descriptive of positive emotion include: "I am excited in a calm way"; "warm excitement"; "I seem to be more alert." In contrast, those used for negative emotions indicate more intense arousal—for example, "my blood pressure goes up"; "there is an excitement, a sense of being keyed up, overstimulated, supercharged"; "my heart pounds." There is some evidence, then, that the degree of arousal associated with positive emotion is frequently somewhat less than the degree of arousal associated with negative emotion, just as the preceding discussion suggested.

The "completion" hypothesis

Although the positive emotions typically may be experienced less intensely than the negative emotions, there is little question that, on occasion, they are characterized by a high degree of arousal and are experienced very intensely. The "control hypothesis," for reasons discussed above, would seem to be inadequate to account for these instances. We may propose, however, that there is another class of circumstances that satisfies the two conditions we have suggested as being necessary for positive emotion (i.e., that the stimulus event must be interrupting, but it must also be perceived by the individual to be benign). These are circumstances under which the stimulus event is unexpected and interrupting, but a meaning analysis of that event subsequently reveals that it largely facilitates, rather than interferes with, the completion of ongoing organized behavior sequences or, more likely, higher-order plans.

Two subclasses of events appear to fall into the general completion category. The first of these has been suggested by many emotion theorists (e.g., R. S. Lazarus, 1968; Mandler, 1975) and are stimulus events that suddenly and unexpectedly *remove* the presence of a stimulus that has previously interrupted an organized sequence or plan. The happiness attendant upon learning that a biopsy has revealed a cancerous growth to be nonmalignant, or that an oncoming tornado has swerved in another direction, or that the IRS has decided not to audit one's books, are examples.

The second subclass of circumstances in the completion category that is important in the production of positive emotions may be hypothesized to be stimulus events that permit the unexpected and sudden completion of organized sequences or higher-order plans that have *not* been previously interrupted, but which, rather, have been proceeding "on schedule" as expected. Here, the event allows the last response in an organized behavior sequence (consummatory, in many cases) to occur earlier in the sequence than expected, or it permits the goal of a higher-order plan to be reached before it was anticipated. The occurrence of such an unexpected "premature" event should be interrupting of the performance of the remainder of the sequence or of lower-order plans and should produce ANS arousal, thus satisfying a sufficient condition for emotion. The event also should receive a meaning analysis since it is unexpected and does not fit into existing cognitive structures. Again, the arousal signals the individual that "something important has happened," and the individual must do a meaning analysis to answer the questions "What does this mean for the future?" and, perhaps, "What have I done right and how would I do it again?" The postulation of this second subclass of completion events is congruent with the emotion theorist Carr's (1929) speculation that "joy is awakened by the sudden and unexpected attainment of a highly desired end."

To illustrate this class of completion events, imagine a couple who have devised a careful plan to purchase a house in 5 years that involves setting aside and investing a certain amount of money each month. The plan will virtually assure them that at the end of the 5 years they will be able to buy the house of their dreams. If this plan runs off precisely as they expect, we would predict that, upon arriving at the last response in their plan (buying their house), the couple probably would *not* feel a good deal of intense emotion. Rather, they might describe their feelings as "contentment" or "satisfaction" or one of the "cooler" (low ANS arousal) positive emotions. Suppose, however, that, after only two years of their setting-money-aside activity, the house of their dreams suddenly and unexpectedly comes on the market at a price that is exactly equal to the amount in their savings. One would expect that the couple might now feel "joy" or "elation." Their higher-order plan has been interrupted—and, indeed, probably many organized behavior sequences and lower-order plans as well (e.g., as represented by the lease on their apartment). Further, their meaning analysis, which will trace the implications of this interrupting event for all other sequences and plans in existence at the time, may reveal that the accomplishment of these, too, has either been secured or facilitated. For example, they will now have three years worth of monies that may be used to buy the appliances and furniture they expected to have only in ten years time; the house is located near a much better school for their child; they will no longer have to put up with the noisy apartment neighbor next door; and so on. Thus, as the meaning analysis traces the implications of this event, and as the full extent of its interruption of other sequences and plans becomes realized, and as it also becomes realized that these interruptions not only do not signal the couple's personal loss of control over their fate but actually enhance it and their well-being, such intense positive emotions as "euphoria," "joy," and "elation" may well be the result. The interruption, and thus the arousal, the event has occasioned have been considerable, and the favorability of the implications of the event for the couple's welfare also is considerable.

Implications of the completion hypothesis for close relationships

If the completion hypothesis proves to be correct, it has a number of implications for positive emotion as it may or may not be experienced within a relationship, and just a few of these will be mentioned here. First, it is clear that, to be the occasion for the experience of positive emotion for the partner, an individual must either have the power or resources to remove an interrupting stimulus, which carries the unfortunate implication that the partner must be in a state of unpleasant emotion initially (hopefully because of causes other than the individual), or the individual must have the power and resources to help the partner unexpectedly complete ongoing behavior sequences and plans. These conditions seem to adequately describe those in which intense positive emotion (e.g., romantic love) is felt for another; that is, the sudden unexpected realization that another is able and willing to help one fulfill one's most cherished plans and hopes is likely to be associated with intense emotion (see Berscheid, in press-b), and Berscheid & Walster, 1978, for a discussion of this point in terms of interpersonal attraction).

It may be important to note that the emotion should follow the meaning analysis precipitated by the stimulus event (e.g., the unexpected proposal of marriage from the rich, intelligent, and handsome stranger) and the tracing of its anticipated benevolent implications for the future, *not* the *actual* fulfillment of those implications, for, as these unfold in time, they have become expected and should not produce much in the way of hot emotion. (Their subsequent lack of fulfillment, however, should produce quite another emotional story.) This reasoning implies that persons whose cooperation and aid in the facilitation of plans has become expected, for whatever reason, should not be able to produce as much positive emotion in the partner as someone whose help is unexpected.

The above analysis also implies that, for positive emotion to occur in association with the individual, the partner must *have* some plans and, in addition, be unable to complete them as readily alone as with the individual's help. It is interesting to note, in this connection, that it is particularly when people are young that they have many unfulfilled plans for their life, several of which require for their completion the presence and cooperation of a member of the opposite sex. As the years roll by, however, many of the sequences and plans will have run off smoothly and will have been completed with the aid of the partner—a home will have been established; children will have been had, raised, and sent off to college; career and financial security will have been secured. Or, the plans will have been abandoned. It's too late to become a concert pianist, and the chances of ever becoming president of the company are now nil. It appears to be at this point (typically in "mid-life," e.g., Mayer, 1978), when most of an individual's plans cease to exist either because they have been fulfilled or because they have been abandoned as hopeless, that people seem to experience a "mid-life crisis." The crisis seems to be that the person feels "emotionally dead" and life has lost its meaning.

If one has completed or has been forced to abandon one's original plans and if one has no other plans to take their place (or at least none that seem even remotely feasible), then it follows that there are no plans to be interrupted, either by unexpected fulfillment or unexpected failure to complete them. If there are no plans to be interrupted, there should be few occasions for feeling intense emotion—either joy or sorrow—and so perhaps "numbness," "deadness," and "staleness" are the labels that we apply to a life in which there is no emotional experience at all. And it is interesting that the absence of any emotional experiences at all is regarded by many as a negative experience in itself.

Thus, one implication of the foregoing analysis of the determinants of positive emotion is that it is especially people who have plans who can experience joy or sorrow, and, in the case of close relationships, it is especially people who have plans that may be interrupted or fulfilled by the other who can experience either joy or sorrow within the relationship with that other person. Perhaps, then, it is only people who continue to dream throughout their life and who continue to strive for long-term goals who are capable of staying emotionally alive. And, within close relationships, perhaps it is only people who can continue to dream dreams that include their partner who can stay emotionally alive within the relationship. For, as the very plans whose anticipated fulfillment may have occasioned positive emotion at the beginning of the relationship are completed (or abandoned), the partner's potential to interrupt (or help complete) the plans vanishes. Unless other plans are put in their place, emotional investment in the relationship should lessen, other things being equal.

At least one other implication of the foregoing analysis of the determinants of positive emotion ought to be mentioned, and that is how difficult it is to "engineer" the experience of positive emotion for ourselves. To do our part in the pursuit of happiness and joy, we can devise plans, but the interruptions—good and bad—are left to the gods and other people. We can't throw our own "surprise party," for then it's not a surprise, not an interruption, and not an occasion for much emotion. We can chase after the butterfly, but, when our plans to catch it are skillfully executed and the prize is in our grasp, satisfaction and contentment may be our only emotional rewards. Only the butterfly itself can give us joy by unexpectedly lighting on our shoulder.

The subject of positive emotion and its implications for close relationships should not be left without the reminder that the above analysis applies only to the hot and intense emotions and the prediction of their occurrence. It is not directed toward the cooler emotions and feelings, such as those often engendered by a superb dinner, a fine bottle of wine, and the cozy warmth of an evening shared before the fire. These are undoubtedly the staff of life for many satisfying long-term close relationships.

The Course of Emotion Over Time

Some emotion theorists and investigators, most notably Solomon and Corbit (1974), have become interested in predicting the course an emotion will take over time. Solomon and Corbit do not attempt to explain when it is one will experience an emotion, nor do they attempt to predict the hedonic sign of the emotion that is initially experienced; rather, their "opponent-process theory of motivaton" is addressed to the temporal dynamics of affect. Specifically, given the occurrence of an emotion of a positive or negative hedonic sign, their theory attempts to predict: (1) the intensity of that emotion over time with continued presence of the stimulus whose onset precipitated the emotion; (2) the hedonic sign and intensity of the emotion that is experienced upon the offset of the stimulus; and (3) the intensity of the emotion that is experienced upon subsequent and repeated onsets and offsets of the stimulus.

Solomon and Corbit's theory assumes that many emotional states are automatically "opposed" by central nervous system mechanisms that, over time and with repeated onsets of the emotion-producing stimulus, reduce the intensity of the pleasant or aversive feelings it initially produced. These opponent processes are assumed to strengthen with use. Thus, with repeated onsets over time of a positive emotion-inducing stimulus, for example, the opponent process associated with it should become increasingly strong, so that the stimulus that initially produced a "hot" emotion now only produces a "cool" one. Using the instance of the stimulus onset of a loved person, for example, the initial onsets (or encounters with the person) may be associated with intense love, euphoria, and joy. Over time, however, and after repeated encounters with that person, Solomon and Corbit predict that the "opponent process" should increasingly dampen the initially strong and intense emotion, and interaction with the loved person now should be associated with the cooler feelings of contentment and comfort. Offset of the stimulus should reveal the full extent of the opponent process, and, since it presumably takes time for the opponent process to develop in strength, removal of the loved person after only a few onsets (or encounters) should be associated only with a mild feeling of loneliness. After much interaction with the loved person, however, the opponent process should by now be so strong that when the loved person is removed, the individual should be plunged into a very strong opposing emotional state, such as grief.

The temporal course of emotion that Solomon and Corbit envision is generally compatible with that which follows from Mandler's view of emotion and with the analysis we have presented in previous sections insofar as it engages this issue, although the mechanisms and dynamics presumed to be responsible for the course of emotion over time obviously differ (see Berscheid, in press-b, for a discussion of this point). It is also agreeable with a number of other theorists' views of the temporal fate of strong emotions, especially the positive emotions. Sigmund Freud (1930/1961), for example, commented, "One feels inclined to say that the intention that man should be 'happy' is not included in the plan of 'Creation.' . . . When any situation that is desired by the pleasure principle is prolonged, it only produces a feeling of mild contentment" (p. 23). Brickman and Campbell (1971), in their analysis of happiness, observe, "While happiness, as a state of subjective pleasure, may be the highest good, it seems to be distressingly transient" (p. 287). Klinger (1977), too, observes, "'Highs' are always transitory. People experience deliriously happy moments that quickly fade, and all attempts to hang on to them are doomed to fail" (p. 116). This fact, if it is a fact (and there is no evidence to the contrary), has a number of implications for the probable course of emotion within close relationships and for a number of emotional puzzles often associated with such relationships.

The temporal course of romantic love

One of the features of romantic love (referred to as "passionate" love in Chapter 7, "Love and Commitment") that is often remarked upon is its swift onset, as opposed to the slower growth of the milder emotions and feelings associated with friendship, such as liking. Given that romantic love is usually conceived of (and experienced as) an intense emotion, it makes sense in terms of everything that is known about emotion that its onset *should* be relatively swift and sudden. The supposition that emotions are precipitated by unanticipated events also helps explain another phenomenon that investigators have regarded as somewhat curious, and that is that simple liking, developed and maintained over a long period of time, infrequently results in the emotion of romantic love; that is, more and more liking does not seem to culminate in the emotion of passionate love (Berscheid, in press-a). That a mild feeling (such as liking) does not progress into an intense emotion (such as love) makes sense in terms of what is known about emotion. Thus, the analysis of emotion presented here supports the views of several interpersonal attraction theorists (see Berscheid & Walster, 1974) who have suspected that mild feelings of liking and the intense emotion of love are not simply different quantitative values on the same continuum of attraction, but that they are qualitatively different phenomena with different antecedents.

There is another aspect of the emotion of romantic love that has interested investigators and concerned laypersons. Compared with the stability over time of such milder positive feelings toward another as liking, the emotion of romantic love seems to be distressingly fragile. As a 16th-century sage poignantly observed, "the history of a love affair is the drama of its fight against time." Most who have considered the matter tend to agree on empirical grounds, if not upon a consideration of the dynamics of emotion, that romantic love is a fleeting phenomenon. For example, most authors of marriage and family texts warn that romantic love should be considered temporary and that, although the experience of this intense emotion may be the occasion for a marital contract, it is bound to dwindle after long interaction. Romantic love theorist Reik (1944) also warns that the very best a couple intensely in love can hope for after several years of marriage (or frequent interaction under other auspices) is a warm "afterglow." This view receives support from such studies as that conducted by Blood (1967), who examined the course of romantic love over time in American marriages contracted on the basis of love ("love matches") and in arranged marriages between Japanese couples. He found that "love matches indeed start out hot and grow cold" (p. 86). The arranged marriages appeared to show an increase in the indicants of romantic love shortly after marriage, but then the usual decline over time.

To many, the prospect of an "afterglow" seems distinctly pallid and uninteresting compared with the flame of the fire itself, but, much as one wishes to hold on to the intensely positive emotions, to retain the euphoric state, there is little that is known about the underlying dynamics of emotion and of its antecedents (even apart from such empirical data as Blood's) that suggests this is possible. The afterglow of comfort, contentment, and affection that Reik describes as the best outcome possible under the circumstances is congruent with what is known about the role of arousal in emotion and, in turn, the role of surprise and interruption in arousal. It should be noted that the fact that intense emotion is not maintained—not even intense positive emotion that may be fervently desired, coddled, and encouraged—stands in stark opposition to many popular beliefs and aims, particularly with respect to close relationships and especially marital relationships. As Klinger (1977) comments:

> The people of Western society, about whom we know most, value and strive after elevating emotional experiences—elevated pride, love, lust, and joy. . . . Society holds out to all young people the prospect of over-powering elation, elation both now-and-then and forever after. Contentment is all well and good, but it cannot hold a candle to elation for sheer emotional power. (p. 117)

To strive for contentment and serenity, rather than joy and euphoria, seems to be an incentive for wise men and philosophers—not ordinary folk.

Finally, it may be observed that the emotion of romantic love brings with it a further irony. Even assuming that the couple intensely in love manages over time to successfully navigate through the shoals of discord and conflict and to arrive safely in the harbor of an "afterglow," there is a further price that will be ultimately exacted from them if our analysis of emotion in close relationships is correct. If both partners have integrated their lives with one another, they will suffer the costs of grief and depression upon dissolution of their relationship through death, if not before. The ripping apart of their bonds of interdependence should result in great pain, pain whose intensity may be far out of proportion to the intensity of the positive emotion they enjoyed when the bonds were intact.

Conflict-habituated interactions

Another interesting emotional puzzle often associated with close relationships, especially marital relationships, is that of couples whose emotional experiences within the relationship appear to be predominantly negative. The puzzle, of course, is why such relationships continue despite the fact that the interaction seems to be characterized primarily by arguments, fights, needling, and hostility.

That such couples are not unusual is suggested by the work of Cuber and Harroff (1966), who identified "conflict-habituated" couples as a prevalent type of marital relationship. Cuber and Harroff quote the husband in one such relationship that had endured 25 years as follows:

> You know, it's funny; we have fought since the time we were in high school together. As I look back at it, I can't remember specific quarrels; it's more like a running guerilla fight, with intermediate periods, sometimes quite long, of pretty good fun and some damn good sex. . . . It's hard to know what it is we fight about most of the time. You name it and we'll fight about it. . . . You called them arguments a little while ago—I have to correct you—they're brawls. . . . When I tell you this in this way, I feel a little foolish about it. I wouldn't tolerate such a condition in any other relationship in my life—and yet here I do and always have. . . . (pp. 44–46)

One interpretation of the stability of such relationships, and hardly a satisfying one, is that such couples are "masochistic." This sort of interpretation simply labels the interaction and provides little understanding of it. Further, this label may be misleading. It suggests that the couples enjoy the fights, that this is a positive emotional experience for both of them rather than the negative emotional experience the outside observer perceives it to be.

Two observations might be made about such couples with respect to the intensity and quality of emotion they experience in their relationship with one another. First, although the brawls may appear to an outside observer to be horrendous, to be the occasion for very intense negative affect, one cannot help but question whether such observations are valid. If the quarrels have become habitual, then there should be nothing about them that is serving as an "interruption," or as the occasion, then, for intense emotion. To the contrary, such behavior sequences often seem to be highly organized and integrated and, ironically, appear to run off very smoothly toward their usual conclusion. It is not necessary to assume that these couples are actually "enjoying" the behavior sequence; it does seem reasonable to hypothesize, however, that these interactions are—despite their surface content—emotionally neutral. And, oddly enough, these couples, who are often labeled as "incompatible" couples, may be very compatible with one another in that their fighting behavior sequences are very well meshed. Thus another irony associated with these couples is the prediction one should have to make should one partner fail to rise to the other's "bait" as he or she usually does, or to pursue the argument in the anticipated way. A loving "turn-the-cheek" response to the barb might be the occasion for an emotional experience on the part of the needling partner, since the other's reaction should be totally unanticipated as well as interruptive of the behavior sequence the partner has begun. One wishes, then, that during one of these habitual brawls, the physiological responses of each of the partner could be monitored. One might be surprised to see that each, in contrast to the ferocity of their overt behavior, is relatively quiescent internally.

A second observation one might make with respect to these couples, and which leads to the same conclusion, is that these battling sequences are not only not interruptive, but time has shown them to be threatening neither to the relationship itself nor, apparently, to each person's vital interests of comfort and welfare. A brawl at the beginning of a relationship may seem to be the end of the world—it signals that the relationship is in jeopardy and that all of one's higher-order plans with respect to the other person may be interrupted. When it is the perennial argument about where Christmas should be spent, it clearly does not carry such threatening overtones.

Finally, of course, even if the sound and the fury in such relationships *does* represent the frequent experiencing of strong negative emotion, it may be the case that the couple is highly interdependent and facilitative of each other in areas not represented by the quarrels. Such interdependence may preclude their dissolving the relationship despite its heavily negative emotional content.

Effects on the Relationship of Emotion Experienced Outside the Relationship

Every relationship is embedded within its own special social and physical environments. The causal conditions these environments represent have important implications, both direct and indirect, for the relationship and for the emotion experienced within it. It is, in fact, the great change in the causal conditions represented by the environments surrounding close relationships that has been identified as primarily responsible for the decrease in the longevity of many of them, as is exemplified by a rise of 700 percent in the divorce rate since the turn of the century (Berscheid & Campbell, 1981). As opposed to the relatively harsh physical environments and the relatively constricted and limited social environments typical of most persons at the beginning of this century (Gadlin, 1977), most close relationships now are embedded within benign physical environments and within mobile and varied social environments. These undoubtedly make substitute response alternatives in the face of interruption far more possible than they were even a few decades ago. The response alternatives for half of the partners in marital relationships, the women, especially, have increased enormously within the past few decades. When both partners in a relationship have response alternatives, it is far easier to "agree to disagree" and to fulfill one's plans without the aid (or the interference) of a partner. Other chapters in this volume discuss the effects these widespread social and physical conditions are likely to have on the probable course and duration of the relationship. Here, we simply point out that their effect on the relationship can be expected to be direct in terms of the experience of emotion in the relationship and its consequences. As the individual perceives more and more alternatives for the completion of sequences and plans that have been interrupted by a partner, the duration and intensity of emotion experienced upon interruption ought to be less, and, of course, the tendency to take advantage of these response alternatives ought to increase.

The social and physical environments in which the relationship is embedded also have indirect implications for emotion as it is experienced in the relationship. As outlined in Chapter 2, the causal conditions that impinge on a relationship govern the origin and development of the relationship, and changes in these conditions are responsible for changes in the relationship. This causal linkage is as true for changes in the emotional tenor of the relationship as it is for other changes. Changes in the causal conditions that impinge on the relationship, as may be represented by changes in the individual's job, in health, and so on, may cause considerable interruption to that individual's highly organized behavior sequences as well as have important implications for the possibility of completion of higher-order plans. Thus, these environment-inspired interruptions may be the occasion for emotion for the individual. Such emotional events may be conceived of as occurring "outside" the relationship, but, when the couple's relationship is characterized by many highly meshed sequences and plans, then these events also become the occasion for emotion for the partner. For example, if the individual's performance of responses in a sequence is interfered with by external events (e.g., job demands), then that person's inability to perform his or her responses in the appropriate form and at the appropriate time in the sequence necessarily interrupts the partner's sequences and plans and may become the occasion for the partner to experience emotion "within" the relationship.

It is perhaps ironic that it is relationships that are close and harmonious, that have developed many meshed sequences and plans, for which changes in external causal conditions ought to have the strongest emotional reverberations within the relationship. There is little evidence on this point, but what little there is does support the hypothesis that couples whose relationship is close are more vulnerable to changes in external conditions, at least insofar as they may create emotional turmoil within the relationship, than are couples who are not. Cuber and Harroff (1966), for example, speak of men and women in "intrinsic marriages" (a term that satisfies most of our conditions for a close relationship) and observe that "some of these pairs actually have had more conflict than typically occurs for couples in Utilitarian Marriages. Partly it is simple mathematics. There are more numerous points of contact, hence more potential for conflict" (p. 143). They also observe that, for these marriages, "perhaps the darkest shadow is: 'It's hell when it ends,' whether by death or because 'something went wrong.' Despite the closeness, the empathy, and all of the other solidifying forces, Intrinsic Marriages have proved for some to be vulnerable" (p. 142). We would simply comment that it may not be "despite" the closeness, but rather because of it, that both partners in these relationships bear a special emotional vulnerability, not simply to the vagaries of their own fate and the causal conditions that determine it, but to the external forces that impinge upon their partner and to which they also are hostage.

Jealousy

There are many close relationship phenomena of interest that are associated with changes in external causal conditions. Perhaps one of the most notable of these in the realm of emotion is the effect of a third person upon the relationship and the emotion of "jealousy." The analysis of jealousy presented below is intended to illustrate how external events may produce emotional chaos within a relationship, and how, paradoxically, the turmoil may be greatest when the relationship is a close one.

Interest in the dynamics and "treatment" of jealousy has steadily increased in the past decade or so (e.g., Clanton & Smith, 1977). Much that has been said on the subject implies that jealousy is an "abnormal" emotion, that the "green-eyed monster" afflicts only the chronically inadequate and insecure neurotics among us, and that the emotion of jealousy itself is far more toxic to a close relationship than the events that precipitated that emotion. It is thus also often implied that jealousy is an emotion for individuals to avoid if possible and, if not, to suppress or exorcise (Berscheid & Fei, 1977). Only one comment on the prevailing view (but see Pines & Aronson, 1982) will be offered here, and that is that our delineation of what constitutes a close relationship, as well as the analysis of the human emotional system presented in this chapter, makes it clear that to ask individuals who perceive that a threat exists to their behavior sequences and plans not to experience negative emotion is akin to asking the sun not to shine.

Although very few systematic data exist on emotional "triangles" (but see M. Hunt, 1969; Neubeck, 1969), the analysis we have developed in this chapter enables us to make a number of predictions about the circumstances under which the emotion of "jealousy" (probably a mixture of "fear" and "anger") ought to be experienced, and with whom, and in what intensity. It seems clear, for example, that the occasion for jealousy in the relationship is, at minimum, represented by: (1) an interruption (or threat of interruption) to one partner of sequences and plans caused by events in the partner's chain, (2) whose partial causal source, in turn, is perceived to be outside the relationship in the form of another person, object, or activity (such as work or a hobby), (3) but whose causal source is also perceived to lie within the partner (i.e., the partner is perceived to have the power to change the causal conditions these events represent for the relationship).

Some relationships (and some partners within relationships) should be more vulnerable to jealousy than others. For example, and ironically, it is individuals who are in close relationships in which sequences and plans are highly meshed who ought to feel most keenly the disruptive threat of external events in the partner's physical and social environments. Again, it is probably relationships that are not close, and not characterized by meshing, that should be most able to tolerate the development of new relationships (or new activities, or unilateral "growth" experiences, as they are sometimes called) on the part of each partner; these should not disrupt the status quo of the relationship, including its current emotional tone.

It should further be the case that the severity of the emotional reaction to the partner's development of a relationship with a third person that is perceived to interrupt (or have the potential to interrupt) the individual's sequences and plans ought to be, like all emotional responses, to some extent dependent on the individual's perception that alternative behaviors are available that will allow the completion of the interrupted sequences and plans without the partner's cooperation. A "jealous" reaction, then, ought to be most probable and most severe for persons who have a great number of sequences and plans meshed to sequences in the partner's chain; for those who perceive that few substitute alternatives are available for their completion; and for those who perceive that the external person represents a very high threat of interruption.

With respect to this last, it should be the case that it is those external persons who are perceived to represent a particularly good "mesh" with the partner's current sequences and plans who ought to be most threatening to the individual, other things being equal. That is, external persons who are perceived to have the resources to permit completion of many of the partner's plans, as well as a behavioral repertoire that would allow them to easily mesh their current behavior sequences to those of the partner, ought to occasion the most alarm. Under these circumstances, the third person represents an important substitute alternative for the partner (in addition to representing opportunity for the fulfillment of currently unfulfilled plans).

In relationships that are not close, in which the individual has many sequences and plans not meshed with the partner and in which these are generally unfacilitated, it is clear why a third person who is perceived to be likely to be facilitative ought to be inviting to the individual. There is, however, the question of why, within a relationship that is truly close, one of the partners would initiate and maintain a relationship with a third person simply to satisfy largely the same sequences and plans that the current partner now does. We might speculate that, in truly close relationships, this is probably not the precipitating factor for the initiation of a new relationship on the part of one of the partners. Rather, the partner most likely perceives, as is the case in the initiation of most social relationships (Berscheid & Graziano, 1979), that interaction with the third person will facilitate certain specific and limited sequences and plans that heretofore have not been facilitated (or have been interrupted). From our previous discussion, one can predict that it is precisely these sequences and plans that ought to dominate the partner's awareness and consciousness (rather than those currently being fulfilled). Further, the perception that another may facilitate their completion ought to be associated with the experience of positive emotion. For example, the person who has been unable to share his interest in collecting American pottery is likely to experience much positive emotion upon discovering a woman who not only shares his interest and expertise but appears to be able and willing to facilitate his aim of developing one of the finest collections extant.

It might be noted that such an event may have no implications whatsoever for the "goodness" or "badness" or "closeness" or "noncloseness" of the original relationship, if one assumes that our sequences and plans generally outrun our partner's behavioral repertoires and/or the partner's ability and resources to help us toward the completion of our plans. It seems reasonable to suppose, in other words, that there is truth in the popular cliché that one single other person cannot satisfy all our needs. And, if our previous analysis is correct, it is the unfortunate case that our partner may satisfy a goodly number of our needs and help us toward the completion of many of our plans, but, if this person has done so for quite some time, these occasions are unlikely to produce positive emotion toward the partner. The point is simply that the experience of a strong positive emotion toward a third person with whom one has just established a new relationship need not necessarily have implications for the goodness or badness of the relationship of the original pair. At least not initially.

It is the case, however, that the emergence of the third relationship represents a new causal condition for the original relationship, an event that may have severe emotional implications for both persons in the original relationship and implications for the course of the original relationship as well. We may hypothesize, in fact, that it is just such a situation and its frequency that makes the question Can you love two people at once? one of the most popular among laypersons (Berscheid & Walster, 1978). The preceding analysis of the antecedents of emotion makes it quite clear that it is very possible to "love" two people at once (although most probably the term *love* has different referents in the two cases). At the least, we may easily assume that an individual's experience of positive affect (whether in the form of cooler positive feelings or hotter positive emotion) with respect to both individuals may be quite genuine. Confusion on this issue seems to be largely the result of the assumption that if an individual initiates a relationship with a third person and experiences positive emotion within that relationship, then the original relationship must necessarily have been unfulfilling or have been characterized by negative affect. As previously noted, this need not be the case, but the effect of the new relationship on the old may be the same as if it had been.

Consider, for example, the following scenario (not entirely unusual if popular and anecdotal reports may be believed): One can suppose that P and O's relationship is very close. Their future plans extensively involve each other. Home, children, family, their economic future—all represent well-meshed sequences and plans. Further suppose that all of these sequences and plans have been running off smoothly, as expected and as intended, for some time. Suppose, also, that, when asked for their verbal report of the affect they experience in the relationship, both P and O report that they "love" each other. (We might also suppose that little affect is experienced in the relationship on a daily basis, because there is little or no interruption experienced.) Additionally suppose, however, that P, the wife, has always had an unfulfilled plan to develop a career. The husband, O, has not stood in the way of the completion of that plan (and so there has been no interruption or conflict in the relationship on that basis), and, yet, O has not had the resources to help P develop the career of her choice and so the relationship is not interdependent in this area.

Onto this stage enters a third person, Mr. X, who (miraculously to P) not only has the resources but the willingness to help P develop her career. P and X may develop some interdependence on this basis. It is not a close relationship, because the interconnections are limited to events in P's chain that have to do with the development of her career. But because X represents an unexpected means of completing P's plan, one may predict that P should experience a good deal of positive affect for X (perhaps in the form of gratitude, or joy, or happiness).

At this point, then, P should be aware of experiencing positive emotion within the relationship with X and, yet, may be experiencing very little emotion (positive or negative) within the relationship with O since there is little interruption of sequences or plans at this point (and, of course, O should not even be dominant in P's thoughts for nothing O is doing or not doing requires attention). The positive emotion felt for X, and seeing the relationship as one that facilitates an important plan, however, should lead P to approach X more and more, and depending upon other factors, the relationship may broaden to other areas. For example, if there is a good deal of overlap between the sequences and plans P currently has meshed with O and the sequences and plans X can facilitate, then P may easily, even without awareness, start substituting P–O sequences and plans with X (e.g., simple sequences, such as eating lunch together or discussing personal experiences together).

The very fact that the P–X relationship is broadening or that the number and diversity of causal interconnections are increasing should by itself not be the occasion for emotion for P; that is, these simple substitutions should not prompt additional positive feeling toward X on P's part. However, if and when P starts broadening the relationship with X by substituting behavioral routines previously shared with O, O ought to be "interrupted." In fact, the sheer amount of time P now spends with X on career development, if nothing else, ought to interrupt O if their relationship has been close. Further, O ought to be having some new thoughts and emotions about P, her career, and this fellow X. At the least, O might be expected to press for the resumption of the old sequences involving P. Because of limitations of time and energy, however, P may not be able to do this without interrupting her new sequences and career plan, and O's efforts at resumption of the old relationship are now interruptive and the occasion for negative emotion for P within the original relationship. Unfortunately, this may occur at the same time P is experiencing positive emotion for X. In fact, in terms of P's current emotional experiences, X may be looking very good and O may be looking very bad.

And, so, P may now decide that the relationship with O is not very good. It is disruptive and unpleasant, and it is steadily worsening as O, as inducement to get P to resume their old relationship or in retaliation for her interruptions, now stops facilitating many of P's old sequences and plans (e.g., taking the children to school in the morning). "Alienation," and the "spiraling of negative emotion" within a relationship of which Waller (1967) speaks, may be under way. In any event, both P and O should start looking for available substitutes for interrupted P–O sequences and plans. If X does substitute for many of the interrupted routines P had with O, P should experience a willingness to terminate the P–O relationship.

The point to be made here is simply that the P–O relationship was a close one initially, that it was not necessarily characterized by negative emotional overtones, but that the new relationship with X did manage to destroy the original relationship. It would seem not to be the case, then, that no third person can threaten a "good" marriage. The Achilles' heel of any relationship lies in the sequences and plans that each partner, for one reason or another, simply cannot facilitate for the other. When a third person can not only facilitate those plans, but also the sequences and plans that the original partner has been fulfilling, one can expect that the fabric of the original relationship should be severely weakened by the emergence of the third relationship.

It is probably more frequently the case in close relationships, however, that a new relationship with X, initiated and maintained because it facilitates certain limited previously unfulfilled plans, cannot fully and adequately substitute for all of the many old meshed behavior sequences and plans shared with O. If, as P explores the relationship with X and is unsuccessful in attempts to substitute P–O sequences, and if O continues to interrupt P in the old sequences in an attempt to get P to give up the third relationship and the new sequences, P is in a situation in which she (or he) is likely to experience great emotional trauma and turmoil; *both* O and X are now necessary for the maximum facilitation of P's behavior sequences and plans, the old and the new, but O (and perhaps X, also) may not be willing to maintain a "limited" relationship that would allow P maximum facilitation. P then loses (suffers interruption) if either relationship is terminated.

CONCLUSION AND COMMENTS

In this chapter we have tried to illustrate the importance of preserving a conceptual distinction between the closeness of a relationship and the magnitude and quality of the emotion that is likely to be experienced within it. When emotion is used as a classificatory variable of relationship closeness, some interesting and important questions concerning emotional phenomena within relationships are masked and others appear unnecessarily mysterious and paradoxical. We have attempted to relate some of what are now generally believed to be the causal antecedents of emotion to the properties of interdependence characteristic of close relationships, with special attention to degree of "emotional investment" each partner may have in the relationship.

The analysis here has focused entirely upon the "hotter" emotions as they may be experienced within the relationship and has set aside the "cooler" feelings and thoughts and appraisals that undoubtedly play an important role in the affective character of relationships and in many behaviors of interest that occur in them. We have done so not only for the practical reason of space; in addition, and as previously mentioned, feelings are currently thought of as simply "little emotions" whose antecedents and consequences are much the same as the "bigger" emotions. Further, whether they are, or whether it is truly useful to view these milder affective phenomena in this way, it is the case that the affective mathematics people use to review and combine their affective experiences within a relationship are not known, nor is it known how the experience of big emotions and little emotions or their cognitive combinations (if such are made) relate to certain behaviors of interest within relationships. These questions, then, constitute matters for considerable future investigation.

Finally, it should be noted that space considerations also have precluded discussion of the implications of the analysis offered here for a number of other special problems associated with emotion as it occurs in close relationships. One of the most important of these is the matter of individual differences in the experience of emotion stemming from differences in the development of an emotional language, innate constitutional differences, differences in the chronic ingestion of drugs that block ANS discharge, and so on (for a discussion of this issue, see Berscheid, in press-b) and the consequences of these differences for the relationship. Another is the treatment of couples whose presenting symptoms are largely emotional ones. This, too, is discussed elsewhere (Berscheid, Gangestad, & Kulakowski, in press), but it may be observed that there is little in the analysis presented here that is not generally compatible with such an approach as offered by Ellis' (1974) "rational–emotive" therapy. In addition to these, we hope the present analysis will shed light on a number of other fascinating emotional problems associated with close relationships, most of which lie waiting for the attention of investigators.

CHAPTER 5

Power

TED L. HUSTON

Bierstedt (1950) has observed that "in the entire lexicon of social concepts none is more troublesome than the concept of power. We may say about it only what St. Augustine said about time, that we all know perfectly well what it is—until someone asks us" (p. 730).

In this chapter, we venture to say what power is. In doing so, we seek to disentangle the varied uses of the concept of power that have been employed in the literature on close relationships. The complexity of the phenomena associated with power creates difficult conceptual and methodological problems for the scientist. Nonetheless, because people have a good intuitive understanding of "power," as Bierstedt observed, researchers have been able to study it fruitfully without solving—or even recognizing—the many problems of conceptualization.

A basic reason for the confusion surrounding the concept of power is that the construct has been used to denote and explain several different aspects of interpersonal influence. In the present chapter, we will be careful to distinguish among three related terms: influence, dominance, and power.

Gratefully acknowledged are the efforts of Rosemary Blieszner, Gilbert Geis, Susan McHale, Margaret Ray, and Elliot Robins, who read various drafts of the manuscript and provided useful suggestions. Preparation of this chapter was facilitated by a grant from the National Institute of Mental Health (MH 33938-02).

Influence is a descriptive term, referring to instances in which events in one partner's chain are causally connected to events in the other's chain. Thus, the occurrence of influence events and the effects of such events on the partners represent the essence of the close relationship. We would not say a relationship is close unless two people have influence on each other for a relatively long period of time.

Of course, the extent to which people influence their partners in close relationships varies in magnitude. Within a particular relationship, influence may be symmetrical or asymmetrical in varying degrees. When influence is asymmetrical over a broad range of activities, we may meaningfully speak of *dominance.* Dominance produces a hierarchical arrangement of the partners, which is reflected in differences in such aspects of relationship as freedom of movement, the utilization of resources, and rights and responsibilities.

The influence that occurs in a relationship involves various phenomena. The student of power is particularly interested in a certain class of influence, a class that includes only influence that is under the person's control and, being so, is exercised by the person in pursuit of various concerns. These exercises of influence reveal a person's power. *Power,* specifically social or interpersonal power, refers to the ability to achieve ends through influence. Persons who fail to achieve their own ends (or worse yet, consistently defeat them) may be influential, but they are not generally considered by social scientists to be powerful. As we will see, the notion of power as "ability" has important implications for its assessment. Like other abilities, it is not always exercised; when exercised, it is not always successful; and, even when successful, its magnitude may not be fully evident unless it is pitted against a counterforce (resistance to influence) of appropriate strength.

Influence and power can be located in our general framework in the manner shown in Figure 5.1. Influence refers to the description of interaction in terms of the effects of events in one person's (P's) chain on events in the other's (O's) chain and vice versa. Dominance is also a descriptive term, used to characterize asymmetry in influence over a broad range of activity. Power, in contrast, is an explanatory concept used to account for the portion of influence that each person generates intentionally (i.e., deliberately) in pursuit of particular ends. P's power is reflected in (and inferred from) P's intentional influence, and the same is true for O. If P's power is greater than O's over a broad range of their activities, this difference would frequently be reflected in a dominance relation between them.

As shown in Figure 5.1, the concept of power serves to abstract the common relevance of a variety of causal conditions to the intentional influence that occurs within a relationship. P's ability to influence O is determined by such conditions as cultural norms about whether P's or O's views are to be given priority in arriving at decisions, the degree to which P and O control resources or information, and their relative amount of love,

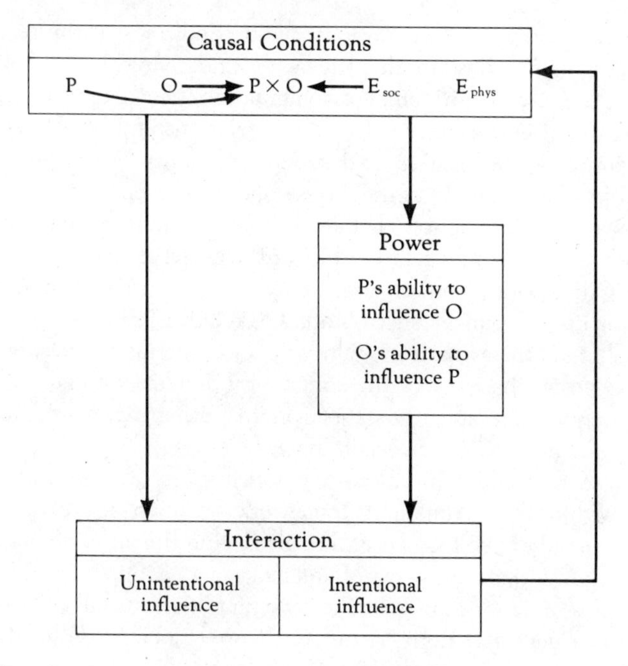

FIGURE 5.1
Analysis of power within relationships. Power reflects those causal conditions that enable the persons, P and O, intentionally to influence each other.

commitment, and dependency. Similarly, O's ability to influence P resides in one or a combination of the various causal conditions. In each case, the ability is derived from certain causal conditions, but also, in each instance, that ability is possessed by the person and is his or hers to control for whatever purposes become important. Figure 5.1 shows power as a unitary concept in linking the causal conditions to interaction. However, as we will see later in the chapter, different types of power can be distinguished, with each type serving to link a particular set of causal conditions to a particular component of intentional influence.

In Figure 5.1, the unintentional influence in a relationship is shown, like power, to derive from a broad set of causal conditions. The specific causal conditions, of course, may be different for the two classes of influence, and, indeed, they probably are. Moreover, even though power is conceived as bearing upon intentional influence, its possession also increases the powerful person's likelihood of exerting unintentional influence (see Lippitt, Polansky, Redl, & Rosen, 1952). There is no parallel concept in the social science literature that links causal conditions to unintentional influence in the way that power is used to link them to intentional influence.

The arrows connecting the various causal conditions in the rectangle at the top of Figure 5.1 illustrate that the factors underlying observed patterns of influence generally act in complex combinations. For example, the norms pertaining to power that have emerged in a relationship (a P × O condition) may be causally anchored in more distal cultural norms, in the resources of P and O, or in a combination of these types and other types of conditions. The arrow drawn from interaction back to the causal conditions reminds us that the influence activity in a relationship will itself serve to modify the causal conditions that control power and influence. An initial asymmetry of influence, for instance, may affect P's and O's relative resources in a way that serves to sharpen the asymmetry. Similarly, P–O interaction may give rise to new P × O norms that provide the dominant individual with greater freedom and that require the subordinate person to anticipate the desires of the dominant one and to behave accordingly.

Social science conceptualizations of power in close relationships all can be analyzed within the rudimentary framework set forth above. Theory and research generally have focused on understanding the antecedents of power and its manifestations in terms of influence and dominance. Theories of power in close relationships refer, at least implicitly, to all of the features identified in Figure 5.1. Both "resource" theory (Blood & Wolfe, 1960) and "ideological" or normative theory (Rodman, 1967), for example, focus attention on the causal conditions that underlie power and look for power to be reflected in influence and dominance patterns. French and Raven (1959) differentiate types of power in terms of both their antecedent conditions and several distinctive types of influence. Emerson (1962) gives particular consideration to the feedback connections from immediate P–O interaction to P and O causal conditions. Specifically, Emerson analyzes the ways in which participants in a relationship react to, adjust to, or otherwise respond to the power balance or imbalance in a relationship.

The framework shows how power can be analyzed in terms of its antecedents, its manifestations, and its consequences. The remainder of this chapter consists of an elaboration of the basic features of the framework and their linkages. We begin at the bottom portion of Figure 5.1 by considering matters of describing interpersonal influence and dominance. In the second section, we turn to the middle part of Figure 5.1 and the particular problems associated with measuring power. We move in the third section to issues pertaining to the linkage between the top and bottom portions. Specifically, we will examine ideas about the causal conditions that relate to power (the bases of power), especially as these conditions account for influence between husbands and wives.

We finally deal with aspects of the feedback connection between influence processes and the causal conditions (shown by the upward arrow in Figure 5.1). The fourth section reviews research that has attempted to determine how the influence a parent exercises over a child changes the child's personality. The fifth and final section deals with how the influence and power exercised in a relationship act to change causal conditions relevant to subsequent influence. This discussion will concern the entire system shown in Figure 5.1 and explore circular causal effects in which causal conditions relevant to power undergo change.

DESCRIPTION OF INTERPERSONAL INFLUENCE

Most generally, we may say that influence occurs when the events in one person's chain are causally connected to events in the other's chain. Causal connections are indicated by the fact that something changes in O's chain as a consequence of some event in P's chain. This idea is similar to that offered by French and Raven (1959), who suggested that influence occurs when P–O interaction results in "psychological change" in O. Psychological change was defined broadly to include any alteration in the person's psychological system, including the person's cognitions, affect, and behavior. The P–O interaction was also given comprehensive meaning and included everything from O's awareness of P's presence to direct interaction.

A brief exchange between a "dominating" husband and an "acquiescent" wife can be used to illustrate the basic phenomena of interpersonal influence. Figure 5.2 portrays the sequence of actions, cognitions, and affect, and the connections among them. Our depiction describes the interaction of the husband and wife at a relatively molar level; a more detailed description might include information regarding such things as facial expressions and body orientations.

The scene takes place in the couple's home, where the husband and wife are getting ready to go to a party. On the behavioral level, the interaction unfolds as follows. The wife takes her yellow sundress from her closet while her husband watches. The husband says firmly, but pleasantly, "You really ought to wear your blue outfit." The wife hesitates a moment and then responds by returning her sundress to the closet and taking out her blue outfit. At the same time, she asks her husband why he doesn't like the sundress. Her voice is somewhat strained, her movements a bit tense. He hesitates before responding and then replies that her blue outfit would be more appropriate. She begins to put the blue outfit on, and he leaves the room, thus ending the episode.

In this interaction, the behavior of each partner has influence on the other. This is shown in Figure 5.2 by the arrows that cross from one partner to the other. These arrows mean that the behavior of one partner affects—stimulates, facilitates, interrupts—the chain of thoughts, actions, or emotions of the other. Or to put it differently, an arrow drawn from one partner

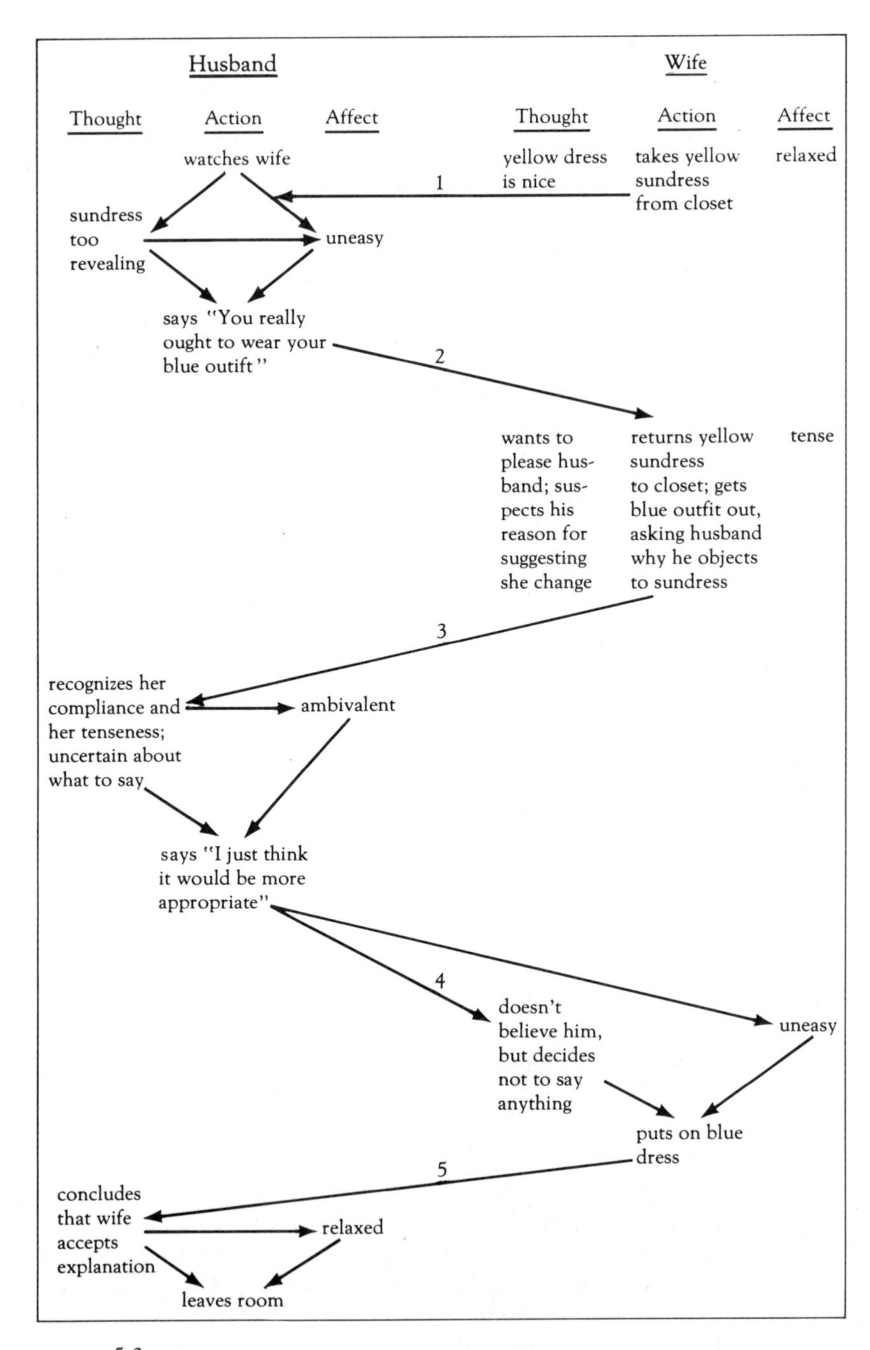

FIGURE 5.2
An episode of interaction between a husband and wife in which the several interchain causal connections (numbered 1 to 5) illustrate interpersonal influence.

to the other indicates that the probabilities of particular thoughts, actions, or emotions have been modified by the behavior of the person from whom the arrow is drawn. The episode shows five instances in which the events in one person's chain have influence on subsequent events in the other person's chain (these are numbered in the figure). Many complex configurations of influence are possible (see Chapter 2), and only a few of the possibilities are illustrated in this episode. Consider, for example, the husband's suggestion that his wife wear her blue outfit. His statement is shown to be jointly determined by her behavior, his awareness of her actions, his thought that the sundress is too revealing, and his feelings of uneasiness.

It is also important to keep in mind that this episode occurs in the context of other episodes and that what happens partly reflects what has gone on before and the participants' ideas about the future. The husband's thoughts and reactions, for example, may reflect insecurities based on what occurred the last time his wife wore the sundress to a party, or, alternatively, they may stem from a general need on his part to control his wife.

Each event in the episode is shown in the figure to be causally connected to other events, with the effects of one event on another often being mediated and indirect. The feelings of the wife and husband at the end of the encounter reflect not only the events that immediately preceded them but also previous events both in the episode and earlier. The husband's relief can be traced to such earlier events; similarly, the wife's uneasiness developed after the husband expressed his desire for her to wear the blue outfit and was maintained by subsequent events.

The intricate intertwining of actions, thoughts, and emotions illustrated in Figure 5.2 provides an approximation of a naturalistic portrayal of influence. Given the complexities of the interactional give and take suggested by this example, how can influence be described systematically? How can the complicated interplay between the two persons' chains be separated so as to describe each partner's influence on the other? We will attempt to answer these questions by first considering what such a description might cover in principle. This consideration is made possible by focusing our analysis on a single unit of influence, which we will call the "influence sequence," and identifying its important features. Subsequently, we will shift to a consideration of the practical problems encountered in assessing influence, most of which arise in the course of identifying the causal connections illustrated in Figure 5.2 and in assembling information about regularities in influence.

The Basic Unit of Influence

The smallest interaction segment that permits meaningful analysis of influence is a P–O–P sequence of events. This is illustrated in Figure 5.2 by the husband's suggestion to the wife that she wear the blue outfit, her subsequent replacing the sundress and taking out the blue one, and the effects of that compliance on the husband. This sequence is defined by the interchain connections labeled 2 and 3 in the figure. Influence is involved because, as the arrows show, the husband's suggestion caused the wife to do something she would not have done without his suggestion. Had it been otherwise, the switch in dresses would have been shown to be causally linked to earlier events within the wife's chain or to events from external sources.

The influence sequence is shown schematically in Figure 5.3, where the originator (P) of the sequence—the influencer, or, more neutrally, the actor—is shown to affect the target (O), who, in turn, affects the actor. The $p_1 \rightarrow o_1$ connection constitutes the *influence,* but it is necessary also to consider the subsequent $o_1 \rightarrow p_2$ connection in order to specify the *consequences* to P of the influence and thereby to understand such things as P's intention and the success and failure of P's influence.

The two-step influence sequence is the minimal segment necessary for discussing basic features of influence, but it is possible to identify more extended sequences. For example, in Figure 5.2 we may examine the sequence that begins with the husband's suggestion and ends with the wife putting on the blue dress and the consequences of that action for him. This sequence spans several interchanges as the husband's influence on the wife's final behavior is mediated through several intervening connections, going from 2 through 3 and 4 and ending with 5. For simplicity, it is convenient to disregard the intervening connections and simply specify the net effect of the initial suggestion on the final change of dress. This simplification moves the analysis of influence to a more macroscopic level.

In any complex interaction, such as that shown in Figure 5.2, many successive influence sequences can be discerned. In the example above, connections 2 and 3 can be identified as pertaining to influence and consequence, respectively. In describing the husband's influence, we might also examine the sequence constituted by connections 4 and 5. Similarly, the

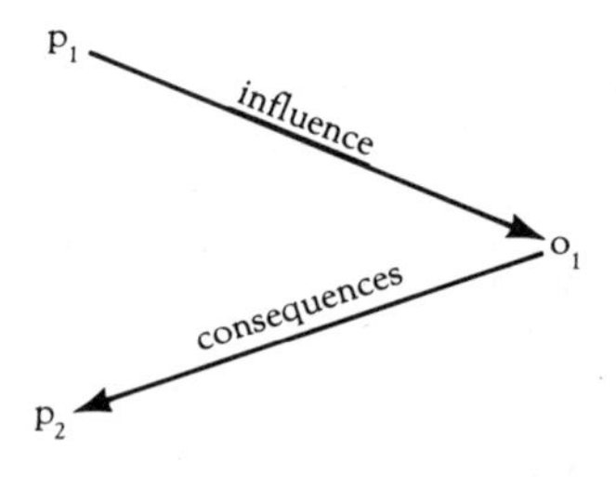

FIGURE 5.3
Schematic representation of the influence sequence showing the influence of person P on person O and the consequences to P of the influence.

wife's influence is reflected in connections 1 and 2, a sequence in which she induces his suggestion through her initial choice of dress, and in connections 3 and 4, a sequence in which she induces him to explain his objection to the sundress.

It is an inherent property of interaction in close relationships that influence sequences overlap. In the example, the wife's influence defined by connection 1 results in the husband's influence defined by connection 2. Thus, she influences him to influence her. This overlap provides the kernel of truth in the psychologist's joke about the rat in the learning apparatus who boasts about its control over the human experimenter. In any closely connected and symmetrical interaction of the sort shown in Figure 5.2, it is difficult to disentangle the influence of the two people involved. The influence of P is dependent on the counterinfluence of O and vice versa. The separation of the two persons' influence is easier (in the sense that there is less overlap in the sequences) when either (1) the two chains are less closely connected, or (2) the influence is asymmetrical, going mainly from P to O and not the other way. In such cases, it is possible to isolate sequences in which P's initiation (referring to the portrayal in Figure 5.2) comes "out of the blue," in the sense that it is caused by prior events in P's chain or by events outside the relationship and, therefore, is little dependent on prior interaction with O. In closely connected symmetrical relationships (and, therefore, in many close relationships), it is useful to identify events in O's chain that instigate or provoke P's influence. The specification of the interchain antecedents of the sequence extends the analysis backward to show a kind of reactive influence (see Figure 5.4). Whereas the influence sequence (as shown in Figure 5.3) suggests that P initiates the influence and that the influence is proactive, directed toward introducing something new in O's chain, the sequence of interchain events shown in Figure 5.4 suggests a

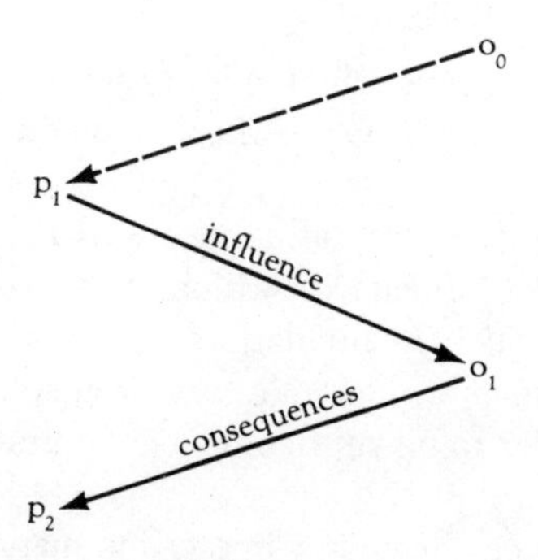

FIGURE 5.4
Schematic representation of reactive influence. Because an event in O's chain affects P (broken arrow), P attempts to influence O.

more reactive kind of influence directed toward modifying something pre-existent in O's chain.

The most important features of influence sequences can be listed briefly:

Content

What kinds of events (see Chapter 2) are involved? What is the influence about? In what realm of psychological and behavioral activity does it occur? Is O changed in behavior, thought, or affect? What kind of behavior, thought, or affect is involved? The influence in Figure 5.2 had to do with the wife's attire for a particular social occasion. Other instances might involve other aspects of her behavior, other situations the couple enters, various aspects of the wife's attitudes (say, about her husband or about some political issue), the information or skill she possesses, and various joint decisions (about major purchases, for instance).

Actor's intention

Is the influence intentional or unintentional? If it is intentional, to what end is it directed? Sometimes the influence reflects a deliberate attempt on P's part to induce some desired change in O. On other occasions, P influences O unwittingly, the influence being an incidental consequence of their interaction. To label the influence as intentional, we must assume that P has some idea about the consequences for self of the change produced in O (i.e., the consequences connection in the influence sequence in Figure 5.3). Intentional influence implies that P desired certain changes in O's chain more than other changes (or no change) and that, in view of these preferences (and P's understandings about consequences), P acted so as to produce the changes.

Mode of influence

The mode of influence refers to the kinds of causal connections between P's and O's chains (see Chapter 2). While there are many ways in which an actor can effect change in a target, our present analysis concentrates on modes of influence that involve direct causal connections between P and O, rather than indirect methods of influence in which the social and physical environment is manipulated. Such manipulation might involve, for instance, an individual arranging chairs in a certain way or encouraging people to sit in particular chairs in order to make them feel comfortable and conversational (cf. French & Raven, 1959).

Direct influence can be effected by impression management, by appeals to norms, or by any of a number of types of explicit messages from P to O. Several systems have been developed to categorize such messages. Tedeschi and his colleagues (Tedeschi, Schlenker, & Lindskold, 1972) differentiate messages in terms of whether they involve threats, promises, warnings (predictions of bad consequences of specific actions) or "mendations" (predictions of good consequences). Falbo and Peplau (1980) recently had 200 college students write essays on how they believe they get their romantic partner to do what they want. The researchers classified these statements into 13 categories that were then ordered within what emerged as a two-dimensional space. The first dimension had to do with the extent to which the influence mode was direct versus indirect. Direct tactics included such behavior as telling, asking, or talking to the partner; indirect tactics included the use of suggestions, hints, and withdrawal. The second dimension concerned the extent to which the tactics were interactive, ranging from unilateral to bilateral. Unilateral tactics included taking independent action and withdrawal; bilateral tactics included bargaining and efforts at persuasion.

The influence modes described above are clearly not exhaustive inasmuch as they include only what people say they do when they wish intentionally to influence their partner. Other modes, such as those used in unintentional influence, involve paralinguistic cues and nonverbal communication by facial and body cues (Mehrabian, 1972).

Magnitude of influence

Magnitude of influence is assessed by reference to changes in the target that result from the actor's behavior. An intervention (or series of interventions) by an actor may have weak, limited, and short-lived effects on the target. This would be the case, for example, when a person's comment about the weather stimulates a minimal reaction, such as a shrug. At the other extreme, an act (or series of acts) may have strong, diverse, and far-reaching effects. Physical violence in the course of a heated argument between a husband and wife may affect the recipient in extensive and enduring ways.

The magnitude of an actor's intervention thus can be assessed along three basic dimensions, the first two of which have been described in Chapter 2. These are strength, diversity, and temporal reach.

Strength of influence refers to the degree to which an intervention produces change in the target. The greater the amount of change induced in the probability of an event, the stronger the actor's influence on that event. Because of presumed target resistance, acts that transform or reverse ongoing behavior usually have greater strength than acts that set in motion new lines of behavior, and the latter have greater strength than acts that merely augment or reinforce existing behavior.

Diversity of influence refers to the number of events and range of types of events affected by the actor's intervention. In our husband–wife scenario, we distinguished among three categories of events—cognitions, actions, and affects—but, as is noted in Chapter 2, these categories are merely suggestive; other distinctions, as well as further refinements, could be made.

Temporal reach refers to the number of subsequent occasions on which the response is different from what it otherwise would have been or, to put it more precisely, to the number of times and length of time the probability of the event has been altered (cf. Cook & Flay, 1978). An act may have large immediate effects, but little or no long-term impact. Alternatively, the immediate effects may be negligible but the delayed ones considerable. Delayed effects will be detected, of course, only if they are encompassed within the time span of the influence sequence subjected to analysis. (As noted in Chapter 2, influence that is evident after a long delay implies a change in one or more causal conditions.) If the husband's suggestion about the sundress causes the wife to reject it as her attire for subsequent parties, it is reasonable, even necessary, to assume that the influence has produced a change in the stable causal conditions constituted by her relevant attitudes or beliefs (regarding the husband and his likely reactions, her dress, and so on).

Consequences for actor

The change induced in the target may be facilitative of the actor's intrachain sequence of thought, action, and affect—promoting the actor's interests, providing satisfaction, and reducing costs. Alternatively, the change in O may be disruptive for P, preventing or inhibiting P's ongoing activities, increasing his or her costs, or causing tension and discomfort. P may be said to have positive influence if, on balance, the consequences for P are desirable (facilitative) and to have negative influence if the consequences for P are undesirable (interfering). Influence can be described as positive or negative whether or not it is intentional. If P's influence is intentional, then the consequences can be described in terms of success or failure. Determination of success or failure requires a comparison of the change actually induced in O with the change P tried to bring about. Failure includes the case in which O does not change in the desired direction or changes in a manner opposite to what P intended.

Assessing Influence

Influence can be described, in principle, in terms of the sequence features delineated above. Needless to say, the interchain connections pictured in Figure 5.2 are not always visible to the participants or to others observing the interaction. There are two kinds of evidence of their existence, and, since analysis of influence must detect these connections, it has to rely on one or both kinds of evidence. The first detection procedure employs observers, either outsiders or the participants themselves, who make *subjective judgments* about the influence of P's activities on O. The second approach, which, for lack of a more precise term, will be referred to as *objective,* involves a fairly mechanical recording of interaction events by a machine or by observers. The causal importance of P's activities is assessed by detecting statistical regularities between them and O's events.

Subjective evidence

Subjective evidence regarding influence is provided by human observers who witness sizable portions of the P–O interaction and make judgments as to when an event in P's chain has caused an event in O's chain. Attribution research (see Kelley & Michela, 1980, for a review) suggests the cues that are likely to be used in these judgments. Ordinarily, event o will be seen as caused by event p if o follows closely after p, does so regularly, and if o otherwise occurs rarely. Thus, an unexpected change in O's behavior will be seen to have been caused by P's intervention (Layton & Moehle, 1980).

Several investigators have measured influence in terms of the apparent contingency of O's events on P's activities. The influence may be seen as either incidental or intentional, depending on the context and on whether the mode of influence appears to be under the actor's control and whether the consequences can be coded as successful or unsuccessful. Lippitt et al. (1952) studied "behavioral contagion" (unintentional influence) in a summer camp, with contagion being defined as "the spontaneous pickup or imitation by other children of a behavior initiated by one member of the group where the initiator did not display any intention of getting the others to do what he did" (p. 37). They inferred the existence of contagion when O's behavior both followed and closely resembled P's activity and when P had made no discernible attempt to influence O.

Lippitt et al. (1952) distinguished behavioral contagion from "direct influence," which was defined as a social interaction in which "one child consciously and deliberately tries to get another child to do something, in such a way that the research observer is aware of the intent" (p. 41). Direct influence was inferred when the behavior that P attempted to induce corresponded to the subsequent behavior of O. Lippitt et al. used these data to derive for each individual an index of "manifest power," which was defined, following Goldhammer and Shils (1939), as the percentage of influence attempts that are successful.

Investigators studying influence in husband–wife relationships (Bahr & Rollins, 1971) and in small groups (Mayhew, Gray, & Richardson, 1969) have assessed direct influence in a fashion similar to that used by Lippitt et al. (1952). Bahr and Rollins (1971), for instance, trained observers to note any direction, instruction, suggestion, or request by one partner intended to control or modify the behavior of the other. The consequence of each attempt was coded as either accepted, ignored, or rejected, according to the response of the recipient. From these data, they derived an index for each partner of "relative marital power," which was defined as the ratio of the partner's accepted influence attempts over the total accepted influence attempts for the couple. Mayhew et al. (1969) also had observers note influence attempts and code them as to whether or not the other complied. They, too, proposed an index that describes each actor's power as the proportion of the total compliance in the group obtained by the actor. Estimates of overall influence based on the frequency of successful influence attempts overlook the magnitude of the effects P's and O's actions have on each other. The fact that one "win" may overwhelm twenty "losses" suggests caution needs to be exercised in drawing conclusions about overall patterns of influence without assessing the magnitude of the influence that occurs. The circumstances under which power can be seen as reflected in such factors as the "success rate" will be considered later when we undertake a conceptual analysis of power.

Subjective judgments of P–to–O causation can be affected by irrelevant cues. Event o will be seen as caused by event p if p is a plausible cause or is particularly salient, even if there was no causal connection. Explicit messages sent by P that elicit compliance are more likely to be noted by an observer than is compliance elicited by indirect or subtle means. There is some evidence that the perceptual salience of person P—as produced, for example, by providing observers of a conversation with varying opportunity to observe P's activity compared to O's—affects the degree to which P is perceived to exert influence (Taylor & Fiske, 1975). Biased estimates of influence in a particular setting also may result from attributions regarding P's overall potency as an influence agent, as well as the observer's ideas about norms, roles, and the usual impact of particular modes of influence. With regard to the latter, several investigators interested in influence have counted instrumental (or directive) acts, using the coding scheme developed by Bales (1950), and have made the reasonable (but certainly fallible) assumption that the frequency of instrumental acts (Kenkel, 1957), or ratio of instrumental to noninstrumental acts (Caputo, 1963), reflects amount of influence.

Global assessments of influence require the observer to abstract from memory the extent to which particular types and combinations of behavior occur. Such assessments require people to indicate how frequently P gives advice, makes suggestions, praises, or provides information to O; and how often P responds to O's communications in particular ways. Schweder and D'Andrade (1980) have shown that such ratings are systematically biased by the semantic similarity of the items on the rating scales. Types of behavior that are seen as similar are rated as occurring at rates more equal than is in fact the case; the reverse is true for behaviors semantically dissimilar.

The foregoing suggests sources of inaccuracy in subjective judgments of interchain causation. Subjective judgments are influenced by such factors as plausibility, salience, errors in detecting the order of events, erroneous conceptions of what is typical (base rates) for a particular target person, and ideas about the potency of the influence agent or the mode of influence. On the other hand, subjective judgments have certain advantages. Observers' understandings of what is common and uncommon behavior, or common and uncommon interchain sequences of behavior, make it possible for them to detect influence within a single influence sequence (as would be evident when a person reacts with alarm to having a gun pointed at him, or when a *sotto voce* comment to a young man about the position of his pants zipper makes him blush). In contrast, the procedures described below, which rely on statistical analyses to detect influence, involve recording large numbers of interaction events in order to calculate stable conditional probabilities. These procedures require prior specification of relevant p's and o's, and therefore they are less likely to pick up unanticipated instances of influence than are approaches that rely on subjective judgments.

If the participants themselves are used as observers, events of a private nature (thoughts, feelings, etc.) not detectable by others or by machines can enter into the analysis. Moreover, in their own reports, participants can describe causation that reaches back to prior episodes and thereby can expand the temporal scope of the analysis of influence activity.

Objective evidence

Objective evidence of influence is provided by recording events in the two persons' chains and detecting statistical regularities between them. Changes in O's chain that indicate causality by events in P's chain are formally specified by comparisons of certain conditional probabilities. If we focus on any particular type of event p in P's chain and another subsequent class of event o in O's chain, the influence of p's on o's is indicated by the difference between (1) the probability of o occurring in the absence of p and (2) the probability of o occurring following p.

Patterson (1974a, 1977) illustrates how these probabilities can be employed to infer P–to–O influence, using a sample of interactional data involving a highly aggressive 6-year-old boy with other members of his family. The "target" of influence (O) was the boy, and the particular o events of interest were those reflective of the boy's "hostility" (whining, yelling, expressing disapproval). The designation of particular acts as hostile was not based on a direct assessment of the consequences of the acts for P, but rather on a subjective judgment made by the researchers regarding their likely consequences for P. Human observers recorded these hostile o events and many types of p events (including their source—mother, father, and sister) on a time line divided by 6-second intervals.

Patterson's causal analysis was objective in the sense that influence was inferred from statistical regularities between observed p and o events. Calculation was made of the overall base rate of a particular type of o event and the probability of that type of o event's occurrence in the 6-second interval after specific types of p events had been enacted by each of the family members. A comparison of the base-rate probability of o events and the conditional probabilities following particular types of p events enables identification of p's that, as antecedents to o, either facilitate it (increase its probability) or inhibit it (decrease its probability).

This analysis revealed that particular p's differed in their effects on the boy, depending on who enacted them. Thus, disapproval by the father increased the likelihood the boy would behave in a hostile manner, whereas the same disapproval by the mother served to inhibit this hostility. Other p's, such as a family member talking to the boy, had a uniform effect on the likelihood of the boy behaving in a hostile manner, regardless of whether the source was the father, mother, or sister. Finally, still other p's, such as ignoring or indulging the boy, apparently had no influence on the boy's hostile behavior.

Patterson (1973, 1977) also used his data to study the reactive influence shown schematically in Figure 5.4. The analysis described above examines the effect of particular p's on the *initiation* of hostile behavior, whereas the analysis of reactive influence considers the p's that accelerate or decelerate hostile behavior already initiated by the boy. Events that significantly increased the likelihood of consecutive hostile o's were defined as "accelerating" stimuli; those associated with lessened probability were labeled "decelerating" stimuli. Patterson began this analysis by examining p's that, following a hostile act by the boy (o_0), either increased or decreased the likelihood that the next act recorded for the boy (o_1) would also be hostile. Calculation was made of the overall base rate for the recurrence or persistence of hostile activity in adjacent 6-second time frames, and this rate was compared to the rates of o's recurrence (or persistence) when various p events were sandwiched between o_0 and o_1.

These formulations handle many of the conceptual problems involved in describing influence. They require information only about the time of occurrence of each p and o over a period long enough to provide frequencies adequate for meaningful statistical statements. They cover the possibility that P's influence may either promote or inhibit some particular behavior. They detect causal connections without reference to intentionality, the similarity between P's acts and O's changes, and the consequences of the change for P. If various time lags are used, they encompass both instances in which P's influence is immediate and in which it is delayed. Thus, they can deal, at least in some ways, with the "temporal reach" aspect of magnitude.

The principal problems in the application of these formulations lie in the sampling of settings and the specifications of p's and o's to be recorded and analyzed. Inferences regarding generalized influence patterns require careful attention to sampling modes of influence, the settings in which influence occurs, and types of target events. There is also some problem in selecting the time lag to be used in analysis, and it is difficult to establish causal connections between p events and o events that are separated in time. Gottman (1979) shows how the spectral density function obviates part of this problem, but, in general, delayed effects are difficult to detect unless they are particularly strong and persistent and unless the investigator has considerable data at hand with which to estimate the relevant probabilities.

Patterns of Influence and Dominance in Close Relationships

Social scientists studying close relationships frequently are interested in general patterns of interpersonal influence. Once we have determined the influence of P over O and of O over P, the balance of influence between the two can be described. A dominance relationship exists between P and O if, over a wide range of activities or large portions of P–O interaction, one partner consistently exercises more influence than the other. By "more influence," we mean not only the frequency with which P and O exert influence, but also the magnitude of influence measured in terms of strength, diversity, and temporal reach. Determination of patterns of influence requires the investigator to disentangle influence sequences, a task which involves deciding on the particular sequences to be included in the analysis and then ascertaining the extent to which P's events are influenced by O's previous actions and vice versa.

Gottman (1979) has demonstrated how dominance in husband–wife interaction can be ascertained by calibrating the degree to which each partner's observed affect is predictable from the other's previous affect. Asymmetry in predictability reflects dominance, with the partner whose affect is less predictable being dominant. Gottman videotaped and coded husband–wife interactions in terms of the sequence of affective displays, with each "floor shift" (change in speaker and listener role) being the unit of analysis. The data were subjected to an analysis based on time-series procedures (phase spectrum analysis), which allowed determination of whether the affect evident in verbal and nonverbal communication while one partner had the floor preceded or followed the affect when the other was speaking. This analysis allowed specification of whether the affect at one time predicted the immediate affective expressions during the next floor shift; it also allowed assessment of the extent to which the affect at one time predicted the affective tone when the partner had the floor several shifts after the first affective expression. The former type of influence sequence, based on the analysis of contingencies among adjacent floor periods, was used to determine whether one partner dominated fluctuations in the *immediate expression of affect;* the latter was used to ascertain whether one or the other partner tended to set the *mood* of the discussion. The analysis allowed Gottman to classify couples in terms of whether they were egalitarian, husband dominated, or wife dominated and whether the influence pattern changed, depending on whether consideration was being given to immediate affect or to lasting moods.

Influence in a close relationship stretches over a long period of time. As a consequence, accurate estimates of the overall pattern of influence require knowledge of the history of interaction between the interactants. The observable actions of the partners in a particular episode (or set of episodes) are generally influenced by events that have occurred earlier in their relationship. Thus, low interdependence may result from an avoidance of the highly influential partner by the one who has had less influence.

More importantly, asymmetrical influence, early in a relationship, may produce adjustments in the attitudes and behaviors of the partners. The subordinate partner may tend to conform spontaneously to the wishes of the dominant partner and thereby reduce the need for the dominant one to exert overt influence (Friedrich, 1963; Nagel, 1975). Thus, the dominant partner may make few overt attempts to influence the subordinate partner in a particular situation. Indeed, in some situations, the subordinate partner might exert more influence. The idea that asymmetrical influence, or dominance, produces a hierarchical arrangement between the partners will be discussed more fully when we take up the idea that influence processes change causal conditions relevant to subsequent influence.

POWER

This section deals with the middle part of Figure 5.1, the notion of power as the ability to influence. By likening power to the idea of "ability," we do not wish to imply that power is necessarily, or even primarily, anchored in the individual attributes of the powerholder. Individual characteristics, such as Machiavellianism or need for dominance, may cause a person to have power, of course, but a person's ability to influence also may be strongly affected by such things as cultural conditions (e.g., social norms), vulnerabilities in the other (O characteristics), and previous agreements between the partners (P × O conditions). Moreover, a person's ability to influence may be general, or it may vary considerably, depending on the changing relevance of power-related causal conditions.

As we noted at the outset of this chapter, the construct of power has been developed out of an interest in that portion of influence that is under the actor's control. In the context of close relationships, power refers to P's ability to influence O when P wishes to do so. The controllability of influence is important, both to students of relationships and to the participants themselves, because it indicates the extent to which persons can make things go their own way. If there were not conflict of interest in close relationships, neither theorists nor the partners would be concerned about power.

Influence would still be worthy of attention, however, because it is through influence that close relationships are formed and changed and because influence has to do with the effectiveness with which partners are able to pursue their common goals. Power is important whenever the partners have incompatible goals or when they have different ideas and propensities in connection with their common goals, as when they have contrary preferences about when and how to attain them.

Problems in Assessing Power

Describing power entails special considerations that pertain to its being conceived of as an ability. Power is the potential for effecting changes in the world (cf. Lewin, 1951a). To put it more formally: "*Social power* is (a) the potentiality (b) for inducing forces (c) in other persons (d) toward acting or changing in a given direction" (Lippitt et al., 1952, p. 39). This view of power is consistent with the way power has been conceptualized in both social psychology (cf. Cartwright, 1959; French & Raven, 1959) and sociology (cf. Blood & Wolfe, 1960; Rollins & Bahr, 1976; Safilios-Rothschild, 1970; J. Scanzoni, 1979c; Szinovacz, in preparation; Weber, 1947).

If power is defined as an ability, its assessment brings forth problems similar to those that occur in the measurement of other abilities. Abilities have to do with the potential to accomplish some purpose, and, as such, they can be gauged only when a particular set of circumstances converge. Let us examine a simple situation. How would we assess a child's spelling ability? (1) We would induce the child to try to spell a number of words. The full extent of the child's ability would be apparent from performance only if the child tried its best to spell each word. (2) We would judge the child's success on each word. This assessment would be a simple matter because there are common standards for the correct spelling of most words. (3) We would give the child words to spell ranging from easy to difficult and determine the most difficult ones it could spell successfully. The level of difficulty of the hardest words the child could spell would become the child's score, indicating its spelling ability.

The principles of measuring a person's power are identical to those for ascertaining a child's spelling ability. However, at each step, it is a complex matter to determine whether the conditions for good assessment are fulfilled (cf. Turk, 1974). Before considering some of the available measures of power, it is useful to examine issues specific to (1) ascertaining whether a person is trying to exert influence, (2) gauging the success of attempted influence, and (3) calculating the amount of resistance an individual is able to overcome.

Trying to influence

To assess power, it is necessary to distinguish between when an individual is attempting to influence another and when such influence is unintentional or incidental. How do we know a person is trying to exert influence? In natural social situations, we are often able to "see" the intent in the nature of the action. Recall the husband–wife episode portrayed in Figure 5.2, which centered on which dress the wife would wear to a party. The husband's statement about the greater appropriateness of the blue compared with the yellow dress would seem to indicate he was trying to influence her choice of dress. We would be even more certain of our inference that the husband's behavior was directed toward inducing change in his wife had she not changed immediately following his assertion and if he had then made new or more intense appeals. P's persistent and directed efforts, which cease only after a particular change has occurred in O, are taken to indicate that P has tried to exert influence. Trying is thus characterized by equifinality—that is, invariance in the ends sought, coupled with variability in the means or influence modes (cf. Heider, 1958).

The sundress episode portrays two kinds of information that allow one to identify whether a person is trying to exert influence. The first of these has to do with P's choice of influence mode. Direct requests, demands, or suggestions, for example, are generally seen to indicate P is trying to influence O. The husband's verbal statement was directed toward the wife; his desires regarding the wife's behavior were clearly evident in his comments. On the other hand, the wife selected the yellow dress with no evident intent to provoke comment by the husband. Lippitt and his colleagues (1952) differentiated intended influence from "behavioral contagion" (the spontaneous pickup of behavior) partly in terms of whether P's activity implied that P was trying to produce a particular change in O. Judgments were made by observers who looked for the kinds of directive behavior noted above. While this procedure is far from infallible, investigators studying intended influence in husband–wife relationships (Bahr & Rollins, 1971; Caputo, 1963; Kenkel, 1957) and in small groups (Mayhew et al., 1969) have assessed attempted influence in a similar fashion.

The second kind of information that identifies trying pertains to the motives of the people involved. From knowledge of the husband's thoughts just prior to his assertion, we are aware that he did not like the idea of his wife wearing the yellow dress. It follows, then, that his comment about the blue dress can be seen as an attempt to dissuade her from wearing the yellow one. Frequently, in both natural and laboratory situations, similar information is available from which to infer the motives of the parties involved. We use either our common knowledge about what people in general want, or we determine what two specific people want, either usually or at a particular time. If the motive we impute to a person is consistent with a change produced in the partner, we have a basis for concluding that the person was intentionally trying to exert influence.

In laboratory situations, investigators (e.g., Strodtbeck, 1951) require participants to form an individual opinion and then through discussion arrive at a group position. By selecting issues on which the group participants disagree (Caputo, 1963) and by inducing the members to become committed to their individual position, it is possible to stimulate them to try to influence each other. Sometimes participants in close relationships, however, may not try to influence their partner in such situations. They may not try because they have concluded that they are not apt to be successful. They believe they are powerless (or at least less powerful than their partner) or perhaps fear that a direct test might prove them to have insufficient power. Along these lines, if people wish to maintain a self-image of being efficacious (Greenwald, 1980; Jones & Pittman, 1982), they may try to exert influence only when they believe they have a reasonably good chance of success. It is a moot question in such situations as to whether the individual lacks power.

To summarize: Trying means making an effort to induce change in another. Such an effort is most easily identified when P (1) makes an overt attempt to induce change in 0, (2) persists in making the effort in spite of O's resistance, and (3) is known to have motives consistent with the nature of the changes sought in O.

Success

Power has to do with a person's ability to change another individual in a particular way. There is no external criterion here, as there is for correctness of spelling. Success is entirely a matter of P's inducing the change in O that P sets out to induce. To gauge success, then, we first need to know what P's intention was. Success occurs when the effect produced corresponds with what P desired.

From P's point of view, assessment of success involves a comparison of two features of the influence sequence identified earlier: P's intention and P's assessment of the consequences of O's actions for P. From an outsider's perspective, success can be gauged by (1) knowledge of P's goals and the degree to which O's behavior subsequent to an influence attempt is goal supportive or reinforcing; (2) reference to the connection between the intentions evident in P's actions (mode of influence) and the congruence of O's subsequent behavior with those apparent intentions; or (3) observation of P's reaction to O's behavior subsequent to an influence attempt. With regard to the latter, success may be evident in P's becoming relaxed or by a redirection of P's effort. Failure, on the other hand, may be accompanied by signs of distress, such as tension and emotionality, or by a redoubling of effort.

The change P desires in O must remain in place from the beginning to the end of the influence sequence under consideration in order to consider issues of success and failure. P's commitment to an end or set of goals, coupled with evidence of trying, are necessary for assessing success. In interactions where participants' ends are readily modified by feedback from the other, the assessment of success is not possible. In some instances, an actor may have more than one equally salient goal (as might be the case when an individual both wants to persuade his or her partner and yet maintain a sense of harmony). To the extent that feedback from the other modifies the salience of the actor's goals, it is difficult to assess success and failure. In such circumstances, it is possible, of course, to ascertain whether influence is positive (produces favorable consequences) or negative, but it is arguable whether such information is relevant to the assessment of power. Power assessments require P to set out intentionally to produce a given effect or range of effects on O.

Resistance overcome

Power is gauged by the degree of challenge that P is able to surmount. Just as the full extent of a child's spelling is only knowable by the most difficult words the child is able to spell, a person's power is evident only in the amount of resistance the individual is able to overcome. Several sorts of data can be used to estimate the level of resistance. Resistance sometimes can be seen in O's response to P's efforts, as when O argues with P, hesitates, or otherwise pulls against O. Knowledge of O's likes and dislikes, either at a particular time or in general, forms a basis for inferring likely resistance. Information that other P's have tried without success to influence O in a particular way can also be used to gauge the amount of resistance. In some close relationships, there is little resistance to influence. Close relationships are formed and maintained, at least in part, as a consequence of the compatibility of the partners (see Chapter 8, "Development and Change"). Similarly, partners in close relationships may change as a result of the resolution of conflict in ways that make them more compatible. When partners are highly compatible—both in the ends they seek and the means they prefer to use to achieve such ends—few occasions crop up for power to be displayed. Both partners are apt to find little need either to try to influence the other or to resist the other's influence. These circumstances make it difficult to see power in operation.

Assessing Power

Assessments of power in close relationships have examined how partners resolve conflict-of-interest situations. Our discussion of assessment procedures is broken down into those that rely on subjective judgments of who prevails in conflict-of-interest situations and those that use objective assessment procedures. Then, we consider some reasons behind discrepancies between subjective judgment and objective assessments and suggest ways the two approaches can be combined to provide a more solid basis for making attributions of power.

Subjective evidence

Research concerned with attributions of power date from the Lippitt et al. (1952) study of influence patterns in a boys' camp. The investigators had campers rank each other in terms of how good they were at getting the others to do what they wanted them to do. Several studies, beginning with Heer (1958), have asked participants in close relationships to provide a retrospective general indication of the relative success of the partners in influencing each other in conflict-of-interest situations. Heer (1962), for instance, asked working wives: "When there's a really important decision on which you two [meaning the wife and her husband] are likely to disagree, who really wins out?" (pp. 65–66). Such an estimate may involve a complex calculation, which begins with memory for the relevant decisions and the outcomes of the disagreements and which requires the wife to combine information to come up with an overall assessment. It is not surprising, given the nebulous nature of this kind of task, that, when Turk and Bell (1972) asked husbands, wives, and children who generally prevails, they arrived at the same opinion only about half the time.

Another approach to assessment, used by Peplau (1979), has dating partners indicate how they believe particular hypothetical conflicts of interest would be resolved. The balance of power is indicated by the proportion of times the conflict is said to be resolved in one or the other partner's favor. The following situation typifies those used by Peplau:

> You and ________________ are trying to decide how you as a couple will spend the weekend. You really want the two of you to go out with some of your friends, but ________________ wants just as strongly for the two of you to go out with some of [his/her] friends. Obviously you can't go out with both sets of friends at once. Who do you as a couple decide to go out with?

The women generally believed that they would be the ones to give in, while their boyfriends were about equally likely to believe the conflict would be resolved either way. Peplau's results showed, in general, that men were perceived as more likely to prevail than women, but this perception was more evident in the women's responses than the men's. One explanation for this divergence lies in the possible disinclination of men, for reasons of modesty or out of a sense of morality, to acknowledge greater power. The overwhelming majority of the men (87 percent) in the study professed an egalitarian philosophy, and consequently they may have been motivated to avoid the suggestion that conflicts would ordinarily be resolved in their favor.

Data gathered by Olson and Rabunsky (1972) serve to caution those who might take at face value subjective reports of how conflict is apt in the future to be resolved or how in the past it has been resolved. These researchers brought husbands and wives into the laboratory and had them resolve a number of issues on which they were led to hold different positions. Prior to the discussion, the researchers asked the participants to predict whether their own view or that of their spouse would prevail should there be a difference of opinion. Two to three weeks after the issues had been resolved in the laboratory session, the participants were asked to recall whose view actually had prevailed. Findings indicated that neither predictions of power nor retrospective reports were accurate. Both were related, however, to respondents' ideas about which partner had the legitimate right (the authority) to make each of the particular decisions.

Other investigators have tapped subjective aspects of power in indirect ways, such as by asking wives to report who makes each of a number of decisions (e.g., Blood & Wolfe, 1960) or by asking children to indicate "Who's boss, your mother or father?" These procedures do not inquire directly about the extent to which the partners get their way in conflict-of-interest situations. Implicit in the Blood and Wolfe (1960) procedure, if it is to be viewed as a measure of power (rather than an indicator of authority), is the idea that the partners have equally strong interests in controlling each of the decision domains. The partner who manages to control decisions in most domains, accordingly, is regarded as having the most power. Turk (1974) suggests that this assumption is dubious, at best, and researchers using the technique (e.g., Centers, Raven, & Rodrigues, 1971; Safilios-Rothschild, 1967) have shown that estimates of the relative power of husbands and wives vary, depending on the decision domains tapped and whether the respondent is the husband or the wife. Safilios-Rothschild (1969), for example, found that the wives in her study attributed more power in decision making to their husbands than their husbands acknowledged for themselves. Part of the discrepancy may lie in the fact that family decisions are made in several steps, extended in time, and it is possible the husbands and wives thought of different aspects of the overall process when they came up with their global reports.

Consumer researchers (H. L. Davis & Rigaux, 1974; Wilkes, 1975) have attempted to identify phases of family decision making and to determine the points at which husbands and wives have influence on products and services the family purchases. Wilkes (1975) suggested that family influence in decision making can be broken down into four phases: (1) Who was responsible for recognizing the need or problem? (2) Who was responsible for acquiring information about the purchase alternatives? (3) Who made the final decision as to which alternative should be purchased? (4) Who made the actual purchase of the product? When decisions are broken down into phases, it becomes evident that the roles of the husband and wife change from one phase to the next. Moreover, this research showed that husbands and wives within particular families hold similar perceptions about their relative influence for a given phase of the decision process. What remains to be determined, however, is whether they aggregate information about the phases differently in coming up with overall estimates.

A procedure for measuring power developed by Szinovacz (1981) overcomes some of the ambiguities of interpretation that result from asking only who controls what decision areas. Szinovacz asked married partners to indicate whether they would like themselves or their spouses to carry out various family-related decisions, such as those regarding expenditures for food, the choice of the car, and purchases of furniture. Szinovacz also had them report who actually carried out each of the decisions. From a comparison of the desired with the perceived actual pattern, Szinovacz derived scores indicative of the degree to which the respondents felt that they had been able to implement their desires. This approach improves upon those that focus exclusively on global reports of actual decision-making patterns, but ambiguities of interpretation remain. Both partners are likely to be able to implement their desired pattern without experiencing interference from the other if the partners have quite similar ideas about who ideally ought to make what decisions. High consensus regarding ideals, then, can produce for each partner a high correspondence between the ideal and the actual. To correct for this consensus factor and get a more direct measure of power, it would be necessary to identify decisions in which the partners' ideals conflict and to assess the extent to which each partner is able to implement his or her ideals in regard to these issues.

Apart from questions of accuracy, however, subjective judgments of power are useful because they tap the experiential world of close relationship partners. The power-related activity of partners may be understandable, at least in part, by reference to the partners' ideas about how much power they have in the relationship. Thus, beliefs about power may constitute causal conditions that affect power usage, an idea that will be discussed in a later section.

Objective assessment

Objective assessment devices attempt to establish empirically whose position prevails in particular conflict-of-interest situations. As Kelley (1977) notes:

> Power is taken to mean something like causing the final decision, and . . . it is usually assumed to be a personal property. Thus, the attribution of power requires

> detecting instances where one or the other person can be inferred to be responsible for the final decision. Such instances exist when, before the discussion, two persons have opposing preferences. Then, the decision itself can be compared with those opposing prediscussion preferences and is credited to the person to whose initial preference it conforms. In other words, the investigator assumes that the person whose initial preference more closely matches the final decision influenced the other to change toward him. (pp. 105–106)

Strodtbeck (1951) initiated this line of research when he introduced the "revealed difference technique" to study power. His task required husbands and wives to identify three reference families with whom they were well acquainted. The spouses were then separated and asked to indicate for each of the the 26 statements which family most fit the statement. Thus, they were asked to designate the family having the happiest children, the family which is most religious, ambitious, and so on. After both partners had individually marked their choices, they were brought together, shown their disagreements, and, in cases where they disagreed, asked to agree on a choice. Strodtbeck's (1951, 1954) general procedure has been refined and adapted by a number of students of power (e.g., Caputo, 1963; Hadley & Jacob, 1973, 1976; Kenkel, 1957; Olson, 1969; Olson & Rabunsky, 1972; Szinovacz, 1981; Turk & Bell, 1972).

Perhaps the most sophisticated effort at refinement has been accomplished by Olson (1970), who designed a version of the revealed difference technique that brings couples together to discuss 18 vignettes pertaining to issues usually of relevance to young marriage partners. For 12 of the 18 vignettes, husbands and wives read versions of the vignette that are slanted to induce them to adopt differing views regarding who was at fault and how the problem ought to be resolved. The remaining 6 vignettes are not slanted differently for the spouses so the partners generally reach the same conclusions. Olson (1970) assessed the split-half reliability with 200 couples and found a moderate correlation (Spearman-Brown, $r = .41$); when only 9 of the 18 vignettes were used, the reliability was even lower ($r = .26$).

The low internal consistency suggests that, in general, neither partner typically prevails when they disagree. This finding could be true for a number of reasons. For one, if partners have about equal amounts of power, who prevails in one discussion would not predict who will prevail in any other. Second, the power balance may differ depending on the topic of disagreement. Thus, one partner may consistently prevail when there is a disagreement regarding how to allocate discretionary funds, whereas the other might prevail when the issue pertains to child care. Third, not all issues will be of equal importance to both participants. And fourth, as Szinovacz (1981) notes, the more powerful partner in intimate relationships probably refrains from exercising power consistently over all task situations. Indeed, choosing when to win is an important element of the effective exercise of power.

The congruence of a group decision with the initial view of one rather than the other partner in and of itself does not provide a clear basis for making an attribution of power. A person may simply have a change of mind independently of the efforts of the other. Or, as noted above, partners may differ in their level of commitment to their positions prior to the discussion (Turk, 1974). For these reasons, it is necessary that the investigator gather information about the decision process itself with an eye toward establishing the causal connections between the actions of the people involved and the joint position they adopt.

Strodtbeck (1951), in his original study, did just that. He recorded his conversations and found that, in general, the spouse who talked most won most of the disagreements. To gain further insight into the interpersonal processes involved, Strodtbeck examined cases in which there was a significant difference in the number of acts initiated by the marital partners. The spouse who talked the most tended also to be the one who most frequently asked questions, stated opinions, carried out an analysis of the issue, and made rewarding remarks. In contrast, the person who talked less tended to vacillate between agreements and disagreements, as well as to engage in aggressive acts designed to deflate the spouse's status. "Taken together," Strodtbeck (1951) notes, "these characeristics suggest the passive agreeing person who from time to time becomes frustrated and aggresses" (p. 473). In this same study, Strodtbeck found cultural differences in the way influence as exerted. Navajos did not attempt to reason with their partner, but rather exerted influence by reiterating their position and by imploring their partner to go along with them. Mormons and Texans drawn from a farming community, on the other hand, tried to resolve their differences through reasoned argument.

Subsequent research has not been directed toward identifying the connection between interpersonal events within a particular discussion and the outcome of that discussion. Instead, investigators have assessed the connection between who prevails, in general, and their behavior in group discussion, also in general. Turk and Bell (1972), for instance, had families engage in two types of decision-making tasks. The first task was based on a procedure developed by Kenkel (1957) in which families had to decide how they would spend a gift of $300. Power was assessed by the proportion of items to be purchased that each family member initially suggested. A second index used was based on the revealed difference technique. Families discussed two issues, and power was ascertained for individual family members on the basis of the extent to which their prediscussion views were different from those of the others in the family and whether they were adopted as the family's position. For each index of power, correlations were run to determine the extent to which the relative power of the husband and the wife was reflected in various features of the discussion, such as who initiated more conversation; who seemed to lead the discussion—as would be indicated by giving suggestions, stating opinions, or confirming the opinions of others; and who was more likely to interrupt other family members. Turk and Bell (1972) found no connection between any of these features of the way the participants took part in the discussion and their power as measured by whose position prevailed, using either of the two types of decision tasks.

Hadley and Jacob (1973, 1976) also have attempted without success to identify the patterns of interpersonal behavior that account for power as measured by the extent to which one spouse's prediscussion view prevails over the other's in arriving at a group position. In the more comprehensive of their studies, Hadley and Jacob (1976) had family groups (consisting of a husband, wife, and teenage son) discuss five issues. Prior to the group discussions, each family member was asked to answer two questions regarding each issue, with each question having five ordered alternatives. On each issue, the individual responses for one of the questions but not the other were revealed to the group prior to discussion. The first procedure is similar to Strodtbeck's (1951) revealed difference technique, and Hadley and Jacob (1976) suggest that the procedure is tantamount to asking family members to say who has the power in the family. The connection between the initial positions of each family member and the final group decision is within the public domain. The procedure in which individual responses are not shared allows the participants to define their own choices as the family's. The rank order of power of the husband, wife, and son on the measure using the revealed difference items was unrelated to the rank order of power using the items in which the initial differences were not revealed prior to the discussion.

Hadley and Jacob (1976) were interested in correlating the way in which the participants took part in the discussions with their power, as measured by whose views generally prevailed. They assessed the total time in seconds that each person talked during the entire set of discussions; they also recorded the total number of times each successfully interrupted another speaker, as well as the ratio of successful interruptions to total attempts at interruption. These measures were seen, following the lead of Strodtbeck (1951), Chapple (1962), Mishler and Waxler (1968), and others, as indicative of the relative degree to which the participants were able to control the discussion. The data, however, showed no connection between talking time, successful interruptions, and ratio of successful to total interruptions and power assessed by either the revealed difference technique or the technique in which the differences were not revealed.

The general failure to uncover the interpersonal mechanisms through which influence is exerted has led some students of power (e.g., Hadley & Jacob, 1976; Olson, 1977; Turk & Bell, 1972) to suggest two distinct types of power—"outcome" power, which is measured in terms of whose view prevails in conflict-of-interest situations, and "process" power, which is assessed by reference to patterns of interpersonal events. Such a solution effectively sidesteps the issue of establishing the validity of either assessment procedure as a measure of power. We would insist, instead, that assessment devices should include information that correlates the interpersonal activities that occur in discussion with the outcome. The discovery of such correlations serves to verify that the change in position was caused by the activities of the person whose position was initially similar to the group position.

Given this view, it is necessary to consider why many of the studies that have attempted to discover such correlations have failed to do so. Most of the research has examined patterns of behavior of couples across discussions as such patterns covaried with the overall extent to which one partner prevailed in getting the other partner to adopt his or her view. Olson's (1970) data showing that who prevails is not consistent from one discussion topic to the next suggests the problem of combining behavior of people across discussions without attention to whether they prevail. Earlier, we mentioned several reasons for the inconsistency in who prevails, and these reasons suggest that the behavior of the participants will also change. Suppose that, in a particular discussion, one participant is more committed to his or her prediscussion position; if so, we might see little resistance to the persuasive efforts of the committed partner, even though the acquiescing person might generally be the one who is powerful. Sometimes the individual who makes the greatest attempt at control is in a weak position. The powerful person, in contrast, may sometimes prevail simply by stating a position. The search for general connections between influence modes and decision outcomes is difficult for another reason. Some couples may characteristically use power assertion tactics much like those used by the Navajo in Strodtbeck's (1951) study; others may generally attempt to resolve their differences didactically. Depending on the mix of couples in a particular study, results could conceivably show a positive, inverse, or no correlation between such factors as talking time and the congruence of the decision outcome with the predecisional positions of the partners.

These observations suggest that future research might begin with careful and detailed probes into the interpersonal events that occur within particular episodes in which partners have differing initial points of view and resolve their difference in favor of one or the other. Such work might not only observe actual interpersonal patterns but also ask the participants to analyze the happenings in terms of influence. This approach may identify types of key events that seem to make a difference, and then it can be determined through more systematic work which types of events are generally important and which ones might sometimes be important, depending on the issue, the type of power involved, and the like.

Reasons for discrepancies between measures

It is not clear to us that one can speak of the relative accuracy of the various methods. This is possible only if one can specify one measure as a "criterion," as in practice Olson and Rabunsky (1972) do. But one investigator's criterion will be another's predictor. For example, why is not the person's belief about who is boss an important "criterion"? It certainly should be if it reflects causal conditions that have implications for understanding the entire history of influence in conflict situations even though it does not predict certain "outcome" or "decision" events assessed in the laboratory. Global reports of power, as noted earlier, are known to be correlated with the respondents' beliefs regarding the appropriate authority structure of the family (Olson & Rabunsky, 1972; Turk & Bell, 1972). For this reason, it has been suggested that reports of actual influence are systematically biased. However, it is possible that when people believe in the legitimacy of a patriarchal or matriarchal authority structure, the belief is causally linked to actual influence patterns.

The decision-making episodes that researchers observe in the laboratory are embedded within a larger decisional context. Olson and Rabunsky (1972) found that the resolutions of issues as they occur in the laboratory are not predictive in any sense of the decisions couples subsequently report having implemented. If we can take these data at face value, we would have to conclude that the laboratory decisions are merely provisional and insist that couples be followed until decisions are actually implemented. The laboratory situation does have an artificial quality to it. Couples ordinarily discuss a series of decisions at one time, and, therefore, the discussions probably take on a hypothetical air.

Laboratory situations also are artificial in the sense that nontypical causal conditions impinge on the interaction (O'Rourke, 1963). The couple is "on stage," and, as Safilios-Rothschild (1970) notes, this limits the types of influence that are apt to occur. Some techniques would not be observed in the laboratory for any of the following reasons:

> (a) because of their intimate nature (e.g., sexual relations or affectionate behavior); (b) because of the "optimum" timing required for their application (e.g., "when she is in a good mood," or "when things have turned out very well at work"); (c) because the application of technique requires the performance of some special tasks (such as cooking some special food that he likes or buying her a hat or blouse . . .); or (d) simply a long time and repeated application (e.g., "nagging" or repeating arguments till [the spouse] gets tired). (p. 546)

The presence of observers also may result in the partners attempting to present themselves in a particular fashion, a fashion that might be different from that evidenced in more private settings. The sex of the observers has been shown to affect patterns of influence in close relationships (Kenkel,

1961). Laboratory settings also remove the couple from their familiar surroundings, particularly the social environment in which they typically interact. Thus, if one partner frequently exerts influence by forming social coalitions (with other family members, for example), this partner's influence is apt to be underestimated in the laboratory. These considerations, taken together, suggest some of the reasons why laboratory solutions to conflict subsequently may be reversed.

Given the problems with laboratory-based data, the temptation to gravitate toward self-report measures is high. We should not be unmindful, however, of the interpretive problems associated with reports of power provided by the participants. Little is known regarding the ways in which people come to conclusions regarding who has the power. Do they think back to particular episodes in which their wants and their partner's wants conflicted, remember who prevailed, and then tally the "wins" for each partner to come up with an overall estimate? Accurate reports of power not only require memory for whose position won out, but also awareness of when the changes that occurred were the result of social influence rather than some other cause. Our earlier discussions of subjective report procedures suggest that they are subject to numerous biases, some of which have to do with faulty memory, others with misattributions of causation, and still others with the way people aggregate events and come to general conclusions. Further discussions of these issues can be found in Chapter 11 on research methods. It should be recognized that the topic of power can arouse the sensitivities of close relationship participants. It is often an important matter to those involved as to who has the most power, and many people are exquisitely aware that others evaluate them by reference to whether they are involved in a relationship of equals or whether they have more or less power than their partner.

Bases of Power

In our discussion of Figure 5.1, we suggested that power is a construct that serves to abstract the common relevance of a variety of causal conditions to the intentional influence that occurs within a relationship. Several bases of power can be distinguished in terms of the causal conditions involved.

French and Raven (1959), in their now classic paper on power and influence, identified five such bases: (1) legitimate power, (2) reward power, (3) coercive power, (4) expert power, and (5) referent power. Later, Raven (1965) added a sixth type, informational power. This differentiation is just one of many that have been developed (e.g., Jahoda, 1959; Kelman, 1961), but it has proved durable and useful because it brings together important ideas about the bases, modes, and consequences of power.

We will consider each of French and Raven's bases in turn:

Legitimate power

Legitimate power is defined as "that power which stems from internalized values in [O] which dictate that [P] has a legitimate right to influence [O] and that [O] has an obligation to accept this influence" (French & Raven, 1959, p. 159). To the extent that P's and O's ideas are culturally derived, they can be seen as representing an incorporation into the relationship of cultural representations. Presumably, as P and O interact, each discovers the extent to which the other understands the authority structure in the same way. With time, they will work out their own set of understandings. These understandings may reflect a compromise between each of their own initial ideas. P × O norms, then, may be derived from traditions within the larger culture, or they may evolve out of interaction. Husbands and wives incorporate from cultural traditions certain beginning assumptions and then work out between themselves a division of labor (see Chapter 6, "Roles and Gender"). The division, once established, has a moral force to it, and the influence is evident in the patterning of behaviors, beliefs, and attitudes of the partners. When participants deviate from agreed-on norms, they are reminded of their responsibilities. If the reminding in and of itself is enough to bring them into line, then the force behind the compliance exemplifies legitimate power. The mode of influence when power is anchored in legitimate authority will frequently involve reiteration of the rule. With time and experience in the relationship, P and O eventually learn the rules so that few occasions crop up in which it is necessary to reiterate them.

The fact that a particular pattern of behavior is consistent with culturally traditional patterns does not necessarily imply that legitimate power is in operation. Such patterns may be maintained by the administration of positive and negative sanctions. When the reasons for compliance are internalized beliefs about how people occupying particular positions ought to think and behave, however, then compliance is based on legitimate power and it ought to occur independent of whether the other person can observe it and directly reinforce it. Legitimate power allows the individual to exert influence in those domains in which the participants agree that such influence is proper and right. Within a particular close relationship, the partners may have an equal or unequal overall amount of legitimate power, or they may have distinctive domains of authority.

Reward power

Reward power is based on P's ability to reward O. The magnitude of P's reward power depends on the extent to which P is able to produce positive affective events or to facilitate O's goal achievement. Reward power also depends on whether an individual can remove unpleasant events or circumstances that block O's goal achievement. Rewards can be mediated directly—as when P gives O attention, approval, or sexual favors—or they can be indirect—as when P controls access to a commodity O would find rewarding.

Some behaviors have widely generalizable rewarding effects. Expressions of approval generally reward the recipient (at least more so than disapproval). But even the rewarding value of approval depends on such things as its basis (in fact or fiction), the status of the source, the self-confidence of the recipient, and the apparent sincerity with which it is given (Jones & Wortman, 1973; Mettee & Aronson, 1974). A person who is deprived of approval is apt to be more rewarded by it when it is given (Gewirtz & Baer, 1958), and, similarly, approval is more likely to be rewarding when an individual is locomoting in a social environment that provides approval begrudgingly. These observations resonate to the general point that the ability of one individual to reward another may be anchored in any number and combination of P, O, P × O, E_{soc}, and E_{phys} conditions.

Reward power occurs when an agent induces a target to comply after (1) indicating one way or another that the agent would like the target to behave in a particular manner and (2) promising (either directly or implicitly) that, should the target comply, rewards will be forthcoming. Given that the term *power* implies that the target normally would not do what the agent wants the target to do, French and Raven (1959) suggest that the stability of the changed behavior pattern elicited by the powerholder depends on the powerholder's opportunity to observe whether compliance continues to occur. Should the behavior that is initially elicited through promise of rewards prove intrinsically rewarding or to be connected to states of affairs that are rewarding, then the behavior will become independent of the agent. Parents are known to use reward power to induce their children to do something with the belief that, once they try it, they will find it rewarding.

Rewards may induce public compliance but fail to produce change in behaviorally relevant attitudes. This result might be particularly common if the rewards consequent to compliance are particularly salient. Reward power that is exercised implicitly, with a sense of subtlety, and over a long period of time, ought to result in friendly interaction, a desire on the part of the target to exhibit the rewarded behavior, changes in the target's attitudes consistent with the public behavior, and the target's identifying with the agent.

When these changes in the target occur, the influence process will "go underground." What we mean by this is that the exercise of reward power will have altered conditions in the relationship and in the target in such a way that an observer watching the couple interact without knowledge of the history of their relationship might not see the influence process in evidence. The individual subject to reward power will have internalized the relevant contingencies governing the willingness of the powerholder to withhold or confer rewards. When this happens, the powerholder may not have to cue the target about acceptable behavior, but rather the target will learn to anticipate the powerholder's desires and spontaneously conform with them in anticipation of thereby receiving rewards. The exercise of power in previous episodes carries over into the behavior of the target in the episode being observed. Thus, we observe events that reflect the temporal reach of previous influence. As we mentioned earlier in this chapter, such carry-over effects ordinarily imply a change in P, O, or P × O causal conditions. A change in the private attitude of the target illustrates a change in a causal condition that results from the previous exercise of reward power.

This manifestation of reward power was recognized in the social science literature as early as the mid-1930s (cf. Friedrich, 1963; Nagel, 1975), and it has been central to many subsequent conceptions (Gamson, 1968; Oppenheim, 1961). Nagel (1975), in his book on the descriptive analysis of power, goes so far as to suggest that power in ongoing organizations usually operates through implicit influence, or "rule by anticipated reactions."

Coercive power

Coercive power is present when P can mediate punishments for O. Coercive power depends on P's ability to cause negative affective events or to otherwise block or interfere with O's goal achievement. Punishments include behaviors indicating disapproval, verbal abuse, and physical violence—actions that run counter to or interrupt O's organized chain of activity. The degree to which particular behaviors are punishing depends on many of the same considerations outlined above regarding rewards. Disapproval, for instance, is punishing in different degrees, depending on what it is for, who it comes from, and the like.

In exercising coercive power, the agent indicates, sometimes subtly, what is wanted of the target, warning at the same time that noncompliance will result in punishment. The threat of physical abuse need not necessarily be made explicit for an individual to have coercive power. Demands emanating from a physically imposing and aggressive individual, for instance, may be complied with because the agent's ability to use coercion is clearly evident.

The use of coercive power has some general similarities to reward power. The use of reward power also involves the agent's providing some indication of what is wanted of the target but, instead of threatening punishments for noncompliance, promising rewards for compliance. Another important similarity between reward and coercive power, as described by French and Raven (1959), is that maintenance of the change (i.e., compliance) depends on the continued observability of the behavior by the agent. In the case of rewards, however, the individual subject to the powerholder's influence ought to maximize the visibility of compliant acts in order to elicit rewards and because the agent, over time, acquires reward value as a consequence of secondary reinforcement (Lott & Lott, 1968). With regard to coercive power, in contrast, the individual will try to avoid showing behavior to the agent. Coercive power, then, in order to be effective, requires that the agent threaten the target with punishment for both noncompliance and concealment (Kelley & Ring, 1961; Ring & Kelley, 1963).

Referent power

Referent power is based on the identification of O with P. By identification, French and Raven (1959) meant that O feels a sense of oneness with P or desires such an identity. O's identification with P can be established or maintained if O behaves, believes, and feels as P does. It is possible, then, that P may have the ability to influence O even though P may be unaware of this potential power. Referent power may be the preferred basis of influence in close relationships, as is illustrated by the following observation wistfully made in a recent novel by a man to a woman he has loved for a long time: "I never had, or wished for, power over you. That isn't true of course. I wanted the greatest power of all. But not advantage, or authority" (Hazzard, 1980, p. 333). The context of the quote suggests that the speaker had in mind power based on the woman's attraction to him—her desire to be, do, and feel like him. People adopt the suggestions and ideas of people with whom they identify, and the man wanted to have power based on the woman's feeling such an identification with him.

Since the individual subject to referent power desires to be similar to the powerholder, the possession of such power reduces the frequency with which the agent will attempt to influence the target by direct means. Moreover, when disagreements between partners arise, referent power increases the readiness of one person to accept the position of the other. To the extent that participants in close relationships identify with one another, their referent power will reduce the frequency with which they are involved in open power struggles. Extreme disagreements, of course, erode referent power.

Expert power

Expert power is based on O's attribution of superior knowledge to P. This type of power may be based on reputational information (E_{soc}), as when people in an individual's social network indicate that a particular person is knowledgeable about x, y, or z. Or it may be based on credentials, such as a medical or law degree. In marriage, it is typical for each spouse to have spheres of expertise in which each can gain the other's ready acceptance without any need to provide reasons or explanations. In a traditional marriage, for example, the husband may unhesitatingly accept the wife's advice about where to buy steaks for the office barbecue, and she may similarly accept, without question, his recommendation as to the best route for her to take in driving to visit an aging relative. "Wherever expert influence occurs," according to French and Raven (1959), "it seems to be necessary both for [O] to think that [P] knows and for [O] to trust that [P] is telling the truth (rather than trying to deceive [O])" (p. 164).

A target person subject to an agent's expertise would change behaviorally and would also change in beliefs; thus, the changes induced by the expert are likely to carry beyond the immediate context within which the agent exercises influence. The use of expert power frees the agent from the task of surveillance that would be necessary if reward or coercive power were utilized.

Informational power

The sixth type of power resides in information P possesses rather than in the P–O relationship or O's attributions to P. The informational content of P's communication must enable O to gain a new understanding of some problem. Thus, P may point out contingencies of which O has been unaware and thereby enable O to see a situation in a different light. In the example above, if the husband explains the pros and cons of various alternative routes so that the wife herself recognizes the preferred one, we would say that he has exercised informational rather than expert power. Rather than saying "I'm doing this because that's what my husband suggested and he knows best" (expert influence), the wife would say 'I'm doing this because I understand why it is best" (informational influence).

A conflict in which the outcome is dependent on informational power would have the elements of a rational debate, with each partner presenting "packages" of information for the other to consider and each attempting to ferret out the meaning, relevance, and importance of the information that the other provides. Spouses may specialize in the domains within which they acquire information, and hence each will have convincing material available to buttress their views regarding some matters more than others. For example, a wife who is knowledgeable about automobiles will be able to explain to her husband the correctness of her diagnosis of a problem with their car. Informational influence is likely to produce change in both cognitions and behavior, and the changes are apt to persist beyond the immediate situation in which the information is imparted.

The French and Raven (1959) typology is instructive to students of power in close relationships for several reasons. First, it draws attention to the idea that different modes of influence are connected to the types of power. Persons possessing expert power merely need to state their position on issues about which their expertise is relevant; individuals who have reward or coercive power will offer promises or threaten the target; those whose power rests with the information they can bring to bear on a matter will try to influence the target by explaining their views.

Second, the causal conditions that account for power depend on the type of power involved. Thus, physical size may have much to do with coercive power, for instance, but have little bearing on expert power. Third, the types of power differ in magnitude of influence. Informational power changes the cognitive system of the target (an O condition), and, therefore, the changes endure and are not dependent on the surveillance of the agent. Reward power and coercive power, according to French and Raven, have more impact on behavioral compliance than on private attitudes. Fourth, the types of power differ in their versatility. Expert and legitimate power both have circumscribed domains; referent power, in contrast, has a broad domain. Fifth, the types of power are causally interdependent. Reward power can increase referent power, while the employment of coercive power ordinarily reduces referent power.

These observations lead to the conclusion that the consistency with which one partner rather than the other prevails when they have conflicting interests will differ depending on the magnitude of each partner's power, the type of power each has available for use, and the types of conflict situations they confront. An individual may consistently prevail in some kinds of situations, but not others, and, depending on the situations sampled by the researcher, a misleading portrait of the power structure of the marriage could be drawn. The fact that modes of influence differ by the type of power being exercised also needs to be taken into account when an effort is being made to correlate features of the influence process with the outcome of the conflict.

In the next two sections, we shall examine two major literatures bearing on power and influence and relate them to our discussion of the bases of power. We will begin with a consideration of power in the marriage relationship and then turn to research covering influence in parent–child relationships. It is useful to consider these two literatures in juxtaposition because they focus on different aspects of power. The research on power in marriage considers the connection between causal conditions and power as evidenced in marital decision making. Thus, the focus is on the downward arrows in Figure 5.1. The research on parent–child relationships, in contrast, has been devoted primarily to the consequences of the exercise of parental power on the development of children. This research draws attention to the fact that the type and extent of parental power has an impact on the development of the child, as shown in Figure 5.1 by the arrow from the bottom of the figure (the influence processes) back to the top (the causal conditions, which include the child's abilities, propensities, and so on).

AUTHORITY, RESOURCES, AND MARITAL POWER

Blood and Wolfe (1960), in their well-known book *Husbands and Wives*, presented two points of view regarding the origins of marital power, and much of the subsequent thinking and research has attempted to refine these viewpoints and gather data bearing on their relative merit as explanations of marital power situations: (1) The *normative* perspective posits that marital power is anchored in culturally derived ideas about authority and that such power is reflected in who is given the right and responsibility for making various group-relevant decisions. This perspective suggests, then, that marital power is largely a consequence of what French and Raven (1959) would refer to as legitimate power. (2) The *resource* perspective suggests that the resources of the husband and wife undergird their amount and relative degree of power. Resource was defined earlier by Wolfe (1959) as "a property of one person which can be made available to others as instrumental to the satisfaction of their needs or the attainment of their goals" (p. 100). He went on to note that resources can be used positively or negatively, suggesting with regard to the latter that a resource can be used to "contribute to a deprivation of needs or may act as a barrier to the attainment of goals" (pp. 100–101). A person possessing great physical strength, for instance, is capable of inflicting harm, thus depriving others of their need for safety. "Resource" refers to any property that can be used to satisfy or frustrate needs or move persons further from or closer to their goals. Skills, competence, status, and knowledge all may be resources that bear on power. Resource theory seems broadly enough conceived to include several types of power, including reward, coercive, informational, expert, and referent power.

Authority

Blood and Wolfe (1960) used data gathered from wives in the Detroit metropolitan area to ascertain whether the balance of power in marriage is better understood in terms of normative prescriptions conferring legitimate authority to the husbands or in terms of the resources the husbands and wives can draw on. The authors suggested that the American family has a patriarchal tradition and that, even though it has moved toward greater egalitarianism, there ought to be subgroups within the society that adhere to the earlier tradition. They assumed the patriarchal tradition would still linger among farm families, recent immigrants, Catholics, the uneducated, and the old. However, their data did not show that pattern. These groups were no more patriarchal than the others in their sample.

Resources

The attempt to relate resources to the balance of power was more successful. Blood and Wolfe (1960) identified variables they thought relevant to the relative degree to which a husband and wife control resources. Rather than focus on particular attributes tied to specific types of resources—for example, those relevant to reward, coercive, and informational power—they argued that particular demographic variables (e.g., education, income, employment status) are related to resource control and dependency. They reasoned, for example, that the employment of the wife should allow her to develop resources (financial, informational, skills) that would not ordinarily be developed in the home and to make contacts outside the home that provide alternative sources of satisfaction and hence reduce her dependence on her husband. Employment, then, would seem to increase the extent to which the wife possesses reward, coercive, and informational power. Depending on her job and how she uses her power, she might also increase in expert and referent power vis-à-vis her husband. In general, then, employment increases her resources (a P condition) as well as the alternatives she has to her husband for achieving satisfactions (the result of which is a lessened sense of dependency, also a P condition, but anchored in E_{soc} conditions). Such employment, because it increases the wife's value to her husband, ought to increase his dependency on her (an O condition). Any or all of these changes in conditions may serve to tie the employment status of the wife to her power in her marriage relationship.

The wife's employment status was found by Blood and Wolfe (1960) and others (Burić & Zečević, 1967; Kandel & Lesser, 1972; Lupri, 1969; Michel, 1967; Safilios-Rothschild, 1967) to bear on her power in marriage, but the data regarding the matter are not entirely consistent. (Bahr, 1975; Blood & Hamblin, 1958; Weller, 1968). Several other variables that Blood and Wolfe (1960) identified as resources and that they found to relate to power have not fared particularly well. The husband's education, occupational status, and income have been positively related to his power in studies conducted in the United States (Blood & Wolfe, 1960; Centers et al., 1971; Kandel & Lesser, 1972), France (Michel, 1967), Germany (Lupri, 1969), and Denmark (Kandel & Lesser, 1972). In contrast, studies conducted in Yugoslavia (Burić & Zečević, 1967) and in Greece (Safilios-Rothschild, 1967) have found an inverse connection between the same variables and the power of husbands relative to their wives.

Authority Versus Resources

To resolve these contradictory findings, Rodman (1967, 1972) proposed that relative resources determine marital decision making only in cultures that are rather egalitarian. The cultural context (E_{soc} conditions) allows (in egalitarian societies) or prevents (in patriarchal societies) resources (P and O conditions) from having relevance as causal conditions for marital decisions. In strongly patriarchal societies, norms confer authority to the husband independent of the spouses' relative resources. In these societies, such variables as education and occupational prestige are correlated with the husband's exposure to liberal ideas that run counter to belief in traditional patriarchy; hence, his high socioeconomic status is inversely related to his reported power. In Western cultures, however, these same properties are seen as "resources" on which he draws in order to exert influence on his wife.

This analysis of power in Western societies would seem to suggest that power is about equally apt to be on the side of the husband or wife and that cultural causes are of minimal importance in accounting for the distribution of power. But nearly all researchers since Blood and Wolfe (1960) have found that husbands generally wield more power than their wives. Of course, one possible reason for this pattern lies in the husband's greater physical strength (Komarovsky, 1967), but most theorists have suggested that a number of intertwined cultural conditions are also likely to be involved (e.g., Gillespie, 1971). Resources have generally been conceived in socioeconomic terms, both by researchers and spouses; since men are given greater opportunity to acquire such resources, their greater power can be attributed to their favored position within the culture (Gillespie, 1971; Safilios-Rothschild, 1976). To carry the argument further, it has been suggested that it is no accident that the resources men possess more than women are the ones valued in the culture as a whole. Walster and Walster (1976), following Marxist reasoning, suggest that "over time, the powerful persons who control community resources will evolve a social philosophy to buttress their right to monopolize community goods; and, over time, the entire community will come to accept this justification of the status quo" (p. 34, italics omitted). This view suggests that one way for women to improve their power position in marriage would be to obtain more of the socioeconomic resources themselves.

Other Factors in Marital Power

Another way for women to obtain more decision-making power does not involve attacking the socioeconomic basis of the husband's authority. Safilios-Rothschild (1976) has shown with married couples, and Peplau (1979) with dating couples, that the balance of love or involvement within a relationship has an effect on power in decision making. This finding illustrates Waller's (1938) "principle of least interest," which suggests the less interested partner has more control over a relationship. Power based on emotional attachment (i.e., referent power) would seem to be one way women might succeed in eroding the husband's power, but unless major changes in the structure of Western society occur, the power balance in marriage is likely to remain on the side of the husband.

There are other societal supports for the asymmetrical power in marriage. Many of these have been articulated by Gillespie (1971), who has tried to demonstrate that socialization processes, legal constraints, and suburbanization all favor a male-dominated marital relationship. The reader is referred to Szinovacz (in preparation) and J. Scanzoni (1979c) for excellent current reviews of the marital power literature.

This analysis of power in marriage can be used to underscore several important points. The first is that the variables actually measured (e.g., education of the husband) are assumed to be correlated with a number of causal conditions (e.g., exposure to nonpatriarchal notions about family structure; knowledge of the husband regarding issues of relevance to family decision making), which, in turn, are said to account for the reported patterns of decision making. Second, particular variables may be connected to more than one type of power. Thus, education may affect legitimate, informational, and expert power and thus impact on the balance of power in marriage. Third, there is not a consistent relation between many of the variables used in power research and either the causal conditions to which they are correlated or the type of power involved. Thus, in some cultures, education of the husband is positively related to the husband's power, whereas, in other cultures, it is inversely related. And, finally, the causal conditions underlying power are themselves causally interdependent. The resources of the husband and wife, as Rodman (1967, 1972) suggests, have more impact on the distribution of power in cultures that do not have strongly patriarchal norms.

POWER AND INFLUENCE IN PARENT–CHILD RELATIONSHIPS

The influence that occurs in close relationships affects the psychological development of the participants. It is not surprising, then, that most of the work on the long-term consequences of patterns of power and influence has been conducted by developmental psychologists interested in parent–child relationships. This section attempts to illustrate changes in how researchers have conceptualized influence processes in parent–child relationships over the past three decades.

We begin with the research program initiated by Sears and his colleagues in the early 1950s (Sears, Maccoby, & Levin, 1957), which set in motion intensive efforts to discover the socialization consequences of particular styles of parenting. We will then compare and contrast the Sears et al. work with a second approach developed by researchers using an operant behavior modification approach (Bijou & Baer, 1961). The third approach, articulated by Bandura and his colleagues (Bandura, 1962; Bandura & Walters, 1963) elaborates on the first two in emphasizing that much of what children learn from parents is not wholly the result of explicit teaching or training. The fourth approach moves away from viewing parents as agents and their children as targets of influence to a focus on reciprocal influence processes in parent–child relationships. In the key conceptual paper that initiated this line of research, Bell (1968) proposed that much of what parents do with regard to their children is reactive rather than proactive. Adequate analyses of parental influence, accordingly, require an analysis of chains of interaction and a separation of the influence the parent has on the child from that the child has on the parent. Social scientists using this general model have begun not only to examine reciprocal influence on a behavioral level but also to consider ways in which the personalities of both the parent and the child are influenced by their interaction.

The four approaches emerged in roughly the order discussed here. It should be made clear at the outset that the later approaches do not replace the earlier ones, but rather they bring attention to issues that had previously been poorly articulated or underresearched. Each of the paradigms remains viable, and each continues to attract the interests of researchers.

Parental Style of Influence and Its Effects on Children

The work of Sears and his colleagues (1957) focused on the deliberate attempts of mothers to influence their children. The basic approach was to make inquiries of mothers regarding their style of parenting and then to correlate this material with ratings of the children's behavior. Most of the questions Sears et al. asked had to do with parental reactions to children's behavior, and thus their work dealt primarily with the kind of reactive influence illustrated in Figure 5.4. The child does something that elicits, at least among some parents, an intention to do something either to reduce or to increase the likelihood of the behavior recurring in the future. The general framework was grounded on basic learning theory principles, with particular attention being given to parental reactions as shapers of the behavior of children. The parent was identified as the agent of influence, the child as the target of the influence efforts. The idea behind this model is that the parent has an assigned role in the culture to socialize the young.

The parent was seen as relatively preprogrammed to deal in particular ways in response to particular behavior exhibited by the child. Thus, some mothers reportedly used primarily "love-oriented" modes of influence (e.g., praised the child for compliance, withdrew love as punishment), while others used "object-oriented" techniques (e.g., offered tangible rewards as incentives, used physical punishment). Sears et al. were interested in the connection between parental socialization practices and children's personality development. The style of parenting was assumed to modify the child's potentiality for engaging in particular kinds of actions, such as dependency behaviors and aggression. Parents who said they used love-oriented techniques, for instance, also had children they saw as likely to confess or feel guilty after misbehaving. An important assumption behind the Sears et al. work was made explicit by one of the authors of the original monograph when she wrote: "When we select aspects of parental behavior for study, and try to relate these to measured characteristics of the child, we usually measure what we believe to be reasonably pervasive, reasonably enduring 'traits' of the parent and the child" (Maccoby, 1961, p. 368).

Although subsequent work has shifted away from examining differences among parents in the way they attempt to shape the behavior of their children, the rationale behind the original work remains convincing. Parents do attempt in an intentional manner to modify the likelihood of their children's behaving in particular ways. The task for the researcher is to extract from the flow of parent–child interactions those sequences in which the parents make direct attempts to influence their children, to categorize the modes they use, and to assess both the short- and long-term consequences of their efforts on the personality development of their offspring. This type of research has been done, of course, but the results have proved difficult to summarize (cf. B. Martin, 1976, and Steinmetz, 1979, for recent attempts) because they seem to depend on such factors as how variables are measured, the behavior settings in which the data are gathered, and the characteristics of the parents and children (e.g., age and sex) studied. This problem suggests that greater attention needs to be given to the causal conditions that are tied to the choice of influence modes and their likely success. It is not surprising, for instance, that "reasoning" with children as a tactic of influence produces different results depending on the mental capacity (or informational power) of the parent and the child.

Operant Behavior Modification Approach to Parental Influence

The operant behavior modification approach to social learning (cf. Bijou & Baer, 1961) is based on Skinner's (1953) reinforcement principles. The central assumption of this approach is that the behavior of children can be shaped and controlled by varying environmental reinforcement contingencies. This approach has many similarities, at a conceptual level, to that taken by Sears et al. (1957). The focus is on examining the relations of various reactions on the part of the influence agent to the frequency with which the target exhibits particular behavior patterns. The trainer (or parent) reacts to the behavior of the target (the child) according to a preplanned schedule, reinforcing or punishing some behaviors and ignoring others. Here the attempt is to identify responses that serve to increase or decrease the frequency with which desired and undesired behaviors occur, respectively. Thus, the operant behavior modification approach and that taken by Sears et al. (1957) are similar in their focus on intentional influence and in their concentration on the kind of reactive influence shown in Figure 5.4.

The ways in which the two approaches differ center, in part, on the divergent purposes of the research. Sears and his colleagues were interested in individual differences in the ways in which real parents behave in natural environments and in identifying whether these differences were connected with socialization outcomes. The operant behavior modifiers, in contrast, are primarily interested in designing effective ways for agents to influence the behavior of target individuals. Operant behavior modifiers, then, are concerned with helping parents—or others who are interested in encouraging or discouraging particular behaviors on the part of children—to be more effective, or powerful. The major advantage of the operant behavior modification approach is that the researcher has considerable control over the parental behavior pattern and is not forced to rely on parents' summary reports of now distant behaviors. At the same time, the data are based on limited time samples of interaction. Another difference is that parental behavior is generally coded in terms of reinforcements and punishments, whereas Sears and his colleagues made more qualitative distinctions in the behavior of the parent. Behavior modifiers working within the operant tradition have focused on the use of reward and coercive power rather than on the other types of power identified by French and Raven (1959). A final difference between this approach and the researching investigating parental styles of influence is that behaviorally oriented researchers are not interested in personality development. Operant behavior modifiers assume that the stability of the changes they induce depends entirely on the subsequent reinforcement contingencies.

Patterson (1979) is one of the principal behavior modifiers who has attempted to analyze parent–child relationships in naturalistic settings. He contends, on the basis of his data, that the rate with which children exhibit particular types of behavior depends more on the characteristics of the stimulus environment than on reinforcement contingencies. Thus, variations among children in the rate of aggressive behavior are shown to be affected more by the density in the natural environment of the stimuli that control its initiation and maintenance than by variations in reinforcement contingencies. The controlling stimuli, of course, are determined by previous reinforcement contingencies. Patterson's (1977) intervention strategy involves first identifying the controlling stimuli and then training parents to modify the frequency with which the controlling stimuli occur in the natural environment of the child. Stimuli that are strongly associated with the likelihood of particular behaviors are said to be "powerful," and, according to Patterson, a parent can gain power by modifying the frequency with which such powerful stimuli are present in the environment.

Behavior modifiers, as a group, have been particularly interested in the "functional" analysis of behavior rather than in constructing models of parenting styles, and, as a consequence, their analysis focuses on particular units of behavior rather than on the people who engage in the behavior. For this reason, the treatment of parenting seems detached from the parenting role as it is culturally defined. The principles developed are presumed to be relevant regardless of whether the analysis is of parents and children, researchers and laboratory animals, or husbands and wives.

Incidental Influence

The idea that children acquire behaviors from observation rather than as a consequence of selective reinforcement was given prominence in the social learning theory developed by Bandura and his colleagues (Bandura, 1962, 1977; Bandura & Walters, 1963). Their framework maintains the focus of the previous approaches on the parent as the influence agent, but it adds to issues of explicit power and control, consideration of the myriad ways in which parents (and others) influence children unintentionally.

According to Bandura (1962; Bandura & Walters, 1963), the patterns of behavior children observe in models provide the sources for the children's behavior. Bandura's initial formulation did not specify any conditions that would be necessary for learning to result from observation, but later, in a restatement of modeling theory, Bandura (1977) proposed four mediating processes. The extent to which the model's behavior is understood, and hence imitable, depends on whether the child: (1) attends to the model; (2) recalls the model's activity; (3) is motivated to perform the activity; and (4) has the necessary skills to carry out the activity.

Children are exposed to a number of models in the course of a typical day, but they do not imitate all of the behaviors they see. How does a child select models to imitate and what considerations regulate when and where the imitation will take place? Bandura, Ross, and Ross (1963b) found that the likelihood that a child would imitate a model depended on the consequences of the observed behavior for the model. The model's behavior imparts information about ways of behaving that are efficacious (or not), and this information is used by the observer as a basis for deciding whether to imitate the observed behavior. Bandura and Huston (1961) found that children are more likely to imitate a model who had previously behaved toward them in a warm and nurturant manner than one who had been distant and non-rewarding. The extent to which an individual possesses reward power also seems to relate to the inclination of children to be imitative (Bandura, Ross, & Ross, 1963a). This finding fits the results of earlier research conducted by Lippitt et al. (1952), which showed that boys in a camp tended to imitate members of their peer group who had reputations as being powerful, and these people, in turn, were observed in the camp to be effective in getting their own way.

Although reinforcement was not seen as necessary for initial learning through observation to occur, Bandura and Walters (1963) suggested that it plays a major role in the continuation of behavior. People acquire behavior through observation, perform it if they have an incentive, and maintain it if it is reinforced. Parents thus serve as a source of the behavior of their children, and, depending on the nature of their power and the control they have over resources, they also function as agents of reinforcement.

Bandura's approach, then, differs in emphasis from the operant approach in assigning a lesser role in socialization to intentional influence. The social learning theory of Bandura also differs from the more operant approaches in that he is interested in analyzing social development—that is, changes in the dispositions of children to act one way rather than another.

Bidirectional Influence

Bell's (1968, 1971) analysis highlighted how children—particularly infants—influence their caretakers. The behavior of caretakers was said to depend more on the activity of infants than vice versa. Bell's writing shares with Bandura's work a focus on behavioral processes and incidental influence. The distinctiveness of Bell's contribution lies in its emphasis on the bidirectional nature of influence in parent–child relationships. An example of Bell's (1968) point is his insistence that parental behavior is generally altered by distress signals sent forth by the infant.

During the 1970s, a number of investigations followed after Bell by focusing on the ways in which parents' behaviors are influenced by the activity of their infant offspring (cf. M. Lewis & Rosenblum, 1974; Stone, Murphy, & Smith, 1973). Infants were seen as learning a sense of control as they discover that by varying their own behavior they can alter the behavior of their caretakers. As children mature intellectually and begin to be able to anticipate consequences, the experience of incidental influence provides a basis for their development of strategies of intentional influence.

Most of the work examining bidirectional influence in parent–child relationships has concentrated on patterns of behavioral interdependence, but some efforts recently have been made to discover changes in parents' personality that result from interaction with their children. This work complements the tradition of research examining the impact of parents on the personality development of their offspring. A study by Patterson (1981), for example, found that mothers involved in a cycle in which they use coercion to control the hostile behavior of their sons have lower self-esteem than do mothers who are not involved in such a cycle. While this evidence, in and of itself, does not show that the low self-esteem of the mothers resulted from the interactive pattern, Patterson found that mothers' esteem rose subsequent to intervention that moved the mother and child away from the hostile–coercive cycle.

There has also been an increased recognition that the interaction of the parent and child depends on elements in the immediate social context. This may foretell a trend toward causal analysis of parenting behavior. The behaviors of fathers toward infants, for instance, have been shown to vary when the mother is present as compared with when she is absent, and this variation has been linked to sex roles (Parke, 1979b). A coordinating context for this work is provided by differentiating between first-order and second-order effects in family relationships. First-order effects pertain to the impact of any one family member on any other, while second-order effects concern the way in which the relationship of a specific family member with a particular other member is affected by the first person's relationships with others in the same family. For example, the husband–wife relationship affects the relationship of the mother with her son. Bronfenbrenner (1979) shows the ways in which family members are connected through social relationships to people and institutions in the larger culture such that the behaviors of people inside the family are affected by these outside influences. Bronfenbrenner's analysis attempts to connect distal causes, such as conditions in the cultural setting, to patterns of interaction in the family setting.

The efforts of developmental psychologists to analyze parent–child relationships are more differentiated than this brief historical overview of recent trends in theory and research suggests. Our purpose in presenting this material, however, has not been to provide a thorough review, but rather to illustrate how the paradigms conceptualize and assess influence processes. It can be seen that the approaches differ in their emphasis on intentional as compared with unintentional influence, their focus on unidirectional as compared with bidirectional patterns, and their concern with the effects of influence on the changes in personality and attitudes of the participants. Each approach focuses attention on a different combination of aspects of the influence that occurs in close relationships.

CHANGES IN CAUSAL CONDITIONS RELEVANT TO POWER

The power of participants in close relationships may change during their involvement. Such changes logically indicate a change in one or more of the relevant causal conditions. A father who used coercive means of control with his young son finds himself no longer able to employ such tactics successfully once the boy has physically matured. A wife who enters the work force is able then to exert more influence on her husband in conflict-of-interest situations. A daughter gradually becomes increasingly persuasive in discussions with her parents about her use of her free time.

Changes in causal conditions relevant to power can be exogenous or endogenous to the dyad. Exogenous changes refer to those that come from outside the relationship and impinge on it, while endogenous changes arise out of interaction of the partners with one another. Largely exogenous are changes in cultural norms regarding legitimate authority, decline in physical or mental resources resulting from health problems, and repeal of legal statutes supporting a patriarchal power structure. J. Berger, Cohen, and Zelditch (1972) have shown how exogenous status characteristics have an impact on influence processes in decision-making groups. Endogenous changes include norms or "contracts" that emerge through interaction. They also include changes in power-relevant resources that result from a particular distribution of power within a relationship. Endogenous changes are identified in Figure 5.1 by the arrow from the bottom rectangle (interaction) to the top rectangle (causal conditions).

Although it is useful to distinguish exogenous from endogenous conditions that affect power, it must be noted that the two sources of causality are frequently interlinked. For example, a husband may learn over time to rely on his wife in household matters; and she, in turn, may increasingly rely on him regarding financial matters. Such a pattern, which did not exist at the beginning of the marriage, evolved over time. At the start of the marriage, each partner may have been equally competent, but various exogenous reasons led the wife to set aside her career goals in favor of the husband's. A number of cultural conditions (which are exogenous to the interactions between the partners) may have determined how the couple reached this decision. Such conditions include cultural traditions regarding the way in which particular matters ought to be resolved, the differential pay scale for men and women, and pressures from family and friends. Regardless of the initial reasons for the decision, other consequences follow that serve to link the employment pattern to subsequent patterns of influence.

People involved in the work world may develop pools of information and interpersonal skills that bear on their ability to exert influence in the marriage relationship. A homemaker also may develop particular skills based on the homemaking experience. Once these different areas of experience are in place and mutually understood, each partner acquires expert and informational power (and perhaps legitimate authority) over particular spheres of decision making. Thus, a linkage among causal conditions is evident: Social conditions (E_{soc}) bear on the initial decisions of spouses, which then result in each developing a different set of skills and resources (changes in P and O conditions); through time, each partner begins to trust the other's judgment regarding particular matters (emergent P and O conditions), and they develop a set of understandings (emergent $P \times O$ conditions) that give power to each in different decision areas.

This pattern was discussed in our earlier examination of marital power and needs no further elaboration. At this point, however, we turn to an analysis of the ways in which causal conditions relevant to power are changed as a result of interaction. Two types of change can be distinguished: (1) changes in causal conditions that reduce the likelihood that power will be exercised or tested and (2) changes in the amount of power an individual possesses.

The first kind of change is illustrated by what Emerson (1962, 1964) refers to as "cost reduction" processes. According to Emerson (1962), the exercise of power results in the target's engaging in behaviors that are inconsistent or only partly consistent with the target's attitudes and values. To continue to believe one thing and do another entails psychological costs. "Cost reduction," writes Emerson (1962), "is a process involving change in values (personal, social, economic) which reduces the pains incurred in meeting the demands of a powerful other" (p. 35). Changes in values, in turn, are apt to result in an increase in the extent to which the individual spontaneously complies with the desires of the powerholder (Friedrich, 1963), thus reducing the need for the possessor of power to display it.

Mazur (1975) suggests that, early in a relationship, partners test each other's power, and, with time, a status relationship emerges reflective of the balance of power within the dyad. Once the power balance of the relationship is mutually understood, a pattern of leading and following emerges. In relationships in which power is asymmetrically distributed, the participants are likely to be able to answer queries regarding the balance of power with ease, and they are apt to have a clear comprehension of their "place" in the relationship. Levinger (1959) also has shown that the perceptions partners have of one another shift as their relative power becomes evident as the result of the influence pattern in the relationship. Persons who find they cannot get their way eventually give up trying, presumably because the effort is costly and the returns minimal. Falbo and Peplau (1980) provide data indicating that people who feel they have more power than their close relationship partner report using strategies of influence involving persuasion and reasoning; in contrast, people who feel they have less power than their partner indicate they attempt to exert influence by telling their partner their needs, withdrawing, or expressing negative feelings (pout, threaten to cry).

The effective exercise of power produces change in the target. Some of these changes make "power displays" on the part of the powerholder less necessary. Conflict can be a means through which P and O become more compatible. Compatibility can be increased by changes in one or both partners. Increases in compatibility (as measured, for example, by changes in P and O preferences regarding important matters) reduce the likelihood that either partner will feel a need to exert influence on the other. The number of ways in which conflict in close relationships can be terminated is further explored in Chapter 9 on conflict.

One consequence of being in the "down" position in a relationship is a reduced opportunity to develop power-related resources. The powerholder may be able to control access to circumstances that would increase the other partner's power. Husbands are said sometimes to resist the desire of their spouses to work because such involvement would reduce their wives' economic dependence, which, in turn, would alter the balance of power in the relationship. An initial asymmetry in power may be sharpened with time, providing it results in the more powerful member gaining greater opportunity to develop power-relevant resources.

At the same time, exercise of asymmetrical power can prove unsatisfactory, perhaps because the individuals are ideologically uncomfortable with such a pattern and strive to achieve a better balance of power. Deutsch (1975) suggests that aysmmetry of power undermines affectional bonds in close relationships. To the extent that partners recognize this danger, and to the extent it is true, they may seek to reduce any asymmetries that become apparent.

Emerson (1962) seems to suggest that imbalance of power usually is uncomfortable, at least for the person who has the lesser amount of it. Equity theory (cf. Walster, Walster, & Berscheid, 1978), in contrast, postulates that power imbalances are not likely to be disturbing as long as the asymmetry in power is a reflection of asymmetry in "inputs." Along similar lines, sociologists have posited that asymmetries of power consistent with the cultural ideology are less likely to produce psychological tension in the participants than patterns that deviate from cultural norms.

CONCLUSION

The analysis of power enlightens a wide variety of human interactions. In any close relationship, there inevitably occur times at which the interests of the parties are in conflict. How such conflicts are resolved is often a function of power. In this chapter, the aim has been to clarify the ingredients of an adequate conception of the nature of power, to illustrate the ways power affects features of close relationships, and to differentiate among such conceptually interrelated concepts as influence, dominance, and power.

We have chosen a broad-gauged approach to power for several reasons. First, we thought it important that the domain of inquiry be set forth as a context for analyzing the research and theoretical work already done on power in close relationships. This idea reflects our view that a coordinating context for power research will facilitate further efforts at understanding how the antecedents, processes, and products of power might be interconnected. Work has focused on particular aspects of power, studied at a particular time in the history of a relationship, but, until research designs recognize that manifestations of power change as the result of P–O interaction, limited progress is apt to be made in understanding it.

Another reason we have cast a broad net is based on our view that we do not yet know how distal antecedents (e.g., cultural forces) are tied to more proximal antecedents (e.g., personal dispositions, social processes) and how these combine to have an impact on power. The conceptual model, by design, provides a context for examining a variety of types of antecedents of power (e.g., qualities of P or O, cultural norms). It also leaves open the connection between the mode of influence the actor chooses and power, the assumption being that the mode varies depending on the bases of the actor's power, the extent to which the target resists, and the like. We also wished to draw particular attention to the idea that the consequences of the exercise of power go beyond the immediate situation and that the influence that occurs within a relationship alters causal conditions relevant to subsequent influence.

Research that examines forces behind power at many levels and that investigates its personal, interpersonal, and societal consequences would seem necessary before its many and complex nuances will be fully appreciated. To understand power, as Bertrand Russell (1937) wrote, is to comprehend the essence of the human condition.

CHAPTER 6

Roles and Gender

LETITIA ANNE PEPLAU

We often describe a close relationship by depicting its characteristic interaction patterns. One young man portrayed his dating relationship this way:

> I play the reassuring, protective father. She is the faithful, dependent child. Her faith and dependency are a form of reassurance and support for me. But there are days when I would like to come to her, as she comes to me, . . . to tell her that I was hurt because my roommate didn't ask if I'd made Phi Bet . . . or that I didn't think that my professor liked me anymore—but I could never bring myself to talk about such sentimental drivel even though I wanted to. (Cited in Komarovsky, 1976, p. 165)

This description highlights regularities in the interaction patterns of the two partners.

Other descriptions of relationships emphasize consistencies in the individual actions of the partners. Thus, a married woman described her husband's activities and her own in their marriage in these terms:

> [His] life is a lot easier; there's no doubt about it. He gets up in the morning; he gets dressed; he goes to work; he comes home in the evening; and he does whatever he

I am most grateful for comments on earlier drafts of this chapter by Steven L. Gordon and Jean Atkinson.

> wants after that. As for me, I get up . . . get dressed . . . fix everybody's breakfast; . . . clean up the kitchen; . . . get the children ready for school; . . . take [the baby] to the babysitter. Then I first go to work. I work all day; I pick up the baby; I come home. . . . Everybody wants me for something but I can't pay them any mind because I first have to fix dinner. Then I do the dishes; I clean up; I get the kids ready for bed. After the kids are finally asleep, I get to worry about the money because I pay all the bills and keep the checking account. (Cited in L. Rubin, 1976, p. 101)

These two examples illustrate consistent patterns that recur over time within a particular couple.

It is also apparent that the same consistent patterns occur across many couples as well. On any given Saturday night, thousands of young American couples can be found on a date, interacting in somewhat similar ways. In many couples, for example, the boyfriend will pick up his girlfriend in his car, take her to a movie and pay for her ticket, hold her hand during the show, talk about recent events at school or work, and initiate greater physical intimacy on the way home. There will be variations in this pattern across couples, but enough commonality to identify a social pattern.

An important task for close relationship researchers is to describe and explain both consistency and variation in relationship patterns such as these. As Biddle (1979) noted:

> It is a fact that human beings behave in ways that are to some extent predictable and consistent and that their behaviors are similar to the behaviors of others who share identities with them and appear in similar contexts. To study and explain these behavior patterns (or roles) is a key problem of the social sciences. (p. 334)

This chapter examines relationship patterns in couples and families. The term most commonly used by social scientists to refer to consistent relationship patterns has been "role." Like power, commitment, and other basic concepts analyzed in this volume, the role concept has been used in a variety of different and contradictory ways. Despite this problem, however, we find it preferable to use the familiar term *role* rather than to resort to a neologism.

This chapter has two major parts. In the first, we analyze the concept of role and contrast our conceptualization with previous uses of the term. We conclude our general discussion of roles with an examination of the types of causal conditions affecting these patterns.

The second part of the chapter examines sex roles in heterosexual relationships. Gender is one of the most basic social categories around which roles are organized. In American society, girlfriends and boyfriends, wives and husbands, and fathers and mothers behave in somewhat different and characteristic ways. No analysis of heterosexual close relationships can be complete without a discussion of gender-linked patterns. Separate sections briefly review research on gender patterns in dating and marriage; present some of the current explanations for gender specialization in marriage; and consider the consequences of gender specialization for the couple, the individual spouses, and their children.

THE DESCRIPTION OF ROLES IN CLOSE RELATIONSHIPS

The analytic framework put forth in Chapter 2 underlies our analysis of the role concept. This framework emphasizes the importance of distinguishing between the description of regularities in intrachain and interchain patterns and the analysis of the causal conditions that produce these regularities. We use the term *role* in a descriptive way to refer to consistent patterns of individual activity (e.g., behavior, cognition or affect) within a relationship. The causal conditions affecting roles include personal expectations, shared dyadic goals, cultural norms, and other individual, relational, or environmental factors that create, maintain, and change role patterns.

Although the distinction between the *description* of roles and the *explanation* of the causal conditions affecting roles follows logically from our conceptual framework, this distinction has not typically been emphasized by previous role theorists. It has been common, for example, to define roles as shared expectations for behavior, rather than as the behavior itself. A major issue tending to blur the distinction between the description and the causal analysis of roles is the identification of the dividing line between the phenomena that constitute a role and the proximal causal conditions that are intimately associated with the role. Suppose, for example, that a husband regularly brings home his paycheck to support his family and that both he and his wife believe that the husband should be the family breadwinner. The husband's behavior is clearly part of his marital role. But, should the spouses' expectations about the husband's behavior be considered part of the role or a causal condition producing the regularity in the husband's behavior? To emphasize the distinction between description and explanation, we will consider expectations to be a causal condition, not a part of the role itself. But it should be recognized that many role theorists have not followed our course. Indeed, it is common in both lay and scientific thinking to identify a relationship phenomenon by pointing to a presumed causal condition, for example, to define roles in terms of social rules for behavior or other hypothesized causal conditions. We wish to avoid this confusion of description and causal explanation.

The Nature of Roles in Close Relationships

We use the term *role* to refer to a consistent pattern of individual activity that is directly or indirectly interdependent with the partner. Two aspects of this definition require explanation: what it means for a person's activities to be "consistent" and what it means for such activities to be "directly or indirectly interdependent with a partner."

Consistency of role patterns

Having said that roles are consistent patterns of individual activity, we need to specify the nature of this consistency or regularity. The first and perhaps most obvious type of consistency is the *repetition* over time of the same activity. For example, we may observe that day after day a mother feeds, plays with, and talks to her child, and we may conclude that these behaviors are part of her maternal role. However, the consistency of a role does not depend exclusively on the temporal recurrence of similar behaviors. Unique or infrequent events can also be part of a role. For example, giving birth or rushing an injured child to the emergency room can be components of the maternal role. Here, role consistency derives from *cognitive conceptions* (a causal condition) that provide meaning and coherence to diverse behavioral events. Thus, birth might be construed as the key event in initiating the mother–child relationship, and seeking medical help might be seen as symbolic of the mother's continuing love and concern for her child. One's conception (either lay or scientific) of the mother–child relationship and one's beliefs about the mother's motives give coherence to these acts and link them to other recurrent patterns. Role consistency can derive either from temporal repetition or from cognitive conceptions of a role. An implication is that the identification of role patterns may require both repeated observations over time and knowledge about important causal conditions, most specifically people's cognitive conception of their roles.

Direct and indirect interdependence

Not all consistent patterns of individual activity constitute a role. In the case of marital roles, for example, we might intuitively suggest that neither a husband's typical way of brushing his teeth nor a wife's interactions with her bowling league partners are part of their marital roles. Our analytic framework provides a clear criterion for determining which consistent individual activities are part of a particular role—namely, that the individual's activities be interdependent with the partner, either directly or indirectly. Thus, roles involve mutual influence between two or more people.

Perhaps most obvious are roles in which each partner influences the other directly through face-to-face interaction. In Chapter 2, we describe this kind of direct influence as *direct interdependence.* Couples commonly develop characteristic patterns in many domains of interaction—typical greeting rituals, particular styles of fighting, coordinated household routines, and so on. In the example cited at the beginning of this chapter, one couple developed a pattern of interaction in which the boyfriend had the role of "protective father" and the girlfriend had the role of "dependent child."

Figure 6.1 illustrates direct interdependence. In this schematic representation, P's role activities (e.g., being the "protective father") include P's intrachain events and connections (e.g., P's thinking about his girlfriend and wondering what to say next), P's actions toward O (e.g., P's giving advice to his girlfriend), and P's responses to O's actions (e.g, P's answer to O's question). The roles of P and O are comprised of the individual things the persons do, think, and feel. It should be obvious that this figure simplifies the nature of roles in several important ways. Although most roles in close relationships involve lengthy and repeated interactions, the figure presents only part of one interaction. Further, the consistency of P's and O's activities is assumed but not illustrated. The main point of the figure is to show that P's and O's activities are directly interdependent as indicated by the interchain connections in the figure.

In addition, Figure 6.1 also shows that any given interaction may itself influence the causal conditions affecting the dyad. For example, today's interaction may influence the partners' expectations about their future interactions or reinforce a shared habit pattern. Links from the specific

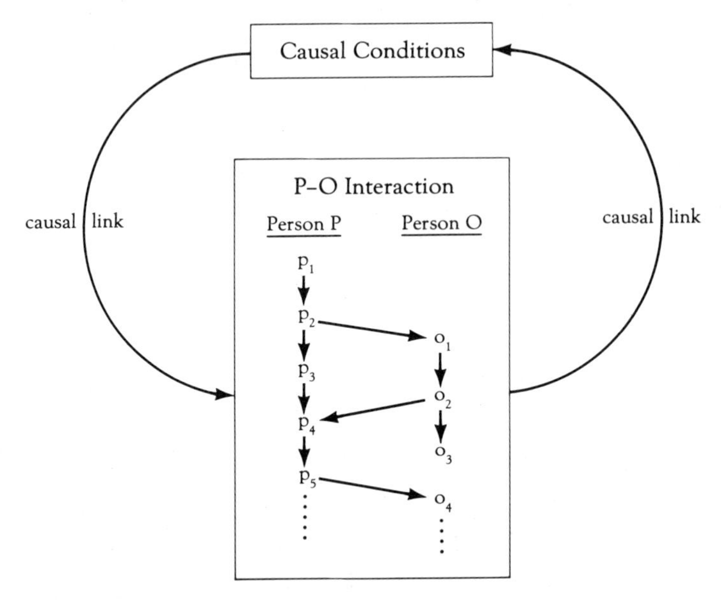

FIGURE 6.1
Direct interdependence. In this interaction sequence, P and O are directly interdependent, as shown by the P-to-O and O-to-P interchain connections. The elements of P's role shown here include P's intrachain events and connections, P's actions toward O, and P's responses to O. As the causal links indicate, the activities of P and O affect and are affected by the causal conditions of the dyad.

interaction sequence to various causal conditions indicate ways in which a particular occurrence of an interaction pattern may have more enduring influence on the individual partners or on other causal conditions affecting the dyad. It is often through such causal links that consistent activity patterns or roles are established in a relationship. Figure 6.1 further shows that the occurrence of a particular interaction sequence is itself affected by the causal context of the dyad, For example, P's perception of his partner (e.g., as weak and helpless) and P's personal dispositions (e.g., to be assertive) may affect his role. Figure 6.1 thus shows the reciprocal links between roles and causal conditions.

A less obvious type of interdependence occurs when the activities of one partner influence the other only indirectly. In Chapter 2, we refer to this kind of indirect influence as *indirect interdependence.* In describing the roles of husband and wife, for example, people often refer to *solitary* activities that are elements of marital roles. One spouse's homemaking activities (e.g., cleaning the house, doing the laundry) or breadwinning activities (e.g., driving to work, balancing the checkbook) are usually construed as part of marital roles because these activities affect the other spouse. In such cases, the influence is not through face-to-face interaction, but rather through changes in the causal conditions affecting the dyad. Thus, one person's homemaking activities may influence the family by affecting their physical environment (e.g., by providing a clean, safe and congenial setting for interaction) or their social environment (e.g, by writing letters to friends who invite the couple to dinner). Similarly, marital roles may also include interactions with *third parties* that affect the causal conditions of the couple. Thus, the husband's visits with relatives and the wife's interactions with business clients may influence the causal context of their marital relationship.

Figure 6.2 illustrates indirect interdependence. The figure shows situations in which the solitary and social activities of the partners do not have immediate interchain causal connections. Rather, P's activities alone and O's interactions with Q affect the causal conditions of the P–O dyad, which in turn influence the partners individually or jointly or both. When people comment that one person's behavior is done "on behalf" of another or is "functional for" the other partner, they are often referring to situations such as these. Finally, Figure 6.2 also shows that these role activities not only influence causal conditions, but are themselves affected by the causal context of the relationship. Thus, the married couple's financial need may prompt the wife's return to paid employment and set the stage for her interactions with clients.

We have seen that an essential feature of roles is that these patterns of individual activity involve either direct or indirect interdependence with a partner. It should be emphasized that most roles in close relationships involve both types of interdependence. For example, marital roles typically include

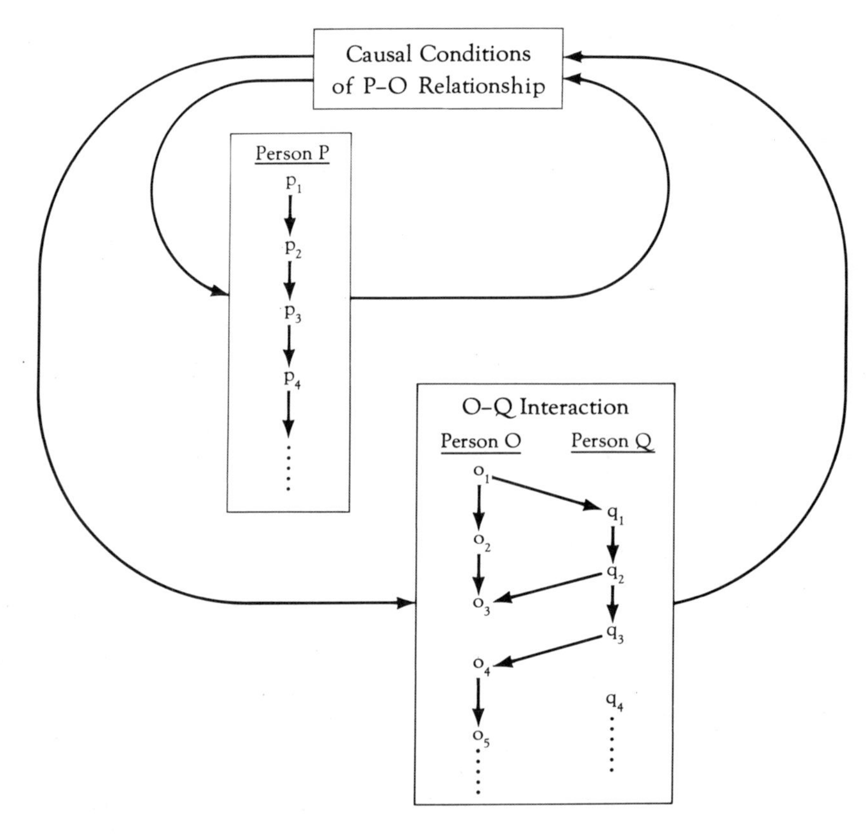

FIGURE 6.2
Indirect interdependence. Two types of indirect interdependence are illustrated. P's solitary activities (e.g., mowing the lawn) affect the causal conditions of the P–O relationship, which in turn influence partner O. The interaction between O and Q does not directly include P, but influences P by affecting the causal conditions of the P–O relationship.

face-to-face interaction between the spouses (direct interdependence) and both solitary activities and interactions with third parties that affect the causal conditions of the relationship (indirect interdependence).

The Phenomena of Roles: Behavioral, Cognitive, and Affective Elements

Roles are comprised of patterns of behavior, cognition, and affect. The connections between behavior, cognition, and affect are often subtle and different to disentangle. We can conceptually distinguish a mother's feelings of love for her child, her thoughts about the child's welfare, and her nurturant behavior, but, in the stream of experience, such components are closely interwoven. So, too, are the elements of roles. As noted in Chapter 2, the important point from our perspective is that intrachain and interchain sequences can include multiple strands of activity. Although individual researchers may decide to focus on a single type of event, such as behavior or cognition, our conception of roles does not preclude investigation of any particular element.

Behavior

Behavioral patterns are the most obvious and visible elements of roles. Behavioral patterns have two major features. The first concerns the *kinds of events* that occur in a relationship—the content of what partners do and say. For example, whereas friendship roles may largely involve conversation and recreational activities, marital roles may include a more diverse range of behaviors, including sexual activities, homemaking tasks, and shared involvement in childrearing. As noted earlier, the content of role behavior is not necessarily directly interpersonal. Cooking meals and washing clothes are part of a marital role if these activities influence a spouse at least indirectly. Second, behavioral patterns also involve the *distribution or division of activities* in a relationship. In friendship, for example, both partners may typically perform similar activities but take turns in doing them. In traditional marriage, spouses often adopt a pattern of greater specialization in which the partners engage in different but coordinated activities.

Cognition

Cognitive processes are part of the specific events and activity sequences that comprise a role. As R. H. Turner (unpublished manuscript) has noted:

> The unity of a role cannot consist . . . simply of the bracketing of a set of specific behaviors, since the same behavior can be indicative of different roles under different circumstances. The unifying element is to be found in some assignment of purpose or sentiment to the actor. (pp. 32–33)

For example, baking a cake can be a part of a parent role if the cake is intended for a child's birthday party or part of a worker role if the cake is intended for a bake sale at the office. Role phenomena include the actors' moment-to-moment cognitions—interpretations of behavior, thoughts about goals, and so on.

In planning or thinking about their activities, people often take a partner into account. For example, as a husband drives to work, he may think, "I mustn't forget our wedding anniversary. I hope I get the new raise so we can afford the trip Susan wants so badly." Meanwhile his wife may be deciding to make chocolate mousse for dessert "because it's Peter's favorite." Cognitive processes are also important in interpreting and reacting to the actions of a partner. Watching her husband do household chores that he dislikes, a wife may think, "He's doing it for me; he really cares about making our marriage work." Partners often seek to make causal attributions about the thoughts and feelings behind a particular action. Dyadic behavior has symbolic importance to participants. What partners do and say cannot be totally separated from their perception and interpretation of these actions. Cognitions are thus an important part of the phenomena of a role. It should be noted, however, that it is sometimes difficult to distinguish the momentary cognitions that are part of a role from the more enduring cognitive causal conditions (e.g., attitudes or values) that influence role patterns.

Affect

Emotional experiences are also an important part of the roles in close relationships. We know intuitively that feelings of love or obligation can motivate our behavior toward a partner, and that specific actions can lead us to feel anger, pride, or guilt. Chapter 4, "Emotion," suggests that intrachain event sequences that are causally connected to the partner are the basis for a person's emotional investment in a relationship. Thus, roles—consistent patterns of individual activity in a relationship—can be sources of emotional investment and so can set the stage for emotional experiences based on interruption.

Other analyses of emotion have suggested several ways in which affect may be connected to specific activity patterns. In *emotional testing* (S. L. Gordon, 1979), people try to gauge their own feelings and those of a partner. For example, knowing that intense emotional arousal is not continuous in close relationships, people may use infrequent events as indices of their underlying feelings. Thus, sorrow at being separated from a partner or renewed sexual passion on a vacation may be interpreted a signs that passionate love is really still alive. Similarly, partners may scrutinize each other's behavior for signs of feelings toward each other. In *emotion management* (Hochschild, 1979), people strive deliberately to evoke, modify, or suppress their feelings. Parents may attempt to heighten feelings of affection for their newborn, and try to put negative feelings out of mind. In *display management* (S. L. Gordon, 1981), gestures are designed to express or conceal particular feelings. We may show gratitude by writing a thank-you letter or conceal anger behind a pleasant smile. In these and other ways, affect comprises an essential component of roles.

Features of Roles: Diversity, Specialization, and Complementarity

Having discussed in a general way the core phenomena that comprise a role, we will consider three properties in which roles may vary. These are diversity, specialization, and complementarity.

Diversity

Roles vary in the diversity and complexity of the activity patterns that they encompass. The smallest unit comprising a role is a single consistent or regular action sequence in a relationship. Benne and Sheats (1948) used this fairly narrow conception of role in describing what they called "functional roles" in groups, including such roles as opinion seeker, harmonizer, and aggressor. Folk conceptions also include simple roles of this sort, as when parents tell their children not to be a "cry baby," "tattle tale," or "bully."

Somewhat more comprehensive roles consist of a cluster of related behavior sequences. For example, in an analysis of interaction patterns in families, Zelditch (1955) offered these contrasting descriptions of the roles of task specialist and socio-emotional specialist. The task specialist is the "boss-manager . . . the final court of appeals, final judge and executor of punishments, discipline and control over the children of the family" (p. 318). The socio-emotional specialist is the "mediator, conciliator of the family [who] . . . soothes over disputes, . . . is affectionate, solicitous, warm, emotional to the children, . . . is the 'comforter,' the 'consoler,' is relatively indulgent" (p. 318). Thus, task and social roles are comprised of clusters of related behavioral patterns.

At a still greater level of complexity or comprehensiveness, some roles refer to many diverse types of activity patterns performed by a single person. For example, when one speaks of the roles of husband and wife, one implicitly refers to the wide range of behavior patterns that each spouse engages in as part of the marital relationship. Thus, the wife's role might include being a socio-emotional specialist, a homemaker, a companion, a sexual partner, and so on; the wife role is a composite of many distinct activity patterns.

Specialization

Role specialization refers to the existence of consistent differences or asymmetries in the roles of individuals in a dyad or group. This property can be illustrated by comparing two families. One family has a high degree of specialization: The husband always makes the decisions, initiates lovemaking, and is the sole provider; the wife always follows her husband's decisions, responds to his sexual overtures, and is the sole homemaker. In contrast, a second family accomplishes similar tasks with little specialization: Husband and wife share decision making and alternate initiating sex, both have paid jobs, and both do homemaking chores. In the second family, diverse behavior patterns occur, but are not consistently associated with one particular actor. Similarly, in one work group, particular members might specialize in the roles of task and socio-emotional leaders, whereas, in another group, individuals might alternate performing these roles. It is also useful to recognize that roles may be specialized in some areas and not in others. For example, one study of married couples (Toomey, 1971) found that specialization in the performance of domestic tasks was unrelated to specialization in decision making. Roles can be arrayed on a continuum from highly specialized to unspecialized, depending on the extent to which specific activities are consistently associated with one actor rather than another.

Complementarity

Role complementarity refers to the coordination of roles in a dyad or group. Our distinction between roles based on direct and indirect interdependence suggests that there are two distinct types of complementarity. One type refers to the meshing of individual activity sequences in face-to-face interaction and is similar to the notion of mutual facilitation discussed in Chapter 2. The gracefully coordinated movements of experienced dance partners illustrate this form of complementarity. A second type of complementarity concerns coordination in accomplishing shared goals or functions or in "managing" the causal conditions influencing a relationship. Traditional marital roles, in which the husband specializes as provider and the wife as homemaker, are complementary in this sense because both spouses contribute in essential ways to the well-being of the dyad. In most close relationships, both types of complementarity are likely to occur.

The most obvious examples of complementarity may be instances in which specialized roles mesh in some beneficial way. However, specialization is not a prerequisite for either type of complementarity. For example, two tennis partners have complementary roles in that their actions must be coordinated for a game to proceed smoothly, yet both partners may engage in nearly identical behaviors during a game. Similarly, in a marital relationship, husband and wife may engage in the complementary roles of sick person and nurse, but alternate who performs each role depending on fluctuations in their health. Complementarity can occur either through specialization or through temporal alternation.

We use the term *complementarity* to refer to the meshing of individuals' roles. It should be noted, however, that the same term is sometimes used quite differently. Some role theorists use complementarity to refer to the normative expectations or rules affecting roles. In this sense, complementarity concerns reciprocity in the rights and obligations of role partners. For example, in traditional marriage, the wife is expected to perform housekeeping tasks and is entitled to receive financial support; the husband is expected to provide financial support and is entitled to receive housekeeping services. One partner's obligations are the other's rights and vice versa. From our perspective, this usage describes a causal condition influencing roles, rather than a feature of the roles themselves. Chapter 8, "Development and

Change," discusses another use of complementarity in describing the meshing of the personality needs of partners in a close relationship.

Previous Perspectives on Roles

The framework of Chapter 2 leads to a perspective on roles that differs in two major ways from perspectives prevalent in the social science literature. First, we are eclectic in our view that roles consist of diverse types of activities, such as behavior, cognition, and affect. Traditional role conceptions have often defined roles more narrowly in terms of either behavior or cognition. Second, we employ the concept of role in a descriptive way to refer to consistent activity patterns in relationships. Although we are interested in understanding the causal conditions that influence roles, our definition of role does not include an explanation of the origins of these patterns. In contrast, most previous role definitions incorporate assumptions about the causal conditions affecting relationship patterns. We have attempted to avoid this merging of description and explanation, preferring to regard the two as important and separate aspects in an analysis of interaction patterns. In this section, we briefly describe three major role perspectives and relate them to our framework. (See extensive reviews of role concepts in Banton, 1965; Biddle, 1979; Biddle & Thomas, 1966; N. Gross, Mason, & McEachern, 1966; Heiss, 1976, 1981; J. H. Turner, 1978; R. H. Turner, 1962, 1970.)

Structuralism

The key idea in the structuralist perspective is that societies recognize certain social categories or positions, such as wife and father, and have norms (i.e., rules or prescriptions) about how individuals in these positions should behave (e.g., Heiss, 1981; Linton, 1936; T. Parsons, 1951). Roles are thus defined as culturally based norms for the behavior and characteristics of people in a given position in the social structure. Roles are relatively standardized and impersonal, applying to all occupants of a particular position. Roles exist prior to and separate from individuals. The process of socialization ensures the perpetuation of these cultural patterns across generations; individuals learn and conform to conventional roles created by society. Structuralists have also called attention to the fact that roles are interconnected. The family, for example, can be seen as a social system composed of positions including husband/father, wife/mother, son/brother, and daughter/sister—each with its own rights and obligations. Roles involve reciprocal sets of norms for members of a given social system.

Our framework leads to several observations about the structuralist perspective. First, structuralists define roles as norms for social conduct. Roles thus tell people what they should do and feel, but roles are not in themselves behavioral or affective. From the structuralist perspective, behavior is construed in terms of role performance, and the focus is on evaluating the match between roles (i.e., expectations for behavior) and actual performance. Second, structuralists do not use the role concept to describe actual patterns of behavior but rather as a causal explanation for the existence of such patterns. The structuralist explanation defines roles as a component of the culture, a feature of the social environment (E_{soc}) that gives rise to observable consistency in behavior. It is not surprising that anthropologists and sociologists concerned with issues of social order should take a feature of the culture as their focus. Essentially, structuralists ask why there is consistency in the relationship patterns of many people within a culture and variation in patterns across different cultures. Their answer is that each culture creates its own distinctive norms for relationships. These norms influence actual behavior through such processes as socialization, conformity, and internalization.

The structuralist perspective is useful in calling attention to an important causal condition influencing roles and in highlighting the interconnectedness of roles in a social system or group. From our framework, however, structuralism has two major limitations. By focusing on norms, it fails to describe actual activity patterns. Equally important, structuralism provides an incomplete analysis of the origins of relationship patterns. Cultural norms are only one of many factors producing consistency in roles. Our own analysis includes a broader range of causal conditions.

Interactionism

The interactionist perspective proposes that people create and negotiate roles in the course of social interaction rather than merely playing out predetermined cultural scripts (e.g., McCall & Simmons, 1978; Shibutani, 1961; Stryker, 1980; J. H. Turner, 1978; R. H. Turner, 1962, 1970). Interactionists recognize the existence of cultural norms for behavior but note that these norms are often vague or inconsistent and so cannot provide an adequate guide for smooth interaction. For example, there are many possible ways of being a wife or husband, and each couple arrives at its own unique pattern. Interactionists thus focus on the active process of "role-making" rather than on the passive adoption of cultural scripts or "role-taking" (R. H. Turner, 1962).

Interactionists emphasize the cognitive meanings and understandings that evolve from and organize social interaction. During interaction, individuals develop a conception of the self as actor and of the other people involved. Such conceptions enable individuals to develop a "plausible line of action" for the self and to predict how the partner is likely to behave (McCall & Simmons, 1978). In addition, members of a dyad develop unique shared norms for their interaction, which are only partially influenced by cultural norms (Shibutani, 1961).

Interactionists focus on the process by which roles are developed, rather than on a more static conception of roles, and thus have seldom given a precise textbook definition of roles. Two interrelated ideas can be detected, however: (1) Roles are fairly broad conceptions that each participant holds about the nature of their interaction. In this sense, roles are an individual (P or O) causal condition that influences actual behavior. (2) Roles are shared dyadic norms that emerge from the process of interaction; roles are thus a dyadic (P × O) causal condition that determines individual behavior. Interactionism shares with structuralism the view of roles as normative expectations, although interactionism sees norms as situationally negotiated rather than culturally determined (Hilbert, 1981). The association of these two ideas about roles (i.e., as cognitive conceptions and as norms) in interactionist writing is understandable. In the course of interaction, knowledge about how a person is likely to act takes on a normative quality of legitimate expectation (R. H. Turner, 1968). People believe that others should continue to act in the future as they have acted in the past.

The interactionist perspective offers useful insights about roles, most notably in emphasizing the active part that partners play in shaping activity patterns in their relationship. From our framework, however, interactionism has two limitations. First, the emphasis is largely cognitive; the description of roles focuses on role conceptions and shared norms to the relative neglect of actual behavior or affect. Second, interactionism tends to blur the distinction between describing interaction patterns and explaining their causal origins.

Behavioral approaches

A third perspective equates roles wth observable behavior patterns: Roles are what people typically do and say (e.g., K. Davis, 1949). A recent proponent of this position is Biddle (1979), who defined roles as "those behavior patterns characteristic of one or more persons in a context" (p. 58; cf. R. H. Turner, 1970, p. 214). Biddle justified his behavioral approach by arguing that a focus on observable behavior patterns permits greater conceptual rigor and avoids the common problem of incorporating assumptions about causality into the definition of roles. Biddle proposed that behavior patterns are the phenomena of interest and relegated cognition and affect to the realm of causal factors accounting for behavioral regularities.

From our framework, this perspective has the advantages of including behavior as a component of roles and of carefully separating description from explanation. However, the behavioral perspective's exclusion of affect and cognition as components of roles is a serious limitation. As we noted earlier, the consistency of roles is not merely a statistical matter of repetitive behavioral sequences but is also a social construction based on cognitive interpretations of behavior.

Our framework suggests several ways in which previous role perspectives have been limited. They have often blurred the distinction between describing the phenomena of roles and explaining the causal conditions affecting roles. They have defined roles in fairly narrow ways (e.g., as only behavioral or only cognitive). They have explained consistent patterns of activity in terms of a limited set of causal conditions (e.g., cultural norms or shared dyadic norms or individual cognitive conceptions). Our framework permits a more comprehensive description and causal analysis of roles.

THE CAUSAL ANALYSIS OF ROLES IN CLOSE RELATIONSHIPS

The basic questions for a causal analysis of roles are how consistent activity patterns arise in relationships and why they take the particular form that they do. This section sketches in a general way the types of causal conditions that can create, maintain, and change roles. A detailed analysis of all factors influencing roles in close relationships is beyond the scope of this chapter. Instead, we attempt to sensitize the reader to the full range of causal factors affecting roles. Following the framework laid out in Chapter 2, we give separate consideration to environmental, personal, and relational conditions.

Environmental Conditions

Roles are influenced by a variety of factors in the social and physical environment. Clear examples of social influences are provided by dramatic cross-cultural differences in relationship patterns (e.g., Brain, 1976) and by historical changes in relationships in our own society (e.g., Degler, 1980). Two basic questions raised by such evidence are why regularities are found among members of a given culture and why differences occur between members of different cultures.

Anthropologists and sociologists interested in such matters have pointed to the influence on roles of cultural attitudes about relationships. Particular emphasis has been given to social norms for role behavior, for example, to widespread *cultural norms* about such relationships as marriage and parenthood. Indeed, as noted earlier, structuralist analysis equates roles with these cultural norms. It has also been suggested that *cultural stereotypes* (i.e., widespread beliefs about the typical behaviors and characteristics of people in a particular social position, such as husbands or doctors) and *cultural values* (e.g., about the importance or usefulness of marriage and children) affect role patterns. The basic idea is that cultures develop unique guidelines for relationships that create consistency in the relationships of members of that culture.

The specific mechanisms through which social and cultural factors affect roles in a particular dyad have seldom been analyzed explicitly (see J. H. Turner, 1978, pp. 353–358). The presumption has been that these generalized social attitudes are communicated to individuals through a process of socialization involving parents, peers, schools, the mass media, and other social institutions. Individuals learn these social attitudes and to some extent adopt, internalize, and conform to these social patterns. Thus, elements of culture become part of the individual's personality (Bates & Harvey, 1975). For example, prior to marriage, boys and girls learn attitudes and values about marriage from their culture and acquire those skills and traits that are considered "appropriate" for their sex. As a result, individuals entering marriage have congruent expectations about the nature of marriage and have specialized skills and interests. Although some individual differences in exposure to cultural beliefs occur, the process of socialization is thought to perpetuate general cultural patterns across generations. One implication is that regular patterns should appear right from the start of a relationship since partners bring a "blueprint" with them and have only to put it into practice.

Although the social environment has profound influence on relationship patterns, several limitations of social influences have been noted. First, although some relationships, such as those in the military, are strongly governed by norms, other relationships, such as friendship, are not. For many kinds of relationships, cultural norms are fairly vague, rather than literal or explicit. Thus, as interactionists emphasize, partners in a relationship may have to create their own norms for conduct. Second, the presumed widespread cultural consensus about norms may not always exist. Especially in times of social change, divergent and contradictory social norms may coexist in a society. Third, the mere existence of cultural norms and attitudes does not prove that they actually cause observed interaction patterns. In a recent critique, Hilbert (1981) suggested that cultural norms are typically invoked after the fact as justifications of behavior "to clear up confusion, sanction troublemakers, instruct others in the ways of the world, and so forth" (p. 217). That actors are aware of cultural norms does not mean that such norms actually control their behavior. (Nor does lack of awareness of social norms necessarily indicate that norms do not affect behavior.)

The physical environment can also influence roles by determining the tasks that individuals must accomplish and by affecting the conditions under which social interaction occurs. Marital roles are often shaped by the necessity of providing food, shelter, and clothing. The degree of privacy available to young lovers can influence their interaction. Another illustration is the impact of technology: Anthropological studies suggest that there is less male–female role specialization in cultures based on hunting and gathering or on horticulture than in societies based on farming or industry (Basow, 1980).

Personal Conditions

Roles are shaped by the individual characteristics that partners bring to a relationship and that develop as a result of interaction. In the sociological literature on roles, much emphasis has been given to individuals' role conceptions—their generalized and relatively enduring impressions or summary beliefs about the nature of their relationship. Role theorists postulate a basic tendency for people to shape and organize their experience into coherent cognitive portraits:

> A concept such as the father role is not primarily a category devised by an investigator to describe the order he observes. Rather, it is a conception which the subjects of his investigation hold and which organize their behavior in situations involving fathers. (R. H. Turner, 1957, p. 131)

R. H. Turner (1979/1980) has used the term *working role conception* to refer to individuals' understanding of their own roles and those of others. Turner emphasized that actors conceive of the self and others not simply in terms of a catalog of typical behaviors, but rather in terms of characteristic goals and means. Thus, a role conception consists of a person's perception of the relationship and beliefs about the characteristics and motives of the self and the partner.

Role conceptions give meaning to interaction and help individuals to appraise their own behavior and their partner's behavior. Thus, a man might believe that working hard at his job, taking care of his health, saving money, helping his children with schoolwork, and spending weekends at home are all important parts of his father role because all contribute to the welfare of his children. The unifying theme among these diverse activities derives from the motives of the actor. Another illustration of how a cognitive conception might affect role behavior is provided by an expert on childcare in a discussion of crying behavior:

> If you pick up the baby when it cries and it stops crying, you can view the baby either as "socially responsive" (it made a demand and was satisfied when it was answered) or you can see it as "exploitative and spoiled." There is a tendency to construct a fantasy about [a baby] from the way that you perceive the crying and then to handle the baby accordingly. (Hotchner, 1979, p. 61)

Cognitive schemas provide a guide for interpreting and evaluating roles in close relationships.

Other relatively stable personality characteristics can also create and maintain consistent activity patterns. The impact of such personal characteristics is perhaps most obvious in considering differences among couples who share a common social and physical environment. For example, to explain why variation occurs among white middle-class American couples with similar social backgrounds, investigators might look to the more unique qualities of the individuals involved.

The conceptualization of individual characteristics depends in large part on the theoretical orientation of the researcher. For example, social exchange theorists might conceptualize diverse personal attributes in terms of "interpersonal resources" (see Chapter 5, "Power"). A more psychodynamically oriented investigator might emphasize individual needs or motives. A nonexhaustive list of possible individual characteristics that can affect interactive roles would include

abilities	perceptions
age	physical attributes
attitudes	physiological capacities
cognitive schemas	psychological needs or motives
expectations	role conceptions
gender	self-concept
goals	self-esteem
interests	skills
knowledge	traits

Such factors are invoked to account for the fact that partners have relatively stable preferences or dispositions that provide a motivational underpinning for their interaction. As Chapter 5 indicates, partners often influence the course of their interaction both deliberately and unintentionally; partners try with varying degrees of success to structure interaction in accord with their own preferences. The partners' repertoires of abilities, skills, and knowledge also influence the nature of their roles. For example, children's friendships are influenced by the maturity of the children's mental and physical abilities (Z. Rubin, 1980).

Although there is much agreement that individual differences in personal characteristics influence roles, there is considerably more controversy about the origins of these personal characteristics. The issue here concerns tracing the chain of causality further back to locate the more historical causes of individual differences. In this instance, causal analyses depend heavily on prevailing theories about the origins of individual behavior and dispositions—

on theories of personality development. Most social science work has emphasized processes of socialization and the extent to which individual differences result from the person's learning history. There can be little doubt that many personal characteristics influencing dyadic interaction are heavily affected by socialization. At the same time, there is also clear evidence of biological influences on dyadic interaction. For example, certain basic sex differences in reproduction and lactation can have important effects on heterosexual roles, especially in cultures in which birth control and bottle feeding are not available. Current thinking has outgrown simplistic debates about whether behavior is caused exclusively by learning or by biology. Our own framework is neutral with regard to the relative causal importance of nature and nurture.

Our framework does, however, suggest several observations about individual causal factors. Although we have discussed personal characteristics as a cause of interaction patterns, it should be clear that the direction of causation is reciprocal. Experience in close relationships can shape and change attitudes, values, and other personal characteristics—sometimes further increasing the consistency of role patterns. Second, as with environmental factors, there is a need for better demonstrations that specific personal characteristics actually do influence relationship patterns; plausible assumptions in this area need to be subjected to empirical test. Third, there is a need to bring greater order to the current plethora of concepts for personal characteristics. Our understanding of the origins of roles will be enhanced by advances in personality theory.

Relational Conditions

As discussed in Chapter 2, relational conditions arise from the conjunction or relation between the partners' characteristics, such as their similarity in attitudes, difference in level of education, shared values, or complementary personality patterns. Some relational conditions are based on preexisting personal characteristics that partners bring to a relationship; other relational conditions emerge from joint interaction. Relational factors may be most obvious when we try to explain why an individual behaves quite differently in different relationships. Thus, Susan's dating relationship patterns with John and Steve may be quite different, even though she presumably remains the "same person" with both partners in the "same" overall social environment. The distinctiveness of these two relationships presumably results from the unique match between Susan and each of her two boyfriends. Two examples illustrate possible relational influences on roles. These concern the effects of shared interpersonal norms and the development of interpersonal habit patterns.

Interpersonal norms

A major explanation for the regularity of interaction in dyads is provided by the concept of norms. In a detailed discussion, Thibaut and Kelley (1959) defined a norm as a "behavioral rule that is accepted to some degree by both members of a dyad" (p. 147). Norms can arise in two major ways. First, norms emerge in the process of interaction over time, a point stressed by interactionists. Explicit and implicit norms may result either from trial and error or from negotiation. Norms are thus products of interaction that then influence the character of subsequent interaction. Because norms result from the process of interaction, they can be idiosyncratic to the particular couple involved. An implication of emergent interpersonal norms is that the regularity of role patterns should increase over time as partners evolve their own norms for interaction. Second, norms may be imported into a relationship from the larger social environment. For example, in growing up, people learn many cultural norms about proper behavior in marriage; if these social norms are accepted by both partners, they may provide the basis for dyadic norms.

Several explanations have been offered for the existence of norms in dyads and groups. Most emphasize that norms improve effective dyadic functioning. Typical is the comment of MacIver and Page (cited in Thibaut & Kelley, 1959, p. 134) that, without norms, "the burden of decision would be intolerable and the vagaries of conduct utterly distracting." Thibaut and Kelley proposed that norms are a solution to problems of interdependence that arise in dyads, and, thus, the content of norms reflects the nature of these problems. So, for example, in marriage, norms may develop about how spouses provide for the economic security of the family, divide domestic responsibilities, and spend joint leisure time. Norms are not necessarily the best possible solution to relationship problems, but are presumably an adequate solution. Norms increase the predictability of interaction and so minimize uncertainty about what to expect from one's partner. According to Thibaut and Kelley (1959), one of the most important functions of norms is to reduce the necessity for the exercise of direct interpersonal influence. "Norms provide a means of controlling behavior without entailing the costs, uncertainties, resistance, conflicts and power losses involved in the unrestrained, *ad hoc* use of interpersonal power" (p. 147). Further, norms reduce the costs of interaction and increase dyadic cohesiveness by fostering facilitative interactions, cutting the costs of communication, and ensuring that important tasks are accomplished.

From this perspective, the consistency of activity patterns in a particular couple is based on the existence of fairly stable dyadic norms. The similarity of patterns across couples requires a somewhat different explanation. Two possibilities may occur. First, some types of couples (e.g., newlyweds, college roommates) may face similar problems of interdependence and so develop similar norms spontaneously. Second, external factors, such as cultural norms for relationships, may affect members of many dyads, producing cross-couple regularities.

The concept of shared norms is not without its critics. Bates and Harvey (1975) suggested that "the concept of norm sharing . . . can and does lead to the notion that norms exist and operate external to the actors who possess them. Some sociologists talk about the norms of a group as though they exist in social space apart from the members of the group" (p. 55). Bates and Harvey are thus arguing against the reification of norms as a third party in a relationship. Instead, they propose that all norms are behavioral rules located within a person. Two people may be said to "share" a norm if they agree about the norms that apply in a particular situation; sharing is thus a synonym for consensus about norms. Sharing presumes some degree of communication between partners to establish the existence of consensus. We also note that, although the existence of interpersonal norms encourages consistent patterns of individual activities in a relationship, roles can exist without partners agreeing about the norms for their relationship. In such cases, individuals might act on the basis of their own personal expectations or dispositions. It seems likely, however, that lack of normative consensus usually leads to conflict in a relationship, and that pressures exist for partners to reach some minimal "working consensus" about their roles.

Interpersonal habits

Whereas norms are a form of cognitive control of activity in a relationship, interpersonal habits develop from reinforcement contingencies that may not be consciously recognized by those involved. Waller and Hill (1951) described dyadic habit patterns as follows:

> The nexus of interaction which is a family may be viewed as a set of intermeshing, mutually facilitating habits. The married pair start with their separate systems of habit. . . . After a time they form interlocking habit systems by modifying old habits and forming new ones; the interlocked habit systems are a great deal more stable than the habit systems of the individual could ever be and rest on a different set of psychological mechanisms . . . [specifically] the habit of adjusting to the situation created by the real or imaginary demands and expectations of others. (p. 328)

Waller and Hill also emphasized that spouses' habits are mutually beneficial. This may often be the case, as in a married couple's efficient morning routine of who gets up first, who uses the bathroom when, who makes coffee, and so on. But other dyadic habits can be dysfunctional. An illustration is the "rejection–intrusion" pattern in some distressed couples (Napier, 1978). One partner, typically the woman, seeks closeness and reassurance, whereas the other, typically the man, desires greater separateness and independence. When the woman's repeated bids for affection are rebuffed, she feels hurt, rejected, and misunderstood. As a result of the woman's behaviors to increase closeness, the man feels intruded upon and engulfed and withdraws from interaction. Such a dyadic pattern can repeat itself habitually in a couple. Couples who develop such upsetting patterns often feel confused about the habit and wonder why they act as they do. One goal of therapy may be to help couples to understand and break out of such habitual dyadic cycles (see Chapter 10, "Intervention").

In summary, we have briefly surveyed several classes of causal conditions that can influence roles. Our discussion has been illustrative rather than definitive. Three issues in the development of roles deserve note. First, in many analyses of roles, causal explanations have been offered post hoc without a clear demonstration that the hypothesized causal link does in fact exist. In future research, the identification of potential causal conditions needs to be augmented by empirical verification.

Second, we have treated various causal conditions separately. In fact, causal links are more complex than we have suggested. For example, cultural norms influence roles, but the reverse can also occur. When role patterns change among many couples, as in the recent increase in American mothers working for pay, general cultural norms about maternal employment also change. Any particular relationship pattern is usually sustained by a web of interconnected causal factors. Isolating the effect of any one causal condition is often difficult.

Finally, our discussion has presented a rather static image of roles. Individual activity patterns are seldom rigidly scripted; variations occur within broad patterns of consistency. Nor are established roles set in concrete. Changes in any single causal condition can produce changes in a role. Such events as the arrival of a new baby, a wife's entry into the paid labor force, a husband's serious illness, or the family's starting therapy can change role patterns.

GENDER PATTERNS IN DATING AND MARRIAGE

We turn now from a general discussion of roles to a more focused examination of sex-linked roles in dating and marriage. Gender is one of the most basic elements affecting the patterning of activity in close relationships. We begin by reviewing research findings about sex differences in relationships and then discuss efforts to create composite portraits of role patterns through the development of marital role typologies.

Gender Differences in Dating and Marriage

Sex difference research has examined both the nature of interaction patterns in relationships and individual characteristics of partners that may influence their interaction. Empirical findings thus provide information about the nature of gender-based roles and about some of the causal conditions influencing these roles. Although we will highlight the description of sex-linked roles, we will also refer to relevant causal conditions as it seems appropriate. (For a more detailed literature review, see Peplau & Gordon, in press; Deaux, 1976.)

Falling in love

College men and women appear to differ in their beliefs or ideologies about the nature of love (Peplau & Gordon, in press; Z. Rubin, Peplau, & Hill, 1981). Men are more likely to endorse "romantic" beliefs, such as that true love lasts forever, comes but once, is strange and incomprehensible, and conquers barriers of custom or social class. Women are more likely to be "pragmatists" who say that we can each love many people, that economic security is as important as passion, and that some disillusionment usually accompanies long-term relationships. When it comes to actual experiences in love, however, the pattern changes. On standardized measures of the intensity of feelings of love, young men and women appear to love their partners equally (Z. Rubin et al., 1981). But, in dating relationships, women are more likely than men to report emotional symptoms of love, such as feeling euphoric, having trouble concentrating, or feeling as though they are "floating on a cloud" (e.g., K. K. Dion & Dion, 1975). It is unclear whether these findings represent actual sex differences in the experience of love or women's greater willingness to reveal such symptoms to researchers.

Self-disclosure

The sharing of personal feelings and information is often considered the hallmark of an intimate relationship. Folk wisdom suggests that men are less expressive than women, but empirical studies reveal a more complex picture (see review by Peplau & Gordon, in press). It is useful to distinguish preferences about disclosure from the level and content of actual disclosure. There is considerable evidence that women *prefer* greater self-disclosure in relationships than men. In actual interaction, however, a norm of reciprocity often encourages similar levels of disclosure between partners; the *amount* of actual disclosure may thus represent a compromise between the preferences of both partners. In marriage, equal disclosure between spouses is common. However, when asymmetries in disclosure do occur, it is typically the wife who discloses more (e.g., Hendrick, 1981). Recent studies of college dating couples (e.g., Z. Rubin, Hill, Peplau, & Dunkel-Schetter, 1980) have found few overall sex differences in level of disclosure, suggesting that younger educated couples may be moving away from the traditional pattern of silent men and talkative women. Finally, even when men and women disclose equal amounts, sex differences have been observed in the *content* of their self-disclosures. For example, men are more likely than women to reveal their strengths and conceal their weaknesses (e.g., Hacker, 1981).

Language and nonverbal communication

Sex differences in communication are evident not only in what the sexes reveal verbally to each other but, perhaps more importantly, in how they interact nonverbally. Research (see Deaux, 1976; Henley, 1977) has found consistent sex differences in the use of language and nonverbal behavior. For example, men do more verbal interrupting, claim greater personal space, initiate more touching, and are poorer at decoding nonverbal communication. Unfortunately, few of the studies in this area have explicitly investigated close relationships. Two exceptions are noteworthy. Fishman (1978) analyzed spontaneous conversations in heterosexual couples. Women appeared to be more supportive of male speakers than vice versa. Women asked three times as many questions as men and were more skilled at using "mm's" and "oh's" to indicate interest and attention. Noller (1980) found that wives were better at encoding nonverbal messages than were husbands.

Instrumental activities

Close relationships involve not only the exchange of confidences, but also the accomplishment of instrumental tasks. For a dating couple, instrumental tasks may include planning a picnic or organizing a party. For married couples, instrumental tasks typically include providing for the economic welfare of the family, maintaining a joint household, and raising children. Pleck (1981a) has distinguished between paid employment and family work (i.e., housework and childcare). We will consider each type of work separately.

In recent years, women's participation in *paid employment* has increased dramatically, decreasing men's exclusive role as the family wage earner. In 1950, only 25 percent of married women worked for pay; by 1978, that figure had risen to 48 percent (U.S. Census, 1979). Today, more than half of all married women work outside the home, including many mothers of small children, and this percentage rises annually. Among unmarried younger adults, a dual-worker pattern is often preferred in marriage to the traditional male breadwinner pattern. For example, Peplau and Rook (1978) found that 65 percent of college women and 48 percent of college men said they preferred a marriage in which the wife worked full-time; another 25 percent of students preferred that the wife work part-time. The causes and consequences of this change in marital roles have been the topic of much debate.

There is also clear evidence that husbands and wives perform different types and amounts of *family work*. For example, Blood and Wolfe (1960) found that husbands usually specialized in mowing the lawn, shoveling snow, and doing household repairs; wives usually did the dishes, straightened up the living room, and made the husband's breakfast. The most detailed information about family work patterns comes from time-budget studies (e.g., R. A. Berk & Berk, 1979; S. F. Berk, 1980; Pleck & Rustad, 1980; Robinson, 1977; Walker, 1970; Walker & Woods, 1976). Results of these investigations support two conclusions: First, wives do the bulk of household work and childcare. Second, this pattern of family work is not dramatically altered if the wife also has full-time employment outside the home.

In an illustrative study, Robinson (1977) found that the husbands' total family work averaged about 11.2 hours per week. The amount of time a husband spent on family chores was *not* related to whether his wife worked outside the home. In contrast, wives who were full-time homemakers spent about 53.2 hours per week on family work, and wives employed full-time spent 28.1 hours per week on family work. Thus, employed wives spent more than twice as many hours on family work as did their employed husbands. The consequence is that employed wives have significantly less free time than either full-time homemakers or employed husbands. Another study (Robinson, Yerby, Fieweger, & Somerick, 1977) found that, in a family with an employed wife and a preschool child, the husband had roughly 339 minutes of "free time" per day compared to only about 221 minutes for the wife—a difference of two hours each day. Women perform the bulk of homemaking and childcare activities, regardless of whether they have a paid job outside the home.

There is some indication that these sex differences in family work may be decreasing. In a recent review, Pleck (1981a) argued that, in the 1970s, women's contribution to family work decreased and men's increased; estimates of the amount of change range from 5 percent to 20 percent for each sex. Pleck suggested that this trend signals an increased convergence in the patterns of paid work and family work for both sexes and that it has reduced the role overload previously experienced by some employed wives. Whether Pleck's view of recent trends will be corroborated by future studies is an important unanswered question. Nevertheless, gender-based specialization in family work remains typical in American marriages.

Decision making and influence strategies

Research on power and decision making in dating and married couples (see Chapter 5, "Power"; Peplau & Gordon, in press) leads to three general conclusions. First, in many couples, men and women specialize in different areas of decision making. In marriage, for example, husbands are more likely to make decisions about the family car and insurance; wives are more likely to decide about meals and home decorating (e.g., Centers, Raven, & Rodrigues, 1971). In dating, boyfriends may have greater say about recreational activities, and girlfriends may have more say about sexual intimacy in the relationship (Peplau, 1979).

Second, attempts to assess the overall balance of power or the dominance structure in relationships (see Chapter 5; Peplau & Gordon, in press) find that many American couples perceive their relationships as egalitarian; these partners report that decision making is mutual or divided equally. When dating and marriage are not seen as egalitarian, however, it is much more often the man rather than the woman who is dominant.

Third, the sexes may use somewhat different tactics to try to influence each other. In one study (Raven, Centers, & Rodrigues, 1975), wives were more likely to attribute "expert" power to their husband than vice versa; husbands said their wives more often used "referent" power, appealing to the fact that they were all part of the same family and should see eye to eye. In a study of dating couples (Falbo & Peplau, 1980), men were more likely than women to report using direct and mutual power strategies, such as bargaining or logical arguments. In contrast, women were more likely to report using indirect and unilateral strategies, such as withdrawing or pouting.

Conflict and aggression

A few studies (see review by Peplau & Gordon, in press) suggest that men and women may respond differently in couple conflict situations. For example, Raush, Barry, and Hertel (1974) reported that in role-play situations husbands more often acted to resolve conflict and restore harmony; wives more often were cold and rejecting or used appeals to fairness and guilt. The researchers speculated that "women, as a low power group, may learn a diplomacy of psychological pressure to influence male partners' behavior" (p. 153). Kelley, Cunningham, Grisham, Lefebvre, Sink, and Yablon (1978) found that both members of young couples expected women to react to conflict by crying, sulking, and criticizing the boyfriend's insensitivity; both sexes expected men to show anger, reject the woman's tears, call for a logical approach to the problem, and try to delay the discussion. In actual dating relationships, partners reported that their own conflict interactions were consistent with these stereotypes. Kelley et al. suggested that men are conflict-avoidant people who find the display of emotion uncomfortable and upsetting, and that women are conflict-confronting people who are frustrated by avoidance and ask that problems be discussed and feelings be considered.

Although Americans like to think of their close relationships in sentimental terms, much physical violence occurs in heterosexual couples (Straus, Gelles, & Steinmetz, 1980). Steinmetz (1978) estimated that roughly 3.3 million American wives and over a quarter million American husbands have experienced severe beatings from their spouse. Although wives are considerably more likely to be the victims of physical abuse, the rates of homicide, the most extreme form of spousal violence, are remarkably similar for husbands and wives.

Reactions to relationship dissolution

The ending of a love relationship is often a difficult and stressful experience. Evidence suggests, however, that men tend to react more negatively to breakups than do women. Summarizing research on the effects of marital disruption on mental and physical health, Bloom, White, and Asher (1979) concluded that "the link between marital disruption and a variety of illnesses and disorders is stronger for men than for women" (p. 192). Divorce is associated with significantly greater increases in the rates of admission to mental hospitals, suicide, alcoholism, and mortality for men than for women. There is also evidence that men may react more severely than women to the ending of a dating relationship. Z. Rubin, Peplau, and Hill (1981) found that boyfriends were less sensitive to problems in their relationship, less likely to foresee a breakup, less likely to initiate a breakup, and tended to have more severe emotional reactions to the ending of the relationship. They concluded that "women tend to fall out of love more readily than men" (p. 825).

Personal attitudes and values about relationships

Much of the current research on sex differences has focused on individual attitudes rather than on actual behavior in dyads. From our perspective, such studies provide information about individual causal conditions affecting roles in relationships. Research (reviewed by Peplau & Gordon, in press) indicates much commonality in what Americans want in close relationships, regardless of their gender. Most people express a desire for a permanent relationship, regardless of their gender. Both sexes value companionship and affection and give relatively less importance to economic security and social status in a relationship. In actual relationships, male–female similarity is usually further enhanced by the selection of a partner who shares compatible attitudes and who is similar in background.

Several sex differences have been found, however. In general, men have more conservative attitudes about roles in dating and marriage; men favor traditional sex-role specialization to a great extent than do women (e.g., J. Scanzoni & Fox, 1980). Women view verbal self-disclosure as more important in a relationship than do men. Among educated young adults, women also show greater concern than do men about maintaining personal independence outside their love relationships by having their own friends or career (e.g., Cochran & Peplau, 1983). Finally, there is some evidence that men and women prize somewhat different qualities in an ideal love partner (Deaux, 1976). Women more often value men's experience, intelligence, and occupational achievements; men more often seek partners who are youthful and sexually attractive. Thus, men and women are likely to enter dating and marital relationships with somewhat different personal values and preferences. These, in turn, may account for some of the gender-based role differences that occur in dating and marriage.

Typologies of Marital Roles

The identification of sex differences provides a starting point for understanding gender-linked roles. But a full analysis must go beyond a simple list of differences to understand the patterning and internal organization of boyfriend–girlfriend and husband–wife roles. The task is thus to identify packages of consistent intrachain activities and their interchain connections. Several issues are important: First, we have seen that there are variations between close relationships in the existence of sex-linked differences; for example, in some couples, both partners disclose equally, and, in others, women reveal more than men. How are we to account for such between-couple variations? A second issue concerns the patterning of different aspects of interaction within a close relationship. How are self-disclosure, power, and the division of labor interrelated—if at all? Finally, how do the actions of each partner—for example, his involvement in work and her involvement in childcare, his verbal inexpressiveness and her talkativeness—mesh or interrelate? The development of role typologies has been one approach to answering these questions.

Typology construction has commonly proceeded on a somewhat intuitive basis, drawing on both empirical findings and the investigator's own understanding of relationship patterns. Typologies usually combine the description of role patterns with assumptions about major causal conditions influencing the patterns. Typologies represent "ideal types" or abstractions that are not perfectly represented in any one unique relationship. Most typologies of marital roles have not been subjected to systematic empirical testing; such testing would be a useful direction for future research. The value of typologies lies in the effort to conceptualize both the diversity of relationship patterns that coexist in contemporary society and the internal consistency of role patterns in a particular relationship.

Researchers have proposed numerous typologies of family roles, typically in an attempt to characterize gender-based role specialization in marriage. In an early work, E. W. Burgess and Locke (1960) contrasted the family as an institution and as companionship. In the institutional pattern, the family is an economic production unit headed by a strong patriarch and based on social norms and laws. In the companionship pattern, the family is based on mutual love and affection, is run by democratic consensus, and has lost its economic function. More recently, M. Young and Willmott (1973) contrasted the patriarchal family and the "symmetrical" family. L. Scanzoni and Scanzoni (1976) identified four patterns in which the relations between husband and wife are that of "owner and property," "head to complement," "senior partner and junior partner," and "equal partners."

Although these typologies represent somewhat different attempts to characterize marital roles, they consistently point to the importance of two basic dimensions: The first concerns the power relations between the sexes, the extent to which the husband is more dominant than the wife. The second dimension concerns the extent of role specialization between the spouses. This includes both activities internal to the couple, such as self-disclosure and housework, and activities external to the couple, such as participation in the paid work force.

We find it useful to distinguish three contemporary patterns of marital roles. Our typology represents a synthesis of existing typologies and draws heavily on Pleck's (1976) analysis of male sex roles. Our typology contrasts traditional, modern, and egalitarian marital roles.

Traditional marriage

In traditional couples, the husband is more dominant than the wife, and there is considerable male–female role specialization. Descriptions of traditional marriages are provided in the work of Bott (1971), Gans (1962), Komarovsky (1967), LeMasters (1975), and L. Rubin (1976). Most of these studies focus on working-class families, and traditionalists may be more common in this group. But traditional marriage is not confined to any one social class.

A happily married British couple interviewed by Bott (1971) illustrates the traditional pattern:

> Mr. and Mrs. Newbolt took it for granted that men had male interests and women had female interests and that there were few leisure activities that they would naturally share. In their view, a good husband was generous with the housekeeping allowance, did not waste money selfishly on himself, helped his wife with the housework if she got ill, and took an interest in the children. A good wife was a good manager, an affectionate mother, a woman who . . . got along well with her own and her husband's relatives. A good conjugal relationship was one with a harmonious division of labor, but the Newbolts placed little stress on the importance of joint activities and shared interests. (p. 73)

In this family, the husband controlled the money; Mrs. Newbolt did not even know how much her husband earned. The Newbolts had separate circles of friends; she socialized with women neighbors and relatives, and he spent time with male friends who enjoyed cycling and cricket. Although Bott tells us little about emotional expressiveness in this couple, it appears to have been limited.

In the traditional marriage, partners believe that the husband should have greater authority than his wife; deference is important, both pragmatically and symbolically. But actual decision-making patterns are often complex, with the wife making decisions about home management and childcare and both partners discussing major family decisions. Nonetheless, the husband retains ultimate control of family decisions. L. Scanzoni and Scanzoni (1976) likened the husband's position to that of a president in a democracy, in which certain powers can be delegated but the chief executive has final responsibility.

Partners in a traditional marriage believe that the sexes should have specialized roles in marriage. This belief is often justified in terms of religious teachings or presumed biological differences between the sexes. The traditional wife does not work outside the home for pay. Prior to marriage, she is supported by her father; following marriage, she is supported by her husband. The wife does not enter paid employment, in part because such activity would reflect negatively on her husband's ability as provider and breadwinner, indeed on his very manhood. It would also be incompatible with the wife's major role as homemaker and mother; wives who want to have a job may be viewed as selfish people who neglect family duties for their own personal benefit. In traditional marriage, the husband does not participate much in homemaking or childcare, since such activities are "women's work." The husband satisfies his main obligation to his family by being the breadwinner; indeed, for some men, duty to family may be a more important motivation for work than the intrinsic interest of the job. Men's and women's work are seen as separate, often incompatible spheres.

Emotional expressiveness tends to be limited in traditional marriages. American society in general emphasizes the importance of marital love and companionship, and few couples are untouched by these cultural themes. Yet, in traditional marriage, many factors hinder open communication and companionship. The widely divergent interests and activities of the sexes may hamper communication. He may not be interested in her talk about baby's teething or new recipes; she may be equally bored by his enthusiasm for sports or politics. Further, traditional men typically believe that men should conceal their tender feelings; masculinity is defined in part by being "tough" and presenting a strong impression to others. Traditional men may not learn how to disclose feelings nor believe that they should. For both men and women, relations with same-sex friends and relatives may be a more important source of companionship than marriage.

Traditional marriage undoubtedly takes many different forms. R. H. Turner (1970) identified three possible elaborations of the traditional role of wife: In the "homemaker role," a wife develops technical expertise in home management and childrearing that permits her to exert influence over her husband's behavior. In the "companion role," a wife cultivates "social graces, personal attractiveness, and personal and sexual responsiveness to her husband, so that she may serve as hostess to his friends and relaxer and refresher to him" (p. 269). In the "humanist role," a wife becomes active in community and volunteer work. Turner noted that the adoption of these roles is affected by social class and education, and that some women adopt combinations.

Modern marriage

In modern marriage, male dominance is muted and role specialization is less extensive (e.g., Blood & Wolfe, 1960; L. Scanzoni & Scanzoni, 1976; M. Young & Willmott, 1973). The Carsons illustrate the modern pattern:

> Jill and Charlie met in a college drama class 22 years ago. They quickly discovered that they both loved hiking and camping, an interest they have shared throughout their marriage. They spent long hours talking about their feelings and planning a future together. Jill respected Charlie's intelligence and logical arguments, and found herself going along with his ideas in most matters. At graduation, the Carsons were married, and Jill worked as a nurse to put Charlie through graduate school. After Charlie got his first teaching job, they started a family and were surprised by the arrival of twins. Jill quit her job to care for the girls. She enjoyed being a full-time homemaker for a while, but went back to work once the children started school. Although the family moved several times to advance Charlie's career, Jill was always able to find new jobs. The Carsons feel that their marriage has improved over the years, and they continue to enjoy many joint activities.

In the modern marriage, the husband's dominance is less evident. Modern couples believe that both spouses should share in decision making, and wives often have considerable influence in some areas. Nonetheless, husbands still tend to take the lead.

Role specialization is less pervasive. The wife has major responsibility for housekeeping and childcare, but the husband is able and willing to help at home. In modern marriage, the wife's paid employment is tolerated or even approved and encouraged, but it is understood that the wife's work is secondary to that of her husband. If a conflict arises between their jobs, the man's career comes first. It is also understood that the wife's work must not interfere with her responsibilities at home. Thus, modern wives typically work for pay before having children and after children enter school. Modern roles blur but do not eradicate the principle that the husband is the major breadwinner and the wife is the major homemaker.

Modern roles emphasize togetherness and companionship. Pleck (1976) suggested that, in the modern marriage, men want emotional support rather than deference from their wives. In their leisure time, modern couples typically prefer couple activities over same-sex socializing with friends. Bott (1971) noted that, among the "joint conjugal" marriages in her study, compatibility was stressed and couples felt that their relationship with each other should be more important than any separate relations with outsiders. Pleck (1976) added that, since modern men do not have close friendships with other men, they channel their desires for companionship into marriage.

Partners in modern marriages recognize that they are, in some measure, redefining conventional marital roles. Bott (1971) reported that her "joint conjugal couples" frequently discussed sex differences, rather than taking them for granted as traditionalists might do. The modern pattern departs in important ways from the traditional pattern, but role specialization is still clearly evident.

Egalitarian marriage

The egalitarian marriage (e.g., Stapleton & Bright, 1976) is best understood as an ideal that some couples are striving for, rather than a common pattern in American life today. At its core, this pattern rejects the basic tenets of traditional marriage: male dominance and role specialization by gender. In an egalitarian marriage, both partners share equally in power, and gender-based role specialization is absent both inside and outside the marriage. M. Young and Willmott (1973) consider this type of marriage "symmetrical" because gender does not determine the division of labor and because the bases of interdependence are similar rather than complementary.

The egalitarian marriage is an attempt to alter the traditional structure of the American family (L. Scanzoni & Scanzoni, 1976). Joint responsibility extends to housekeeping, childcare, and the financial support of the family. The modern idea that husbands "help" their wives with domestic chores is replaced by the concept of equally shared responsibility. Similarly, both spouses typically engage in paid work. A salient value is that both partners' work be considered equally "important"; the wife is no longer the junior partner. Among those contemporary American couples striving for an egalitarian relationship, an emphasis on companionship and sharing is typically important. There is an effort to overcome traditional sex differences in emotional expressiveness.

The central theme in egalitarian marriages is a rejection of the culture's traditional model for marriage. In its place, many alternative patterns for the conduct of married life are possible. Some couples may do housekeeping tasks together, others may take turns, and still others may divide tasks according to personal interests. Similarly, in supporting the family financially, partners may alternate holding paid jobs, experiment with sharing one job, or prefer that both partners have full-time jobs. L. Scanzoni and Scanzoni (1976) view the egalitarian marriage as "an emerging form—a life-style that may very well represent the wave of the future" (p. 237).

Currently, the couples who may come closest to the egalitarian model are dual-career marriages in which both spouses have major commitments to a full-time professional career. Studies of such couples (e.g., Bryson, Bryson, Licht, & Licht, 1976; Holmstrom, 1972; Rapoport & Rapoport, 1976; Yogev, 1981) indicate, however, that, although these marriages are often happy, they seldom achieve a truly egalitarian relationship. For example, Poloma and Garland (1971) concluded that only one of the 53 dual-career couples they interviewed was actualy egalitarian. In all the others, the wife was responsible for domestic tasks and the husband's job was seen as more important. Even for couples who intellectually endorse an ideal of equality in male–female relationships, this goal is not often attained. Some of the reasons why partners who want an egalitarian marriage may have difficulty in eliminating gender-based role specialization are discussed in the next section.

In summary, the three marriage role patterns we have identified represent different mixes of the elements of power and role specialization. Traditional marriage is based on a form of benevolent male dominance coupled with clearly specialized roles. Egalitarian marriage rejects both of these ideas. Modern marriage represents a middle position.

THE CAUSES OF GENDER SPECIALIZATION IN MARRIAGE

Earlier in this chapter, we discussed general issues in the causal analysis of roles. Studies of gender patterns provide a more detailed examination of causal questions about roles. In this section, we consider several explanations that have been offered for the existence of gender-based role specialization in marriage. Although the discussion focuses on gender, it also illustrates the various types of causal conditions that may produce role specialization in a wide range of close relationships.

Personal Conditions

The personal characteristics of partners can influence role specialization in two general ways: First, partners may have similar attitudes that promote differences in their roles. For example, both partners may endorse traditional views of marriage in which the husband and wife are expected to behave in sex-typed ways. Agreement about traditional role prescriptions and their adoption as shared dyadic norms would lead to role specialization. It is likely that role specialization is greater when both partners adhere to a belief in its value. Second, differences in the partners' personal characteristics can create asymmetries in their behavioral patterns. For example, in heterosexual relationships, it is common for the boyfriend or husband to be bigger and older, have more education, know more about cars and finances, have more ambitious career plans, and so on. These asymmetries encourage role specialization. In any relationship in which asymmetries exist—in age, wealth, knowledge, or other personal attributes—role specialization may be more likely. The influence of personal factors is illustrated by discussions of how attitudes and biological dispositions may affect gender specialization in marriage.

Personal attitudes

The degree of gender-role specialization in marriage has been linked to the individual attitudes of the marital partners. Although Americans' sex-role attitudes have become more egalitarian in the past two decades (Mason, Czajka, & Arber, 1976), many people continue to believe that specialized male and female roles should exist in marriage. Evidence linking sex-role attitudes to role behavior comes from two studies (Beckman & Houser, 1979; Perrucci, Potter, & Rhoads, 1978) showing that people with more traditional attitudes report lower levels of husband participation in housework and childcare.

Recently, much interest has been directed toward identifying the specific attitudes that limit men's participation in family work, especially in dual-worker families in which wives have full-time paid jobs. As noted earlier, in dual-worker marriages, husbands typically spend much less time on family work than do their employed wives. Yet, most Americans, including employed wives, report being satisfied with the division of labor in their marriage (Bryson et al., 1976; L. Harris & Associates, 1971; Robinson et al., 1977). A common theme emerging from studies of dual-worker families is the belief that the employed wife's major responsibility should still be as homemaker and that the husband's major responsibility should still be as breadwinner. The comments of a successful woman professor illustrate this view:

> Even though my career is clearly secondary, I don't feel cheated in any way because I want it this way. If I didn't want it this way, I think the marriage institution as we know it . . . would be disrupted and that my marriage wouldn't be a successful one. (Cited in Poloma & Garland, 1971, p. 534)

Even when a wife works full-time for pay, her job is often interpreted as less important than her husband's job or than her own family obligations. Robinson et al. (1977) suggested that there may be psychological benefits in maintaining separate "role territories" for men and women. Both sexes may experience psychological rewards from traditional role performance and may fear loss of these benefits if the activities are shared with a partner. It is also possible that husbands and wives have rather different attitudes about housework and childcare. For example, some women may view their homes as a personal reflection of themselves to a greater extent than do their husbands. Men may be more casual about standards of cleanliness and less disturbed if others observe a "messy" house. Attitudes such as these may contribute to traditional gender-role specialization in marriage.

Biological causes

Some explanations of gender specialization point to the personal predispositions that men and women bring to the marital relationship. A specific illustration is provided by recent discussions of assumed genetic sex differences in parental investment (e.g., Mellen, 1981; Symons, 1979; Wilson, 1975). Sociobiologists and others argue that females have a "biologically-based heightened maternal investment in the child" (Rossi, 1977, p. 24). Men, in contrast, have a lessened investment in parenting—a biological indisposition toward childcare. Sociobiologists explain this sex difference by arguing that humans evolved in ways that tend to maximize the likelihood that their individual genes will survive by being passed on to their offspring. Whereas men produce many sperm, women typically release only one egg per month and then must invest years in pregnancy and nursing. As a result, the most efficient reproductive strategies for the two sexes differ. For men, reproductive success (the survival of one's genes) is enhanced by impregnating as many women as possible and investing a minimal amount of time and energy in the rearing of any one child. For women, in contrast, reproductive success depends on maximizing the chances that a few children will survive to maturity; investment in childcare is a necessity. The ultimate cause of men's lesser participation in family work is thus seen as genetic and may operate through sex differences in dispositions that in turn influence behavior.

The sociobiological view is quite controversial (e.g., H. E. Gross, 1979), and support for the position is indirect. For example, it is noted that, in most nonhuman primates as well as in humans, females engage in considerably more care of the young than do males (G. Mitchell, 1981). Many biologically oriented researchers (e.g., M. McClintock, 1979) acknowledge that actual behavior is influenced by factors other than biology. Symons (1979) and others believe that evolutionary influences are most evident in psychological predispositions rather than in abilities or actual behavior. The debate about how biology contributes to observed gender specialization in marriage is likely to continue for many years.

Relational Conditions

Some explanations of role specialization emphasize the importance of dyadic or relational causal conditions. As discussed earlier in the chapter, shared interpersonal norms and interpersonal habits can influence role specialization in a relationship. Two additional relational explanations for gender specialization in marriage deserve comment: the functional requirements of social systems and the relative power of husbands and wives.

The functional requirements of social systems

It has been suggested that the very nature of group interaction creates the necessity for role specialization. In this view, specialization arises from factors intrinsic to a relationship, not from external forces. The best-known statement of the perspective is found in the work of T. Parsons and Bales (1955). Even though many of their ideas were later rejected by subsequent researchers, the Parsons and Bales analysis provides a useful illustration of this general approach. Parsons and Bales argued that

> the tendency toward differentiation [is] probably not dependent on any gross differences between persons, upon preexisting cultural prescriptions, or upon any particular task demand, although all of these may play their part. The tendency toward differentiation depends basically . . . on the fact that all social systems are confronted wth several fundamentally differentiated problems, and with a limitation of resources which makes it difficult to keep them all solved in short time spans. (p. 300)

Role differentiation in the family or in any other group is thus considered as a special case of more general principles of group functioning. Parsons and Bales believed that role differentiation occurs because all groups must simultaneously accomplish two goals: the maintenance of group solidarity or cohesiveness and the performance of instrumental tasks. As a result, there emerge in all groups two different types of leaders: a task leader concerned with solving instrumental problems and a social leader concerned with maintaining relations among members and relieving group tensions.

Parsons and Bales further argued that task and social roles are incompatible and so must be performed by different individuals. They offered several explanations for the hypothesized mutual exclusivity of these two roles (see also Burke, 1967, 1968). For example, when group members coordinate their activities to accomplish task goals, feelings of frustration, anxiety, tension and hostility commonly arise. Conflicts of interest, reactions to taking directions from the task leader, and other aspects of interdependence create socio-emotional problems. Since the task leader is often the source of tension and the target of hostile feelings, she or he cannot ease group tensions effectively. Hence, a separate social leader is needed. In addition, Parsons and Bales believed that the limited flexibility of adult personality necessitates role specialization: "Society . . . requires a higher order of role differentiation than the normal personality is capable of achieving" (p. 385). Hence, individuals tend to specialize in particular types of roles. All of these factors foster the emergence of two specialized but complementary roles in all groups.

To explain the allocation of individuals to specialized roles, Parsons and Bales had to look beyond the universal processes of social interaction to the specific characteristics that people bring to relationships. In discussing roles in the family, they emphasized biological sex differences in reproduction and physical strength, childhood socialization that builds sex-typed personalities and skills, and other aspects of American society that accentuate male–female role specialization.

The attempt by Parsons and Bales to explain role specialization in terms of universal features of social systems is intellectually appealing. They provided a simple and parsimonious explanation based on the functional requirements of group interaction. Unfortunately, research evidence collected over the past 20 years has failed to support their views.

The assumption that task and social roles are a universal feature of social systems has not been substantiated by research on group interaction (see review by Meeker & Weitzel-O'Neill, 1977). Many factors, including group size, the nature and complexity of group tasks, and differences in the abilities of group members, influence whether these two roles emerge.

Further, when task and social roles do occur, they are not necessarily specialized (e.g., R. A. Lewis, 1972). Studies of group interaction (e.g., Bales, 1970) suggest that task and social roles are actually independent, rather than mutually exclusive. Cross-cultural studies of family interaction (Crano & Aronoff, 1978) demonstrate that the degrees to which parents participate in expressive activities (e.g., childcare) and instrumental activities (e.g., subsistence work) are unrelated. Levinger (1964) argued on logical grounds that "social" specialization is a meaningless notion when applied to a two-person group. Whereas instrumental tasks can be delegated and are subject to specialization, socio-emotional activities necessarily involved two people and so cannot be delegated. Levinger thus proposed that, in the marital dyad, "both spouses are task specialists and neither spouse is a social-emotional specialist" (p. 435). His original research and that of others (e.g., Rands & Levinger, 1979; Raush, Barry, Hertel, & Swain, 1974) have supported this view. Additional evidence against the necessity of task–social specialization comes from studies of same-sex dyads. For example, empirical research on homosexual couples (reviewed by Peplau & Gordon, 1983) has typically found a pattern of role sharing and turn taking, rather than rigid task–social specialization.

Recent research (e.g., S. L. Bem, 1981; Spence & Helmreich, 1978) has also challenged the assumption that adult personality cannot encompass both instrumental and expressive components. Studies of psychological androgyny suggest that a sizeable number of adults incorporate both "masculine" (instrumental) and "feminine" (expressive) elements into their personality.

If specialization along task–social lines is not a given of social interaction, then we must look elsewhere to explain the emergence of specialization in close relationships. Current sociological theory (e.g., Biddle & Thomas, 1966; R. H. Turner, 1968, 1970) does not offer a comprehensive analysis of the origins of role specialization and provides only the most general guidelines. Theorists postulate that, when interaction is both repeated and diverse, specialized roles tend to emerge (e.g., Biddle, 1979). R. H. Turner (1979/1980) proposed that roles must be organized so that they are "functional," in the sense of organizing behavior in ways that effectively and efficiently accomplish group goals. He also asserted that roles must be "tenable"; individuals must experience roles as rewarding and supportive of their self-esteem. Research has not yet specified, however, what constitutes minimal levels of efficiency and viability; it seems likely that these assumed constraints permit wide variation in the nature and degree of role specialization.

The early suggestion by Parsons and Bales (1955) that role specialization follows universal patterns has proved much too simplistic. The origins of specialization are more complex. Specific problems of interdependence may tend to give rise to particular specialized roles, although it seems likely that any particular interdependence problem can have several alternative solutions. The pattern that develops in a relationship is significantly affected by characteristics of individual partners and the social environment.

Power

Another relational explanation for role specialization emphasizes imbalances of power between partners. In this view, the proximal, immediate determinant of role specialization is to be found in asymmetry in the individuals' status or power. Although the chain of causality can be traced to more distal factors that establish power imbalances, the focus is on the consequences of dominance for role specialization. In a general statement of this view, J. Berger, Rosenholtz, and Zelditch (1980) argued that small groups seldom "create a social organization *de novo,* out of the interaction of their members, but instead maintain external status differences inside the group" (p. 2).

Henley's (1977) analysis of patterns of nonverbal behavior and communication illustrates a power explanation. She suggested that power equality in a relationship leads to reciprocity in behavior. Thus, in relationships among power equals, there tends to be mutual touching, reciprocal self-disclosure, equal sharing of physical space, and similarity of conversational attentiveness. In contrast, among power unequals, the more powerful person initiates more touching, receives more self-disclosure, occupies more territory, and interrupts and talks more in conversation. The patterning of interaction is structured by dominance in the relationship. Finally, and most pertinent to our discussion, Henley argued that many of the sex differences observed in heterosexual relationships are similarly caused by men's greater power. Henley's analysis is provocative in that it offers a unified explanation for role specialization in diverse areas of interaction, but more empirical documentation of the causal contribution of power is needed.

A further question for causal analysis concerns the mechanisms through which status and power differences influence roles in relationships. One detailed explanation derives from the sociological theory of expectation states (Berger et al., 1980; Meeker & Weitzel-O'Neill, 1977). According to this view, external or culturally based status characteristics, such as gender, influence interaction internal to a group through the establishment of performance expectations. High-status individuals, such as men, are expected to perform well, are given more opportunities for task performance, and receive greater approval for their behavior. In addition, Meeker and Weitzel-O'Neill argued from this theory that women are less likely than men to try to raise their own status in a social system by their performance, and that women are less likely than men to perceive competitive behavior as legitimate. A demonstration that parallel processes operate in marriage would be useful.

Environmental Conditions

Gender-based role specialization is also influenced by the social environment. In the case of marriage, the formal forces of law and religion have long promoted specialization, emphasizing the husband's role as head of the family and the wife's role as homemaker and mother. Etiquette books and guides to married life offer detailed rules for distinguishing male and female roles (e.g., Andelin, 1963). Analyses of social factors influencing gender specialization in marriage are diverse. For example, the increasing participation of American wives in the paid labor force has been attributed both to economic necessity and to changing social attitudes about women's roles. In this section, we consider the effects of social approval and social networks.

Social attitudes

The behavior and attitudes of married couples are influenced to some extent by social reactions to their conduct. These social reactions may in turn depend on such factors as the social class or educational level of the group, neighborhood, or community. For example, a study of working-class families by Lein (1979) found that men who deviated from traditional family roles by performing childcare and homemaking tasks were often criticized by relatives, friends, and even strangers. Men's peer groups often explicitly ridiculed them for what was perceived as effeminate or weak behavior. One husband of a working wife commented:

> I know I do more than most of the guys I know as far as helping their wives. . . . We talk about it at work. We talk about it when we have a get-together with a half dozen couples, and they say, "What, are you crazy?" We get very personal, you know. The guys want to kill me. "You son of a bitch! You are getting us in trouble," and the wives say, "Does he really?" The men get really mad. (Lein, 1979, p. 9)

Thus, the attitudes of friends and family, as well as more general cultural attitudes about marital roles, may influence the degree of role specialization in a particular relationship. This influence can either increase or decrease role specialization, depending on the nature of the social attitudes.

Social networks

Early studies (e.g., Bott, 1971) suggested that marital role specialization was greater when a couple had a tightly knit network of friends, relatives, and neighbors who knew each other—rather than a more loosely knit network. It was also suggested (C. C. Harris, 1969; Wimberly, 1973) that role specialization was greater when each spouse interacted primarily with a same-sex network. More recent evidence suggests that the effects of network density and sex composition depend on the social norms of network members. Network norms can either promote traditional role specialization in marriage (as in a working-class community) or devalue such specialization (as in a feminist network). Tightly knit and same-sex networks may be more effective than loosely knit or mixed-sex networks in influencing individual behavior toward group norms.

In summary, the various explanations of gender-role specialization in marriage illustrate several more general points: First, gender-linked patterns are affected by a large number of causal conditions; our list is illustrative, not exhaustive. Biological sex differences are only one of many explanations for observed differences between the roles of men and women. Hence, to say that a particular role pattern is linked to gender provides only a first clue as to the actual causal conditions producing the pattern. Second, causation can occur on various levels simultaneously; different explanations need not be mutually exclusive. Third, one of the difficulties in assessing the relative contributions of various causal factors is that they may all operate to produce the same effect. For example, both biological and social factors may work in the direction of limiting men's participation in family work. As a result, situations in which at least one of the presumed causal conditions varies in directionality are particularly interesting. Thus, comparisons of couples in communities that encourage versus discourage gender specialization in marriage might be especially informative.

THE CONSEQUENCES OF GENDER SPECIALIZATION

Recent changes in the American family, such as the ever-increasing proportion of employed wives, raise important questions about the implications of new patterns for individuals and their relationships. Some popular writers suggest that more egalitarian relationships will ruin our sex lives and destroy the family, but others claim that equality will improve the quality of marriage for both spouses and save the family as an institution. In this section, we consider some of the implications of gender-based role specialization in marriage for the couple, for the individual partners, and for their children. It is useful to bear in mind that marriages that have little gender specialization are not necessarily *un*differentiated or *un*specialized. Rather, nontraditional relationships may develop specialized roles based on such factors as skills or interests.

Consequences for the Couple

Traditional role specialization has often been justified in terms of its assumed benefits to the marital relationship (e.g., T. Parsons & Bales, 1955). It has been argued that specialization by gender is an efficient way to run a family, that the separation of homemaking and breadwinning activities reduces potential competition and conflict between spouses, and that the mutual dependence produced by specialization increases the stability of the marital union. Empirical evidence pertinent to these hypothesized dyadic consequences is limited.

One argument, that traditional gender-role specialization is an *efficient* way to organize family activities, seems plausible. What is lacking, however, is evidence that gender specialization is any more (or less) efficient than a role-sharing pattern or than specialization based on factors other than gender. Researchers have neither attempted to assess systematically the degree of efficiency of family activities nor to compare traditional, modern, and egalitarian couples on this dimension. Time-budget studies (e.g., Robinson, 1977) indicate that employed wives spend about 25 fewer hours per week on housework and childcare than do full-time homemakers, suggesting that employed wives may actually be more efficient in their use of time. It is also possible that employed wives change their standards or methods of homemaking, rather than increasing their efficiency. More generally, it might be questioned whether, beyond some minimal level, efficiency is the most important criterion for evaluating family roles.

A second hypothesis is that gender-based role specialization is beneficial because it reduces *competition* and conflict in a relationship. If this hypothesis is true, we might expect greater competition among dual-worker couples than among traditional couples. Available evidence generally contradicts this position, however. For example, Holmstrom (1972) found little competition among the dual-career couples she studied. Oppenheimer (1977) has argued that a wife's employment can actually benefit families by enhancing their status in the larger community. Although wives typically earn less than their husbands, the wife's income can "provide a functional substitute for upward occupational mobility on the husband's part, or [compensate] for a husband's relatively low earnings compared to other men in his occupational group" (Oppenheimer, 1977, p. 404). Oppenheimer also argued that there are many alternatives to role specialization as ways of preventing competition and conflict between spouses. For example, spouses may work in different occupations or in different work settings.

A final hypothesis is that traditional role specialization increases *marital stability*. As a general principle (Thibaut & Kelley, 1959), partners are more likely to stay in a relationship if they depend on each other for important rewards and have no alternative sources of these rewards. Gender specialization, in which the wife depends on the husband for financial support and the husband depends on the wife for homemaking services, is one possible basis for marital interdependence. But other bases of dependence also exist, such as strong feelings of love and attraction or the sharing of pleasurable joint activities. There is evidence (reviewed in J. Scanzoni, 1979b) that traditional role specialization, particularly women's economic dependence, does contribute to marital stability. Wives who are financially dependent are less likely to seek a divorce than are employed wives who are financially independent (e.g., Hannan, Tuma, & Groeneveld, 1977). If one values the permanence of marriage at all costs, traditional role specialization may therefore appear beneficial. However, an alternative interpretation (J. Scanzoni, 1979b) is that the asymmetry of economic dependence in traditional marriage puts wives at a power disadvantage that can force them to stay in an unsatisfying relationship. Increased economic independence could permit women to bargain more effectively to improve the quality of their marriage, and so avoid divorce, or enable women to escape from a hopelessly unrewarding relationship. For many Americans today, marital satisfaction is a more important goal than merely staying together.

Consequences for Partner Satisfaction

Gender specialization can influence individuals in many ways, affecting, for example, their self-concept, personality, or economc resources. Our discussion focuses on marital satisfaction—partners' evaluations of their happiness with the marriage. The literature on this topic is large and often inconsistent (see reviews in Aldous, Osmond, & Hicks, 1979; Laws, 1971; R. A. Lewis & Spanier, 1979; Peplau & Gordon, in press). In general, most husbands and wives report that their marriage is satisfying, and spouses' happiness ratings are positively correlated. Differences between the sexes, when they do emerge, are small.

There is little evidence that traditional sex-role specialization enhances marital satisfaction. In a study of British couples, Bott (1971) found no association between marital satisfaction and the degree of role segregation. Similar results were obtained in a study of middle-class American families (Rainwater, 1965). Some evidence has linked role-sharing in marriage to greater enjoyment of couple activities (Rapoport, Rapoport, & Thiessen, 1974) and to reporting fewer serious problems in marriage (Rainwater, 1965). One reason for these mixed findings may be that people's global assessments of marital satisfaction are based not only on their actual experiences but also on their aspirations (Komarovsky, 1967). Couples with clear-cut specialization of husband–wife roles may expect little interaction or sharing between spouses and judge their marriage on that basis. More generally, traditional and nontraditional couples may use different yardsticks in assessing marital success.

Most satisfaction research (see reviews cited earlier) has examined specific aspects of marital roles, rather than global measures of degree of role specialization. Many studies have found that the greater the husband's occupational success and income, the greater the marital satisfaction of both spouses (R. A. Lewis & Spanier, 1979). Recently, Aldous et al. (1979) suggested that this relationship may actually be curvilinear, with extremely low and high occupational success by the husband detracting from the enjoyment of marriage. The impact of the wife's employment is more controversial. Overall marital satisfaction is probably highest when both partners are satisfied with the wife's employment status (Lewis & Spanier, 1979).

Satisfaction in heterosexual relationships is significantly associated with the balance of power or decision making. Studies of married couples (e.g., Blood & Wolfe, 1960; Centers, Raven, & Rodrigues, 1971; Rainwater, 1965) have generally found higher levels of satisfaction among both male-dominant and egalitarian marriages, and lower satisfaction among female-dominant marriages.

The specific pattern of interaction that a couple adopts may be less important to satisfaction than whether the partners agree about the pattern. Several studies (reviewed in Hicks & Platt, 1970; Lewis & Spanier, 1979) document the importance of role consensus or agreement between the marital expectations and behavior of spouses. It seems obvious that an ardent feminist who desires shared roles in marriage will be happier with a partner who supports these views than with a staunch traditionalist. Disagreement between spouses about marital roles is a major source of potential conflict and dissatisfaction.

Several older studies (reviewed in Hicks & Platt, 1970; Laws, 1971) found that marital satisfaction was significantly linked to the wife's ability to perceive her husband as he perceives himself and to conform to his expectations—but not vice versa. Laws (1971) referred to this phenomenon as the norm of wife-accommodation and explained that "an accommodative (or empathic, or considerate) spouse contributes to *anyone's* marital satisfaction, . . . and the social norms decree that it shall be the wife's role" (p. 501).

Consequences for Children

Is it "good" for children if parents interact in distinctive, sex-typed ways? Ultimately, an answer to this question rests on assumptions about the processes of personality development and the desired outcomes of childhood socialization. Pleck (1981b) has offered a detailed and provocative analysis of social science models of sex-typing and personality development. He contrasts the "role identity" paradigm that has dominated both lay and scientific thinking with a newly emerging "sex-role strain" paradigm.

The role identity paradigm, found in the work of T. Parsons and Bales (1955) and others, rests on two basic ideas: First, to be psychologically mature as males or females, individuals must develop a secure sense of sex-role identity, manifested by having the psychological characteristics culturally defined as appropriate for their sex. It is not sufficient for individuals to know their biological sex; rather they must psychologically "validate" or "affirm" their sex-role identity through exhibiting sex-typed traits, interests, and behaviors. Second, the development of sex-role identity is often a risky, failure-prone process, especially for boys. Because sex-role identity is a learned outcome, it is susceptible to faulty socialization. Parental role specialization is seen as essential to the development of adequate sex-role identity in children. If parents deviate from traditional roles, the paradigm holds, their children will not develop properly and will suffer from an array of "problems," including aggressiveness, learning difficulties, homosexuality, and delinquency. Pleck (1981b) examined the theoretical and empirical support for this paradigm and found it largely inadequate. Data suggest, for example, that individuals who are highly sex typed may actually function less effectively than those who are less rigidly sex typed.

Pleck believes that traditional sex roles can create problems for individuals, but he interprets such difficulties from an alternative, sex-role strain paradigm. In this view, sex-role development involves conformity to culturally defined sex roles. Individuals sometimes experience difficulties because cultural role definitions are internally inconsistent or are incompatible with the person's own temperament and interests. Traditional sex roles can be unnecessarily confining, and overconformity to traditional roles can be dysfunctional, as when male assertiveness is expressed as physical violence.

The work of Pleck and others challenges traditional wisdom about the goals of sex-role development and the importance of parental sex-role specialization. Whereas the establishment of a secure, sex-typed identity was once considered the ideal goal, today some social scientists propose that the ability to transcend sex roles and to behave more flexibly is preferable. The extent to which personal identity must and should be linked to gender is now under debate. This reassessment of goals also calls into question how parental roles influence children. Although much more will undoubtedly be written on this matter, there is no clear evidence that highly specialized parental roles contribute to the psychological well-being of children.

Our discussion of roles has examined relatively broad and comprehensive patterns in close relationships. Although the concept of role has been important in the social sciences, the term has been used in diverse and often contradictory ways. Our conceptual framework has helped to clarify different usages of the term and has provided the basis for our own analysis of roles. We have found it useful to employ the role concept descriptively to refer to consistent patterns of individual activity in a relationship. We thus distinguish the role concept, which describes observable regularities in a relationship, from various causal conditions that create, maintain, and produce change in role patterns.

This chapter has also considered the more specific topic of gender-linked roles in contemporary American dating and marriage. Gender specialization is a prominent feature of many close relationships. Thus, a careful description and causal analysis of gender-linked roles not only illustrates the application of more general notions about roles but also addresses central aspects of heterosexual relationships.

CHAPTER 7

Love and Commitment

HAROLD H. KELLEY

In a much publicized court case, Michelle Triola sued Lee Marvin for a share of the financial assets he had acquired during the 6 years they had been living together unmarried—a relationship he had broken off. Triola claimed a right to a portion of the assets on the grounds that she had served in every way as a faithful wife and had foregone a promising singing career to do so. In the course of the trial, Triola's lawyer asked Marvin whether he had ever loved her "even a little bit." Marvin's reply, hardly surprising at that point in the proceedings, was that he had not had the kind of love that involves "deep regard for the other person, truthfulness, loyalty, fidelity, and a tremendous sense of selflessness toward the other person." In a St. Valentine's Day commentary on this exchange, Zick Rubin, a prominent researcher on love, observed:

> At bottom, this case is not a matter of love but of commitment. Lee Marvin may, in fact, have loved Michelle Marvin a great deal. . . . But loving someone is not a commitment to love and support that person forever. (*Los Angeles Times*, February 14, 1979)

In this chapter, we consider the two concepts that Rubin contrasts. Our main purpose is to show how love and commitment relate to one another in heterosexual relationships (dating, engagement, marriage, cohabitation, and so on). On the one hand, we will see that the phenomena to which these concepts refer have much in common, and, in many cases, love and commitment are intimately related. On the other hand, we will see that love and commitment can and should be distinguished. An important point is the one that Rubin makes: Love for a person may involve no commitment to maintain a relationship with that person or to promote their welfare.

THE CONCEPTS OF LOVE AND COMMITMENT

There can be love without commitment, as illustrated by the Triola–Marvin example, but there can also be commitment without love, as in the marriage that endures even when, as the saying goes, "the love has gone out of it." And, of course, there can be both love and commitment or neither. The conceptual relation between the two is shown in Figure 7.1. A distinction is made between (1) the type of causal condition promoting interaction between two persons (the positive factors that draw them together versus all

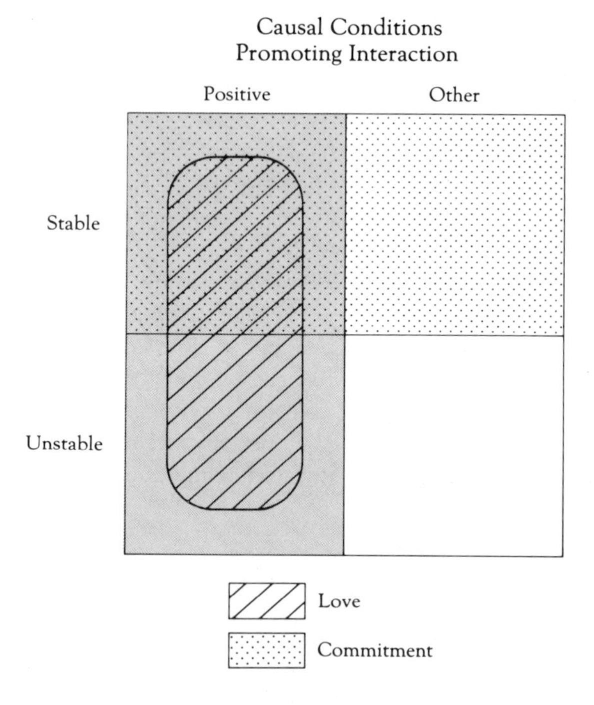

FIGURE 7.1
The conceptual relation between love and commitment. Among all the causal conditions that promote interaction, love *refers to a major subset of the positive conditions and* commitment *refers to all of the stable conditions.*

other causal conditions, such as constraints and external forces) and (2) the stability of these conditions (stable versus unstable). The phenomena of love are found in the diagonally lined area of the figure, love reflecting a particular subset of the positive factors that draw and hold people together. These positive factors are stable in some cases but unstable in others. The phenomena of commitment are found in the stippled area of the figure. As we will see, the concept of commitment summarizes *all* the factors that act stably to promote and maintain interaction between individuals, including the stable aspects of love.

The Importance of Love and Commitment

Love and commitment are members of a large class of global concepts that researchers have found useful in characterizing relationships and, particularly, in thinking about what draws individuals together to form and maintain relationships. This class of "attraction" and "cohesion" concepts (to use the most general terms) includes such other items as attachment, involvement, loyalty, dependence, and affection. All of these are intimately associated, both in the common person's thought and in scientific analysis, with the notion of "closeness." As we will see in this chapter, a relationship characterized by both love and commitment can be expected to be "one of strong, frequent, and diverse interdependence that lasts over a considerable period of time" (see Chapter 2).

There are a number of reasons for giving special attention to love and commitment as representatives of the class of attraction and cohesion concepts. For one thing, they frequently occur in the questions people ask about their close relationships: How can I tell if I'm in love? How can I be sure my partner loves me? What can I do to get my partner to make a commitment to our relationship? A second reason for focusing on love and commitment is that their analysis takes us deeply into the personal and interpersonal processes involved in the formation and continuation of close heterosexual relationships. Both phenomena require the identification of the initial personal and social causal factors that shape the interaction between P and O in ways that then give rise to the P × O causal conditions (e.g., understandings, joint possessions) that draw and bind the two together. The causal loops we identify in the study of love and commitment constitute some of the most important processes that underlie the general course of development and change described in the next chapter.

A final benefit to be derived from the analysis of love and commitment is that it illustrates vividly the problems that arise when complex phenomena with rich everyday associations are dissected for scientific purposes. A brief elaboration of this point will set the stage for what follows.

Problems of Conceptual Analysis

There are certain general problems of conceptual analysis common to love and commitment that must be understood before we consider them in detail. These problems become apparent when we compare the various meanings given these concepts by different investigators, and even when we examine their dictionary definitions. Each term is applied, and legitimately so, to such different categories as actions, processes, states, and dispositions. For example, it is possible to speak of commitment as an action (Kiesler, 1971: a pledging or binding); a state (Webster's *Collegiate Dictionary*: the state of being obligated or emotionally impelled); or a disposition (Leik & Leik, 1977: unwillingness to consider any exchange partner other than the current one). Similarly, it is possible to speak of love as an individual process (Webster: feeling passion); an action of an individual (Webster: caressing, fondling); a state (Berscheid & Walster, 1978: a state of intense physiological arousal); a disposition (Z. Rubin, 1973: an attitude toward a particular other person); or an interaction between individuals (Webster: copulation). This is not to suggest that actions, processes, states, dispositions, and such can be neatly distinguished. In some conceptions, dispositions and actions blur together. For example, the "contents" of Rubin's "attitude" of love, a disposition, include behaviors (ignoring the loved one's faults; spending time just looking at the loved one) and feelings (misery when the loved one is absent). Definitions in terms of states may refer to a more or less stable condition (H. S. Becker, 1960: commitment is a position from which it is not easy to extricate oneself) but also to temporary states or processes (Berscheid & Walster, 1978: passionate love as a state of intense physiological arousal). Further complications arise because process definitions may call attention to the initiating events (the pledging or bonding that constitutes commitment) or the consequent course of events (a consistent line of activity in the face of inconsistent pushes and pulls).

The analysis of close relationship phenomena described in Chapter 2 helps us sort through this apparent confusion of definitions and ideas and see the order that underlies it. Consider Figure 7.2, which describes the general analysis of the dyad and its causal context. The figure shows the causal conditions (e.g., personal, interpersonal, social, physical) that, on the one hand, affect the events in the dyadic interaction and, on the other hand, are affected by those events. Within this broad framework, the analysis of any phenomenon entails separating out a particular cluster of interaction events and causal conditions that are thought to be causally interlinked in significant ways and identifying the nature of those linkages. As the history of this interconnected cluster is traced back, the analysis identifies earlier significant events, both within and outside the relationship, that are responsible for the present causal conditions and causal loops. As the implications

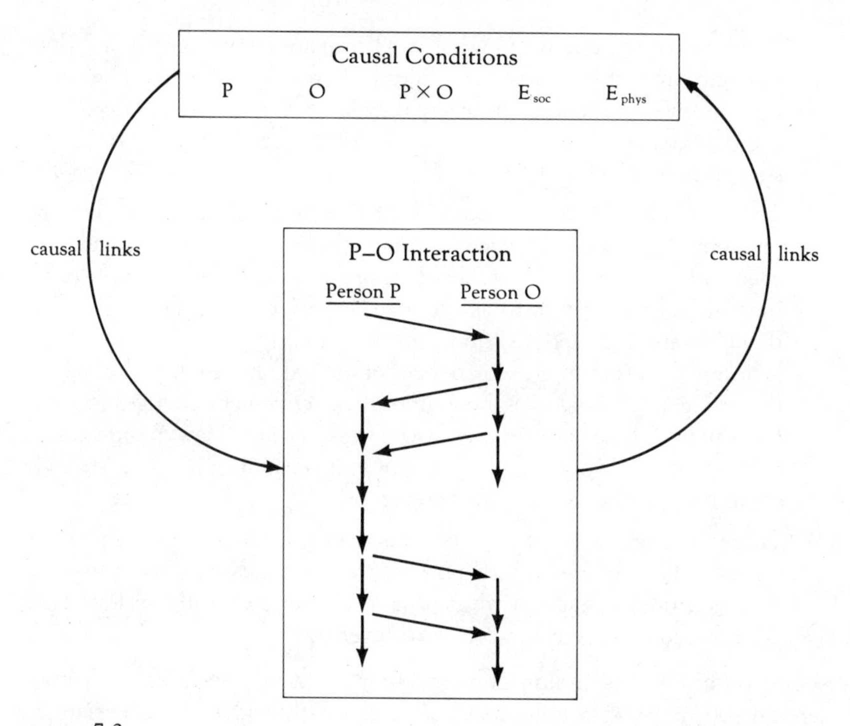

FIGURE 7.2
The causal context of dyadic interaction. This diagram suggests that the analysis of such a concept as love *or* commitment *must identify the interaction phenomena encompassed by the concept, the causal conditions thought to be linked to those phenomena, and the temporal course of the interaction, both past and future, that is implied by the causal loop between causal conditions and interactions.*

of this system are spun out for the future, the system predicts the probable changes in the causal conditions and in their connections to internal events.

We find each of the above components incorporated in the various conceptions of love and commitment. Any particular conception or theory of love or commitment has associated with it a cluster of ideas that includes the following four components:

1. Certain *observable phenomena* are identified—certain interaction phenomena, including actions and feelings, that are the characteristic manifestations of love or commitment. These phenomena are usually specified in answer to the question "How do we know it exists?" and are an important part of the answer to the question "How do we measure it?"
2. There are notions of the *current causes* believed to be presently responsible for the observed phenomena. These causal notions are often given

in answer to the question "What is it?" the answer being that "it" (love or commitment) is an attitude, a state, a position, and so on. The causal notions differ particularly in where they locate "it"—in the person(s); in the environment (situation, position); or between the two persons (e.g., shared understandings).

3. There are ideas about the *historical antecedents* of the current causes and phenomena. These antecedents include causal conditions that had their effects at an earlier point in the development of the relationship (e.g., the cultural ideology of romantic love that affected the persons' earlier thoughts and interactions) and the causal loops by which earlier interaction events (promises, experiences of arousal, and so on), themselves the product of initial causal conditions, have wrought changes in these conditions. These conceptions answer the question "How did it come to pass—that the two persons are deeply in love, that only one is strongly committed to the relationship, and so on?"
4. There are ideas about the *future course* of the phenomena, given the causes and processes currently under way. These ideas provide answers to the questions "What can we expect in the future? Will the love last? Will the commitment bind the two together?"

Given this understanding of the cluster of ideas associated with conceptions of love and commitment, we can see that it is not necessary to choose among actions, processes, states, dispositions, and so on in thinking about them. In fact, it is wrong to make any choice of this sort. Any comprehensive conceptual analysis must encompass all these phenomena. Love and commitment must be described by reference to interaction phenomena that can be understood in terms of both contemporaneous and historical causes and that have implications for the modifications of various causal conditions and, therefore, for the future course of the relationship.

The various analyses presented by different scientists may differ in the particular part of this cluster used as a starting or focal point. It is in focusing on different components of the cluster that people—common people and scientists alike—use a concept with seeming inconsistency and divergency. However, all analyses (or any that can claim comprehensiveness) will explicitly specify all four facets of the phenomena.

Goals of This Chapter

For each concept—love and commitment—this chapter will identify the components described above. In neither case will an attempt be made to provide a comprehensive review of the literature. Rather, these will be two exercises in analysis, designed to illustrate how such complex phenomena as these can be dissected and structured in terms of the classes of concepts outlined in Chapter 2. Considerably more attention will be given to commitment than to love, for the reason that love has been more thoroughly analyzed in recent writings (e.g., Berscheid & Walster, 1978). Although the analysis will undoubtedly have relevance for other relationships (e.g., love for parents, commitment to work), our focus will be on love and commitment as they occur in heterosexual relationships.

LOVE

Common and Scientific Usage

Love has, of course, been in the common language far longer than in scientific usage. Furthermore, it is an important and widely used term in the common person's vocabulary as one person declares love for another and as the course of love is discussed and thought about. It is in the category of highest frequency in the Thorndike–Lorge word count (1944) and is one of the most frequently referenced words in Bartlett's *Familiar Quotations.*

At present, the scientific usage of *love* is rooted in common usage. It is the common person's description of the phenomena of love that pretty much defines the specific interpersonal events on which our analysis focuses. Furthermore, it is the common person's account of love that largely guides our conceptions of its causes and its dynamic course.

On the one hand, this dependence of the scientific analysis on common understanding is as it should be. People's beliefs about love are part of the causal conditions governing the behavior and feelings that occur in love. People's beliefs about the nature of love affect when they attribute love to themselves and to their partners. For example, if love is understood to be a powerful and irrational attraction, then the young lovers who are drawn to each other despite strong opposing reasons have a basis for concluding "We must be in love." Conceptions of the dynamics of love have similar consequences. The advice of "Dear Abby" that, if it's true love, it will last, becomes a prescription for drawing apart to see if each one's feelings persist. And, of course, people's conception of their own love constitutes an important criterion for what they do toward their partner and for how they respond to the partner's actions. For example, felt "love" may be a basis for making an open pledge to the partner and for responding in a self-sacrificing manner to the partner's interests. In these and other ways, common conceptions of love are among the causal conditions that shape its course and expression. The scientific study of love must take these conceptions into account, and they in turn must be traced back to the more distal causes—the cultural models and norms of love—that they reflect.

On the other hand, we must recognize the limitations of common understanding. These limitations include a failure to differentiate phenomena that are frequently associated, a tendency to reify process (e.g., to treat emotions as things), a tendency to locate causality within persons rather than between them, a proclivity for linear rather than circular causal analysis, and being satisfied with a causal analysis that provides a sufficient account rather than requiring a specification of necessary causes. Perhaps most important is the fact that reliance on the common usage of love, as either a descriptive or explanatory concept, limits our analysis to a particular time and place. A history- and culture-free conceptualization is only possible (if indeed, it is possible at all) when we identify the essential phenomena of love and understand the ways in which their underlying reality constrains their interpretation.

The Phenomena of Love

It has almost become a tradition to begin an essay on love with quotations from philosophical and literary sources. It is no longer necessary to do this. We need not learn what Plato, Shakespeare, deRougemont, et al. thought about love because we now know a good deal about what our contemporaries think about it. And it is, after all, contemporary love and loving behavior in which we are interested. Here, we will briefly review the major research findings about the thoughts, behaviors, and feelings associated with love.

Thoughts of love

Figure 7.1 shows love to reflect a portion of the positive factors that bring and keep people together. Prominent among these factors are positive attitudes. We are indebted to Z. Rubin (1970, 1973) for information about how the thoughts and expectations deriving from attitudes of love differ from those associated with other positive attitudes. Specifically, his research was oriented toward distinguishing between love and liking. He drew up separate lists of the thoughts and expectations associated with each attitude, had a number of judges check his hunches by sorting the items into two categories, and finally gave a number of the most differentiating items to some 200 college students. They used the items to express their thoughts about a lover versus a platonic friend of the opposite sex. A factor analysis was then conducted to identify the items best representing each type of attitude and to form two scales, the Love Scale and the Liking Scale. A second study provided further evidence of their distinctiveness, revealing that they were differentially correlated with global reports of being in love and were used differentially to characterize an opposite-sex dating partner and a same-sex friend. Thus, Rubin's research showed that it is useful to separate attraction (positive interpersonal attitudes) into loving versus liking. In doing so, the research also revealed the thoughts and expectations associated with love.

An examination of the Love Scale suggests that it has four main components. The first is *needing:* The person in love has a powerful desire to be in the other's presence and to be cared for by the other (e.g., "If I were lonely, my first thought would be to seek ______ out") and expects it would be difficult to get along without the other ("If I could never be with ______, I'd be miserable"). A second component is *caring:* The person in love anticipates wanting to help the other ("If ______ were feeling badly, my first duty would be to cheer him/her up"; "I would do almost anything for ______"). Beyond needing and caring, the Love Scale includes items reflecting willingness to establish mutual *trust* through exchange of confidences and willingness to *tolerate* the other's faults. These features of love—needing, caring, trust, and tolerance—are to be contrasted with the thoughts associated with liking. The latter consist largely of evaluations of a person as mature and competent and a willingness to endorse the person for responsible positions. Perhaps the Liking Scale might better be characterized as a measure of "respect."

Relative importance of care, need, and trust

A recent study by Steck, Levitan, McLane, & Kelley (1982) indicates that the different thoughts that constitute the components of the Love Scale may not be regarded by common people as equally important indicators of "love." Their method consisted of constructing different patterns of responses to the Love Scale (or to derivative versions of it) and asking undergraduate students to rate the degree of love they thought each pattern expressed. Thus, one pattern revealed that the person taking the Love Scale had agreed strongly with the need items and only mildly with the care items, whereas another pattern revealed that a different person had agreed strongly with the care items and only mildly with the need items. By appropriate comparisons, it was possible to determine which type of item contributes most to the impression that a person's answers reveal love. The results confirm an hypothesis suggested by Kelley (1979), that care plays a more important role than need in judgments of love. In other words, even though two patterns of answers to the Love Scale yield the same total "love score," if one shows strong caring and the other, strong needing, the former will be judged to reveal greater love. It was found incidentally that need plays a more important role than care in judgments of "attraction." Trust is less important than either care or need in judgments of love and attraction but plays an important role (equal to that of care) in judgments of "friendship." The three components affected judgments of "liking" in pretty much the same way they affected judgments of love. These results suggest that, contrary to the implication of Rubin's research, loving and liking may not be as sharply differentiated in people's minds as are the other concepts (e.g., love versus attraction versus friendship).

The great importance of caring in college sophomores' judgments of love may, of course, be time- and culture-bound. After all, "love" can be indicated by anything, depending on what meaning it is given in a particular society. However, it may be an important commentary on the contemporary middle-class usage of *love* that *caring* is its most distinctive component. The value placed on caring may constitute a somewhat unrealistic standard for relationships based on "love" insofar as our partners' capacities to set aside their own interests and to "care" for ours are limited.

Behaviors of love

Since Rubin's important work, Swensen (1972) has provided more detailed evidence about how love is expressed. This research is noteworthy in revealing the manifestations of love in a variety of close relationships, for example, in relationships with parents, siblings, and same-sex friends. However, here we will focus on the behavior toward the opposite-sex lover or spouse. Swensen first had a large sample of respondents list the ways they behaved, the things they said, and the feelings they had that differentiated their relations with persons they loved from those with mere acquaintances. Nearly 400 items derived in this way were then given to four samples of respondents (some 1,200 in all) who used them to describe their relationships with various loved ones. Factor analyses were conducted to reveal the component features of love. The results were moderately consistent over the four samples. Love for the opposite-sex partner was expressed (and more intensely so than was love in other types of relationships) in the following ways:

1. Verbal expression of affection.
2. Self disclosure: revealing intimate facts.
3. Nonmaterial evidence of love: giving emotional and moral support, showing interest in other's activities and respect for their opinions.
4. Feelings not expressed verbally: feeling happier, more secure, more relaxed when other is around.
5. Material evidence of love: giving gifts, performing physical chores.
6. Physical expression of love: hugging and kissing.
7. Willingness to tolerate less pleasant aspects of other: tolerating demands in order to maintain the relationship.

It is clear that Swensen's results reveal roughly the same range and types of interpersonal events as does Rubin's research. Again we see manifestations of needing (4, above), caring (3 and 5), trusting (2), and tolerance (7). In addition, there are verbal (1) and physical (6) expressions of love. These last two are generally the explicit ways of expressing the other components, especially needing and caring. There is, of course, an important difference between Swensen's and Rubin's results. Rubin's factor analysis, made to show the contrast between loving and liking, suggests that needing, caring, trusting, and so on form a single, homogeneous cluster of thoughts and behaviors *relative to* the different thoughts and behaviors characteristic of "liking." In contrast, Swensen's factor analyses, made exclusively within the "loving" domain, identify these same features as being distinguishable components of the broad phenomenon of love. The consistent factor patterns that Swensen is able to identify reveal that love is characterized more by needing for some persons, more by caring for other persons, and so forth. If love were equally composed of all components for all persons, the analyses would have yielded only one general factor. That the analyses yielded several factors is an important clue that love takes different forms. Some examples of these different forms are outlined later in a discussion of models of love.

Feelings of love

In his study of the "language" of emotion, Davitz (1969) had 50 respondents recall times when they had experienced "love" (as well as a number of other emotions) and then go through a lengthy checklist indicating which statements described the experience. We may take the 12 items checked by 35 or more of the respondents as indicating the typical feelings associated with love. These feelings include a number related to the components noted above: needing (desire to touch, hold, and be close to the other); caring (wanting to give of oneself to the other, wanting to be tender and gentle with the other); and trusting (sense of trust and appreciation). In addition, the consensus includes reports of (1) a warm inner glow, feeling optimistic and cheerful; (2) feelings of harmony and unity with the other, a belonging with the other from which other people are excluded; and (3) total concentration on—an intense awareness of—the other.

A number of Davitz' respondents, but not a majority, describe experiences of arousal, that is, excitement, quickening of the heart beat. The experience of arousal in love is further delineated in K. L. Dion and Dion's (1973) analysis of some 240 undergraduates' reports of their romantic love experiences. A brief schedule devoted to emotional symptoms yielded a single factor on which all six items had substantial loadings. The symptoms were (1) feelings of euphoria, (2) feelings of depression, (3) daydreaming, (4) difficulty sleeping, (5) agitation and restlessness, and (6) inability to concentrate. These symptoms seem mainly to consist of the subjective experiences of arousal and mood swings. The fact that they are encompassed by a single factor suggests two conclusions. First, there are variations among individuals in the degree of arousal and affective ups and downs they experience in their love relationships, and second, if a person experiences one element of this set, he or she tends also to experience the other ones. Once again, our attention is drawn to the possibility that love takes different forms, the evidence here pointing to different degrees of arousal and affective cycles.

Dion and Dion provide additional evidence of variations between persons in their feelings of love, in the results of a second factor analysis of 23 bipolar scales used to describe the subjective experiences of love. The results suggest that people differ in whether they experience love as "volatile" (changing, short, mysterious); "circumspect" (subtle, cautious, passive); "rational" (logical, systematic, rational); "passionate" (sensual, physical); or "impetuous" (sudden, fast). The implication for our scientific analysis of love is that we must decide whether to regard it as a single entity that can be experienced in a number of different ways or as a number of different phenomena, each of which generates its own unique pattern of feelings and behavior. The different models of love (described below) imply the latter.

Sexuality and love

What is the role of sexual desire and behavior in the love that exists in heterosexual relations? There have been various allusions to sexuality in what we have already reviewed, for example, in the references to the warm feelings and arousal that accompany love and to the strong desire to make intimate contact with, to possess, and to be fulfilled by the other person (the needing component of love). However, in general, neither writers on the subject nor common people treat sexuality as a *sine qua non* of love. Evidence on the common conception is provided by Forgas and Dobosz (1980). In judgments of 25 different heterosexual relationships, undergraduate students were found to distinguish between those involving "sexuality" and those involving "love and commitment." The former are exemplified by brief sexual encounters in contrast to long-lasting nonsexual or platonic relationships. "Love and commitment" relationships are exemplified by long-lasting marriages, a young marriage after an extended courtship, and even "love at first sight" that results in a brief but intense relationship followed by engagement. These relationships are contrasted with brief flirting or sexual encounters. For sake of completeness, we may note that Forgas and Dobosz' respondents also discriminated between the different relationships in terms of their desirability, the undesirable ones including not only counternormative relations (extramarital affairs, affair between teacher and pupil) but also one-sided ones and relationships entered into for extraneous reasons (marriage after unwanted pregnancy, dating relationship maintained to impress peers).

The result obtained by Forgas and Dobosz, that love and commitment appear in combination as one of the dimensions of discrimination between relationships, appears to conflict with the opening theme of this chapter, that they are partially independent, though overlapping concepts. The problem is resolved when we examine the relationships provided to the respondents to judge. These relationships did not include either examples of long-term relationships that finally break up (such as might, for a time, involve love, but without the stability of commitment) or examples of relationships that endure without positive attitudes (such as a marriage from which love has gone). In the absence of such examples, it is not possible for the respondents to draw distinctions between love and commitment. The reason for the omission of such examples is that the list of relationships was drawn largely from the personal experiences of undergraduate students. In this study, as in similar analyses, the results are constrained by the particular sample of stimulus materials employed and must be interpreted in that light. The occurrence of love and commitment on the same dimension does not necessarily reflect, as Forgas and Dobosz (1980) suggest, a "deep-seated Western misconception that romantic love is the only true path to an enduring and permanent relationship . . . " (p. 296).

With respect to the matter of sexuality and love, Forgas and Dobosz' results show that undergraduates believe that two persons can interact sexually without there being any love between them and, moreover, that there can be long-lasting, close relationships between sexually mature males and females without any overt sexual activity. However, these findings must not be taken to mean that sexual drives do not often play an important role in the development of heterosexual love or that ordinary people (in this case, undergraduate students) are unaware of this role.

The biological functions of heterosexual relations are procreation and ensuring the procreativity of one's offspring. From an evolutionary perspective (Mellen, 1981), it is inconceivable that the relations between men and women in their years of sexual maturity would not be promoted and sustained by sexual drives and behavior. Ultimately, our social–psychological analysis of close heterosexual relationships must take account of the biological requirements. It seems likely that the various components of love—caring, needing, trusting, tolerance—can profitably be viewed from an evolutionary perspective, as relationship-binding dispositions that make possible and sustain the activities by which the continuance of each partner's genes is ensured. We cannot expect an evolutionary perspective to figure explicitly in the ordinary person's conceptions of love. However, it is reflected in their needs for intimacy (including sexual relations with members of the other gender) and in their susceptibility for developing caring and supportive attitudes toward the sexual partner and the shared offspring.

The Causal Conditions of Love

The contemporaneous causes of love, which accont for its manifestations at any given point in time, can be characterized in terms of P, P × O, and E causal conditions. Z. Rubin (1973) conceptualized love as "an attitude that a person holds toward a particular other person" (p. 212). This characterization of what love "is" amounts to the assertion that the manifestations of love are caused by a factor located within each person, that is, by a P causal condition. As Z. Rubin (1970) explains it, ". . . love is an attitude . . . involving predispositions to think, feel, and behave in certain ways toward that other person" (p. 265). Love is differentiated from other positive attitudes by the particular thoughts, feelings, and behaviors to which the individual with attitudes of love is predisposed, these being indicated by the Love Scale.

The notion that love is caused by P factors has the important implication that there need be no necessary relation between P's love for O and O's love for P. Love may be quite unilateral, as in "unrequited love" or "love at a distance." Rubin presents evidence consistent with this notion in his finding that the correlation between the scores of the male and female members of dating couples is a modest .43. Thus, it is possible for there to be a sizeable discrepancy between the two persons in the extent of their love for each other. This, of course, does not preclude the possibility that, as relationships move toward greater seriousness and commitment, beyond the dating stage at which most of Rubin's respondents were found, the two persons' attitudes of love typically become equally intense.

A variant of the conception of love as an attitude views it as a special P × O causal condition, reflecting the interplay between particular properties of the loved one, on the one hand, and special vulnerabilities or needs of the lover, on the other hand. This causal hypothesis is illustrated by the common beliefs that people fall in love only with certain kinds of partners (e.g., those with characteristics like their parents, those who express needs the lover has repressed, and so on) and that people tend to make the same "mistake" over and over again as they fall in and out of love with successive partners. This type of P × O effect is represented by the complementary need hypothesis of mate selection, discussed in Chapter 8, "Development and Change." One doubts that statistical studies of complementarity between married persons are likely to capture the many and varied ways in which one person can have special appeal and significance for another as a potential love partner. This common belief has gone untested and by its very nature—with its emphasis on the unique stimulus value that one person can have for another—may be exceedingly difficult to test.

A different P × O causal analysis views love as a relationship condition, that is, as a factor emergent from the interaction between P and O. For example, Swensen (1972) describes love as "a positive, enjoyable, constructive interpersonal relationship" (p. 87). This relationship is not studied directly but is observed only in its consequences. These consequences are the objects of Swensen's research, that is, "the things people who love each other do for each other or say to each other, or say they feel about each other" (p. 89). This view implies that persons in love not only hold positive attitudes toward each other but also have a strong sense of the shared nature of their attitudes. Accordingly, each one will be inclined to talk about "our love" rather than "my love," and to perceive their love as involving reciprocity of feelings. Swensen describes evidence consistent with this view. His respondents perceived there to be a high degree of similarity between the ways in which they expressed love for their partners and the ways in which the partners expressed it in return.

Both of the P × O causal analyses imply that the love between P and O is likely to be experienced by them as unique—as something never experienced before in other relationships. The stimulus person who possesses the set of features to which the lover is especially attuned will arouse new and unusually intense feelings in the lover. The pair of persons who evolve their own special pattern of trusting and caring will experience their relationship as involving distinctive interaction. They will have a sense that the satisfactions they derive from each other are unique and wholly unavailable in other relationships. This sense of uniqueness is the basis for Foa and Foa's characterization (1974) of love as the most particularistic resource involved in interpersonal exchange. "It matters a great deal from whom we receive love since its reinforcing effectiveness is closely tied to the [particular] stimulus person" (p. 346). With experience in heterosexual relationships, the perceived uniqueness of love based on P × O factors is likely to decline. The individual is likely to learn that the experience of love is repeatable with different partners—that a number of partners possess the magical characteristics or that intimacy can be successfully worked out with a variety of partners. With this realization, the person will move from what K. L. Dion and Dion (1973) identify as the idealistic view of love, that there's only one real love for a person, to a more pragmatic view, that real love need not be limited to a particular partner.

The E_{phys} and E_{soc} conditions that affect the love between P and O have not been the subject of much scientific attention. Common lore attaches importance to both types of conditions but includes contradictory views of their operation. Environmental impediments are believed to promote love (the notions that absence makes the heart grow fonder and that love is spurred by interference from parents and family), but they are also believed to inhibit or erode love (the notion of "out of sight, out of mind"). There is probably some truth to both views, depending on the nature of the impediments and the type of love involved. For example, the obsessive thought about the partner that seems to accompany, and perhaps to promote, passionate love (see below) may be enhanced by a certain degree of separation from the partner. Other relevant E_{phys} and E_{soc} conditions affect the degree to which the pair has the time and privacy necessary to sustain intimate interaction. Employment and economic conditions, housing arrangements, the dyad's social network, and other such causal factors undoubtedly affect the ease with which the needing, caring, trusting, and tolerance that constitute love can be maintained in the close relationship.

Models of Love

Our analysis of the phenomena of love does not reveal how the various thoughts, behaviors, and feelings may be causally and historically interconnected. For example, needing the partner may precede tolerance for the partner, the latter developing as a means to maintain the relationship. Trusting the partner may precede the development of high levels of needing, as we permit ourselves to become highly dependent only on partners who, we believe, can be trusted not to exploit our dependence. Our brief review of the causal conditions of love suggests some of the contemporaneous determinants of the phenomena of love but does not deal with its development and change over the course of time. All of these aspects—causal conditions, interaction events, contemporaneous connections, and developmental changes—are brought together in a "model" of love. A model consists of a package of notions about the causal conditions of love, their effects on interaction, and their temporal course, as they are changed through external events and are modified by the interaction itself. A model implies a scenario of the initiation of a love relationship and its most likely paths over time.

We have seen that love has different components and may take different forms. The single word *love* refers to different phenomena—to different clusters of caring, needing, emotion, and so on. Consequently, in both common lore and scientific thought, there are a number of different models of love. It is important to realize that these models are *not* alternative, competing views of a single phenomenon, "love." Rather, they are conceptualizations of different phenomena, each of which has historically been termed *love.* The different models are addressed to the major forms or types of love. They imply that one person's "love" for another should always be qualified as to the type or combination of types it involves.

The commonsense models of love vary historically (Gadlin, 1977) and across cultures. Evidence on the latter point is provided by Triandis (1972) in a broad study of the perceived antecedents and consequents of various phenomena. Although the evidence is not entirely appropriate for our present purpose inasmuch as heterosexual love is not distinguished from love in other and more general senses, the study is interestingly suggestive of possibly differing views of heterosexual love. Data were obtained from over 1,200 young male students in Greece, southern India, Japan, and the United States. All four samples gave "affection" as an antecedent of love and "friends" as a consequent. Beyond that, there was little commonality across the four samples. The conceptions of the Japanese and United States samples were rather similar and are particularly interesting in that a number of perceived antecedents of love were identical with its perceived consequences, namely, positive affect (happiness, joy) and trust. This identity suggests that the common models in these cultures may include an understanding of circular causal connections between interaction events and causal conditions. That is, positive affect and trust are seen both to promote love and to be promoted by it. The consequents reported by the United States sample also included items related to the caring component noted earlier: concern for others, sacrifice, and goodness. These features were not especially common in the other three samples. This finding implies that one of the components we have been emphasizing in this chapter may not be of universal importance. The causal models in the Greek and Indian samples seem more linear, linking sexual attraction to "marriage" (a consequent not frequently mentioned in either the U.S. or Japanese sample). Scenarios of love with unhappy endings may be more common in the two Asian countries inasmuch as "sorrow" was a commonly reported consequence for the Indian sample and "disappointment" for the Japanese sample.

Lee (1977) describes several conceptions of the causes and course of love that he believes are common in Western culture. One example is that of *erotic love.* An important causal (P) condition in this case is the person's clear image of the ideal physical type that most strongly attracts him or her. There is an instant appreciation of the extent to which the beloved fulfills the ideal, and, once this appreciation occurs, the lover is eager for rapid disclosure, sexual intimacy, and frequent interaction. A second example is that of *ludic love,* which occurs for the person disposed to like a variety of types of partner and to avoid commitment, and who easily switches from one partner to another. The ludic lover doesn't "fall" in love but calmly fits it into a daily schedule. Both one's own and the partner's feelings of involvement are kept at a low level, and a relationship is ended when it is no longer diverting. A third pattern, *storgic love,* flows out of a prior friendship based on common interests and activities. This initial P × O condition gives rise to the further one of mutual commitment, and the latter tends to be followed by sexual self-disclosure.

Lee's work identifies various P and P × O causal conditions that, according to people's accounts of their personal experiences with love, give rise to various forms and scenarios of love. His writings are important in calling attention to the fact that any one of these conceptions of love may become an ideology that is used to justify and sustain certain social institutions and practices relating to courtship and marriage. Lee highlights the need to understand why one ideology of love is predominant at a particular social–historical juncture and is then displaced by another one.

Like common thought about the matter, the scientific study of love is characterized by a number of different models, referring to different forms that love may take. The three to be outlined below represent some of the major variations in scientific conceptions of love. These will be described in sharply contrasting terms in order to highlight differences between them. In reality, heterosexual love is usually some mixture of the phenomena and causal conditions described by these models—a mixture that shifts over time as the relationship develops and the lovers undergo change.

The model of passionate love

Berscheid and Walster (1978) describe passionate love as a "state of intense absorption in another . . . a state of intense physiological arousal" (p. 176). (This kind of love is sometimes referred to as "romantic" love. However, the term *passionate* is preferrable because "romantic" has also been commonly used simply to refer to heterosexual love in contrast to filial and kin love.) The model of passionate love emphasizes the need-related features of the phenomena. For example, Driscoll, Davis, and Lipetz' (1972) conception of passionate love overlaps with Rubin's Love Scale in including affiliative and dependency needs, feelings of exclusiveness, and absorption, but not the caring, trusting, or tolerance features. Driscoll et al. also include physical attraction, passion, and idealization of the partner.

It is the state of arousal and the conditions giving rise to it (sexual maturity, unmet sexual needs, and availability of an attractive partner) that play a central causal role in this model. Obstacles to the formation of the relationship and to sexual fulfillment within it are thought to increase the intensity of passionate love through promoting fantasies about the partner, sustaining the state of arousal, and creating feelings of unity against outsiders. Some evidence that parental interference stimulates rather than deters passionate love is presented by Driscoll et al. (1972). Even irrelevant sources of arousal may give rise to the appropriate feelings if other conditions are favorable (Dutton & Aron, 1974).

The temporal course of passionate love is described in Chapter 4, "Emotion," (where it is referred to as "romantic" love). Its onset is thought to be sudden and its duration brief. The person has little control over the process inasmuch as it is basically a matter of emotion rather than of deliberation and choice. Berscheid (1980) has observed that the common belief that this type of love is uncontrollable makes it useful as a justification for otherwise unacceptable behavior, such as marital infidelity. The instability of passionate love locates it in the lower left portion of Figure 7.1, as a spectacular example of the unstable positive factors that draw people together but do little to keep them that way. By itself, passionate love is not a sound basis for a permanent domestic relationship. The abrasions of ordinary living dull the idealization of the partner, and regular sexual gratification and growing predictability of the partner can reduce high levels of arousal. However, passionate love may provide the conditions under which other forms of love, such as those described by the following models, may develop.

The model of pragmatic love

This model emphasizes the trust and tolerance features of love. The description of conjugal love given by Driscoll et al. (1972) provides an apt characterization of pragmatic love. It is the type of love that occurs between mature adults and is common in lengthy relationships, such as marriage. It evolves from interaction that both persons find satisfying and from which they gain mutual trust. They learn to tolerate each other's idiosyncracies in order to sustain the relationship and the need gratification it provides. Similarly, there is caring for the partner—attention to the partner's needs and efforts to assure their satisfaction—in order to obtain and maintain reciprocation of caring from the partner.

As compared with passionate love, relationships of pragmatic love develop more slowly and under greater control by the two persons. The passionate lover is likely to blurt out his or her feelings, revealing the dependence on the partner and not holding anything back. Pragmatic relationships develop with greater deliberation and self-control. Rather than "falling" in love, each person can be said to let or make himself or herself become attracted to the other one (Kelvin, 1977), and this process is regulated by cues indicating the mutuality of the growing dependence. Accordingly, pragmatic relationships are likely to be those in which the two persons' feelings of love are relatively equal and in which each person sees their behaviors of love as being reciprocated by the partner. Furthermore, pragmatic relationships are likely to be equitable ones, each person feeling that what they get out of the relationship relative to what they put into it (their outcome-to-input ratio) is roughly equal to their partner's outcome-to-input ratio. This balance presumably results from a loose kind of negotiation in which inputs and outcomes are adjusted to the two persons' mutual satisfaction. However, it must not be imagined that this negotiation can be very explicit. On this point, Blau (1964) has important things to say in his "excursus on love" (pp. 76–85). He notes that an important part of what each person seeks in the love relationship is the rewards provided by the partner's spontaneous positive evaluations. To have value, they must come from a person whom one respects, they must be well informed, and they must be given spontaneously. Thus, the partner's words "I love you" have high reward value if the partner is loved, if they are said when, after trusting disclosures, the partner knows the real "you," and if they are spontaneous rather than coerced. The latter condition puts severe constraints on the negotiation process by which expressions of love become exchanged inasmuch as the authentic expression cannot be commanded or elicited through threat or wheedling.

Pragmatic love undoubtedly depends on compatibility between the P and O factors that the two persons bring to their relationship (similar interests, matched or complementary needs, e.g., Centers, 1975) and supportive E_{soc} and E_{phys} conditions. However, it seems likely that the more important causal conditions are various $P \times O$ causes that emerge from and grow with the interaction: e.g., shared awareness of mutual trust, mutual understanding that each is truly known by the other and is fully appreciated and respected, and dyadic agreements about sharing and loyalty. In contrast to the passionate model's emphasis on a special person (i.e., a partner with special characteristics), the pragmatic model emphasizes a special process whereby these $P \times O$ causes evolve (Schwartz & Merten, 1980).

Recent scientific studies of love provide a number of indicators of the presence of relationships that conform to the pragmatic model. Consistent with the model is Walster, Walster, and Traupmann's evidence (1978) that dating relationships perceived to be equitable are expected to be more stable and are more likely in fact to endure. This finding suggests that young adults assess the mutuality of their growing dependence and do not pursue the relationship if the reciprocity becomes too attenuated. This interpretation is supported by evidence from C. T. Hill, Rubin, and Peplau's (1976) longitudinal study of dating couples: Those who initially reported unequal levels of involvement in the relationship were more likely than others subsequently to break up. We may imagine that the nonmutual relationship is broken up in a kind of implicit bargaining, with the less dependent person leaving it for a better alternative or the more dependent person leaving it after finding that his or her position is too insecure and exploitable.

The self-regulation aspect of pragmatic love assumes that people do not let themselves become "in love" unless the implied dependence is accompanied by trust. Some evidence consistent with this view is provided by Berscheid and Fei (1977). Persons who are not sure whether they are in love report a moderate degree of dependence on their partners but high insecurity about the partner's feelings toward them. In contrast, persons who report being in love have high dependence along with low insecurity about the partner. The pragmatic model implies that, for the latter respondents, love followed the feelings of security. The self-regulation aspect of pragmatic love is also suggested by K. L. Dion and Dion's (1973) findings that young people characterized by feelings of internal locus of control (that they control their own fates) are less likely than other persons to have experienced "romantic" love and to have experienced their heterosexual attachments as being mysterious and volatile.

The model of altruistic love

This model emphasizes the caring component of loving attitudes and views caring behavior as intrinsically motivated rather than performed to elicit similar behavior from the partner. Altruistic love is usually thought to be epitomized by "mother love," in which the mother sacrifices her own interests and is wholly oriented toward promoting the child's welfare without any thought of reciprocity. This model assumes that altruism also exists in some heterosexual relationships (or to some degree in many such relationships). Some authors view the altruistic model more as a cultural ideal than as something actually achieved in heterosexual relationships (e.g., Lee, 1977). However, other authors (e.g., Z. Rubin, 1973) see altruism (caring) as commonly occurring, along with needing and dependence, in the development of mature love. The latter view is consistent with Kelley and Thibaut's logical demonstration (1978) that, in relationships of high interdependence, there are mutual benefits to be derived from behavior that is responsive to the other person's interests as well as one's own and that is motivated by considerations of joint welfare.

Clark and Mills (1979) draw a general distinction between communal and exchange relationships. As applied to heterosexual relations, the former would correspond to the altruistic model of love and the latter to certain features of the pragmatic model. Their work is important in suggesting the manner in which a person demonstrates a communal orientation (altruistic attitude) toward the partner. The key idea is that any gift or help given the partner must be dissociated from the partner's prior help or gifts and from any expectation of such help or gifts in the future. Accordingly, a gift or help should not immediately follow the partner's, should differ from the partner's (to avoid any appearance of being a repayment), and should not be followed by any suggestion of how the partner might benefit you. The general principle is that altruistic benefits to a partner are geared solely to the partner's needs and involve no consideration of one's own needs, whether past, present, or future.

In enduring relationships, it is very difficult to distinguish altruistic behaviors from helpful behavior motivated by expectations of eventual reciprocity. The rules for showing altruism, such as those suggested by Clark and Mills, are well known to ordinary people and therefore afford the basis for favorable self-presentations that may misrepresent a person's true motives. There is much to be gained from convincing our partners that we are attuned to their interests and willing to put them before our own. And there is even more to be gained, for many persons, from convincing themselves of their beneficent motives. Even authentic altruists must often be aware of the potential benefits of their sacrificial actions and often become uncertain in their own minds that those benefits do not figure in their motives. Of course, the best evidence of altruism is its persistence in the face of total non-reciprocation. However, such persistence is an unreasonable criterion, whether for purposes of scientific measurement or for the judgment of our associates. The ultimate breakdown of unreciprocated considerate behavior does not necessarily mean that it was not motivated by genuinely altruistic feelings at the outset. After all, there is nothing in the meaning of altruism that requires it perpetually to be tolerant and forgiving.

The mistaken assumption that altruism must be wholly stable reflects the common belief that it reflects a stable P causal condition or, as in the case of mother love, a stable P × O condition defined by the particular mother–child combination. One such causal condition is genetically based altruism. Recent advances in understanding the evolution of altruistic behavior (e.g., as summarized by Dawkins, 1976) hold promise for future insights into altruistic tendencies within human relationships. Even now they suggest that parental altruism toward children is unlikely to be stable over the life of the relationship. Instead, it is likely to vary with such things as the maturity of the offspring and its apparent vigor. The evolution of altruism between father and mother is a much more complicated matter than that of altruism between parent and child or brother and sister. From a genetic perspective, each prospective parent is interested in putting his or her genes into association with other genes that have high viability and in doing so with a mate who is genetically inclined to provide faithful support for the survival of the joint offspring. As the implications of these genetic interests are drawn out in the sociobiological literature, they provide a goldmine of provocative ideas for the close relationship researcher, concerning the occurrence across the animal kingdom of such familiar phenomena as sexual attractiveness, female coyness, and father desertion.

Love in Close Relationships

In actual close relationships, love is typically a blend of the different forms described by the preceding models. Different degrees of needing, caring, trust, and tolerance are involved, and the various behavioral, cognitive, and affective features of these components are not always entirely consistent. One may greatly need the partner but not be able to bring oneself fully to trust him or her. A person one deeply cares for may have important characteristics that are difficult to tolerate. Ambivalent feelings about the loved one are probably more common than consistent feelings. Although loving can be distinguished from liking or respect, the latter type of positive attitude is also usually important in close heterosexual relationships. Rubin's original research revealed that liking for one's dating partner is positively correlated with love for him or her. However, the correlation is low, so the person one loves may not always be a person one respects. Lack of respect is one of the common causes of complaints about heterosexual partners (Kelley, 1979).

Our conceptions of love undoubtedly change with our experience in successive relationships. A case study presented by Schwartz and Merten (1980) is an exceedingly useful illustration of how one person's model of love may differ from that of the partner, may not be understood by other people, and may undergo change and crystallization over a series of encounters. Similarly, the blend of the different forms of love that characterizes a particular relationship undoubtedly changes if it lasts for any considerable period of time. As the relationship develops, initial passionate love may afford the setting within which trust and understanding develop and the relationship can take on some of the features described by the pragmatic model. Altruism may be promoted by the birth of children or by the onset of various problems (illness, dislocation of residence) that stimulate awareness of commonality of fate and the necessity of mutual support.

Few would deny that love, in its various forms, is close to the core of human existence. It is through its complexity, diversity, and successive changes that we are afforded unending fascination with love.

COMMITMENT

Among authors writing about commitment, there has been great unanimity about the phenomena to be accounted for by the concept. However, the crucial features of the causal conditions underlying these phenomena have managed to elude clear specification. It will be useful here to show how these features of causal conditions are implicit in prior uses of "commitment" and to explain the previous failures to identify them. It will also be useful to point out how the failure to specify these key features has led some authors to focus on certain common though nonessential aspects of commitment and, as a consequence, to develop definitions and measures that overlap in confusing ways with other established concepts.

The Phenomena of Commitment

When we say a person is committed to something, we mean that he or she is likely to stick with it and see it through to its finish. A person committed to an activity is expected to persist in it until the underlying goal is achieved. A person committed to a relationship is expected to stay in that relationship, "through thick and thin," "for better and for worse," and so on. Thus, commitment focuses our attention on the relationship property of *duration.* Commitment is to be measured and analyzed primarily in order to help us understand why some relationships endure while others do not.

The anchoring of commitment in the phenomena of behavioral consistency or, in the case of relationships, in stability of membership, is agreed on almost without exception by the various social scientists who have written about the concept. In his now classic paper on commitment, H. S. Becker (1960) observes that commitment is used "to account for the fact that people engage in consistent lines of activity" (p. 33, italics omitted). In a subsequent paper (1964), he describes commitment as "an approach to the problem of personal stability in the face of changing situations" (p. 41). In M. P. Johnson's recent review of commitment (1978), it is clear that the different phenomena for which the concept is invoked include continuation of a line of action once begun and stability of social relationships. Rosenblatt (1977) incorporates the stability notion in his definition of commitment as "an avowed or inferred intent of a person to maintain a relationship" (p. 72–73, italics omitted). Rusbult (1980) specifies that commitment is inversely related to the probability that a person will leave a relationship. Research on the measurement of "organizational commitment" (Steers, 1977) is motivated by the hope that it will be a better predictor of turnover in employment than are other measures, such as job satisfaction. In their concern about stability of membership, some authors (e.g., Berscheid & Walster, 1978) focus on instances in which a person maintains membership even though certain factors favoring membership (e.g., its attractiveness) are weak, and others (e.g., Leik & Leik, 1977) focus on instances in which membership is maintained even though the factors favoring *non*membership (e.g., alternative relationships) are strong.

Beyond endorsing the relevance of commitment for behavioral consistency and stability of membership, the writings about the concept show little uniformity. As we will see, some authors focus on certain current causal conditions and others emphasize events that, historically, in the development of a relationship, are important in promoting commitment. Some writings specify relationship properties in addition to stability that can be linked to commitment, such as control over group members (Kanter, 1972) and consistency of inputs to the group (J. Scanzoni, 1979a). Distinctions are sometimes made between types of commitment, all of which are relevant to membership stability but which differ in their other consequences (e.g., voluntary versus involuntary commitment: both affect stability, but they have different implications for willingness to work to improve the relationship).

The present analysis will focus on commitment as a more or less unitary concept related to membership stability, leaving for later a brief discussion of its types and its other consequences. Our analysis is directed at the critical features of the causal conditions underlying membership stability, something alluded to in various ways by previous writings on commitment but made explicit by none. To understand these features, it is necessary now to develop our own understanding of commitment.

The Causal Conditions of Commitment

A distinction must first be made between the causal conditions that promote a person's remaining in a relationship and those that promote his or her leaving a relationship. For simplicity, we may refer to these as the *pros* and *cons* of membership. A number of analyses of these causal conditions have been presented in the commitment literature, usually drawing on Lewin's analysis of the forces acting on a member of a group (e.g., M. P. Johnson, 1978; Levinger, 1976) or from the derivative analysis of Thibaut and Kelley (1959) of the factors determining a person's attraction to and dependence on a group (e.g., Berscheid & Walster, 1978; Rusbult, 1980). On the *pro* side are *all* the causal conditions that act to keep a person in a relationship. These conditions include love for the partner, the desirable activities and status the relationship makes possible, the costs that would incurred on leaving it (e.g., legal fees, negative social sanctions), feelings of obligation to sustain the relationship, and so on—in short, all the benefits and constraints that push the person toward and hold the person in the relationship. On the *con* side are *all* the conditions that act to push or draw the person out of the relationship, such as the psychological costs (effort, anxiety) experienced with the partner, the attractiveness of alternative relationships that are precluded by the present one, internal and external pressures to experiment with such alternatives, and so on.

We may assume that the magnitude of these pros and cons of membership fluctuate over time. Each causal condition varies in its strength (sexual needs wax and wane, relationship costs vary, and social conditions change), and the resultant of all the pros and cons varies accordingly. We may further assume that the person remains in the relationship as long as the pros outweigh the cons. This is not to say that the person is constantly monitoring the pros and cons and drawing a balance between them. Such balancing may or may not be part of the process, as discussed below in the analysis of proximal versus distal causes. Nor is it assumed that a person leaves a relationship when the cons momentarily outweigh the pros. This decision depends on the person's psychological perspective and the time span over which the pros and cons are aggregated (also discussed below). Here we simply presume that, in general, membership is sustained as long as, over the course of their varying magnitudes, the causal factors promoting membership are more effective than those promoting disruption.

If we are correct in asserting that the central phenomenon of commitment is stability of membership, then it follows that the key feature of the relevant causal conditions is the consistency with which, over time and situations, the pros outweigh the cons. In other words, if we shift from the phenomenon of commitment to its conceptualization in terms of underlying causal conditions, we must refer to the person's being in a causal system that stably supports continuing membership. At this point, a metaphor from statistics is useful. *If membership is to be stable, the average degree to which the pros outweigh the cons must be large relative to the variability in this difference.* Stability of membership reflects two outputs of the system of causal conditions relating to membership: (1) the average level of the pro–con difference and (2) the variance of this difference. Thus, different cases of stability can be distinguished, for example, one in which the average pro–con difference is small but the variance is also small, so the difference never becomes zero or negative; another in which the variance is large but the average level is so high that, again, the difference never dips below zero, and so on.

This conceptualization of the features of the causal conditions underlying commitment is simple and, even, obvious. However, it reveals the ways in which prior conceptualizations of these features have been overly limited. Some have focused exclusively on the average level of the pro–con difference, and others have (implicitly) focused exclusively on the variance. For convenience (and *not* to propose a new concept), the pro–con difference will hereafter be described as the person's "adherence" to the relationship.

Conceptions relating to the average level of adherence

The conceptualization of commitment in terms of average level of adherence is illustrated by the work of Rusbult (1980). This investigator conceives of commitment as being a positive function of (1) the outcomes (rewards minus costs) the person has experienced in the relationship and (2) the investments he or she has made in the relationship (e.g., the time, money, and affect put into developing the relationship), and a negative function of (3) the attractiveness of available alternatives to the relationship. Rusbult shows that these factors are related as predicted to self-report measures of commitment, which include such items as estimates of the probable duration of the relationship and feelings of attachment to it. This work shows that the higher the average level of adherence to a relationship, the more stable it is perceived to be. Rusbult gives no attention to possible variations in this level over time or situations. The implicit assumption is that variance is unrelated to average so, the higher the average, the less likely it is that adherence will ever become zero or negative. This assumption is not unreasonable, but it must be made explicit, and consideration must be given to possible cases in which it is not justified.

Another instance of identification of commitment with average level of adherence is found in Berscheid and Walster (1978). These authors describe commitment as being determined by the person's outcomes in a relationship relative to those attainable in the best available alternatives to the relationship—the Comparison Level for Alternatives. Commitment is distinguished from attraction, which is determined by the person's outcomes relative to his or her expectations about such relationships—the Comparison

Level. Berscheid and Walster note that this distinction enables us to account for instances in which a person remains in a relationship even though it is unattractive. This kind of adherence was one of the phenomena that had led Thibaut and Kelley (1959) originally to make the distinction between the two kinds of comparison levels. More generally, they had differentiated between attraction and dependence, the latter corresponding to what Berscheid and Walster refer to as "commitment," and had used this distinction to analyze the phenomena that distinguish voluntary from nonvoluntary relationships.

It may be noted that Berscheid and Walster have replaced the concept of "dependence" with that of "commitment." Such replacement is a common consequence of conceptions of commitment based exclusively on the average level of adherence: "Commitment" is not clearly distinguished from other existing concepts having to do with cohesion and attachment. One may ask why "commitment" is more appropriate in this context than the concept "dependence" proposed by Thibaut and Kelley. The same question can be raised about Rusbult's conceptualization of "commitment." In both cases, in seeking to account for membership stability, the authors have focused on the *average level* of adherence (which is perfectly well described by "dependence") and have overlooked the other necessary feature, the *temporal variability* in degree of adherence. In each case, we can ask what relevance "commitment," as defined by the authors, has for membership stability. In each case, the answer is that it bears no *necessary* relation to stability. These limited conceptions of "commitment" are relevant to the phenomena of membership stability only on the assumption that the higher the average level of adherence, the less likely it is, over time and circumstances, that the level will become zero.

As to why these authors have overlooked variability in their conceptualization, we may speculate that the omission has been encouraged by the special attention given to the problem of membership maintained without positive attraction to the relationship. This problem is successfully resolved, *though only for a single point in time,* by making a distinction between two different adherence concepts, such as attraction and "commitment." On achieving this partial solution to the problem, these authors seem to have felt that the entire problem of conceptualizing membership stability has been satisfactorily solved. We see that this has not been the case.

Conceptions relating to variability of adherence

The conceptualization of commitment in terms of variability in the level of adherence is implicit in a number of writings. The notion of variability is introduced implicitly by defining commitment in terms of certain causal conditions *that may be presumed to be relatively stable.* Recall our earlier model of membership stability in which stability reflects the fact that the pro–con difference remains consistently greater than zero over time and circumstances. Some authors identify commitment with certain components of the pros that are (or may be assumed to be) highly stable. For these authors, "commitment" is the magnitude or importance of these particular pro-membership causal conditions that are relatively invariant over time.

Several illustrations of this approach can be provided. *The Modern Dictionary of Sociology* (Theodorson & Theodorson, 1969) defines commitment as the "feeling of obligation to follow a particular course of action." Why are feelings of obligation relevant to relationship stability? Presumably such feelings are relevant because they are more stable (or reflect more stable causal conditions) than other types of feeling, such as sexual desire or need for companionship.

H. S. Becker (1960) defines commitment as being "a position in which [the person's] decision with regard to some particular line of action has consequences for other interests and activities not necessarily related to it" (pp. 35–36). The person follows a certain line of activity "for reasons quite extraneous to the activity itself" (p. 40). As examples, Becker describes a person who cannot afford to leave a particular job because of institutional rules (e.g., the perquisites of seniority would be lost on leaving, the person has a stake in a nontransferable retirement fund) or because of psychological adaptations (e.g., he or she has become adapted to the job and is no longer suitable for alternative positions). From these examples, it is clear that commitment is seen to depend on causal conditions that, once in place, are highly stable. It is their assumed invariance relative to other components of the pro-membership causal package (e.g., the satisfaction from the work, the interaction with favorite workmates) that recommends these particular causes for inclusion in Becker's list, rather than their "extraneous" nature per se.

Conceptions of commitment that focus on the more stable components of the pro-membership causal package are consistent with the present conception in highlighting the *variability* of the pro–con balance as an important feature of the causal system underlying commitment. However, the present conception requires that we consider *all* the pros and cons and *both* the average level of the resultant and its variability. This view also leaves as an empirical question, one central to research on commitment, the matter of the relative stability of the various pros and cons of membership. It recognizes, furthermore, that the stability of the various components is a relative matter. Thus, the present conception makes it necessary not only to ascertain the stability of pro-marriage motivation that stems from, say, having children and financial obligations, but also to evaluate the level and variance of this motivation relative to the level and variance of the motives that act to disrupt the marriage (its frustrations, the attractions of alternative lifestyles).

In the present view, commitment exists when the total set of relevant causal conditions stably generates a resultant that is supportive of continued membership in the relationship.

Definitions in terms of process

We may note in passing that a number of authors define commitment in terms of events that create causal conditions favoring stable membership. These definitions, which specify a pledging of oneself or an avowal of intention to sustain an activity or relationship (Kiesler, 1971; Levinger, 1980; Z. Rubin, 1973), draw on the portion of the dictionary definition that describes commitment as an activity rather than as a state. A somewhat similar definition, by Leik and Leik (1977), defines commitment in terms of a process that presumably serves to *sustain* the requisite set of causal conditions once they are in place. These definitions are to be compared with ones mentioned above that specify commitment in terms of a state of the person (e.g., level of adherence) or a particular subset of the causal conditions relating to membership (e.g., causal factors extraneous to the membership itself).

Together, these various definitions illustrate the problem highlighted in the introduction to this chapter. The concept of commitment (and others of its sort) refers to a complex set of components, including observed phenomena, assumed underlying causal structures, and processes that provide a historical account of the existing causes. As various researchers analyze this set, attempting to understand its inner relations, they can enter it at different points. What each researcher identifies as a useful entry point into the tangle of properties, states, processes, and causes becomes "commitment" for that researcher and the thing for which he or she tries to identify the ramifications. Unfortunately, the paths followed from these different entry points do not always converge. Because the phenomena associated with commitment overlap those associated with other concepts (for example, with those of dependence, as we have seen), some authors are led off into an area already conceptualized in other terms and away from the core of commitment itself. Researchers who begin their analyses with certain antecedent processes run a different risk, namely, that the antecedent does not always lead to the assumed consequences. For example, a pledge to maintain a relationship may not always result in the person's remaining in it, either because the pledge itself does not give rise to stable pro-membership causal conditions (e.g., feelings of obligation dissipate as the pledge becomes resented) or because other causal factors overwhelm it. From a conceptual point of view, a definition of commitment in terms of certain historical antecedents clouds over the detailed causal analysis that is necessary for understanding the core phenomenon of commitment, namely, membership stability. The questions

"What is commitment?" and "How does commitment come about?" become confused. What should be a hypothesis about the development of the state or condition of "commitment" (e.g., that certain pledges made under certain conditions enhance membership stability) becomes taken as a statement of what constitutes the essence of commitment.

Subjective versus objective perspective

Little is known about the exact sequences by which relationships are terminated, whether by decision, drifting apart, enforced separation, or other scenarios (see Chapter 8, "Development and Change"). However, it seems reasonable to assume that most of the terminations envisioned in discussions of commitment entail more or less explicit decisions on the part of one or both members of the dyad. This being the case, a proximal causal condition necessary for commitment is the person's relatively invariant *perception* that the pros of membership outweigh the cons. For membership to be stable, it is only necessary that the person's subjective assessments of the pros and cons consistently yield a balance in favor of continuation.

The key role of subjective assessments as proximal causes for decisions to stay or leave presents us with a number of problems. A first set has to do with when and on what basis such assessments are made. Very little is known about these factors, but it seems unreasonable to assume that the pro–con balance is continually monitored and that the relationship is left when the balance first drops below zero. Rather, there must be certain occasions that provide the impetus to assess the relationship, as when attractive alternative partners present themselves or when one's partner begins to express doubts about the association. There are probably great variations between people in how they aggregate (tally up, take account of, weigh) the various pros and cons. Certain aspects of this process are considered below in a discussion of time perspective and aggregation.

A second set of problems has to do with the scientific measurement of commitment. These problems are discussed more fully in a later section, but certain general issues must be considered here. If subjective assessments of the pro–con balance are the crucial proximal causes of membership stability, then may we not base our measurement of commitment exclusively on subjective measures? To answer this question, it must first be made clear that it is the *objective* level and variability of the subjective pro–con assessments that mediates stability. To ascertain the commitment that has characterized a certain person over a particular period of time, we must obtain a systematic record of the person's subjective pro–con assessments as they vary over that period. The person's subjective impressions of the level and variability as provided, say, at the end of the period will not necessarily correspond to the level and variability of the actual day-by-day assessments. To estimate the person's commitment for a particular period in the future, we must estimate the values that person's subjective pro–con assessments will take over that period. We may enlist the person's help in making this estimate, as by asking him or her to predict the level and variability. However, we must take this prediction for what it is—a mere subjective forecast, with an uncertain basis, and possibly badly in error because of failing to account for future changes in relevant distal conditions (e.g., changes in the members, their social and physical environments). The quality of people's predictions of their future assessments is not known at present. It is easy to imagine cases in which an investigator who possesses certain facts about distal causes—facts not available to a member (e.g., about typical consequences of having children, or of aging)—will be better able than that member to predict the future course of his or her subjective assessments.

In these and other ways, we see the advantages of maintaining a clear distinction between distal causal conditions (e.g., being objectively committed through legal and economic entanglements) and proximal causal conditions (e.g., being aware of the entanglements and therefore motivated to make the relationship work.) As M. P. Johnson (1978) points out, an advantage of this distinction is that instances can be identified in which the person is objectively committed but doesn't know it. This situation can give rise to a special scenario of discovery in which initial attempts to discontinue the relationship are terminated as the facts of the situation become known. Johnson notes that this situation constitutes one of the many instances in which attitudes (as measured before realization of the situation) do not predict behavior (as it is finally affected by the situation). Most importantly, to maintain the distal–proximal distinction keeps our attention upon the cognitive and communication processes by which objective situations become part of the subjective life space and, therefore, proximal causes of behavior.

Summary

The preceding analysis provides an abstract causal account of the *phenomena* of commitment. Commitment to a relationship means that a person is in a causal system that stably, over time and situations, supports membership in the relationship. The varius *causal conditions* relevant to membership favor his or her continuation of such membership and do so relatively invariantly over time. The degree of *invariant* support constitutes the degree of commitment. Specified in this manner, commitment is to be distinguished from the level of the pro–con difference at any given time and from the average level of the difference over time. The latter are more appropriately referred to as the person's adherence to or (per Thibaut & Kelley, 1959) dependence on the relationship. (When adherence or dependence is aggregated across the various members of a group, it has been referred to as the cohesiveness of the group (Cartwright, 1968).). As we have seen, commitment is a function of *both* the average level of the pro–con difference *and* its variability.

This conception links commitment most directly to the proximal causes constituted by the person's subjective assessments of the pro–con difference. However, the level and variability of these subjective assessments are affected by distal causes, so commitment is ultimately determined by fluctuations in the distal factors that affect the pros and cons of membership. Insofar as the latter effects are not fully anticipated by dyad members, their subjective expectations about relative stability will be in error. In this case, their expectations must be replaced by a more fully informed understanding (e.g., the investigator's "objective" perspective) of changes in the distal factors and of the distal-to-proximal links.

The Processes of Commitment

We now consider theories and research that identify the processes that promote the person's being in a state in which the causal conditions favoring continued membership stably outweigh those acting against it. These processes include both significant events in the development of the relationship and ongoing sequences of events that continue throughout its course. Our strategy here is to analyze such events in terms of the causal conditions they bring into play.

According to our model of membership stability, to be considered a part of the "commitment" process, an event or sequence of events must (1) raise the average pro–con difference without producing an offsetting increase in its variance, (2) decrease the variance in the pro–con difference without producing an offsetting decrease in its average level, or (3) both raise the average and decrease the variance. In the first case, the process may increase the level of pro-membership factors, decrease the level of con-membership factors, or both.

The effect on the variance of the pro–con difference that results from adding new causal factors is complex inasmuch as it depends on the correlation between the preexisting sources of variation and the new ones. The appropriate formula for the variance of the resultant is

$$\sigma^2_{\text{pre + new}} = \sigma^2_{\text{pre}} + \sigma^2_{\text{new}} + 2r_{\text{pre}\cdot\text{new}}\sigma_{\text{pre}}\sigma_{\text{new}}$$

where the successive terms refer, respectively, to the resultant variance (reflecting both preexisting and new factors), the variance due to preexisting factors, the variance due to new factors, the correlation between variations in the two sets of factors, and the square roots of the variances in the two component sources. The major point to be drawn from this formula is that new factors will *increase* the variance of the resultant *unless* they are totally stable across time and situations *or unless* their variations are strongly *negatively* correlated with corresponding variations in the preexisting factors.

Illustrations can be found for both types of factors. Certain commitment processes generate causal factors that are stable. Such processes are exemplified by interpersonal and public events (marriage vows, entering into joint contracts, public behavior) that bring into play formal and informal social systems (church, loan companies, the circle of family and friends). These social systems are themselves enduring, they tend to have long memories, and they try to promote stable membership in certain types of pairs. If they have an interest in a particular relationship, these systems can be expected to be a stable part of each person's pro-membership causes unless he or she is able somehow to find ways to circumvent their pressures and sanctions.

Other commitment processes provide variations in pro-membership factors that are negatively correlated with variations in the strength of the con factors. These processes involve negative feedback causal systems in which the tendency of a person to be drawn or pushed out of a relationship is monitored and in which increases in the strength of this tendency lead to counteracting increases in the causal factors supporting membership. Such "corrective" systems may involve the partner (who increases the person's rewards when he or she is detected to be wavering in loyalty), the external social system (the family that, on sensing possible defection, takes action to keep the two persons together), or the individual (who, on experiencing the temptations of attractive alternatives, steers away from them and engages in self-persuasion about the merits of the present relationship). An interesting example, relating to processes maintaining membership in a communal group, is provided by Kanter (1968). In Oneida, a 19th-century utopian community, after outside visitors left each day, the group held a ritualistic scrubbing bee to purify the community. Furthermore, those members most exposed to contact with the outsiders were required to submit to a session of group criticism in order to free them from possible contamination. In this case, the corrective measures, to weaken the possible pull of outside influences, were set in motion by the contact with those influences and presumably served to maintain a resultant positive orientation to the Oneida community.

The examples in the paragraphs above illustrate an important point: Commitment processes vary greatly in the time and duration of their activity. Some, like the initial pledge of loyalty, serve to *establish* membership-promoting causal systems. Others may exist throughout the life of a relationship, serving to *maintain* pro-membership causal systems and to link them, over time, to the interaction. The latter are illustrated by the negative feedback systems just described.

What follows is a list of some of the major processes that have been proposed as heightening a person's commitment to a close relationship. It will be seen that these processes vary in many different respects, such as initiating versus maintaining processes (as just described); public versus private processes; processes that affect intrinsic properties of the relationship versus those that affect extrinsic factors; processes that involve voluntary decisions versus those that create commitment without the person's awareness; and processes that generate commitment precipitously versus those that create it in cumulative small increments. It is not argued here that all these processes have the same effects (see discussion on types of commitment below). However, each of them has been advanced as having some relevance to membership stability. From the perspective of our simple model, we may consider how they affect the level of an individual's pro–con balance and its variability over time. Finally, it must be noted that the processes in the following list are not mutually exclusive, nor are they easily distinguished from one another, either conceptually or operationally.

Improving the reward–cost balance of membership

Both M. P. Johnson (1978) and Rusbult (1980) identify satisfaction with the relationship as a factor contributing to commitment, and Rusbult provides evidence consistent with this notion. Similarly, as one process leading to commitment, Hinde (1979) describes the growing preference for the ongoing relationship that is inherent in its development. As partners develop a history of positive interaction, the relationship acquires secondary reward value through its association with past happiness. Furthermore, each person learns the other's ways and mannerisms and develops a preference for continuing to interact with the now predictable partner rather than new, unfamiliar ones. Along the same lines, Rosenblatt (1977) describes the shared memories and patterns of living that become tied to the spouse and that provide a level of shared meaning and comfort no other relationship can provide. Through such processes, a person's love for a partner may create and promote commitment as new bases of reward are experienced and the costs of interaction are reduced.

This view clearly sees membership stability as increasing with the average level of adherence, so we must be concerned about a possible accompanying increase in its variability. However, it seems that the particular new causal conditions referred to in these examples (deriving from mutual accommodations, familiarity, and so on) are relatively invariable ones. They are anchored in the experiential history of the pair and, although subject to some reinterpretation, this history is stably represented in the individuals' memories and affective associations.

In contrast to the processes above, which increase the *intrinsic* benefits of the relationship, there are processes that increase the *extraneous* benefits to be derived from remaining in it. For example, through decisions relating to purchases and employment, a person may find that continuing a marriage is necessary in order to retain possession of a house or to continue to hold a job in an in-law's firm. H. S. Becker (1960) proposed that the linking of extraneous interests to following a consistent line of activity be considered a necessary component of the commitment process. As noted earlier, what seems important about the kinds of interests Becker identifies is the stability of the effective causal conditions rather than their extraneous nature. Becker's examples include generalized cultural expectations (which constrain a person to act consistently in order to avoid getting a reputation for being erratic), impersonal bureaucratic arrangements (which create benefits related to seniority that are lost on leaving a position), and individual adjustment to a social position (which makes the person unfit for other positions). All of these involve pro-membership considerations that are extraneous to the inherent rewards and costs of membership itself, but, what is more important, they derive from causal conditions that are relatively stable in their action.

Irretrievable investments

Among the rewards and other benefits (e.g., avoidance of costs) that a person will lose on leaving a relationship are certain special ones that are associated in the person's thought with "investments" (time, money, effort, and so on) the person has earlier "put into" the relationship. Because of this association, the anticipated loss of these particular rewards or benefits is felt more keenly than would otherwise be the case, and they therefore carry special weight in keeping the person in the relationship.

An example will illustrate the point. If employees will have to give up a pension fund on leaving a company, we might expect them to be equally deterred from leaving whether that fund exists (a) through their own contributions over the years of their employment or (b) through a recent benevolent action by the employer. However, the notion of "investment" implies that the loss will be felt more keenly in the first case, in which the benefit to be foregone is understood to be the result of the persons' own sacrifices, those sacrifices having been made in order to create the fund and with expectations of ultimately enjoying its fruits.

The phenomenon of "investment" derives from the common experience (itself derived from the typical structure of "tasks") that we must put effort into a venture before we get rewards from it. Except for the unusual cases of "good luck," we expect to incur costs first and to reap benefits later. To discontinue an activity without gaining rewards from it is to leave it uncompleted and to have wasted the effort earlier put into it. Most of us believe that effort deserves to be rewarded (Weiner & Peter, 1973) and, indeed, that it will be rewarded (Lerner, Miller, & Holmes, 1976). If we find that we have put a great deal of effort into an activity, we are likely to feel that it must be worth pursuing (Aronson, 1961). We also tend to feel that we have revealed a personal incompetence in an activity unless its rewards are commensurate with the effort we have put into it. These various attribution tendencies, just-world beliefs, and dissonance mechanisms often converge to induce us, following initial irretrievable investments in an activity, to make further contributions to it. These contributions are sometimes greater than those warranted by the returns currently to be expected from the activity, in which case we are aptly described as "throwing good money after bad."

In her study of commitment to utopian communities, Kanter (1968) identified a number of investment mechanisms used to promote membership continuance. These included being required to sign over one's property to the community in order to join it and not being reimbursed for one's work for the community if one leaves it. Both of these rules were found more often in successful communities, that is, those of long duration, than in unsuccessful ones. This evidence is consistent with the notion that investment promotes membership stability. As applied to close relationships, the notion of "irretrievable investment" has been used for the time, energy, emotional costs, self-disclosure, and money that are put into developing and maintaining the relationship (M. P. Johnson, 1978; Rosenblatt, 1977; Rusbult, 1980).

Heightening the social costs of termination

Various processes enhance the stability of membership by increasing the specific costs a person will bear should he or she decide to terminate it. In part, these processes bring into play stable social systems that have an interest in relationship stability, that monitor membership activities, and that negatively sanction attempts to disrupt the association. Other processes, discussed in the next sections, give rise to causal conditions that result in high termination costs mediated by the partner or by self-evaluations.

Some of the social costs of termination involve the loss of one's good reputation among friends, family, and work associates. These costs become a part of the deterrents to termination when, through their public actions, a pair of individuals become defined by their social circle as a "couple." The actions include such obvious ones as public pledges (e.g., wedding vows), but also more subtle ones, such as going places and doing things in public, monopolizing each other's attention and conversation, and wearing jewelry and clothing known to be given by the partner (Hinde, 1979). Having publicly presented themselves as a couple in these ways, if their social environment values stability of their sort of relationship, a pair will find it desirable to maintain their association if they are not to be regarded as irresponsible or undependable (H. S. Becker, 1960). For a person's reputation to constitute a stable constraint on behavior, it is necessary that there be a stable set of "significant others" in whose eyes the reputation matters. Thus, public behavior generates stable causal conditions promoting membership only for persons whose other important relationships are fairly circumscribed and stable.

Understandings between the partners

Hinde (1979) describes the "private pledge" as one of the processes by which commitment develops and contrasts it with the sorts of public pledges and activities described above. As Hinde characterizes the private pledge, it may be an explicit act, such as a promise of loyalty to the partner. However, he notes that such acts are often the culmination of more implicit processes in which each person makes clear his or her intention to continue the relationship. These processes involve sequences of events in which one person changes his or her behavior to suit the other's desires and adopts roles that mesh with the other's, the two work out understandings about what activities they do (or do not do) with each other and about what they should or should not do with outsiders, and they differentiate their relationship from other ones by drawing "we–they" contrasts and using signs with private meanings.

In these and other ways, each person reveals his or her dependence on the partner, acceptance of that dependence, and willingness to make the adaptations it necessitates. With such revelations, each person becomes more willing to let himself or herself become dependent in return. Thus, there develops an understanding between the two that each is attached to the other, that each regards the attachment as stable, and that each fully accepts the long-term implications of such attachment. This mutual understanding represents a stable P × O causal condition inasmuch as it is stably represented in the two persons' thoughts and continues to possess over time the potential for eliciting guilt and recriminations when its terms are violated.

The scenario above envisions a cumulative process in which the partners show parallel increases in their dependence on the relationship and in their open displays of that dependence. In reaching their understanding about these matters, the two are essentially striking a bargain in which each agrees that "I'll stick with you if you'll stick with me." The terms need not, of course, be entirely equal for the two, but, as in all such negotiation, each person acts to avoid being subject to unfavorable terms. So, insofar as commitment is generated by processes of achieving mutual understandings, it will tend to be symmetrical in strength. However, one must not conclude from this thesis that degree of commitment to a close dyadic relationship will always or even generally be symmetrical. The other processes listed here, which also give rise to causal conditions that promote membership stability, are often more effective for one person than the other. As Rosenblatt (1977) observes, the social supports for sustaining a marriage and the accessible alternative partners that weaken marriages are often of differential potency for men and women, depending on cultural differences, demographic factors, and similar distal causes.

A number of authors, among them, Hinde (1979) and Rosenblatt (1977), have suggested that there are causal loops between the two types of processes discussed above. The external, "public" processes tend to stimulate and promote the internal, "private" ones and vice versa. As a pair of individuals become defined by their social environment as a "couple," circumstances are created in which they are encouraged to act as a couple—doing things together, reconciling their different interests, meshing role performances, and so on. As the pair, out of their interaction, develop understandings and express these publicly, through favoring each other over outsiders, referring to themselves as "we" and to their activities and possessions as "our," and so on, their social environment is encouraged to define and treat them as a couple.

Linking membership to the self-concept

The assumption here is that most people are motivated to be consistent in their behavior and attitudes. They tend to try to maintain a consistency, over time, in how they feel, think, and act on important matters. One possible basis of consistency motivation is found in the negative implications of inconsistency for self-esteem. As Kiesler (1971) puts it:

> One is committed because he feels responsible for his past behavior. Change for the committed self would involve not only a new opinion but also some change in his self-view. Since he is . . . responsible for behavior consistent with the new opinion, he must somehow explain that, if only to himself. Many of the explanations are not complimentary: He is stupid, he made a mistake, he acted without forethought, and so forth. [In short,] . . . change for [a] committed subject involves explaining his previous behavior [and] these explanations are largely demeaning to self. (p. 168)

The implication of this view is that pro-membership actions for which a person feels responsible become linked to the self-concept in a way that creates stable motivation to continue such actions and to maintain a positive attitude toward membership.

A different view of commitment in relation to self-concept is provided by Kanter (1972). She describes the "core of commitment to a community" as being a relationship in which "both what is given to the group and what is received from it are seen by the person as expressing his true nature and as supporting his concept of self. . . ." A committed person has a "feeling that the group is an extension of himself and he is an extension of the group" (pp. 65–66). In her explanation of these notions, Kanter identifies processes that "reduce the person's sense of autonomous identity, so that he can have no self-esteem unless he commits himself to the norms of the group . . ." (1968, p. 712). These entail self-mortification through confession, self-criticism, and mutual criticism, and deindividuation through wearing uniform dress and lacking opportunities and places for privacy. Such processes serve to make the self seem inadequate and incomplete unless one adheres to the group. Although these processes are represented in rather extreme forms in the utopian communities Kanter investigated, they may have their parallels in close relationships. For example, in marriages, self-disclosures in the early stages of the association may entail a mild form of self-mortification, and the loss of privacy that comes with joint living may reduce feelings of separate indentity.

Decision, dissonance, and self-regulation

Both Kanter (1968) and Z. Rubin (1973) propose that the dissonance created by decisions made on entering and staying in a relationship may heighten a person's commitment to its continuation. When the person endures costs in order to enter a relationship or makes irrevocable sacrifices for the partner and then finds that the relationship is less then ideal, an unpleasant state of cognitive dissonance is created. Dissonance can be reduced by distorting one's cognitions about the benefits of the relationship, exaggerating its merits, and playing down its drawbacks (Aronson & Mills, 1959; Gerard & Matthewson, 1966). These changes would presumably be stable, on the assumption that the memories of the earlier costs and sacrifices remain vivid and do not become distorted themselves.

A more general view of the contributions of decision processes to commitment highlights the benefits of a life that is free from continual uncertainty and internal conflict. Rosenblatt (1977) expresses the point so:

> One obvious gain from making a high commitment seems to be freedom from having to make decisions. Once one is dedicated to a relationship, the issue of whether it should be continued, the monitoring of alternative possibilities, or questions about how much one can take root in the relationship will no longer be present. This saves energy and allows time for other things. (p. 83)

The various theories of decision processes and their effects on behavior can be traced to Lewin's (1926/1951b) analysis of intentional action. Intentions are formed when we decide to do something. We make such decisions in situations in which, at the outset, a number of different forces act simultaneously on us but in opposing directions. A decision restructures the situation so that the action is controlled by a more or less unidirectional causal system and is, therefore, more certain of enactment. In a sense, an explicit decision to do something is a means of regulating oneself so that a line of action presently undertaken will be carried through to completion despite subsequent variations in the field of forces. According to Lewin, the self-regulating, unidirectional causal system is brought about through restructuring the life space, by reconciling opposing forces, or by isolating one set of forces and suppressing the others.

This conception of self-regulation through managing the causal context within which one acts has been reflected in subsequent theories about postdecisional cognitive processes (Festinger, 1957) and unequivocal behavioral orientation (Jones & Gerard, 1967). Common to these various views is the idea that self-regulation processes serve to help a person maintain a resolve, live with a course of action once it is adopted, and carry it out effectively without being "forced to listen to the babble of competing inner voices" (Jones & Gerard, 1967, p. 181). Such self-regulation is carried out partly by managing the environment and partly by managing one's own attention, memory, and thoughts (see Kelley, 1980, for examples). The management results in a restructuring of the life space and a maintenance of that structure so that forces consistent with the chosen activity are strengthened and those inconsistent are weakened.

The self-regulatory processes relevant to close relationships have not been analyzed or even fully identified. Remarks by Rosenblatt (1977) suggest that, in marriage, these processes may entail controlling fantasies about adulterous activities, reducing flirtatious behavior with opposite-sex acquaintances, and tolerating marital problems. In connection with the last, Rosenblatt makes the interesting observation that, with high commitment, a couple may avoid the discussion and self-evaluation by which marital problems might be resolved contructively.

Psychological perspective and aggregation

If a person decides to enter a relationship after extensive analysis of its pros and cons and accurate anticipation of its probable ups and downs, that decision is unlikely to be overturned by subsequent events. This decision process is to be contrasted with the establishment of a relationship on an impulse of the moment or under the pressure of a particular situation. As impulses or situations change, the balance between staying in and leaving the relationship is likely to change. Thus, the broader the person's psychological perspective on the relationship—the greater the range of its implications the person has considered—the more likely that his or her pro-membership motivation (assuming that to have been the outcome of the considerations) will be stable.

The concept of "psychological perspective" regards commitment as an anchoring of a line of activity in a broad set of causal conditions, including anticipations of future difficulties and expectations about long-range consequences. Once an activity is undertaken with this broad understanding, the person can endure setbacks and frustrations inasmuch as they will already have been discounted. In contrast to activity that draws consistency from its anchorage in a broad causal context is behavior that, being under the control of local, situation-specific causes, varies in its course and intensity. The former is the type of behavior more characteristic of adults and the latter, more characteristic of children (Kelley, 1980). Commitment deriving from breadth of perspective at the initiation of activities or relationships is an essentially adult phenomenon.

The manner in which perspective contributes to membership stability can be illustrated by a simple statistical analogy. Consider first the range of experiences, over the daily and weekly ups and downs of the interaction, to which the person's subjective assessment of the relationship is geared. In answering the question "How satisfactory is this relationship?" what sample of interaction events does the person aggregate in order to "calculate" a balance? A person with broad time perspective will aggregate experiences over a lengthy period, taking account not only of present dissatisfactions but also of long-past good events and anticipated future benefits. This person's orientation to a relationship that is, on the average, satisfactory, will be more stable than that of a person who, with limited perspective, reacts strongly to each discrete experience, finding in the unsatisfactory events reason to terminate what may usually be a good relationship. The appropriate statistical analogue for this difference is provided by the formula for the variance of a running average as a function of the length of the sequence over which the average is calculated. The running average shows smaller variance than do the individual points. Furthermore, the greater the number of points for which the average is computed, the less its variance.

Reducing the availability and attractiveness of alternative relationships

Here we consider processes that enhance continuation of membership through reducing the quality of the possible alternatives to it. One consequence of becoming publicly identified as a "couple," at least in societies with monogamous norms, is that other persons who might have been available as partners now take themselves out of the running and look elsewhere for associations. Alternatives may also become eliminated because, in the course of adaptation to a particular partner, the person loses his or her fitness for other relations. Skills and preferences may narrow down to those appropriate to the given relationship. For example, the professional person who stays home ten years in order to raise children may lose both the skills and the interests necessary to resume an outside career. In these and similar ways, becoming a "couple" and growing into a particular relationship may reduce the accessibility and attractiveness of alternative lines of activity.

Alternatives may also be reduced through actions taken for that purpose by the members of a relationship. As proof of their dedication to the present association, the two may require each other to break off interaction with persons who might constitute alternative partners. The renunciation of these outsiders may also involve downgrading their attractiveness, with elaborations on their negative qualities and unfavorable comparisons with the current partner.

These and similar deliberate renunciation processes were noted by Kanter (1968) to occur in utopian communities. Outsiders were described in negative terms, and contact with them was controlled. As part of the effort to engage each person wholly in the community, he or she was often required to give up special relations with spouse and family. This renunciation appeared to be associated with the stability of membership.

Leik and Leik (1977) propose that the cessation of "monitoring" of alternative partners (observing and interacting with them, sizing them up in comparison with the current partner) be considered the criterion of "commitment." In other words, a person can be regarded as having reached the highest level of involvement in a relationship, namely, commitment, when he or she shows "an unwillingness to consider any exchange partner other than that . . . of the current relationship" (pp. 301–302). This limitation on considering alternatives is seen to grow out of confidence that "in the long run, the current exchange partner will come through" (p. 303). Alternatively, we might regard nonmonitoring of alternatives to be part of a self-regulation process in which one's life is made free of decisional conflicts by focusing attention on what one has and putting out of mind what one might have. Other renunciation strategies, described above, may also play a role in membership-stabilizing self-regulation.

Leik and Leik (1977) pursue the implications of their view by considering when it is that monitoring of alternatives is reinstated and the adherence to the relationship is possibly upset. The existence of the person in a complex social network with involvement in a variety of different associations always poses the possibility that alternative relationships will be brought to the person's attention. Unless the couple is wholly isolated from other contacts, there is always the chance that an outside relation will create a challenge to their current one. This analysis is particularly useful in suggesting the causal conditions (namely, the structure of the social environment) under which a membership-stabilizing process (nonmonitoring of alternatives) is likely to succeed or fail. When the process fails and an alternative is thrust into the person's awareness, there is set in motion a scenario in which the person experiences a new state of conflict and opportunity for choice, and (according to the Leiks) a resolution is finally reached in which the challenge "succeeds or . . . is purposely avoided or nullified" (p. 317).

Research and Measurement

As the general research question for the commitment area, the present perspective suggests the following: How do various causal factors and processes affect the stability of each person's adherence to the relationship? Research can focus on *commitment factors* (various causal conditions thought to be relevant to stability or instability); *commitment processes* (various intra- and interpersonal processes thought to establish and sustain stability); *commitment* itself (a property of each person's relationship to the partner: the average level of his or her adherence relative to the variability in the level); and various combinations of these three. Examples of commitment factors and processes have been provided in the foregoing section of this chapter. The causal factors are assessed directly, for example, by reports of reward–cost balance, investments in the relationship, or understandings between the partners; or indirectly, for example, by observation of events that can be assumed to set them in place, such as marriage vows, contracts, contributions to the dyad's existence, and public behavior. The operation of commitment-preserving processes can be detected through observation and report. For example, an agreement to limit contacts with attractive third persons (an agreement constituting a causal condition) will presumably have some observable or reportable effect on the members' interaction with and knowledge about such third persons.

A common question in commitment research concerns the relative importance of various factors and processes for the long-term stability of the relationship. This question is addressed in Rusbult's research (1980). Commitment was measured by a respondent's ratings of the probable and desired duration of the relationship, degree of commitment and attachment to the relationship, how attractive an alternative partner would have to be before the person would leave the present association, and (weighted negatively) the likelihood the person will end the relationship in the near future. These items can be regarded as single time-point estimates of commitment and, as such, may have severe limitations (see below). Rusbult was interested in how much each of the following causal conditions affected the estimated commitment: (1) the outcomes derived from the relationship (rewards and costs), (2) the investments made in the relationship, and (3) (negatively weighted) the goodness of the available alternative associations. These three causal factors were also measured at a single point in time. To the degree their stability can be assumed to be constant across respondents, their resultant pro-membership value should determine membership stability. Rusbult's data indicated that all the factors were related to estimated commitment, although the cost component of relationship outcomes was less important than the reward component. Thus, Rusbult's study identifies some of the causal conditions whose magnitude is relevant to membership stability. The study assumes, of course, that the single time-point measure of stability, based on estimates of probable duration, is truly reflective of the future pattern of variations in adherence. It may also be noted that Rusbult provides evidence consistent with the overlap between love and commitment, shown in the upper left portion of Figure 7.1: Attraction to the relationship (a positive attitude akin to love) was shown to be determined only by certain causal conditions (the rewards provided by the dyad), whereas commitment was shown to be determined by both rewards and the other causal conditions, presumably the more stable ones, namely, investments and available alternatives.

Research on organizational commitment (Porter, Steers, Mowday, & Bulian, 1974; Steers, 1977) has used records of attendance at work and duration of employment to measure membership stability over time. We may consider membership stability as a measurement of commitment even though the researchers treat it as a measure of the consequences of commitment. The antecedents of membership stability are measured by workers' ratings of their loyalty to the organization, their beliefs in its goals and values, their willingness to exert effort in its behalf, and their desire to maintain membership in the organization. The conceptual status of these measures is not entirely clear, but on the whole they seem to reflect attitudes and beliefs supportive of membership, that is, psychological causal conditions proximal to continued work for the organization. The magnitude of these causal factors is measured at one point in time, and, on the unstated assumption that their stability is roughly constant across respondents (and not correlated with magnitude), this index is expected to predict membership stability. The investigators find some evidence consistent with this expectation. This research, like Rusbult's, suggests some of the causal conditions whose magnitude at a given point in time is related to membership stability.

The necessary directions for future research on commitment are implied by our characterization of these studies. It will be important to determine not only the magnitude of the membership-relevant causes but their stability over time. It will be desirable to distinguish the different factors that work for and against membership and to determine their respective courses over time. Additionally, the analysis of causes must be extended beyond the proximal, psychological ones (evaluations, expectations, intentions) to the more distal ones. The latter include the important events in the association (amount of conflict, frequency of joint pleasure), the objectively available alternatives, attitudes toward the pair's continuation held by significant others, and legal and economic impediments to dissolution.

In fairness, it must be noted that prior research has not entirely overlooked such distal factors. Steers (1977) reports evidence about various antecedents of the attitudes that predict membership stability, including the attitudes toward the organization held by the worker's friends; background characteristics of the person, such as educational level; and the organization's dependability in fulfilling its promises to its employees. Kanter's (1972) study of commitment to utopian communities, described in certain of its aspects above, deals only with distal factors, these being the institutionalized practices of the community that require sacrifices, investments, mortification, and so on. The psychological consequences of these practices, which are presumably the proximal causes of loyalty and conformity, would have been difficult or impossible for Kanter to recover from her historical sources. However, a combination of Kanter's method for identifying significant practices with, say, Rusbult's assessment of current beliefs and attitudes would seem to hold promise for future research.

Highly important will be studies of the waxing and waning, over time and situations, of a person's adherence to the partner. If membership stability is to be of central concern in commitment research (as it has been in the past), there is a clear necessity to assess it by an extended time series of measurements. It will not suffice to have persons project or predict their adherence to the partner. Or at least, this kind of prediction will not be sufficient until we have a great deal of systematic evidence about the origins of such predictions and the conditions under which they are valid. Furthermore, the assessment of variability in adherence can be combined with assessment of stability of various causal conditions to determine which factors account for within-person variations in continuance tendencies. In research that follows close relationships over time, the investigator has a particularly good opportunity to detect circular causal processes. Only through longitudinal work can researchers observe how disturbances in a person's adherence may give rise to corrective, negative feedback, or how, in other instances, there develop exacerbation cycles in which small loosenings of interpersonal bonds elicit sensitivity to association costs, scanning of the social environment for better alternatives, and discounting of termination costs.

Measures of commitment

In the preceding remarks, we have distinguished commitment factors and processes from commitment itself. The latter refers to the average level of a person's adherence to the relationship relative to the variability in that level. To assess commitment, ideally we must do exactly what the above implies: Over a long time period, during which situations and problems vary, we must repeatedly measure the level of each person's adherence to the relationship—the balance of the pro-membership and con-membership factors acting upon him or her. The higher the level of the pro–con difference relative to its variability, the more stable is the causal system acting to maintain membership.

For scientific purposes, we will often want to predict the future stability of certain classes of dyads. And, of course, the members of close relationships will often have the same interest with regard to their own particular dyads. A measure of past commitment provides one basis for such predictions. On the assumption that the variations during the period of observation are representative of those that will occur in the future, the ratio of the average level to its variability will be predictive of the future stability of the dyad. This assumption is not to be made uncritically, either by us or by the partners themselves. For example, a couple's history of weathering past difficulties may seem to afford a better indication of future stability than does the history of a couple for whom everything has been sunshine and fair winds. However, the future may have even worse things in store for the first couple and only continued good weather for the second. Furthermore, a couple's history is not, as our meteorological metaphor suggests, wholly determined by external forces. It is partly of their own making, as they exacerbate problems or avoid and resolve them. So a history of surviving difficulties can be a negative indicator for the future—a sign not of good coping processes but of dangerously weak ones. In short, the prior ups and downs in a person's pro–con balance must be examined closely as to their sources and the processes entailed if their relevance for the future is to be estimated.

These considerations have special relevance to the meaning of estimates of future stability of the relationship, provided by the participants themselves. These estimates are probably based on both recollections of past events and anticipations of future ones. The quality of the estimates is determined by the interpretations made of the past, for example, by whether they veridically reflect the commitment-sustaining mechanisms exhibited in the past; and by the intelligence with which the future is imagined, for example, by whether account is taken of external factors that are not presently salient but will come into play when disruption threatens. These considerations should make us wary of taking at face value respondents' predictions of the future of their relations. More importantly, they suggest fascinating problems for research, having to do with the origins of such predictions and the factors underlying their validity.

In the world of physical objects, if we want to know how long something will last, we sometimes subject it to a strain test. For example, in product testing, sample objects are placed under stress and a determination is made of their breaking points. Something similar to this testing may occur naturally in close relationships. If serious about testing the partner's commitment and doubtful that it has yet been revealed by natural events, a partner can create "test" conditions. As Rosenblatt (1977) suggests, such tests are made by asking for favors that the partner will find difficult to provide or by being personally difficult in ways that test the partner's tolerance. By these and similar means, a person can heighten the factors negative to membership in order to see how resilient and strong are the pro-factors. For ethical and practical reasons, the researcher will not be able to introduce such test conditions. However, the researcher can be sensitive to the existence of such tests, whether they are created by the partner or by natural variations in relationship stressors.

Types of Commitment and Their Consequences

So far we have viewed commitment as an undifferentiated concept. This approach was taken with the thought that it would be better to clarify "commitment" in general, and to devote this chapter primarily to that end, before introducing the complexities of the different types of commitment. We have also focused on one particular consequence of commitment, namely, the tendency for a committed person to remain in a relationship. This focus is not to preclude an examination of the other consequences, such as the resultant conformity, emphasized by Kanter (1972), and the confidence in reciprocation and relationship growth that flows from mutual commitment, as noted by Hinde (1979). The present focus reflects the major purpose of prior writings in the field, which has been to provide a causal account for the fact that a person would continue an association even though it is not particularly satisfying. For example, the concept has been useful to account for the fact that marriages can be unhappy but stable, a fact that is well documented by Cuber and Harroff (1965) and Levinger (1976).

In contrast to this unitary view of commitment is one that distinguishes different kinds and their different consequences. We have seen that the term *commitment* has been applied, and appropriately so, to a number of different phenomena. There is commitment to an activity or course of action (H. S. Becker, 1960; Kiesler, 1971); to a close relationship (Rosenblatt, 1977; Rusbult, 1980); to a community (Kanter, 1972); and to a work organization (Steers, 1977). In the domain of close relationships, one might wish to distinguish commitment to the relationship (i.e., to its continuation) from commitment to the partner—a dedication to his or her welfare whether or not the relationship continues in its current form. A parent may be committed to a child's well-being without desiring continuation of a close relationship with the child, even being willing to forego the latter in the interests of the former.

When we focus on commitment as it relates to continuing a relationship, we have seen that the relevant causal conditions are exceedingly diverse. A number of subclasses of these conditions have been suggested: commitment as personal dedication versus commitment as conformity to external pressures (Rosenblatt, 1977); personal commitment versus structural commitment (M. P. Johnson, 1978); and endogenous commitment versus exogenous commitment (Hinde, 1979). These distinctions differ in their details but converge in drawing a line between (1) continuation that is sought by the members themselves, on the basis of their interaction experiences, and (2) continuation that derives primarily from the outside. The first reflects such causes as the persons' attitudes, private pledges, and understandings, and the second reflects such causes as pressures from family and friends, legally imposed termination costs, and available alternatives. Even if both types of causes are relevant to continuity of association, the processes by which they are maintained and act to promote continuity are obviously different, as are the conditions under which they become ineffectual and no longer sustaining of the relationship. Beyond that, the two sets of causes undoubtedly have different implications for other aspects of the association. For example, Hinde (1979) describes two scenarios of commitment that imply two different classes of cause. In the first, which would seem compatible with endogenous commitment, the two persons actively strive to make the relationship work, to optimize its mutual rewards, and to provide each other the rewards felt necessary to protect the relationship. The second scenario, more compatible with external pressures for continuation, envisions a slackening of efforts on behalf of the partner and taking the relationship for granted in the confidence that "it needn't be worked at."

With further analysis and investigations of commitment, we can expect the membership-sustaining causal conditions to be differentiated along a number of lines and their different implications to be identified. This kind of dissection of an intertangled skein of processes and phenomena is not easy. A precedent can be found in research on the closely related problem of group cohesiveness (Cartwright, 1968). After a long period of research and debate, the problems are far from being solved. However, such differentiated causal analysis is absolutely necessary if the rich variations in processes related to membership stability are to be understood.

LOVE VERSUS COMMITMENT

This chapter has drawn a distinction between love and commitment, but, in doing so, it has recognized the considerable overlap between the two, as shown in Figure 7.1. The phenomena of love—the caring, needing, trusting, and tolerance that are shown in behavior, thoughts, and feelings—reflect the *positive* factors that draw and keep two people together. People in love voluntarily seek out each other and, while love lasts, cleave closely to each other. The phenomena of commitment—the enduring adherence of persons to their close partners—reflect the *stable* causal conditions that draw and keep people together. These conditions include both stable positive factors that comprise part of love and stable extraneous conditions, such as social pressures, felt obligations, and investments, that keep a pair together whether or not they feel positively about their relationship. Insofar as the positive factors—for example, sexual satisfactions, pleasures of joint activities, exchange of consideration and esteem—are stable, love and commitment will go hand in hand. However, love based on unstable causes, for example, on transient passion, will not promote commitment, and commitment based entirely on extraneous considerations will not include love.

Love is properly measured at any given point in time, and the level of love may fluctuate considerably over time. Love involves distinguishable components—caring, needing, and so on, and these may be assessed separately. These several components are related to different developmental courses and current causal conditions. Various combinations of components, history, and contemporaneous causes are envisioned in different models of love, such as the three that were outlined in this chapter: passionate love, pragmatic love, and altruistic love.

Commitment cannot be measured at a single point in time. Conceptually, commitment means that a person is located in a causal system that stably, over time and situations, supports his or her membership in a relationship. The phenomena of commitment can be assessed only by determining over a series of temporally separated occasions the level and variability of the person's adherence to the partner. The level of adherence generated at any particular time by all the pros and cons of membership has only a tenuous relation to commitment. A measure of this sort can be used to *predict* commitment, but cannot reasonably be said to *measure* it. Commitment is most plausibly predicted by assessing the strength of factors that have been found independently to be highly stable. However, as shown by our review of the processes promoting commitment, it is not merely the level of stable *causes* that is relevant to membership stability. Also to be considered are certain *processes* that serve to maintain low variability in adherence, such as the processes of social- and self-monitoring through which the weakening of certain ties to the partner are systematically counteracted by the strengthening of other, pro-membership forces.

Both love and commitment can be distinguished as to their varieties. The different varieties of love, summarized by different models, involve different patterns of phenomena and different histories and dynamics. The different varieties of commitment involve different orientations (e.g., the partner versus the relationship) and different patterns of underlying causal conditions and sustaining processes. The varieties of love and commitment are reflected in the fact that, in natural language, each concept refers to a "fuzzy category."

Like other such categories (Rosch & Mervis, 1975), the categories corresponding to love and commitment include a number of different phenomena that are distinguishable as to their prototypicality. For example, the research of Steck, Levitan, McLane, and Kelley (1982) suggests that, among American youth, the expression of caring is considered to be more prototypic of love than is needing or trusting. The latter are more peripheral members of the category, closer to the unclear boundary that separates love from related concepts. In addition, there is overlap between the fuzzy categories associated with love and commitment, as suggested by Figure 7.1.

The fuzzy and overlapping nature of the categories of referents of love and commitment has afforded the subject matter of this chapter, as we have tried to identify various prototypes of each and to characterize the nonshared elements in their meaning. We may note in conclusion the important implications of the fuzziness and overlap for communication and attribution within close relationships. Expressions of love can easily be confused with expressions of commitment. The display of ardent affection stimulated by passionate feelings can be taken as an avowal of obligation to adhere to the partner. Misunderstandings about a person's love versus commitment can be based on honest errors of communication, on failures of self-understanding, or on the intentional manipulative use of language and behavior to portray false intentions and induce the partner to accept the relationship on spurious terms. Returning to the example with which this chapter opened, Lee Marvin's comments imply that he loved Michelle Triola but had no intention of permanently supporting her. Although we may accept Marvin's distinction, we must not overlook the problems involved in its being accurately communicated to Michelle. We may easily imagine that Michelle honestly believed that Lee was both in love with her and committed to her.

CHAPTER 8

Development and Change

GEORGE LEVINGER

This chapter examines how relationships start, how they are built up and continued, and how they may eventually decline and end. In contrast to Chapter 3, "Interaction," which considers short-run interaction sequences, it is concerned with relatively broad, often diffuse changes over time in relationship properties that result from a pair's personal, environmental, and relational causal conditions.

THE COURSE OF AN ILLUSTRATIVE RELATIONSHIP

In order to focus on particulars, an illustrative case will be referred to throughout this chapter. It is based on interviews conducted separately with Susan and Tom Darber, two New Englanders who were married for 22 years before they were divorced. Their accounts of the early years are rather similar, but they later diverge. We begin with Susan's account:

First Meeting

> I met Tom in Paris in 1955, during my junior year at the Sorbonne, where I had gone to improve my French and learn about another culture. Tom had already

I appreciate the valuable suggestions that I received from Ann Levinger and Elliott Robins.

finished college; he was taking off a year before starting graduate work in architecture at Harvard. Our first meeting was at a party for new students. Someone announced that Tom and I both came from Rhode Island, but I remember that neither of us made any effort to talk to the other. We were both far too eager to meet French students and to stay away from Americans. Did I have any first impressions? Well, I must have noticed that Tom was tall, very good looking, and probably energetic, but he registered very little in my awareness.

Initial Attraction

It wasn't until the Christmas break, 3 months later, that I again noticed Tom. All my new French friends were with their families, and I was suddenly feeling pretty lonely. At dinnertime, I saw Tom sitting by himself at a table in a local bistro, and I asked him if I could join him. I still remember him rising with a mock bow and saying: "*Enchanté, mademoiselle.*" He, too, felt a need for companionship, and we spent a lively evening together. Actually, I was fascinated by him. The wine, the candlelight, and the season of the year must have contributed, but I hadn't had this lively a conversation with anyone since I arrived in France, and perhaps never before.

It snowed that evening for the first time that winter, which utterly delighted me. I had been waiting for snow, especially because I'd planned a ski trip and was leaving the next morning for the French Alps. Tom was staying in Paris to make the rounds of painters and sculptors during his Christmas vacation.

Building a Relationship

We didn't see each other again until I returned 2 weeks later. But something had blossomed. Perhaps it was the romance of Paris in the winter, perhaps it was the many miles we were away from home, but anyway we soon fell madly in love; only the fact that we lived almost an hour apart slowed our courtship. A month later, I moved in with Tom—something that hardly any American college girl would have done back at home in the fifties. That was a splendid winter and spring. We each got involved in the other's interests. I started walking around with Tom into all sorts of galleries and studios. He, despite his initial mockery, learned to ski rather quickly and well.

Where could our relationship go after that? Well, we surprised our parents when we wrote them that we would get married after our return in August and that I planned to finish college in the Boston area while Tom went to Harvard. We couldn't bear the thought of my going back to my old school and being apart a whole year.

Marriage

Tom spoke as follows about their early marriage:

Those early years seemed good to us. Susan got pregnant near the end of her senior year and we had a baby boy the next winter. I had to work hard in school—very hard; in fact, some days I saw Susan and Tommy only early in the morning and late at night. We were both encouraged, though, when I finished with high honors and got a job with a top firm in Boston. During the next 4 years, we had two more babies—two girls following the boy. I was doing well professionally and even attracted some national attention with my designs. In spite of what some colleagues called my workaholic state, we sometimes would take off for a camping trip into the wilds. I had always loved the wilderness, and Susan also learned to appreciate that sort of camping. Sometimes, when I couldn't get away, Susan went off alone with the little kids in our VW bus.

Susan also talked about these years:

Looking back now, it's obvious that Tom spent a lot more evenings and weekends with his work than with his family. He was driven partly by his desire to provide for us, and his early successes fed his ambition. Also, he found the office more orderly and predictable than our rather disorganized household. At the time, I didn't think much about it; I just accepted it as the way things were. We didn't talk much about things. We did, however, enjoy each other intensely in bed. Since we weren't together a whole lot, it was usually easier to ignore disagreements than to work them through; when we were together, we tried to have fun and brush away unpleasant matters. I didn't miss talking with Tom much, though. I was very busy with the kids and I had a lot of friends of my own.

When our youngest was about 8, I decided it was time for me to go back to school. After a while of agonizing about it, I started going to law school in the evenings. Five years later I had a degree.

Let us hear from Tom again:

Deterioration

Susan did rather well in school and made some connections through volunteering for Legal Aid. When she graduated, she was offered a job with a fine law firm and I think she wanted to take it, but I was going through a tough time at my firm. I had taken on new responsibilities as a head partner and needed extra help at home to entertain colleagues and clients. I asked her if she would postpone taking a full-time job and she agreed to wait. Instead, she did volunteer work for Legal Aid and helped out in the law practice of a friend of hers.

Susan described her feelings of that time as follows:

I was relatively satisfied with my decision. I was able to help Tom and at the same time do interesting law work. My volunteer work was well recognized in the legal community. A year and a half later, I was to receive a special award for my contribution to Legal Aid. I was really happy and looked forward to the award dinner, which many lawyers I respected would attend. But that evening, I was very embarrassed when Tom never showed up at the banquet; the chair reserved for him next to me near the center of the head table stayed empty all evening. I was even angrier when I returned home that night to discover him watching TV and to hear him tell me that he had been just too tired to attend the dinner.

That incident brought matters into focus. It forced me to realize how much the two of us had been drifting apart. I realized how much effort it now took me to ask Tom about his work and how rarely he inquired about mine or even listened when I did tell him about an experience. I also realized there wasn't much of a spark in our sex life. I started actually reveling in my extra "space" when Tom was away on a trip. I began to feel resentful about the imbalance in our lives: Tom, at the peak of his career, expecting me to continue making sacrifices while giving me little in return. I became resentful about doing the bulk of the housework; resentful about Tom's taking me for granted; resentful that, after all the years I'd listened to *his* stories and supported his career, Tom now didn't give a damn about *my* career.

Ending

Finally, I figured I'd had enough. I received another good job offer and decided to take it without consulting Tom. I also told Tom that I'd probably want to separate from him. When I told him that, he seemed stunned. He said I was the most important person in his life and asked what he could do to keep me. I was so shocked at his insensitivity, though, that I blurted out: "Nothing. There's nothing you can do to keep me, if you can't even figure out how unhappy I've been."

After I moved into my own apartment, I began to feel a lot better physically. The headaches and backaches I'd been having went away, and I felt good to be on my own. At the same time, I felt bad about the breakup and felt bad for our kids. Sometimes I worry about my future alone. But I think I made the right decision.

In his last interview, Tom remained unclear about his part in the deterioration of this relationship. Although he claimed that he had wanted to change things, he accepted at face value Susan's statement that there was "nothing" he could have done. One aspect of Tom's story worth mentioning is that, on the day of Susan's award banquet, he had lost an extremely important competition for a major design contract he thought he would receive. Even to the interviewer, though, he disclosed only that he skipped the banquet because he was tired. He did not acknowledge any understanding of Susan's hurt or his own sense of insecurity.

On the other hand, Susan herself responded incredulously when the interviewer asked if she thought her success might have threatened Tom. She still saw Tom so far ahead of her professionally as to be invulnerable, and thinking of him as insecure in any way seemed ridiculous to her.

Interpretation

The Darbers' relationship developed uniquely in ways influenced by their own personal and sociocultural backgrounds, but its development also shared characteristics of many other relationships. Its pattern of interdependence developed gradually over time. At their first meeting, neither Susan or Tom was much influenced by the other, and their lives continued independently. Nonetheless, their first impressions were sufficiently strong that they recognized each other 3 months later.

In their separate interviews, Tom and Susan mentioned a variety of additional points. Thus, after their evening encounter at the bistro, each one's thoughts focused repeatedly on the other's activities. For example, while riding up the ski lift during her vacation, Susan sometimes found herself wondering what Tom was doing at the same moment. Tom said independently that he too was occupied by thoughts about Susan at that time. Following their reunion in Paris, therefore, the two discovered that their casual evening had led both of them to consider a more serious relationship. Their temporary loneliness on a December evening had given rise to transactions that led each one to desire increased closeness.

Subsequent weeks and months led to an increased frequency of interaction—to events that diversified and strengthened their interconnections. Both of them experienced many instances of mutual facilitation and few sources of friction or interference. Noticeable, too, was a high degree of symmetry in their feelings.

Susan's moving into Tom's apartment removed one important impediment to their relationship; it made their physical environment more conducive to further progress. This is an example of how changes in a pair's relational connections can modify important causal conditions in their environment. The progress of the Darbers' relationship also encouraged both partners to change some of their personal dispositions—Susan became more interested in art, and Tom became a good skier. These personal changes facilitated further progress, which in turn led to an even greater convergence of their interests. From the standpoint of contemporary norms about courtship, their eventual commitment to marriage was not surprising.

Without discussing the early and middle stages of the Darbers' marriage at this point, let us look at its deterioration and breakup. After Susan attended law school, she became more aware of the asymmetry in her relationship with Tom. It had not bothered Susan when she moved to live with Tom in Paris or when she transferred colleges during her senior year. Nor had she greatly minded Tom's absence from home while he was building his professional career. After she herself achieved professional status, however, she expected reciprocity on Tom's part. When he failed to give it, Susan began to think that their marital pattern was unfair. And, once having focused on the asymmetry of their marital pattern, she found more and more evidence of Tom's egocentrism and insensitivity.

What Susan had earlier considered tolerable and even comfortable, she eventually found intolerable and even outrageous. At the same time, Tom's feelings about the marriage stayed fairly stable; he remained largely oblivious to Susan's disenchantment until it had gone too far. In other words, the major causal conditions affecting Susan's marital role (e.g., her occupational status and her social network) changed over the course of the 22 years, but those affecting Tom's did not, and the two of them could not communicate constructively about this discrepancy. When Susan's final blowup occurred, her perceptions were far out of line with Tom's. It would have taken enormous restructuring to salvage the relationship, a task that neither partner—for different reasons—was prepared to face. Thus their marriage ended. (Incidentally, although the Darbers' terminated their marriage legally, their social relationship did not end. As parents of their children and as members of other overlapping social networks, Susan and Tom continue to come into contact with one another.)

CONCEIVING OF RELATIONAL DEVELOPMENT

Varieties of Relationship

Marriage is, of course, only one of a large variety of close relationships. It would be possible to consider the development of many other types, such as same-sex or other-sex friendships or romances, business partnerships, employer–employee relations, familial or kin relationships, and even the long-standing ties between close neighbors. In each of these cases, the formation of a relationship is marked by an increased frequency in the persons' interaction, a diversification of their shared activities, and a strengthening of their ability to influence one another. In other words, causal interconnections become more frequent, more diversified, and stronger as a relationship is formed.

The nature of a pair's development depends greatly on its type of association. Friends have different aspirations for their bond than do romantic partners; business associates have different interpersonal goals than family members. Two neighbors may end their relationship when one moves away, whereas parent and child are unlikely to terminate their connections no matter how great their physical or psychological distance. Given these great differences, it is beyond the scope of this chapter to chart the course of "relationships in general." Our more modest aim is to examine heterosexual relationships between similar-age adults, which may progress from a casual acquaintance to courtship and marriage and may later face issues of conflict and deterioration or ending.

Postulating a Developmental Sequence

According to the basic causal model introduced in Chapter 2 (Figure 2.5), P's and O's personal, environmental, and relational conditions influence their interaction, which in turn influences those same causal conditions. In an encounter between two strangers, the relational influences are very slight, but they increase substantially as two people become interdependent. Relational development is thus accompanied initially by a growth in the P–O relational causal conditions (e.g., shared norms). Both personal and environmental conditions also change over time. For instance, persons get older and acquire new attitudes or tastes; environments are transformed by changes in residence or employment and by altered patterns of friendship or kinship.

Given the many ways that relationships can develop, is it possible to identify any general long-term sequence? Beyond emphasizing that all relationships have beginnings, middles, and endings, the present chapter will consider five sequential phases that appear pertinent to adult heterosexual pairs, as well as transitions across adjacent phases.

The following five potential phases are postulated:

A. Awareness of or *acquaintance* with another person; this phase may last indefinitely.
B. *Buildup* of an ongoing relationship, an exploration of the extent of the partners' mutual facilitation or interference, their pleasures and problems of connecting with each other; phase B does not necessarily require deliberate effort and can occur imperceptibly.
C. Following a mutual commitment to a long-term relationship, the next phase is one of *continuation* or consolidation of the relationship in a relatively durable midstage, marked by marriage in many pairs.
D. *Deterioration* or decline of the interconnections; like the buildup phase, this phase may develop imperceptibly, at least for some time.
E. *Ending* of the relationship, either through death or through some other form of separation.

This A-to-E sequence is a useful prototype, but only a small minority of all heterosexual relations actually pass through all five phases. Most pairs never get beyond phase A, acquaintanceship. Most of those that do are likely to terminate during phase B. Of the small fraction that commit themselves to an enduring association and thus engage themselves to maintain a long-term bond (phase C), not all find that their high interdependence is necessarily followed by deterioration (phase D). Indeed, many marriages build increasingly strong interconnections over the course of time and end only through a spouse's death. Furthermore, it is possible for relationships to start deteriorating during phase B, a process that may not be noticeable until long afterwards. Finally, the ending of relationships can take a variety of forms, from death to voluntary separation.

Particularly important for our analysis are the *transitions* between adjacent phases or periods. What will lead a pair to move from phase A to phase B, from a casual acquaintance to a significant buildup of their bond (A → B)?

After a bond has been formed, what factors impel partners to commit themselves to a more enduring continuation of their pairing (B → C)? Later, during phase C, what conditions lead to decline or deterioration (C → D)? Finally, given that a relationship has indeed deteriorated, what conditions lead to its ending (D → E)? Questions about such transitions will be examined in the present chapter.

The five-phase sequence as an organizing device

There would, of course, be alternative ways of dividing up a pair's development from initial acquaintanceship to final ending. The present A-to-E sequence, though, is long enough to permit differentiation, but short enough for easy comprehension. Its five phases, and the transitions between them, have several characteristics that should be noted.

To start with, note that different specialists have focused their attention on different parts of the sequence and have neglected others. Social psychologists, for example, have until recently concentrated mainly on phase A, pertaining to initial attraction and impression formation (see a review by Huston & Levinger, 1978). Family sociologists have focused largely on phases B, C, and E (R. A. Lewis & Spanier, 1979). Clinicians have concentrated on phase D, the analysis of distressed relationships (see Chapter 10, "Intervention"). Knowledge about phase-to-phase transitions or change processes has remained fragmentary.

The five developmental phases differ in the extent to which they refer to the pair versus the individual person. Initiation and termination (phases A and E) imply little P–O interdependence; therefore they are often studied from the perspective of the single individual. In the three middle phases, though, relational properties are much more salient, and pair indices are profitable tools for research.

BEGINNINGS

This section will consider, in turn, four aspects of the beginning of a long-term relationship: phase A, the initial acquaintanceship; A → B, the transition from a superficial to a deeper relationship; phase B, the buildup; and B → C, the transition to a more enduring bond.

Acquaintanceship (A)

An acquaintance cannot begin until one person attends to another. P either sees O directly or gets information about O indirectly through mutual friends or other channels. In a "closed field" situation, where persons are forced to be together for an extended time—as when two persons share a seat on a bus trip—the environment makes them aware of each other. In an "open field" situation, such as a bus station, people's attention may be directed to a wide variety of others; in such a situation, one important determinant of P's attention is O's distinctiveness (Berscheid & Graziano, 1979). Only if P attends to O can the various determinants of interpersonal attraction begin to operate (Huston & Levinger, 1978). Before examining impression formation and initial interaction, let us note the influence of environmental and personality factors on encounters and interpersonal goals.

The physical environment—that is, the size and density of a community, its transportation system, its climate, and a host of other factors—affects whom P encounters and continues to see. Whether people live in the city or the country, in the hills or the plains, or in luxurious or impoverished surroundings influences their initial encounters.

If the physical environment sets the stage, the social environment writes much of the script. Our culture defines rules of eligibility for friendship and mate selection, maps paths over which developing relations are supposed to travel, and writes scenes that help shape relationships at different stages. We are all actors in a drama for which a large portion of our lines is already written, even though we may imagine ourselves speaking extemporaneously. Our social environment also consists of our social networks—our kin, friends, co-workers and others. These people influence whom we meet, get to know, and continue to see, and they communicate their approval and disapproval of our associates (Ridley & Avery, 1979).

People's personalities also affect their tendencies to form and maintain acquaintanceships. Both enduring characteristics (e.g., introversion–extroversion) and transient dispositions (e.g., feeling busy or idle) influence their interest in meeting other people. The other person's characteristics further affect the likelihood of one's approach: How good looking does the other appear? How interesting to converse with? How free to engage in conversation (Huston & Levinger, 1978)?

Impression formation

When one stranger (P) attends to another (O), P's intrachain events (perceptions, thoughts, affect) are causally influenced by O's characteristics or actions, but there is no necessary reciprocal effect. Initial impressions are often formed unilaterally. A large literature in social psychology has analyzed the determinants of impression formation. P's impressions are affected not only by O's characteristics (e.g., physical appearance or perceived competence) but also by P's own moods, goals, or values.

One prevalent hypothesis is that initial images govern subsequent interaction and resist disconfirmation. This hypothesis has not been tested beyond studies of early encounters in the laboratory (e.g., M. Snyder, Tanke, & Berscheid, 1977). An important question for future research concerns the conditions under which early impressions are changed in actual relationships: How and when do partners become open to new information that alters their previous perceptions?

Initial interaction

Having become aware of and favorably impressed by O, what leads P to initiate interaction? A combination of environmental and personal conditions are facilitative—for example, spatial proximity and O's personal desirability and eligibility as an interaction partner. Such factors had little effect on Susan and Tom Darber's first encounter, since neither one felt inclined to talk to the other, but these factors did influence their second meeting.

Given availability and desirability, interaction permits each individual to explore the other's enjoyability. If this exploration seems unrewarding to either one, little more may happen between the two—unless outside circumstances keep them together long enough to find each other attractive. If early interaction is mutually enjoyable, a couple may act to alter inhibiting environmental conditions; for example, the Darbers eliminated their irksome commuting by having Susan move in with Tom.

Early interaction may also be transformed in other ways. People often meet each other in one set of roles and later take on new roles. For instance, two college students may first meet as members of the same English class, where they only discuss literature. Only later, when one of them needs an evening companion, is their interaction transformed from that of classmates to that of dates. In the role of dating, new behaviors and new standards of evaluations become appropriate.

From Acquaintance to Building a Relationship (A → B)

How do people move from merely knowing each other to beginning to care for each other? How do they proceed from a casual acquaintance to a deeper friendship? The transition from encounter to relationship is not well documented (see Huston & Levinger, 1978) but will be examined briefly here.

As mentioned earlier, there are many different ways for developing a relationship. Some formative transitions occur gradually, through a slow increase of interactive rewards over costs. One meeting follows another, date follows date, and slowly a mere acquaintance is transformed into an intimate friendship.

In other instances, a dramatic episode marks a breakthrough to an unexpected change in a relationship. Consider the following two examples. In one case, a 27-year-old skier returned from a ski weekend and announced to his parents: "It was wonderful! Lori broke her leg!" Although he was extremely competent in first aid and on skis—and although he had long been interested in Lori—he had always felt insecure around women. Lori's ski accident gave him an opportunity to show his competence and thus to gain the confidence necessary to further his relationship with her.

Another example of a dramatic buildup was an encounter between two previously unacquainted persons in their late twenties, one of whom had placed an advertisement in an underground newspaper. After some preliminary contacts, they made a date to meet at a restaurant. The woman later reported that "as soon as we saw each other, sparks flew . . . . We spoke for hours that first day and I think both of us knew that something was going to happen." The man agreed: "Yeah, after that luncheon, I knew that I was interested in pursuing her" (Berman & Weiss, 1978, pp. 46–47).

In each of these instances, both individuals were ready to build a relationship with someone of the other sex. Each had been looking for the "right" partner and thus was eager to expand on the opportunties for building connections. In either case, different personal or environmental conditions (e.g., one person's previous commitment or unpleasant food at the luncheon) could have led to different interpersonal outcomes.

Transitions from a casual acquaintance to a solid relationship are usually marked by both partners' successful testing of their interchain connections (Altman & Taylor, 1973). Each finds it easy to further the other's goals, and both come to anticipate mutually rewarding future interaction. Their interdependence expands not only with the increased frequency and diversity of interchain connections, but also with increases in the affective strength of those connections. Such developments may, in turn, lead to helpful changes in the partners' environmental conditions (e.g., the partners' proximity) and in their personal dispositions (e.g., their interests or goals), which serve to encourage further progress.

Sometimes two persons have been casual friends for a long time but do not strengthen their connections until external circumstances drive them together. For instance, two female housemates had been living together for half a year without sharing much intimate information. Then each discovered that both were getting increasingly angry at the third housemate's inconsiderateness, and they began to discuss their mutual feelings about her. After they had persuaded the third housemate to leave, they discovered that their collaboration had redefined their own relationship. Dealing with the irritating housemate had transformed their acquaintanceship into a strong friendship. In other words, working together to achieve one particular joint goal often serves to build interdependence with regard to additional goals.

Buildup (B)

There is little empirical literature on the processes whereby interpersonal relations become increasingly interdependent. Altman and Taylor's (1973) research on social penetration was directed at this issue but was mainly confined to the finding that verbal disclosure increased over time in "breadth" and "depth." In their recent reappraisal of the assumptions of social penetration theory, Altman, Vinsel, and Brown (1981) question its implicit assumption "that the development of successful relationships always follows a unidirectional and cumulative path, with ever-increasing openness of people to one another" (p. 7). They now emphasize that relational buildup entails a continual cycling between superficial and deeper contact, a repeated ebb and flow of exchange.

Direct observational data concerning actual change during the buildup phase are scarce. Longitudinal accounts have been confined, with few exceptions, to the descriptions of novelists or biographers or autobiographers, limited by the selectivity that such descriptions entail. One exception is Schwartz and Merten's (1980) careful sociological analysis of a young woman's changing loves and commitments during and after high school in a small Midwestern town. But even this lengthy account, as well as those provided by retrospective studies of larger samples (Bolton, 1961; Braiker & Kelley, 1979; Huston, Surra, Fitzgerald, & Cate, 1981) focused on only a relatively small number of selected time points over the developmental course.

An alternative approach is to obtain data about people's *expectations* regarding behavioral changes in typical relationships. Such an approach was employed by Rands and Levinger (1979). Respondents estimated the probable occurrence of 30 different behaviors in pair relationships varying in sex composition at four degrees of closeness—casual acquaintance, good friends, very close friends, and marriage. Respondents believed that all 30 instances of interpersonal behavior become more probable as relationships increase in closeness, but the rise was perceived to be much steeper for affectively loaded behaviors such as praise, criticism, or affectionate touching. These increases were steeper in male–female than in same-sex pairings.

Gradual versus stepwise buildup

In an early study of the transformation of romantic relationships into marriages, Bolton (1961) analyzed varying forms of relational development. In most of the 20 couples he studied, Bolton found that the process involved a sequence characterized by "advances and retreats along the paths of available alternatives . . . . [by the crystallization of commitments and felt obligations], by the reassessment of self and other, and by a tension between open-endedness and closure" (1961, p. 236).

Altman et al. (1981) suggest that relational development displays two basic dialectic processes. One dialectic is an opposition between stability and change. The second pertains to the opposition between openness and closedness–tendencies to affiliate and reveal versus tendencies to withdraw and remain private. In this regard, Eidelson (1980) has reported findings from two studies that show that temporal changes in pair involvement depend on varying strengths of two friends' respective affiliation and independence motives.

In the buildup of pair interdependence, then, "progress" is not necessarily constant. The solution of one interpersonal problem is often followed by the discovery of another. The unfolding picture looks different at any present moment from the way it looks in retrospect. Unfortunately, we lack data about the extent of that difference or its theoretical implications.

From his study of courtship, Bolton (1961) recognized the existence of "turning points" in relational development, but he emphasized that changes are rarely dramatic or even clearly perceived. He quotes one of his newly married interviewees as follows:

> It is hard for both of us to say when we privately got engaged. The subject would come up time after time, and each time we would be more seriously attached afterwards, until it was just "there." I'm not sure when it occurred exactly. There was no time it exactly occurred—it was a gradual transition. (p. 237)

Other retrospective investigations of relational buildup (Huston, Surra, Fitzgerald, & Cate, 1981; Purdy, 1978) have noted similar gradual changes punctuated by occasional turning points. In Purdy's study, respondents graphed changes in their feelings of pair involvement in two same-sex and two opposite-sex close relationships. Personal involvement reportedly rose more gradually and with significantly fewer upward or downward turns between same-sex than opposite-sex friends. Upturns and downturns in involvement were sometimes linked, respectively, to overcoming external obstacles or to the failure to overcome them. Huston et al. (1981) asked both members of recently married couples to graph from memory the history of their premarital relationship in terms of the changing probability of their marriage. These respondents, too, remembered few abrupt jumps or falls in their feelings about or definitions of their premarital relationship. Recollections of advances and retreats usually blended into the general flow of the pair's development. Turning points (i.e., sharp changes in the graphs) generally referred to discrete events; the recall of a major fight might be linked to a drop in the probability of marriage, whereas the first discovery of mutual intense affection or the couple's first intercourse might be tied to an abrupt rise (Huston, 1981, personal communication).

Whether the actual developmental process is smooth or jagged can be confounded by people's perceptions of how it changes. And their perceptions or recollections are governed by the nature of their conscious thinking. Recent theorizing (J. G. Holmes, 1981; Levinger, 1979b; Newman & Langer, 1977) suggests that people think a great deal about a partner at the beginning of a relationship, when they are comparing the relationship to alternative pursuits, but that continuing such comparisons later interferes with enjoying the relationship. If this theory is true, and no systematic data yet confirm it, this tendency would help account for the paucity of memorable markers in respondents' accounts of later changes in their relationships. Furthermore, it seems likely—and here too we lack data—that changes in the frequency and diversity of causal interconnections are more gradual than changes in their affective strength. Our framework (see especially Chapter 4, "Emotion") suggests that the greater the meshing of interchain connections, the more a relationship becomes predictable and emotionally placid.

For example, both Darbers emphasized the smooth progress of their relational buildup, but each recalled certain important marker events. Susan remembered how touched she was when, soon after her return after Christmas, Tom took her for an outing to a reclusive sculptor he had gotten to know. And Tom recalled being overwhelmed, not much later, when Susan gave him a pair of woolen socks she had knitted for him.

Linear versus circular causal models of buildup

Our framework assumes that the causal conditions that contribute to the building, maintaining, and eroding of a relationship stem from environmental, personal, and relational conditions, but that these factors are themselves affected by consequent pair events. In other words, our model (see Figure 2.5) is one of circular rather than linear causality.

This assumption is basic for considering "filtering models" of courtship and mate selection (Kerckhoff & Davis, 1962; R. A. Lewis, 1972; Murstein, 1970). These models conceive of mate selection as a sequence of decisions by two partners about the goodness of fit between their individual attributes. These models assume that, as two persons get to know each other, they get information about each other through a progressive series of filters or screens. At the start of their acquaintance, such information is limited to the other's physical appearance and social memberships. Later, they obtain information about the other's interactive responsiveness, and, after that, about attitude or value similarity. Still later, partners are said to receive information about each other's degree of "need" or "role" compatibility. Filtering models imply that, at each point in the sequence, each partner may decide either to continue the relationship, if current outcomes and future prospects remain favorable, or to cool it down. Thus, from point to point in the sequence, a partner's attributes are presumed either to pass or to fail the screening; one either increases or decreases his or her personal involvement, or continues the relationship at its present level. Filtering models of mate selection imply that all couples follow a similar causal sequence on their developmental path.

Although these models are intuitively plausible, they have failed to receive general empirical confirmation (C. T. Hill, Rubin, & Peplau, 1976; Levinger, Senn, & Jorgensen, 1970; Z. Rubin & Levinger, 1974). Here we will examine the theoretical weaknesses of such fixed-sequence linear models.

First, filtering models seem to assume that each person brings to a relationship a preexisting bundle of characteristics that, if they can only find a proper match with someone else's bundle, will lead to a properly compatible pairing. This assumption seems questionable. Although personal characteristics are an important determinant of the initiation of relationships, they are subject to alteration as a relationship develops. As two persons get to know each other, it is likely that relational conditions emerge that exert an increasingly important influence on a pair's outcomes (Levinger & Snoek, 1972).

Second, whereas filtering models imply a common set of ways in which different relationships develop, our present conceptualization emphasizes the multiplicity of ways. For instance, couples differ widely in their speed of buildup and their range of considered alternatives; some pairs go from superficial acquaintance to marriage with hardly a thought, whereas others go through a very prolonged and vacillating course of involvement before they decide to get married (Huston et al., 1981). Even among pairs that seem to progress at an average rate, studies of courtship have found multiple pathways of locomotion (Z. Rubin, 1975, personal communication). One pair interacts with great intensity on its very first encounter and only later diversifies its interdependence. Another continues its initial interaction at a casual level for a long time and only much later builds strong interconnections. Different pairs give widely differing accounts of how their relations have developed. Such different stories of progression imply widely different sequences of information availability.

Third, many aspects of relational buildup are neither deliberate nor voluntary. People do not always choose either their partners or their environment. Even if partners feel well suited to each other, circumstances often intervene to separate them. In contrast, two people who decide they are not well enough suited to each other may later revise their feelings in the light of new circumstances. For example, one successfully married older couple reported that they broke off their relationship after a brief but tumultuous engagement (Levinger, unpublished data). Believing that they had too little in common at that point, each person began sadly to explore relationships with alternative partners. Months later, the man was in an automobile accident. At this point the woman felt drawn back to help him recover. And this time the relationship clicked; to everyone's surprise including their own, this couple soon announced their impending marriage.

The notion of filtering, then, refers to important aspects of the developmental process, but it does not follow that filters operate in the linear fashion suggested by existing models. Decisions about the compatibility of a partner's attributes do probably require successive screenings, but such decisions are not irrevocable. Rather, both positive and negative impressions are reviewed and rereviewed. Factors that influence feelings about the relationship favorably or unfavorably are reevaluated in the light of new events and changing causal conditions. Decision processes do not necessarily pass through a single screening; rather, old issues often remain unresolved and resurface repeatedly.

Furthermore, the building of a relationship may be accompanied by a transformation of the criteria for judging satisfactory outcomes (Kelley, 1979; Levinger & Snoek, 1972). Early in a relationship, the other's actions are evaluated primarily in terms of the actor's own self-centered criteria. Later, those same actions may be judged more and more in the frame of the pair's own event history. If that is so, then the screening of the "other's" actions is increasingly translated into a screening of "our" transactions.

Suffice it to conclude that fixed-sequence models of mate selection describe only a part of the overall processes of relational buildup. As currently constituted, they do not account for the effect of interaction on initial causal conditions nor for the nonvoluntary transactions that function outside the deliberate decision-making process. Empirical research on relational buildup can profit from a continuous loop conception of causality.

The assessment of pair compatibility

As implied above, the development of a relationship is accompanied by increasing information about the partners' degree of compatibility or incompatibility. Theoretically, P discovers more and more instances in which O either facilitates or interferes with P's plans or activities, leading to either anticipations of harmony in times to come or doubts about a joint future. Practically, though, people do not always scrutinize their interaction so carefully. Matched, compatible interactions, being less problematic are less likely to be noticed (see Chapter 4's discussion of meshed sequences). Anecdotal data suggest that people assess their relationships primarily when they are trying to decide about making transitions from one to another phase of relationship.

The literature on interpersonal attraction (e.g., Berscheid & Walster, 1978) has identified two different determinants of potential compatibility—that is, P–O similarity and P–O complementarity. Each of these will now be considered within the framework of our present causal model.

Similarity in attitudes or other personal characteristics

P–O attitude similarity is a relational condition. It may exist before acquaintance but can also develop over time. Social psychological research has shown repeatedly that attitude similarity tends to facilitate long-term attraction (e.g., Newcomb, 1961), although the actual dynamics are a matter of some debate (Huston & Levinger, 1978). Chapter 2 (Figure 2.7) has already given a paradigmatic illustration of how P–O attitude similarity can encourage positive interaction over a short sequence.

Nevertheless, P–O attitude similarity does not always encourage interaction (C. R. Snyder & Fromkin, 1980); people at times prefer to interact with others who are different from themselves. For instance, when Susan and Tom Darber first met in Paris, neither wished to talk with another American, each preferring dissimilar, French-speaking students. Only at their second meeting did their similar feelings of loneliness give impetus to their relationship. Subsequently, their relationship blossomed in spite of their dissimilar recreational tastes; and later both of them modified their preferences so as to build their interconnections. Thus, Susan and Tom's degree of dissimilarity in attitudes and interests had neither a fixed meaning nor a static quality.

It appears, then, that the association between partners' attitude similarity and their mutual attraction depends on the degree to which their agreement is positively reinforcing (Byrne & Blaylock, 1963) or is instrumental for furthering the partners' goals (Levinger & Breedlove, 1966). Whether some given degree of similarity facilitates the partners' mutual outcomes depends on the interplay of other causal conditions. In the absence of adequate data about those conditions, we hypothesize the following temporal sequence: At the beginning of a relationship, another's similar interests or attitudes are likely to breed comfort and attraction. Later, the other's continuing sameness may get tiresome and lead to a search for novelty and difference. Still later in a relationship, particularly in times of crisis or confusion, partners are likely again to prize attitude communality and familiar experiences.

The puzzle of pair complementarity

The idea that "opposites attract" has long been popular among citizens and professionals alike. Early sociological research on mate selection largely discredited that notion with regard to dissimilarity in values, interests, or attitudes (see Berscheid & Walster, 1978), but it emerged again in the hypothesis that partners with complementary rather than similar *needs* are especially attracted to each other as potential mates (Winch, 1958; Winch, Ktsanes, & Ktsanes, 1954). Winch argued, for example, that persons high in the need for dominance would better complement (i.e., feel more satisfied with) others who are low in the need for dominance than others having a similarly high need. He listed a long set of additional "complementary" need pairs. Our present theoretical orientation leads us to reexamine Winch's hypothesis and to suggest an alternative formulation.

First consider the general meaning of complementarity. According to *Webster's Dictionary,* to "complement" another means to supply another's lack or "to fill out or complete" another's performance. Many of the most productive human relations are complementary—for example, male and female in the sex act, seller and buyer in an exchange, talker and listener in conversation. To the extent that one person gives what the other wants to receive, the interaction is completable and thus rewarding. Nonetheless, Winch's need complementarity hypothesis has received little empirical confirmation. Studies have usually failed to find that pairs in which partners' need constellations appear complementary progress more readily in their relationships than do pairs with noncomplementary needs (Levinger et al., 1970; Seyfried, 1977; Stroebe, 1977).

One difficulty is that Winch's hypothesis assumed that people's "needs" remain stable over time and across situations—for example, that needs for dominance or nurturance are stable personal conditions. This assumption is dubious; motives are hardly constant over time but rather wax and wane under varying conditions of deprivation and satiation. At best, measures of people's needs describe central tendencies around which need-expressive behavior varies depending on its situational context.

Empirically, it is possible to find evidence for the facilitative effects of need complementarity under limited, task-connected conditions. One well-designed study by Smelser (1961) found that individuals scoring high on dominance worked better when paired with another who scored low in dominance, under a condition in which each was assigned to the corresponding high or low status in a temporary laboratory task, than did any other composition of dominance and work status. Over the longer term, however, it seems dubious that such pairwise role assignments remain permanent, even if their respective personality scores were to remain so. Everyday observation suggests, for example, that even very dominant individuals fail to display that trait under all conditions. Furthermore, dominant individuals seem likely to enjoy similarly dominant partners when they are looking for excitement.

Any successful analysis of pair complementarity, then, must account for the contextual conditions under which partners interact. The social environment is an especially influential aspect of this context. Relationships with third parties strongly affect, and are affected by, the dyadic relationship. A proverbial image pictures the man whose dominance needs are frustrated at work, but who comes home to lord it over his wife and children. Conversely, the passive husband at home is shown going out to a job in which he bosses his employees. Needs for achievement, autonomy, or novelty also find their expression in a variety of associations; thus, they may not serve as reliable predictors of either facilitating or interfering relational events within a particular pair relationship.

Interpersonal complementarity, then, should be conceived as a feature of relational roles played in particular situations (see also Chapter 6, "Roles and Gender"). Such situations are liable to recur repeatedly, but they are not constant over the course of the relationship. It is the *meshing of situation-specific intrachain sequences* that should be considered to assess whether there is high or low complementarity, depending on whether the interchain connections are facilitative or interfering. To understand continuing complementarity, one must assess the continuity of two actors' social and physical contexts, as well as the meanings that they each attach to them.

During the early construction of a relationship, partners do indeed learn about the situations during which their actions are mutually facilitative and about the conditions that contribute to their compatibility. To the extent that their compatible interactions exceed incompatible ones, they will experience satisfaction. Often, they will even infer that their future together will continue just as compatibly. Yet, unless these early situations represent a good sample of the pair's later situations, such inferences are erroneous. More accurate, it seems, are limited inferences about the partner's ability to connect with one's own behavior under particular situational constraints.

From Buildup to Continuation: The Development of Commitment (B → C)

The transition from a phase of tentative "buildup" to one of long-term "continuation" presents important problems for causal analysis. Two basic questions are: What sorts of forces move a relationship from relative instability to relative stability? And what forces tend to inhibit such a transition?

Such questions are treated in considerable detail in Chapter 7, "Love and Commitment," in which commitment is analyzed in terms of membership stability. According to this analysis, high commitment implies that positive ("pro") forces that favor membership continuation heavily outweigh the negative ("con") forces, and that the variability of the difference between pro and con forces is relatively low. This is a general analysis that applies to people's commitments to any sort of pair or multiperson group. Friendship and memberships in groups usually grow with little awareness or publicity (Levinger, 1980; Purdy, 1978), so that the B → C transition in such relationships is often an imperceptible process. Processes of commitment in heterosexual relationships, however, are often more voluntary, more public, and more likely to entail deliberate decision making than those in nonexclusive relationships.

In view of Chapter 7's careful analysis of processes that increase people's commitment to a close relationship—for example, improving the reward–cost balance of membership, making irretrievable investments, heightening termination costs, or forming explicit understandings of partnership—this section is limited mainly to two contrasting illustrations of B → C transition. Both cases refer to movement from a premarital to a marital relationship. One case exemplifies modern Western patterns, in which commitment processes occur after a growing personal involvement; the second illustrates marriage by family arrangement, in which an important commitment process precedes personal involvement.

Commitment following growing personal involvement

One path into a marriage leads from "falling in love" to an eventual mutual commitment. This path was taken by the Darbers during their courtship in Paris. Early in their relationship they discovered a strong mutual attraction. While exploring the bounds of their mutual interests and activities, Susan and Tom built up diverse and strong interconnections that both valued highly. To continue and to strengthen these connections, they later made an explicit pledge, publicly reinforced by familial approval—that is, an engagement to get married. In other words, both removed themselves from the interpersonal marketplace and agreed to limit their closest intimacies to one another.

In this case, commitment developed mainly through the partners themselves, who built frequent, diverse, and strong interconnections during a long sequence of interactive and private events. Both followed culturally determined scripts in building their relationship. Each partner had an image of appropriate ways of getting acquainted, living together, and constructing a joint life as a permanent couple, which in their case facilitated straightforward progress toward marriage.

In other cases, personal and social influences are less facilitative. For example, two potential partners may hold incompatible attitudes about the meaning of marriage. Or their families are antagonistic. Or his friends approve warmly of her, but hers are critical of him. In each such case, the pair's privately developed interconnections are tested. Some relationships that survive initially antagonistic influences may actually be strengthened by the partners' struggle to resolve the problems. Other relationships, probably the majority, are weakened by such interfering forces and fail to reach the point of long-term commitment.

Commitment preceding personal involvement

A very different B → C transition is exemplified by the institution of arranged marriage. In the most traditional form of arranged marriage, the relational buildup is achieved by third parties—that is, parents or matchmakers—without any consultation between the persons to be betrothed. Without the partners' personal choice, but merely with their assent, the social environment creates conditions that close off a bride's and groom's future alternatives.

In the traditional arranged marriage, feelings of involvement and love can only follow the partners' commitment to their relationship. This fact leads to a rather different conception of love. Levenson (1973) writes about such a traditional marriage between his parents, who grew up in working-class Jewish families in Brooklyn and did not meet before their wedding. Levenson reminisces as follows:

> 'Love, schmove!' Papa used to say. 'I love blintzes; did I marry one?' To Mama, love was not passion, or infatuation, or compatibility. She had given birth to ten kids without any of those. 'Love,' said Mama after many years of marriage, 'is what you have been through with someone.' . . . Not that they didn't believe in love. They felt it, but avoided the precise definition that young people demand. Defining it might lead to misunderstanding rather than understanding. (p. 210)

From our present point of view, Mama and Papa did define the meaning of their love clearly, but differently from its contemporary meaning. For them, love was a personal attitude that grew out of their long-time interdependence—the psychological residue of interaction, and no more. "Love is what you have been through with someone." Their commitment to each other remained stable, for they did not seem to consider alternatives to their married state and their "pro" forces were strongly embedded in the network of both spouses' sociocultural and religious obligation.

Blood (1967) surveyed both arranged marriages and love matches in postwar Japan, where the custom of the traditional "interview marriage" was being replaced by Western forms of "personal selection." He found relatively few overall differences in marital satisfaction between these two contrasting paths into marriage, especially if arranged weddings were preceded by numerous (10 or more) "dates" between the spouses-to-be. Nevertheless—in marriages of 5 years or longer—husband's marital satisfaction was significantly higher in couples who followed the traditional paternalistic arrangement than that of their love-match counterparts, whereas traditionally married wives reported significantly lower satisfaction than love-match wives. Most important for the success of both types of Japanese marriages, according to Blood's findings, was the degree of enthusiasm expressed by the spouses' family members: the greater the parents' approval, the greater was the spouses' subsequent marital satisfaction.

Finally, it should be noted that the B → C transition does not necessarily occur at a single identifiable point in time. Its processes overlap with those initiated during phases A and B, and they carry into the middle-stage processes to which we now turn.

MIDDLES

This section concentrates on middle stages in marriage, successively considering phase C (continuation), the C → D transition, and phase D (deterioration). Note, however, that the continuation and deterioration phases merge into each other. Two other important points must be noted. First, marriages differ widely in the nature of their P × O conditions as well as the strength of P's and O's attachment and commitment. In some marriages, both spouses are highly affiliated, the content of their interaction is highly personal, and the frequency, diversity, and strength of their interconnections is high. In other marriages, one or both spouses are disaffiliated, and their relationship is contained mostly by external or institutional constraints. This distinction parallels those made by E. W. Burgess and Wallin (1953) between "companionate" and "institutional" marriages, by Cuber and Harroff (1965) between "intrinsic" and "utilitarian" ones, or by Bernard (1972) between "interactional" and "parallel" marriages. The glue maintaining the relationship depends on the nature of the marriage.

Second, the nature of the spouses' interdependence is likely to affect processes of possible deterioration or breakup. If a couple's interconnections are strongly meshed, then a breakdown is likely to be accompanied by different sorts of events or processes than if the interconnections are weakly developed, and different causal explanations may be appropriate. That is, if P × O interconnections were once very strong, explanations of deterioration may focus primarily on the partners' patterns of social exchange and the decline of their interpersonal attractions. If, on the other hand, they were never very strong, then one's explanation is likely to center on significant changes in the pair's environment.

Continuation (C)

Beginnings of relationships are marked by the partners' experience of novelty, ambiguity, and arousal. In contrast, middles are accompanied by familiarity, predictability, and the reduction of cognitive and emotional tension. The smoother the functioning of a marriage, the less will be the partners' self-consciousness or their ambivalence. Three aspects of the C phase will be treated here: changes in marital satisfaction over time and the impact of two different "critical events"—the impact of the first child and the impact of one spouse's disabling illness.

Changes in marital satisfaction over time

Susan Darber reported that her marital satisfaction rose during the early part of her marriage. The occasional stressful times were outweighed by her pleasure with their growing family and the continuing advance of Tom's career. Only later did Susan realize that those also were years when she and Tom established patterns of not talking about their occasional marital difficulties—patterns of meeting the other's irritations with joking or belittlement and of suffering one's own discontents with resignation. After the children grew older and Susan returned to school, both spouses realized that they now were sharing their feelings much less than during the earlier years. Both, however, accepted this change as a "normal" part of later marriage. How truly "normal" was the Darbers' experience? And, whether or not it was typical, what might have been the contributing factors?

Textbooks on marriage sometimes say that average marital satisfaction declines from the beginning to the end of marriage. For example, Blood and Blood (1978) write:

> The average marriage coasts downhill. The longer it lasts, the lower it sinks. Partly the decline is psychological, an ebbing enthusiasm for the partner. But it is also pragmatic, in the sense that partners become disengaged from one another. Some marriages fall apart. Others not only resist decay, but grow and develop. In between are the bulk of marriages—superficially intact but steadily less cohesive. (p. 547)

This statement reflects findings from a large sample survey of Detroit housewives about 20 years earlier (Blood & Wolfe, 1960) and also seems supported by a variety of subsequent research findings (A. Campbell, Converse, & Rodgers, 1976; Glenn, 1975; Pineo, 1961; Rands, Levinger, & Mellinger, 1981) that suggest that length of marriage is inversely associated with reported satisfaction. For instance, Blood and Wolfe (1960, p. 174) wrote that only one Detroit wife out of four was "still enthusiastic" about the quality of her marital companionship, a second was merely "satisfied," and the remaining two out of four wives were dissatisfied—many of whom said they were living with their husband as relative strangers under the same roof. From an interview study of over 200 affluent Midwestern spouses, Cuber and Harroff (1965) concluded that a substantial proportion of those long-term marriages were "devitalized," many others were either "passive-congenial" or "conflict-habituated," and that only a minority remained "vital" 15 to 30 years after their wedding. Thus the Darbers' drift from an exciting and full to a dull and empty relationship was not atypical.

This apparent decline in marital satisfaction over the course of an average marriage may be interpreted in various ways, ranging from dismissing the finding as a methodological artifact to considering it as a combination of environmental, personal, and relational changes.

METHODOLOGICAL ARTIFACT. To begin with, the evidence for a temporal decline in marital satisfaction is limited. Blood and Blood's (1978) conclusions about change over the length of marriage are based primarily on cross-sectional survey data from spouses of differing ages who represent different birth cohorts and therefore different initial marital expectations and historical experiences. As has been suggested elsewhere (e.g., Spanier, Lewis, & Cole, 1975), basing conclusions about change on cross-sectional findings runs the risk of confounding sheer length of marriage with spouses' age and generation, and the data are also susceptible to sampling and response biases. More convincing support for the decline in marital satisfaction comes from only one retrospective study of individual spouses (Cuber & Harroff, 1965) and a single large longitudinal survey in which engaged couples were followed up at a single time point 20 years later (E. W. Burgess & Wallin, 1953; Pineo, 1961).

Furthermore, there are two sets of contrary data that should be noted. One set has suggested that the long-term decline in long-term marital satisfaction, generally attributed to the impact of children on the spouses' preoccupation, is succeeded by an upswing after the children have left home (A. Campbell, et al., 1976; Rollins & Cannon, 1974). But this finding, too, is based on cross-sectional data and is subject to the same possible artifacts discussed above. A second set of data were recently obtained from another longitudinal study of married individuals (Skolnick, 1981). In an analysis of 82 spouses in their fifties, for whom marital satisfaction data were available at two follow-ups, Skolnick finds "no evidence of decline" (p. 294). Fifty-four percent of those spouses did not change markedly in their marital ratings over the decade; of the remainder, 17 percent declined in satisfaction, whereas 29 percent improved. Although Skolnick's finding differs from Blood and Blood's implication, it refers to a limited time span in her respondents' marriages as well as a sample of fairly conservative individuals of high socioeconomic status. Let us therefore consider explanations for possible changes in marital satisfaction over time.

CHANGES IN ENVIRONMENTAL CONDITIONS. Assume that most spouses are at the time of their wedding near a peak of satisfaction with their relationship. This is referred to as the "honeymoon" period, when a pair is childless and supposedly free to devote themselves to each other. Compared to the external pressures facing them, newly married spouses seem to possess adequate if not abundant resources. If in succeeding years, however, environmental demands rise more rapidly than do the couple's resources, it is reasonable to suppose that their satisfaction with their physical circumstances is likely to decline. For instance, a just-married childless couple may require little physical space; after children arrive, though, their original home may feel cramped. Similarly, their initially sufficient income becomes insufficient after the family expands, which often coincides with the reduced earning ability of the principal childcare giver. Low marital satisfaction has been found especially common in marriages with low income and unstable employment (Cherlin, 1979; Levinger, 1965; J. Scanzoni, 1970). Even in couples with considerable increases in income, marital satisfaction has been found to decline if the high-income husband finds it necessary to spend increasing time away from home to meet added organizational demands

(Dizard, 1968). Furthermore, particularly for husbands, occupational satisfaction has been found significantly linked to later marital satisfaction (Skolnick, 1981); to the extent that occupational satisfaction declines, marital satisfaction will also decline.

CHANGES IN PERSONAL CONDITIONS. Both constancy and change in spouses' personal characteristics may affect the course of a marriage. Some forms of constancy—such as continued caring or acceptance of the other's foibles—are likely to be associated with marital stability and enhancement. Other forms—such as intractable habits or rigid role expectations in the face of changed needs—function so as to irritate or interfere with a satisfying close relationship. Depending on their content and context, inflexible dispositions may affect the marriage either positively or negatively.

Personal change also may function either positively or negatively. Some spouses interpret the partner's changing attitudes, interests, or skills as contributions to novelty and stimulation (Brooks, 1978); others label them as signs of immaturity or unreliability or as threats to the existing relationship. Whether the other's personal changes are met with approval or disapproval seems to depend on the degree to which they are perceived to facilitate or interfere with the existing meshed interconnections. To evaluate their impact requires an assessment of the manner in which they fit the previous and current interchange.

CHANGES IN P–O INTERACTION. The critical question, to which we have few definite answers, concerns the nature of temporal changes in spouses' actual interaction. The bulk of current knowledge is confined to retrospective accounts from individual spouses (e.g., Cuber & Harroff, 1965) that do not directly inform us about changing patterns of husband–wife behavior over time. Thus, if a spouse tells us that his or her marriage remains as "vital" and satisfying as ever, we do not know what has continued to keep it so. Or if another says that marriage has become "devitalized" over the course of time, it gives only limited information (that may not match the other spouse's) about how the relationship functioned at earlier times or the transactions that led to its decline. In the case of the Darbers, we obtained retrospective accounts from both partners—although Tom's story was much less rich than Susan's—but we have no convincing way of getting independent observations on their earlier interaction.

At the present time, there seems to be only one published study that reports sytematic longitudinal interaction data for married couples over a period of years. Raush, Barry, Hertel, and Swain (1974) collected such data on a subsample of 13 couples at three time points: (1) 4 months after marriage; (2) an average of 14 months later (7 months into the first pregnancy); and (3) 4 months after the birth of their child. At each of these times, each couple participated in a series of four conflict role-play improvisations. Couples' interaction styles showed consistency over the three times, but during the wife's pregnancy the husbands took more care to prevent conflict and to reconcile differences than either before or after. The data from this small sample, studied over less than 2 years, yield rather little information about the manner or the degree to which spouses' interaction patterns change over time.

Given the scant longitudinal data about marital interaction, one is forced to make inferences about temporal changes in satisfaction from studies conducted at a single time point. For instance, Gottman (1979) has found that spouses are generally less polite to each other than to strangers; it appears they learn to talk to each other more frankly but also more rudely. Beyond that, most existing interaction studies consist of comparisons between the behavior of "distressed" and "nondistressed" couples (e.g., Gottman, 1979; Jacobson, Waldron, & Moore, 1980; Koren, Carlton, & Shaw, 1980; Raush et al., 1974; Wills, Weiss, & Patterson, 1974). Today's preponderant evidence from such studies suggests that harmonious couples have learned a different form of interaction than distressed pairs. That is, spouses in harmonious couples are found to reciprocate each other's rewarding behavior but not to react in kind, either subjectively or behaviorally, "when the partner delivers punishing stimuli" (Jacobson et al., 1980). In contrast, distressed spouses' behavior is unpredictable when the partner acts positively, but their tendency is to respond in kind to the other's negative actions (Gottman, 1979; Jacobson et al., 1980). It seems, then, that satisfied spouses learn over the course of their relationship that the other's unpleasant behavior is more to be accepted than to be worried about; unpleasantness is more a reason for sympathy than for counterattack. And, given this tolerant reaction, the fundamental compatibility of their interaction is not endangered. Distressed spouses, on the other hand, are far more likely to respond unpleasantly to the other's negativity and thus to escalate the original altercation.

Gottman and his colleagues have observed that highly satisfied spouses tend to listen to each other's expressions of "problems" and to respond with some form of "validation"—that is, a nonjudgmental expression of understanding. Validation encourages the speaker's continued expression, permits the listener to hear more, and also prepares the speaker to listen to the other in turn. Validation is a sign of considered attention. Although it is easier to use in times of calm than of storm, its consistent use may make it into a stable property of interaction. Contrasted to validation is "cross-complaining," in which the other's complaint is not listened to but met with a counterattack (Lederer & Jackson, 1968). Dissatisfied spouses are observed repeatedly to engage in this form of interaction, setting off lengthy chains of reciprocal negativity (Gottman, 1979).

If couples' marital companionship worsens over the years, it is likely that their conflicts are marked more by cross-complaining or passive avoidance than by validation or consideration (see Rands et al., 1981; Raush et al., 1974). In her book on working class marriages, L. Rubin (1976) notes the unresolved friction in a majority of the 50 pairs she interviewed. Most of Rubin's couples had allowed resentments to accumulate, chilling the feelings of the aggrieved, and in turn worsening their future situation. In many of her cases, the sexual relationship had become progressively worse, a condition traceable to deterioration in marital interaction. Consider the following example:

> Wife: I want him to talk to me, to tell me what he's thinking about. If we have a fight, I want to talk about it so maybe we could understand it. I don't want to jump into bed and just pretend it didn't happen.
>
> Husband: Talk! Talk! What's there to talk about? I want to make love to her and she says she wants to talk. How's talking going to convince her I'm loving her? (L. Rubin, 1976, p. 146)

In this case, the wife believes that "talking" will help to restore "love." The husband does not. These two spouses' definitions of talking and of love are vastly different, and thus their causal analyses diverge. The husband equates love with relief from sexual arousal. The wife believes that her sexual arousal is tied to her feelings of appreciation, feelings that need to be restored before she can enjoy sexual play. Neither partner, particularly not the husband, properly understands the other's perspective. By failing to recognize each other's intrachain sequences, these spouses fail to attain interchain synchrony.

If two out of four Detroit wives indeed became progressively disenchanted with their marriage while only one in four grew in enthusiasm (Blood & Wolfe, 1960), the causes may be sought in multiple places, their particular mix obviously varying widely. The main point addressed here is the interplay among those causes.

We can also consider changes in the course of satisfaction or love from another standpoint. Chapter 4, "Emotion," examined how emotion in a long-term relationship becomes tranquil to the extent that the pair's interchain sequences become increasingly well meshed with each individual's intrachain sequences. Early in a relationship, interaction is marked by uncertainly and a high probability of "interruption"; later it is marked by relative security and stability. It is possible to mistake such tranquility as implying emptiness or passivity. Thus, spouses may not fully appreciate the degree of their dependence on each other and so may dream of greater excitement elsewhere. In other words, in a society that emphasizes romance and novelty, long-married spouses may raise their earlier expectations for "a good marriage," and they may become critical of the predictability of their existing relationship. Our present view, however, is that emotional highs are likely to be transitory and not a good index of a pair's closeness.

Nevertheless, it is possible within the confines of a committed relationship for partners to create continued interest and even excitement. Spouses can add challenge by engaging in new undertakings at home or in journeys away from home, or they can change familiar interaction sequences through humor or trying varied sexual practices. To the extent that their joint enterprises elicit the unfamiliar, new interchain meshings will emerge. In creating such new uncertainties, the question from the days of courtship—"Does my partner love me?"—is replaced by "What new surprise will tomorrow bring?"

Coping with a critical life event: Parenthood

The appearance of the first child transforms the family from a dyad into a triad and has many ramifications for the future of a marriage. It is therefore an important locus for the analysis of changes in marital relations. Nevertheless, social scientists have little definite knowledge about "how, why, and to what extent" parenthood exerts influences on family relations (Lamb, 1978, p. 157).

Early studies by family sociologists (e.g., LeMasters, 1957) emphasized the stresses associated with the first birth. The findings of various surveys (e.g., A. Campbell et al., 1976) that indicate that childbearing parents have lower marital satisfaction have implied that "having children" will do damage to a marriage. Nevertheless, other findings suggest that most parents feel that parenthood is more rewarding than stressful. In a review of diverse evidence, Hoffman and Manis (1978, p. 175) conclude that the total picture is more sunny than cloudy: Despite the existence of strains, the majority of parents surveyed, in contrast to childless spouses, say they believe that having children brings a couple closer together.

The latter impression is shared by Susan and Tom Darber, who had prepared themselves well for the arrival of their firstborn. They think they coped relatively well with having children and thoroughly enjoyed their parental experience. Less fortunate couples, particularly those who do little joint planning, are likely to find parenthood far more taxing. If they have little income, the new addition will strain their precarious resources. If their home is cramped, it will become more so, and the increased noise and reduced sleep will raise the level of tension. Teenage parents, particularly those who are premaritally pregnant, are likely to experience parenthood as especially stressful (Furstenberg, 1976); the unplanned pregnancy and birth cut short the time they might otherwise have devoted to building their pair relationship, and the economic consequences produce added job pressures on both parents.

In an important new longitudinal study of "becoming a family," Cowan and Cowan (1981) and their colleagues have focused on couple properties before and after the birth of the first child, as well as each spouse's personal characteristics. The Cowans' study assumes that the impact of the baby will depend both on the quality of the couple's relationship before the birth and on how the couple experiences the transition. Their study of 96 young couples—three groups of 24 expectant couples in differing research conditions and one otherwise comparable group of nonexpectant pairs—investigated the spouses' changing role arrangements, communication patterns, and sense of self. Their preliminary findings suggest that the sharing of household tasks declined substantially from pregnancy to 6 months after birth; most pairs in this sample became "traditional" in their household roles following the baby's arrival. Mothers whose role arrangements were least equal tended to report the lowest role satisfaction, and this also depressed their marital satisfaction. In contrast, mothers whose partners were more involved in household and child care had relatively high role satisfaction. This study's preliminary findings indicate that the majority of young couples experienced more negative than positive changes in their self-esteem, communication, and conflict, but there also were important exceptions to this trend (P. A. Cowan, 1981, personal communication). Couples most "at risk" were clearly those whose prebirth functioning was rated the poorest. In contrast, "well-functioning" couples before birth tended to cope well with the experience of parenthood.

Altogether, then, the "same" objective event of a child's birth has a wide range of impacts on different young couples. To the extent that today's social environment is less supportive of parenthood and that it fragments mothers' and fathers' familial and extrafamilial concerns, young parents are likely to experience greater stresses than the older survey respondents cited in Hoffman and Manis's (1978) optimistic review of the evidence. On the other hand, to the extent that today's norms emphasize the importance of a father's active acceptance of egalitarian role obligations and the couple's frank discussion of interpersonal issues, partners are likely to experience parenthood as a strongly positive marital experience. Further longitudinal research on this topic is needed, which will take into account individuals' prior characteristics as well as couples' previous and current patterns of interaction.

Coping with a critical life event: Serious illness

A spouse's serious illness or disabling accident can also be a critical event in a marriage. Although such an occurrence is undoubtedly stressful, its impact on a marriage can vary greatly. For example, studies show that after a husband's heart attack a minority of marriages deteriorate, a larger proportion remain about the same, and a substantial fraction actually improve (Doehrman, 1977).

Michela (1981) has proposed a model for analyzing changes in marital relationships following a husband's heart attack. He identifies several categories of important variables: (1) the initial conditions of the marriage before the illness, including the couple's social and financial resources and the partners' personal characteristics; (2) the nature of the actual heart attack and its medical treatment; (3) subsequent personal and interpersonal events; and (4) the interpretations that spouses place on the subsequent events and outcomes. Changes in any of the later categories may feed back to events and evaluations in the prior categories, including, perhaps, the recurrence of heart attack.

Michela's model is a good example of a circular causal model. That is, the improvement of a heart patient's physical health may enhance marital interaction, which in turn leads to further improvement in his health. The converse, of course, is also possible. The model calls attention to the linkages among analytic elements in the complex cyclical chain of causal conditions and marital events.

One important aspect of Michela's analysis pertains to spouses' attributions about each other's behavior (Kelley, 1979). A partner's sacrifice is usually perceived as a sign of love, but not always. If the sacrifice is interpreted as motivated by "duty," by external role requirements, or by ulterior goals, it will not necessarily help strengthen the marital bond.

Marriages under stress provide an important opportunity for examining both the vulnerabilities and the strengths of close relationships. Crisis situations are important sources for conflicts of interest regarding day-to-day matters, which allow the partners to display the extent of their concern and affection for each other (Kelley, 1979, pp. 114–116). The analysis of differing couples' responses to similar crisis situations offers a window on how and when close relationships either maintain themselves stably or change in a positive or a negative direction.

From Continuation to Deterioration (C → D)

Between the maintenance of a stable relationship and its noticeable impairment there may be one or more points or periods of transition, occasions when things go from good to bad, from pleasant to unpleasant, from easy to difficult. A major problem in trying to identify such points is that we cannot easily ascertain whether a specific current trouble is just a temporary fluctuation or a prelude to further decline. Furthermore, relationships are systems amenable to regeneration. If partners sense that things are going poorly, they may intervene to identify and solve their problems.

Consider the Darbers' marriage, for example. After Susan had attended law school for a year, she began to notice that Tom took little interest in her new learning. Although she was disappointed by this, she was reluctant to make an issue of it. Looking back today, however, Susan thinks that if she had expressed her feelings at that time it might have changed subsequent events. She now points to that period as signaling a crucial downturn in the marriage, a time when she tried to confine their interaction to Tom's concerns but at the cost of weakening their overall relationship. Early in their marriage, they kept expanding their mutual interests and strengthening their interdependencies; now there was a reversal of that trend, but Susan and Tom failed to notice it until it was too late.

What are signs of a downturn?

A downturn in a relationship may be marked by an increase in interference or a reduction in the strength, diversity, and/or frequency of the pair's interconnections, especially of those that entail mutual pleasure. In addition, a downturn is usually accompanied by one or both partners' feelings that their outcomes have become unsatisfying, either with regard to their own absolute comparison standard or with regard to outcomes obtainable in alternative relationships (Thibaut & Kelley, 1959).

How such a change actually occurs has not yet been well documented in the literature on close relationships. Ordinarily, it would appear, a series of repeated incompatible events is necessary for a marriage to begin to deteriorate. O's actions, or perhaps O's mere presence, must interfere with P's ability to carry out his or her personal plans or activities. Sometimes P's disaffiliation seems to precede O's interference, as occurred in the case of the Darbers. In other cases, O's interference is considered to be the cause of P's disaffiliation. Regardless of how one punctuates such a series of events, though, a negative cycle of actions and reactions ensues. In any event, during this transition, spouses sense a loss of vitality, enjoyment, or trust in their relationship. One potential danger sign, whose dynamics are difficult to study empirically (Murstein, Cerreto, & MacDonald, 1977; Walster, Walster, & Traupmann, 1978), is either partner's worry about mutual equity or the fairness of the marital exchange (J. G. Holmes, 1981; Levinger, 1979b).

Whatever the causal conditions, any particular pair's C → D transition reflects how that pair reacts to the various personal, environmental, and relational influences on its current interactive situation. Some spouses prevent deterioration by confronting instances of irritation or unfairness and handling their conflicts constructively (see Chapter 9, "Conflict"). Other spouses avoid acknowledging such instances and permit their worries to fester silently. Still other spouses meet irritation or perceived inequity with aggression that may heighten the original problem. A recent survey of young married couples (Rands et al., 1981) found that all three responses to conflict were reported by substantial numbers of spouses. Although marital satisfaction was reported to be above average by a sizeable portion of respondents who either avoided conflict congenially or aggressed against their partner warmly, the highest satisfaction was reported by those who were confrontative without aggression; feelings of intimacy permitted confrontation without "escalation" of the problem. Thus, intimacy begets further intimacy; its lack contributes to further estrangement.

The same objective action can elicit very different reactions. An example is taken from a series of interviews by Levinger and his co-workers. Each of two women told about an incident early in her marriage when her husband forgot her birthday. One wife said nothing about it all day but made sure that evening her husband would notice she had received birthday cards from her family. After he noticed it contritely, she teased him unmercifully, but with laughter; since that occasion, she says he has never again forgotten her birthday. The second wife perceived her husband's forgetfulness as no joking matter, but as a sign of his lack of care; she felt too hurt to say or do anything about it. Years later, long after their divorce, she reports that this incident first "proved" that he was insensitive and not concerned about her, and that this event encouraged her soon after to pay attention to another man's interest in her. These very different reactions to these parallel incidents can be interpreted in terms of the two wives' differing personalities, differing social environments, or differing marriage histories, or some combination of all three. Our present perspective would emphasize the possible interplay of these three types of causal condition.

The timing of downturns

Little is known about the timing or sequencing of C → D transitions. Their timing does not necessarily correspond to that of objective life events such as childbirth or a geographical move. In one study of ex-spouses' accounts of their marital dissolution (R. L. Miller, 1982), some respondents said that their relationship began to deteriorate before they even got married, and some said that the estrangement started soon after the honeymoon. In other cases, more like the Darbers, marriages continued compatibly for many years before the advent of interfering causative factors.

Deterioration (D)

Years ago, the family sociologist Waller (1930/1967) offered the following hypothesis about marital breakdown:

> Alienation, once begun, moves on as ineluctably as if a thousand Titans were pushing it and the actors were but puppets in a show . . . . Husbands and wives who have begun to quarrel find themselves unable to stop until it is too late. (1967, pp. 104–105)

Waller's term *alienation* appears synonymous with antagonism. His assertion that once spouses start to quarrel they will be "unable to stop until it is too late" is a hypothesis. This hypothesis—that marital fighting leads to an endless cycle of interchain sequences with negative, perhaps increasingly negative, intrachain consequences—probably fits the experience of some distressed couples (see Gottman, 1979). But, for a variety of reasons, it fails to describe the circumstances of many other couples.

First, as pointed out in Chapter 4, "Emotion," for some couples, quarreling is an important part of the meshed interconnections between their intrachain sequences. To the extent that quarreling contributes to the smooth meshing of interactive sequences, it will arouse little emotion or disaffection. Although observers may consider two spouses' fighting as unpleasant or disruptive, the spouses themselves may view it as entirely normal and not at all upsetting.

Second, a *lack* of quarreling does not necessarily imply that a marriage is going well. As was true for the Darbers, spouses can drift apart by continuing to avoid arguments about serious couple issues. They can become increasingly independent of one another in order to avoid irritation and thereby gradually sever the connections between interactive and intrapersonal sequences. For example, "If he's too busy to go on vacation with me, then I'll go by myself." Of, "If she's not interested in helping decide how our basement should be remodeled, then I'll just have to make the decisions alone." But, while the partners' occasions for joint activities and problem solving diminish, the severity of their interpersonal problems may actually expand.

Third, the two spouses do not necessarily use the same criteria in judging the deterioration of their marriage (Bernard, 1972). It is entirely possible for one spouse to believe that the interaction has become unsatisfying while the other thinks it remains as good as ever. The dissatisfied one either may perceive drops in rewards or may raise his or her comparison level for evaluating interaction. Furthermore, one spouse can believe that current marital tension is temporary, while the other fears it signifies a lasting trend.

Fourth, even if two spouses agree that their marriage is in trouble, they can make deliberate efforts to alter its "ineluctable" character. They may seek to resolve their conflicts either through their own efforts or through outside intervention (see Chapters 9, "Conflict," and 10, "Intervention"). Such change is difficult, of course, even if both partners perceive the problems similarly. Interconnections that now seem maladaptive have usually been shaped over many months or years of interaction. Each individual's intrachain sequences are linked to various external events, persons, or other influences in the social and physical environment that surrounds the marriage. Nevertheless, the beginning of deterioration need by no means signal the eventual ending of a relationship. Rather, a period of distress *may* be a storm before an eventual calm—which perhaps must be preceded by periods of increasing turmoil, steps toward resolution, and the renegotiation of mutual commitment within a new context.

Both the paths and the outcomes of pair deterioration may thus be as varied as those of pair formation. Before examining a few of those paths, let us consider an additional statement about marital deterioration:

> Most marriages begin to fall apart during the honeymoon. The moment the spouses say "I do," the relationship becomes compulsory instead of voluntary. Almost everything within the newly formed joint system is viewed within a new perspective . . . . The change is profound. During the wooing, both people constantly attempt to be as attractive as possible; each tries to exhibit only those parts of himself which will please and capture the other . . . . As the novelty of the relationship begins to wear off, the effort to change one's idiosyncracies in order to be accommodating becomes more difficult. (Lederer & Jackson, 1968, pp. 245–246)

This analysis of marital difficulty, like Waller's, is probably accurate for some couples, even though it touches on only a few of the many-sided issues of pair deterioration. It especially emphasizes the issue of compulsion versus volition. One major difference between a marriage and a friendship is that marriage partners have made a public, externally reinforced pledge to each other, whereas friends generally have not. Hence acts of goodness to the marriage partner can be attributed merely to one's fulfillment of socially structured role obligations, whereas, in friendship, such acts are seen as pure generosity. Thus, marital difficulties may be compounded by each partner's uncertainty of when and how to attribute personal volition versus external compulsion to one's own and the other's actions (Haley, 1963, p. 120).

Lederer and Jackson also appear correct in noting that the beginning of marital breakup can sometimes be traced to the very start of a marriage, for example, when the new living arrangement requires partners to establish new and to revise old meshings between their intrachain sequences. Nevertheless, in many other cases—including Susan and Tom Darber's—the most significant changes do not appear until long after the initial marital patterns are established.

In view of the need for systematic data regarding the dynamics of marital deterioration, R. L. Miller (1982) recently studied the retrospective reports of 40 ex-spouses. Each of these divorced persons, 20 men and 20 women, described the process of their breakup—beginning with courtship and concluding with the present time. Respondents first gave unstructured accounts and then graphed their changing feelings of confidence (e.g., from 100 percent to 0 percent) that they would remain indefinitely married, from before their wedding until the final divorce. Each respondent also rated his or her feelings at four time points between the peak of their feelings of confidence in the marriage (time 1) and the point at which they felt single again (time 4). Ratings were made separately for "attractions" toward the marriage, for "attractions toward alternatives," and for "barriers" against leaving the marriage (Levinger, 1976).

There was no one preponderant explanation or sequencing among the different descriptions of the deterioration process. Some respondents began to be bothered soon after the inception of the marriage, or even before its very start, by their partner's demands or anger or rigidity or thoughtlessness—or by their own negative reactions or interactions. For other ex-spouses, it took a change in jobs, a geographic move, or the arrival of a second child until bothersome interpersonal patterns emerged or before they began to reappraise previously established patterns. In every case, though, these ex-spouses reported a decline in the rewards they experienced from the relationship—either an absolute decline or a decline relative to their changing expectations.

The largest portion of Miller's 40 ex-spouses indicated that the greatest initial change in their marriage was a drop in their feelings of attraction toward their spouse, followed by their attention to the costs of ending the marriage (barriers), followed by their preoccupation with the attractiveness of alternatives. Other ex-spouses reported that the attractiveness of alternatives had risen before they began to worry about the unattractiveness of the marriage itself. Still others believed that the drop in their internal attractions and the rise in alternative attractions had been almost simultaneous. Still different sequences of changes in their commitments were noted by several additional interviewees.

Miller's findings confirm what was suggested earlier in this chapter—that processes of deterioration are not confined to any single time period in a marital relationship. They may originate during the buildup phase and extend through the ending of the relationship. In the absence of additional research data, let us speculate briefly about several different causal processes that may contribute to deterioration.

Changes in P's private judgments

Imagine that a married woman is displeased by her husband's thoughtlessness, such as forgetting her birthday. Her displeasure leads her to act less generously toward him subsequently, bringing about less positive responses in him as well. Although he may for the time being remain unaware of his own changed behavior, she uses the data from that new interchange to confirm her previously negative evaluations, which then pervade her future behavior. This sequence illustrates how changes in one partner can remain uncommunicated to the other. It is not too different from the gradual change reported by Susan during the years before she confronted Tom openly.

Berscheid and Campbell (1981) suggest that the decline in a marriage's attractiveness is today furthered by partners' tendencies to evaluate their relationship and its alternatives more than in previous eras when separation and divorce were socially stigmatized. They propose that such a continuous assessment creates its own costs, which themselves contribute to further unease. For example, a wife becomes uneasy if her reading of her current marital state indicates its "pro–con" level has slipped; even if her reading is quite high, she may worry whether her husband also reads it that way. Furthermore, spouses' expectations of a "good" marriage have increased in recent times. Standards of evaluation have shifted upward so that a placid relationship that formerly seemed acceptable may now, without any change in its actual functioning, be considered unacceptable.

An accelerated version of the "private change" process is when both partners are simultaneously aware of marital difficulties but do not communicate with each other about the problems. Here two private spirals of negativity operate side by side, thus increasing the opportunity for mismatching and interference while preventing constructive communication. This version of marital deterioration represents Waller's (1930/1967) description of alienation. The inability of many spouses to communicate constructively about their problems (Rands et al., 1981; Raush et al., 1974; L. Rubin, 1976) suggests that such spirals are especially difficult to reverse.

Interactive changes

Two different forms of marital communication about interpersonal difficulties can contribute to the deterioration of a relationship. One type of communication, already referred to above, includes cross-complaining (Gottman, 1979), mutual criticism (Koren et al., 1980), and negative reciprocity (Gottman, 1979; Jacobson et al., 1980). In this type of interaction, partners are more interested in justifying their own point of view than in listening to the other. The result is to perpetuate rather than to resolve conflicts (see Chapter 9, "Conflict").

A second type of communicaton is more open. Here one or both are able to understand the other's point of view, which ordinarily would help them deal with conflict constructively. But if their conflict is important and "veridical" (Deutsch, 1973), such open communication only hastens their realization that the entire relationship is in jeopardy.

Third-party influences

Other degenerative influences may be associated with the intervention of the social environment. For example, if P communicates her unhappiness with the marriage to a friend or family member, it is possible that this communication will aggravate her original complaint. In Susan Darber's case, her talks with her friend made her dwell even more on Tom's negative characteristics. Furthermore, such third parties may themselves act icily or angrily toward the other spouse, with potentially damaging effects. If O meanwhile has likewise shared his feelings with others, the deterioration may be hastened.

Note that such degenerative influences from third parties are especially common in today's era and among certain subcultures. Today a woman who complains of her husband's physical abuse may be directed to a shelter for battered women; in former times, she would more likely have been told that she should accept such beatings. Similarly, today's parents may be more sympathetic to a married son's complaints about his wife than would parents in earlier historical periods.

Jaffe and Kanter (1976) have analyzed the important roles of third parties in marital conflicts. Their study of marital conflicts in urban communes found that other commune members acted as "potential audiences, intervenors, and allies" (p. 177). If one's own mate could not meet one's particular needs, others listened to one's story and often stood ready to meet the need themselves, thereby further calling into question the mate's abilities. The presence of third parties, and the support they offered either spouse, often aggravated grievances that might otherwise have remained unvoiced.

Recapitulation

This section has considered the declining attractiveness of a marriage and some of its interrelated determinants. Some determinants, such as either spouse's personal qualities—for example, neuroticism or psychopathy—or the two partners' dissimilar social backgrounds or antagonistic social networks, have not been considered here, but have been reviewed elsewhere (Levinger, 1965, 1976). It was suggested that when marital attractions are high and stable and when the spouses' interconnections are smoothly meshed, spouses focus little attention on alternatives or barriers to breakup. If pleasures recede and displeasures intrude, however, they weaken interdependencies and encourage thoughts about alternatives. A seriously deteriorated relationship is held together primarily by the barriers around it.

Such "empty shell" marriages do not necessarily break up (Levinger, 1976). Despite their diminished closeness, the sum of their attractions and barriers often remains greater for both spouses than the net rewards from any possible alternative. Only if one's outside alternatives begin to outweigh the advantages of maintaining the union do partners take steps to end the relationship.

ENDINGS

In considering the causes and the dynamics of marital breakup, one should note that professionals themselves differ widely in their perspectives. Psychologists and psychiatrists have in the past accounted for divorce in terms of the individual spouses' lack of "marital aptitude" (Terman & Wallin, 1949) or their "neurosis" or "psychosis" (Bergler, 1948). Sociologists and historians, on the other hand, have explained divorce in terms of underlying historical trends in family economics or the culture's increased emphasis on female equality (O'Neill, 1973; J. Scanzoni, 1979a). Social interactionists (e.g., E. W. Burgess & Cottrell, 1939) or systems-oriented clinicians (e.g., Gottman, 1979; Haley, 1963) have seen the causes of divorce in the relationship of the affected partners.

Our own perspective is nearest to that of the interactionists, even though it acknowledges the importance of other determinants. We further note that ending a relationship differs in important ways from either building or continuing one. Forming or maintaining pair connections requires the participation of *both* members; severing them requires the actions or decisions of only *one.* This section therefore considers mainly personal processes, such as reevaluations of relational events, decisions between staying and leaving, and individuals' ways of coping with their past and future.

During the breakup of a marriage, the events and causal conditions experienced during earlier phases are likely to be reevaluated and reinterpreted. Divorced persons generally form "accounts" of occurrences during the marriage and its breakdown that differ considerably from what they experienced before the divorce (R. S. Weiss, 1975). Earlier events are seen in a new light, and they are usually remembered and interpreted differently by two former spouses from the same marriage (Fletcher, 1981). In the case of the Darbers, Susan now recognizes downturn events and empty times in the middle of her marriage that she was unaware of earlier. Tom, on the other hand, does not recall those times; he begins his account of the breakup with a description of Susan's actions after she had already decided the relationship was unsatisfactory.

From Deterioration to Ending (D → E)

While Tom Darber was nearly oblivious to the erosion of their marriage, Susan was distressed after she recognized how far she had drifted away from Tom. Her awareness of having gradually disconnected herself from Tom's influence sometimes gave her a sense of personal strength, but this feeling was mixed with sadness—a sadness for her memories of their earlier love and that Tom's presence now tended more to irritate than to soothe her. Susan thus was led to assess the costs of ending their marriage and to weigh the rewards of alternatives. Except for her talks with one female friend, she struggled alone toward a decision. One issue in her struggle was to resolve her emotional reactions in termination. Other issues pertained to the legal and economic implications of divorce as well as her relations with her children.

Susan's predicament is typical of persons who feel forced to assess the ending of their marriage. During the middle phases, spouses are traditionally expected to remove themselves from the interpersonal marketplace. They do not make detailed assessments of alternative possibilities or calculate the costs of ending their union. If a deteriorating marriage continues to worsen, though, spouses are likely to weigh carefully its remaining benefits and costs and to search for alternative solutions. A 1974 national survey found that 60 percent of all Americans said that divorce is an appropriate solution to a "bad marriage" (Roper Organization, 1974); this percentage is undoubtedly even higher today.

There is relatively little theory or research concerning the processes that operate between deterioration and breakup, but Jaffe and Kanter (1976, pp. 170–171) have proposed a four-factor model to help account for them. Their first factor is "contextual conduciveness," which refers to the "structural effects of the couple's environment [promoting] the weakening of their bond." Second is "systemic strain," that is, "the experience of strains in the couple system from incongruent meanings, needs, or role orientations." Third are the spouses' "generalized beliefs," defined as "the existence of beliefs or values that interpret strain as a relationship problem and . . . . separation as a solution." Fourth and finally, there occur "precipitating events," or circumstances that encourage "friction, protest, and eventually separation."

Note that Jaffe and Kanter's four factors acknowledge the important contribution to the separation process of the social environment (factor 1), of spouses' personal dispositions (factors 2, 3, and 4), and of their relational conditions (factors 2 and 4), as well as the overlap between such conditions. However, aside from Jaffe and Kanter's own interviews and parallel data from a few other studies (Fletcher, 1981; Harvey, Wells, & Alvarez, 1978; M. Hunt & B. Hunt, 1977; Spanier & Casto, 1979; R. S. Weiss, 1975), there exist few research findings that illuminate the D → E transition. The following discussion will therefore touch on only two interrelated questions.

What processes serve to diminish marital commitment?

Chapter 7, "Love and Commitment," considers eight different processes that may contribute to building commitment to a close relationship. In considering the reversal of commitment during the D → E transition, let us recall four of these: (1) improving the reward–cost balance of membership, (2) heightening the social costs of termination, (3) reducing the attractiveness of alternative relationships, and (4) linking membership to the self-concept.

The first three processes pertain to building up the "attractions" and "barriers" and holding down the "alternative attractions," considered earlier in this chapter. Before a dissatisfied spouse can reach a final decision about breakup, it is necessary that his or her net attraction to the relationship become *stably* lower than the attraction to an alternative state (Levinger,

1965) and that costly barriers to termination be overcome. If one's remaining emotional and social interconnections are substantial, the process of reducing the marital commitment is likely to arouse ambivalence and upset (R. S. Weiss, 1975). At times, a person considering divorce will be convinced that the existing marriage is terrible and that any reasonable alternative is preferable; at other times, the remaining marital ties and the obstacles against cutting them will seem far too strong. Until this roller coaster of appraisals and reappraisals is traversed, and the affected individual can hardly know its length or topography beforehand, the decision process is likely to be difficult and distressing.

It should be noted that often these processes are far from symmetrical. In a marriage in which one spouse has already decided to end the relationship, the second spouse has little choice but to accept that decision. For example, Tom Darber felt that Susan's decision left him no opportunity to rebuild their marriage. If the experience of the "leaver" is difficult, that of the "left" partner is even more so (Hagestad & Smyer, 1981, personal communication; R. L. Miller, 1982).

A fourth process of decreasing marital commitment pertains to changes in a spouse's self-concept. This process may hold the key to the final resolution. Whether or not one finally decides to end the marriage hinges on the extent to which one *defines* oneself as an independent individual or a loyal partner. In other words, appraisals of pros and cons, or rewards and costs, are strongly linked to self-appraisals, which are in turn linked to one's relations with kin, friends, and the larger community (Huston & Burgess, 1979; Ridley & Avery, 1979). In Susan Darber's decision, for example, a crucial determinant was her new identity as a career woman and her alliance with other self-assertive female professionals who supported this identity. Tom's inability to accept Susan's new self-concept was probably more important than any other component of Susan's eventual decision.

What is the probability of reconciliation?

Spouses who make plans to end their marriage, including those who publicly file for divorce, do not necessarily carry through that intention. The ambivalence that accompanies such plans leaves considerable uncertainty about the final resolution. Although it is hard to reverse the slide of declining attractions, a final breakup may indeed be forestalled; in fact, an appreciable minority of actual divorce applications are withdrawn by the applicants (Levinger, 1979a). Despite the many studies of divorce, however, there are almost no studies about the circumstances under which plans for divorce are reversed.

An exception is a study that reports on the dismissal of divorce applications at a divorce court (Levinger, 1979a), in which approximately 20 percent of the divorce applications of parents of dependent children were dismissed after filing, and in which a large majority of the dismissing couples reported 6–18 months later that they were still living satisfactorily with their spouse. A sample of 400 such dismissing couples was compared with 400 randomly selected couples who had finalized their divorce during the same 3-year period. The major differences between the dismissing and the divorcing applicants were that the dismissing spouses were more likely to be still living together at the time of filing, and they had higher husband incomes and lower wife incomes. Those results can be interpreted in various ways. Our present analysis would emphasize the greater strength and diversity of the dismissing couples' remaining interconnections. Still-together pairs would presumably retain stronger meshed interconnections than separated ones. Low-income wives would presumably have stronger economic ties to their husbands than higher-income wives. To further illuminate the determinants of decisions during the D → E transition, more research on marriages at the brink of breakup is sorely needed.

Ending of a Marriage (E)

In writing about the stress of divorce, Bohannan (1970) noted the multiple overlapping experiences that are faced by the divorcing individual:

> (1) the emotional divorce, which centers around the problem of the deteriorating marriage; (2) the legal divorce, based on grounds; (3) the economic divorce, which deals with money and property; (4) the coparental divorce, which deals with custody, single-parent homes, and visitation; (5) the community divorce, surrounding the changes of friends and community that every divorcee experiences; and (6) the psychic divorce, with the problem of regaining individual autonomy. (p. 34)

The *emotional* and *psychic* aspects refer mainly to personal conditions of adjustment and readjustment. In contrast, the *legal, economic, coparental,* and *community* aspects refer to important environmental and relational conditions with which each partner must cope. To Bohannan's emphasis on the overlap of these different facets of the divorce experience can be added our emphasis on their causal interrelatedness. For instance, parents who satisfactorily solve their immediate economic problems are likely to handle better their coparental duties, which in turn aids their own psychic readjustment. These interrelated experiences are only recently being studied empirically (e.g., Hetherington, Cox, & Cox, 1977; Spanier & Casto, 1979; R. S. Weiss, 1975). It is evident though, that the final rupture of the well-established connections that have long influenced spouses' intrachain sequencing requires the realignment of a large variety of other personal and environmental connections as well as the building and rebuilding of alternative relational bonds.

The separating individual faces two major tasks. One task is to reorganize the intrachain sequences that formerly were tied closely to those of the partner and to heal the wounds of severing those ties. A second task is to build new relationships to replace the old ones. Both tasks lend themselves to social psychological analysis and research. Here we can examine them only briefly.

Healing processes: Reorganizing old connections

In his interviews with persons who came to his seminars for the separated, R. S. Weiss (1975) found that all experienced extensive separation distress, which usually lasted about a year after the final separation. Ex-spouses interviewed by other investigators (Hagestad & Smyer, 1980; Kohen, Brown, & Feldberg, 1979; Spanier & Casto, 1979) indicate wide variations in the emotional impact of divorce; some respondents, in fact, report an almost immediate emotional uplift. According to Chapter 4, "Emotion," an important determinant of the nature of the emotional impact is the degree to which a person's intrachain plans and behavior sequences are portions of interchain sequences involving the partner. The disruptiveness of a separation is determined by the extent of the interchain connections.

For example, a recent study of mid-life divorce (Hagestad & Smyer, 1980) found that fewer women than men reported negative feelings toward their divorce. For most of the women in the sample, though, marital disintegration was perceived as having been a long, slow process; for most of the men, in contrast, it was perceived as an abrupt event of relatively short duration. In other words, spouses—primarily women—who had begun to disengage themselves from the marriage years earlier, felt more positively toward the actual divorce experience then did spouses who came upon the divorce in the belief that their marriage was still relatively intact.

An emotional disruption is also a cognitive disruption. In trying to reach a new self-definition, one also tries to account for why the marriage failed. Research on attribution processes has begun to examine ex-spouses' reasoning about their marital dissolution. Findings from two studies are that persons attribute more responsibility for the divorce to their former partner than to themselves (Fletcher, 1981; Harvey et al., 1978). Findings from another study (Newman & Langer, 1981) suggest that "interactive" explanations of the divorce are more conducive than "personal" explanations for maintaining an ex-spouse's sense of personal control. Divorced women who attributed their marital breakdown to interpersonal rather than personal factors were found to be more active, more socially skilled, happier, and more optimistic than those who blamed the breakup on their husbands or on themselves. Other recent research on the attribution of responsibility for misfortunes

(Janoff-Bulman, 1979) indicates that victims benefit from focusing on how their own "behavior," but not their "character," has contributed to the occurrence. Reviewing how one's past behavior could have been different in a past event is one way to gain better control over one's actions in future events. Nevertheless, it remains to be learned how people's postdivorce attributions are connected to the specific manner of the breakup or to the magnitude of the marital investment.

Rebuilding processes: New beginnings

Following the Darbers' divorce, Tom and Susan concerned themselves with different issues of readjustment. While Tom went into a long period of private grief, during which he was reluctant to talk to other people about the breakup, Susan occupied herself with building her new life and facing new problems. Tom describes this period of his life as "a real downer," in which he had trouble keeping himself together; Susan, on the other hand, sees this time as "scary, but exciting." Their contrasting reactions paralleled those of the men and women in Hagestad and Smyer's (1980) study. Tom was still working on a psychic resolution of his previous interconnections, while Susan, who had largely worked through that task years earlier, was taking on new challenges.

In their study of divorced persons' steps toward readjustment. Spanier and Casto (1979) found that, in general, "establishing a new life style is more problematic than adjusting to the dissolution of the marriage" (p. 226). Their respondents appeared to be busier with trying to cope with their new economic, social, and sexual problems than with worrying about their past interconnections. Spanier and Casto do not report about the extent to which different persons had disconnected themselves psychologically or socially from their partner before the actual physical separation. It would seem, though, that people who were most ready to deal with new problems had previously settled their old accounts. It also appears that differences in personalities and in the social environment (see Rands, 1980; Ridley & Avery, 1979) affect ex-spouses' abilities to build new connections. Future research on endings, and on new beginnings, ought to encompass information about people's social networks and other aspects of their postmarital environment.

CONCLUSION

This chapter has examined the temporal course of close relationships. It has focused on a variety of conceptual issues associated with their beginnings, middles, and endings. A prototypic five-phase sequence, spanning from initial acquaintance to termination, was postulated to consider various aspects of change and transition. In view of the difficulty of generalizing about all possible pair relationships of different types and contexts, the treatment was limited to adult heterosexual relationships, and the emphasis was put on mate selection, marriage, and marital separation.

The development of other types of close relationships—such as parent–child bonds, friendships, or work relationships—deserves treatment elsewhere. Here it must suffice to note ways in which other types of relations differ from premarital and marital pairings. They differ not only in their duration, diversity, and strength, but also in the nature of their socially structured norms, their members' goals, their exclusiveness, their commitment and love, and in the events associated with their formation and termination. For example, a mother–child relationship begins with a biological connection long before the growth of psychological connections. Its initial asymmetry exceeds what is found in any peer relationship, but it changes greatly with the aging of child and parent. Furthermore, parent–child relationships are not customarily subject to voluntary termination. Same-sex friendships in our culture also differ from cross-sex love pairings in many ways. Friends' mutual concern or interdependence is usually less than that of lovers—although there are wide variations among friendships and among love relations. The norm that people may have numerous close friends without interference among them has permissive implications for the buildup, commitment, and termination of friendships. In contrast, work relationships may be characterized in still other ways. Although work colleagues often exhibit strong influence patterns, often marked by asymmetry, their interconnections are usually narrower than of kin or friends. Work interaction is usually confined to prescribed rules and times and is more likely to involve contractual modes of exchange. The many different types of pair relationships can be described along many different dimensions, and their typical modes of development must probably be charted on very different topographic maps.

This chapter's picture of heterosexual marriage is itself limited. The Darber illustration presents only one of innumerable possible developmental paths. It spotlights a woman who changed her skills and aspirations later in life and a man who did not. An alternative illustration could have focused on a pair in which neither spouse changed much or in which both did so; in either case, their major events and outcomes would have gone differently. More generally our hypothetical A-to-E sequence may leave the impression that relationships "typically" build up to some peak and then deteriorate, even though it has been said repeatedly that there are *multiple* forms of pair development. Buildup can vary from flat to steep to fluctuating; deterioration may never occur or it may start before a public commitment.

The Darbers, furthermore, represent a relationship between two people of high socioeconomic backgrounds who each later achieved professional success. Their marriage differs from those of people in less favored economic circumstances (L. Rubin, 1976; J. Scanzoni, 1970), whose outside alternatives are likely to be less appealing, and whose social settings are more confining.

Nevertheless, our basic model does apply to each of these widely varying relationships and their disparate contexts. Its continuous causal loop represents the essential ingredients of all Person–Other events and changes over time, as well as the interplay among differing sets of properties, conditions, and connections. The development of some relationships, at some times, may appear to be primarily influenced by historical or environmental factors, whereas that of other relationships seems better explained by personal or interactive influences. A task of this and other chapters in this volume, however, is to show how the phenomena of close relationships reveal a dynamic interplay that transcends single-cause explanation or simple linear analysis. In other words, to illuminate Person–Other interdependence over the course of time, we must learn to account for the changing influences among the multiple causal conditions that affect and are affected by such a changing interdependence.

CHAPTER 9

Conflict

DONALD R. PETERSON

A young man and woman in marital counseling were asked to keep detailed records of their quarrels. Here is the wife's account of one episode.

> We were in the car on the way to visit my parents. We had had a tough time getting ready for the trip and were both tired. Paul was in the back of the station wagon reading. I was driving. We were on the Turnpike, and I asked him to move to the side. I wanted to pass and I couldn't see through him. He told me to look out of the side mirror or turn around and look. I'm accustomed to using the rearview mirror and I didn't feel I should have to change my driving habits when he could move to the side a little so I could see. A little later I said, "*Will* you move? I can't see through you." He just sat and glared. Twice more I asked him to move and finally he blew up and told me to pull over so he could drive. Then he told me he'd show me how he could pass without looking in either mirror by looking around. Paul kept on with some more nasty remarks. I told him he was making a big deal out of this and it was ridiculous to fight about and that I thought he really was going overboard about the whole thing. Finally he agreed that he was overreacting and admitted that he had intended to take over driving when we got to this certain point because traffic is heavy there. He decided to tell me to pull over there so I would feel guilty about asking him to move. We made up after he told me that.

The research on which parts of this chapter are based was supported by NIMH Grant MH24698 and an award from the Grant Foundation.

The husband's report is closely similar to the wife's, though his own viewpoints are expressed by somewhat different emphases.

> I was sitting in a position that blocked my wife's view through the rearview mirror. She asked me to move and I became annoyed. I had been reading a textbook and felt it was just as easy for her to use the side mirror rather than disrupt my concentration. I knew we were close to entering a very busy highway and that I should drive, so I asked her, without explanation, to pull over so I could drive. Sometime later I explained why I had taken over (i.e., because of the difficult driving conditions, my wife dislikes driving this particular stretch, rather than irritation over interruption). I admitted at that time that I had "overreacted," and the subject was dropped.

As conflicts go in clinical practice, this one is relatively brief and straightforward. Yet, even so simple an exchange is seen to be an object of considerable complexity when it is described in detail. If the interaction is traced from start to finish, it appears as in Figure 9.1. The arrows in the figure show the most obvious connections among various events in the sequence, but the direct accounts of the participants suggest that many other connections are involved. Besides the direct effects of each person's behavior on the other, delayed internal cognitive–affective processes and indirect situational influences are implied. At one stage, for example, the husband's anger instigates a wish to make his wife feel guilty for "interrupting his concentration." These thoughts and feelings coincide with their approach to a stretch of highway that the husband had intended to drive anyway, and he seizes the occasion not only to take over physical control of the wheel but to punish his wife by "showing her" how to drive.

The episode also suggests the more general issues that appear when conflicts in close relationships are examined in detail. Some relationships are relatively free of conflict. In other relationships, conflicts are frequent and intense. Some conflicts are readily resolved. Others escalate into serious struggles and, in the extreme, may end in separation or physical violence. Some conflicts have destructive effects on relationships. After a damaging fight, "things are never quite the same again." Other conflicts, even serious ones, may have constructive outcomes. Once people have "had it out," their relationship may be more intimate and satisfying than it was before. Some people seem unable to resolve conflicts. The same patterns of dispute occur over and over again, and no new ways of dealing with the problems are created. Other people learn to deal with conflicts in relatively effective ways, and their lives together become more harmonious and satisfying than they were before or during the times of active struggle. How can these different patterns of behavior be described and understood?

The significance of conflict is widely appreciated by lay persons and professionals alike. Anyone who has known the pain of fighting with a

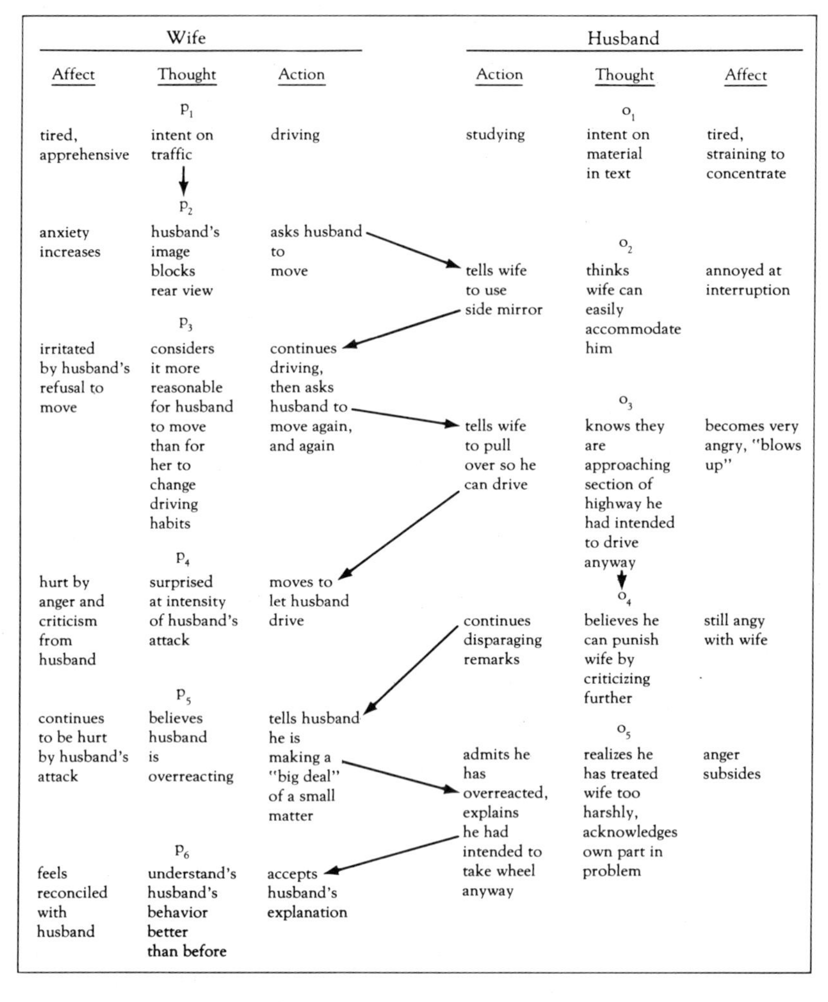

FIGURE 9.1
A quarrel between husband and wife. The designations p_1 through p_6 indicate successive events in the wife's chain while o_1 through o_5 indicate successive events in the husband's chain. The arrows indicate causal connections between the husband's and wife's behaviors.

beloved spouse and any parent locked in battle with a rebellious but vulnerable child know poignantly how much conflicts matter in their lives. The importance of conflict is thoroughly grasped by clinicians, who devote large shares of professional time to clarifying the origins of conflict and helping clients find constructive ways of dealing with conflicts. The topic has been equally intriguing to social scientists, who have developed a large body of research and theory, not only on interpersonal conflict, but on intergroup and international conflict as well.

Although the several literatures on conflict are extensive, they are not as enlightening as one might hope for understanding conflict in close relationships. With a few notable exceptions (e.g., Gottman, 1979; R. L. Weiss, Hops, & Patterson, 1973), the writings of clinicians consist mainly of untested theoretical statements about the origins and therapeutic management of conflict. The scientific literature is based on far more systematic inquiry than that of clinicians, but, until recently, very little of it was derived from the investigation of conflict in close relationships. Conceptually organized within the framework of game theory and methodologically dominated by the rules of experiment, the formal social psychology of conflict today tells mainly about brief interactions among strangers or casual acquaintances in controlled and therefore contrived situations (Deutsch, 1980; Pruitt & Kimmel, 1977). Again, there are exceptions. The work of Raush, Barry, Hertel, and Swain (1974), of Kelley (1979) and his colleagues, and of Gottman (1979) and his colleagues offers prominent examples of systematic research into the natural processes of conflict among closely related people. The concentration of investigative effort on short-term experiments with people who do not know one another well does not necessarily mean that the findings from those studies have no bearing on conflict in close relationships. High degrees of relevance cannot be assumed, however, and even brief consideration of the defining characteristics of close relationships—duration, strength, frequency, and diversity—suggests that they differ in important ways from the kinds of relationships commonly studied in laboratory research on conflict.

By its nature, any close relationship has a history and a future. Family members are related for life. Residuals of past experience and expectations of experience to come affect behavior in the present. For a married couple, for a parent and child, every conflict and every resolution, as well as every failure at resolution, become part of the history whose effects determine the occurrence and course of further conflicts. It seems inconceivable that this conditions could fail to influence the kinds of conflicts people have and the kinds of resolutions they attain.

The interdependence of people in close relationships is great. The influence each person exerts on the other is powerful. They "mean" a great deal to each other, because each person is vitally affected by the other. They "play for keeps" because each person must live with the consequences of his or her own behavior, and those consequences are of central significance in their lives. Although degree of interdependence and emotional intensity should not be confused, close relationships are always marked by strong emotional investment and often by experiences of intense affect (see Chapter 4,

"Emotion"). Compare the passions of a stormy love affair with the feelings of two college sophomores playing Prisoner's Dilemma to gain extra credit for an introductory course. The strong interdependence and the emotional investment characteristic of close relationships raise questions about the resemblance between life struggles and most laboratory investigations.

People in close relationships are together a great deal of the time. They interact frequently with one another. Even a disturbing conflict in an occasional relationship may not matter much in the larger plans of the participants, for both can take comfort in the knowledge that they do not see each other often. Where people are together day by day, however, strife between them can become a matter of consuming concern, especially when conditions of frequency are combined with those of strength and duration. Frequent conflicts in a strongly interdependent relationship expected to last a lifetime can obliterate thoughts of other matters and make attention to other concerns all but impossible.

Interactions in close relationships occur over many different kinds of issues, involve many different kinds of goals, and require various forms of activity for goal attainment. In any close relationship, some issues are likely to be conflictive, while others are relatively free of conflict. Some conflicts are apt to be avoided, while others are engaged. The multiple goals of each individual are likely to be complexly interrelated, and a gain from reaching one goal may entail difficulties in reaching others. The competing and cooperative goals of two or more people in close relationships are likely to be more complicated still, and ambivalence over goal-related behavior is apt to be common.

It is just to order this tangled mass of strivings and activities that the simplifications of laboratory experiment were designed. Indeed, deliberate controls seem needed if the snarls of conflict in close relationships are ever to be unwound. Yet, it must also be apparent, even from this brief discussion, that conflicts arising in the natural course of close relationships and those instigated by controlled laboratory conditions may differ in important ways. Even at this stage of discussion, it should be clear that firm knowledge about the natural processes of conflict in close relationships can be obtained only by studying the natural processes of conflict in close relationships.

This chapter will be devoted to two general topics. First, we will consider the natural processes of conflictual interaction, from the start, through the intermediate stages of fighting and dispute, to the finish, however bloody or genteel. Then, the role of conflict in the development of close relationships will be discussed. We will attempt to identify the conditions involved in shifts from prevailing harmony to patterns of contention, and the other way around, from intense and frequent conflict to relative freedom from conflict. Several developmental outcomes will be discussed in reference to our conception of close relationships.

DEFINITIONS OF CONFLICT

When Fink reviewed conceptions of conflict in 1968, he found widespread disagreement among writers in their definitions of the term. *Conflict* as defined by Stagner (1967) was synonymous with *competition* as defined by Doob (1952). Others regarded conflict and competition as different kinds of struggles (Mack & Snyder, 1957). But, to some, conflict was viewed as a form of competition (Boulding, 1962), while, to others, competition was seen as a form of conflict (Dahrendorf, 1959). To Coser (1954) and Mack and Snyder (1957), only openly enacted struggles were considered to be conflicts. Others (Boulding, 1962; Stagner, 1967) presented "motive-centered" concepts in which states of psychological antagonism were seen as conflictive even without expression in open battle. Efforts to bring clarity out of the confusion often took the form of distinctions among various types of conflict. But these efforts produced a confusion of their own. The typologies of conflict varied as to number of classes, from 2 to 18 (Chase, 1951). They varied as to the basis for classification, which, in some schemes, was the organizational level of conflict (Deutsch, 1973; LeVine, 1961); in other schemes, the logical structure of the conflictual interaction, as in A. Rapoport's (1960) distinctions among fights, games, and debates; and, in still other typologies, the outcome of conflict, as in the distinction between constructive and destructive outcomes suggested by Simmel (1955), elaborated by Coser (1954, 1967), and accepted by most investigators since that time.

It is not clear that conceptions of conflict are any more uniform today than they were at the time of Fink's review, nor even that the conceptions should be uniform. No rules of logic prohibit stating definitions that emphasize different aspects of complex phenomena. Unnecessary confusion on the part of readers may be reduced, however, by noting that different discussions of *conflict* may deal with very different events and processes, and organization of knowledge in the field may be improved if writers say what they mean by *conflict* when they use the term.

For this discussion, *conflict is an interpersonal process that occurs whenever the actions of one person interfere with the actions of another.* In examing conflict in close relationships, we are concerned with interpersonal conflicts, and only indirectly with conflicts among competing action tendencies within one person, or with conflicts among groups or larger social organizations. Any action sequence, as conceived in this book, can include cognitive and affective activities as well as overt physical movements. With Deutsch (1973, 1980) and others, we take *interference* to include not only outright obstructions of activity, but any reduction in effectiveness or benefit or one person's activity that is causally related to the actions of another.

For our purposes, it will be useful to distinguish between structural conflict of interest, that is, any incompatibility between the goals of one person and another, and open conflict, that is, overt opposition between one person and another. It will also be useful, even if operationally difficult, to distinguish between the manifest content of overt conflict and the latent issues that manifest conflict may represent. In keeping with the definition of close relationships used throughout this book, a relationship will be considered conflictual to the extent that conflicts among participants in the relationship occur frequently, intensely, diversely, and over long periods of time.

ANALYSIS OF CONFLICTUAL INTERACTIONS

In Chapter 2, facilitation and interference were identified as among the most important properties of relational interdependence. In the brief discussion that followed, some examples of interference wre given to suggest the wide variety of ways in which one person's behavior may interfere with the goal-directed action of another. Everything from direct physical obstruction to nonverbal innuendos that affect the cognitive and emotional states of partners may qualify as interference. Note was also taken of the frequent asymmetry of facilitation and interference (what is facilitative for one may be interfering for the other) and of the differences between mutual facilitation and the affective positiveness of the relationship (e.g., couples who fight a lot may systematically bring out the worst in each other and so facilitate mutual aggression very effectively, but neither observers nor the battling pair themselves would consider the relationship to be particularly happy).

In Chapter 4, "Emotion," Mandler's "interruption hypothesis" was elaborated to account for affective arousal in close relationships. From that discussion, it became clear that conflict, which is interruptive in its nature, is bound to produce emotion in closely related pairs, and that when strong emotions occur, particularly tense, aggressive, hyperactive emotions, conflict is a likely antecedent. All the conditions that accompany strong emotional arousal are especially pertinent to any discussion of conflict. With these general background conditions assumed, we can proceed to a detailed analysis of the conflict process.

Beginnings of Conflict

Kelley, Cunningham, and Stambul (cited in Kelley, 1979) and Gottman (1979) and his colleagues have interviewed young heterosexual couples to determine the range of problems they report. Results of the studies are consistent in showing extremely wide varieties of issues over which conflict may arise. Kelley et al. needed 65 categories to accommodate the problems identified by the couples they interviewed. Gottman distinguished 85 kinds of conflict situations. The young men and women who took part in the studies reported problems about most imaginable kinds of issues: how to spend time together, how to manage money, how to deal with in-laws, frequency and mode of sexual intercourse, who did which chores, insufficient expressions of affect (not enough affection), exaggerated expressions of affect (moodiness, anger), personal habits, political views, religious beliefs, jealousies toward other men and women, relatives, and the couples' own children. The interviews showed, as have previous studies (DeBurger, 1967; Goode, 1956; Levinger, 1966), that the widely diverse and frequent interactions of marriage provide abundant opportunities for dispute.

Efforts to group problems into clusters vary from one study to another depending on the bases for classification. A particularly interesting classification whose significance wll grow clear in the discussion to follow distinguishes three main kinds of problems as perceived by participants. First are problems related to *specific behavior:* One party complains about something the other does (she leaves her stockings soaking in the sink; he leaves the toilet seat up). Second are problems related to the *norms* governing classes of behavior: Dispute centers on the rules by which interactions are governed (he/she doesn't do his/her part, contrary to agreement, in keeping the house in order). Third are problems related to *personal dispositions.* One person finds the general traits of the other objectionable (he is a messy person; she is sloppy). Braiker and Kelley (1979) propose a hierarchical order for these classes, with influences at the more general (that is, dispositional and normative) levels controlling activity at the more specific levels.

Although married partners, and presumably people in other close relationships, can come into conflict over practically any kind of issue that might engage them, not all of them do experience conflict in the same areas and at the same levels of intensity. The conditions that give rise to conflict in natural settings have not been thoroughly investigated. Systemative knowledge about conflict has been derived mainly from experimental research. Experimentalists set conditions for the conflicts they study as a matter of design, in the interest of learning about the ways people behave under standard, carefully controlled conditions. In making that choice, they have necessarily neglected studying the prior conditions for conflict as these occur in the lives of people outside the laboratory. Clinicians usually obtain reports of conflict from their clients, but the reports are typically retrospective and subject to distortion. Besides, both clinicians and clients are usually more interested in treatment than in facts of history, so questions about the beginnings of conflict are often displaced by more urgent questions of change. Because knowledge about the early stages of conflict is so limited, our account must consist principally of conceptions and conjectures rather than established facts.

Causal conditions predisposing conflict

If open conflict occurs, some structural conflict of interest must necessarily be present. The goals, that is, the valued outcomes, of the participants must be incongruent in some way. In the example at the beginning of this chapter, the wife's main goal was to drive the car safely toward their destination by avoiding collisions with other vehicles. She felt that she might not attain this outcome if her husband's image in the mirror continued to obstruct her view to the rear. The husband's aim was to maintain attention on the book he was studying. He felt that he could not attain this outcome if he had to move. Before any conflict begins, each party is engaged in some kind of goal-directed action (organized intrachain sequence). For conflict to occur, a valued outcome for one person must be incompatible in some way with a valued outcome for the other. This incompatibility appears during interactions by interchain events interfering with the enactment of organized intrachain sequences.

Occasions for conflict will increase to the extent that the goals of the participants are highly valued and incompatible. This statement holds not only for short-term goals of the kind just illustrated, but even more significantly for long-term goals that are perceived by individuals as vital to their personal interests. If a husband cannot have the kind of family life he considers essential except by having a wife whose main preoccupation is with household matters and attention to his needs, and, if his wife cannot reach fulfillment unless she pursues an active career, conflict is inevitable.

When young couples are growing closer and trying to decide whether they would get along well in a lasting relationship, such as marriage, they are often reassured to find that they have many interests, beliefs, and attitudes in common. But the effects of interests, beliefs, and attitudes in predisposition to conflict are not all straightforward. Large numbers of shared interests are likely to lead to more frequent interactions than occur among people with few interests in common, but this is no guarantee that the interactions will be free of conflict, and, if some of the interests are competitive, conflict may arise more often than it otherwise would.

Beliefs and attitudes will play a part in determining conflict only to the extent that they are related to important discrepant goals in the lives of the people involved and lead to incompatible courses of goal-directed activity. A nominal Democrat may marry a nominal Republican and find their political differences a source of mild enjoyment in the games of political debate. If they are professional politicians, however, and their energies and resources are in fact employed in the election of one candidate or another, harmony will be difficult to maintain. A Catholic man may marry a Jewish woman, and both may get along well within their differences, even to the point of attending separate religious services. If they have a child, however, and the mother believes she cannot face Judgment Day unless the child is reared in the Jewish tradition, while the father believes the child's soul, and his own, can be saved only if the child enters the Catholic Church, conflict is assured.

To understand the effects of interests, attitudes, beliefs, values, and other such characteristics in predisposing people to conflict, it is necessary to know not only what the characteristics are and how similar or different they are, but how they are linked to significant goals and goal-directed behavior.

Some conflicts arise mainly through interference with habit patterns, rather than incompatibility of basic goals. For example, some people are disposed to rise early, cheerful and eager to greet the day. Others, who may have stayed up late the night before, are drowsy in the morning, especially before the first cup of coffee. If a "night owl" marries an "early bird," considerable friction may result. In general, conflicts over habits seem less likely to be severely disruptive than those linked to central, long-term goals, but, if unacknowledged and unresolved, conflicts of habit can be relentlessly irritating.

Generally, affection and hostility have appeared at the poles of a single dimension in many factor analyses of interpersonal attitudes (Carson, 1969); Swensen, 1973), and it will be surprising if most measures of hostility are not related to the frequency and intensity of overt conflict among closely related people. It seems even more likely that specific resentments on the part of each partner toward the other will contribute to conflict. As grievances accumulate and resentments grow, very little provocation may be needed to set a quarrel going.

It is quite possible, however, that the associations between affection and hostility are more complex than it may seem. Stambul (1975) found no correlation between content dimensions of "love" and "conflict-negativity" as reported by the young couples she examined. Orden and Bradburn (1968) found scores on "satisfactions" and "tensions" in a "marriage adjustment balance scale" to be uncorrelated, though both variables were positively correlated with self-ratings of marital happiness. Wills, Weiss, and Patterson (1974) found negligible correlations among pleasurable and displeasurable instrumental and affectional behavior as rated by spouses, though systematic associations were found between some of the behavioral tendencies and global ratings of marital satisfaction.

The causal conditions identified so far are mainly relational, P × O conditions, though some are linked to the personal dispositions of the participants. Other causal conditions are more clearly located in the individual characteristics of P and O. Some people appear generally more combative than others, and the presence of one openly hostile member of a pair would seem to guarantee some conflict. On the positive side, a partner devoted to another in a thoroughly selfless way would seem likely to overlook provocations that might draw a less loving mate into bitter arguments.

Among causal influences in the physical environment, it seems reasonable to suppose that high situational stress and scarce resources, such as unceasing noise and severe poverty, would set conditions for conflict. The effects of such general cultural conditions as social class on conflict and other aspects of interpersonal relationships appear to be profound (L. Rubin, 1976). The surrounding network of other people in the immediate social environment can also play a part among the general causal conditions that engender conflict. A young couple living with parents would almost surely be influenced by that situation, and the amounts and kinds of conflict would likely be affected by the demands the parents made and the supports they offered to the younger people in their home. Here, as elsewhere, however, it does not seem profitable to indulge in elaborate conjecture. We need to compare people who fight often with those who do not, and to study not only the ways they resolve conflicts but the conditions under which conflicts begin. We have all known married people who seem doomed by their backgrounds to lives of conflict. They have active, not easily reconciled careers. They differ in religion, politics, race, and age. Yet, some such people appear to get along marvellously well. We need to find out how they do it. Others with "everything in common" fight like badgers. We need to find out why.

Initiating events

In the course of any conflictual interaction, general predispositions to conflict of the kind mentioned above must be converted into overt conflict by some precipitating event. For a structual conflict of interest to erupt into open conflict, behavioral expressions of incompatible interests must occur. The goal-directed activity of one person then interferes with the goal-directed activity of another.

When husbands and wives are asked to write detailed accounts of "significant" interactions in their daily lives, a substantial proportion of the reports concern conflicts (Peterson, 1979). And, when the events said to precipitate conflict are examined, four conditions are frequently described. The first is *criticism*. Interactions that begin as enjoyable or task-oriented exchanges are frequently converted into conflicts by verbal or nonverbal acts on the part of one person that are perceived as unfavorable or demeaning by the other.

> We were playing golf. Both of us were rather tired and hot . . . . Neither of us was having a particularly good game and were irritated about that basically. Then, when Steve missed two putts, I said facetiously, "You're not concentrating." He got ticked off at the criticism and snapped back with something meaning mind your own business.

It does not seem to matter much whether an act is intended to be critical. The important condition is that the behavior is interpreted as critical. The remark or act leads to a feeling of injury and unjustice on the part of the offended person, who is then likely either to retaliate with some form of aggression or withdraw.

A second event that commonly initiates conflict is *illegitimate demand.*

> I asked Jim if he would do the dishes—to get them out of the way—while I fixed the handle of the teapot so we could have some tea. Negative vibrations from Jim, who immediately suggested we do the dishes together, implying that I should wash. I asked if he would like to wash them and he said, "Well, I would if it were just the dinner dishes, but there's all that other stuff in the sink." I had been cooking all afternoon for a camping trip and Jim didn't like the way I left pans and things around. General manifestations of disgust from both participants at this point.

A demand of any kind by one person upon another is apt to be interfering and therefore apt to set conditions for conflict. The demands that seem especially likely to produce serious conflict, however, are those that are perceived as unjust, that is, outside the normative expectations each person holds toward the other. The intent of the person making the demand and the "true" legitimacy of the demand are irrelevant, as far as the precipitation of open conflict is concerned. The key condition is that the person upon whom the demand is made perceives it as unfair.

A third kind of action initiating severe conflict is *rebuff.* One person appeals to another for a desired reaction, and the other fails to respond as expected.

> I came home from work and went to sit on Bert's lap. I kissed him and began to tell him something that had happened to me during the day. Bert said it was time for the news. End of encounter.

The person suffering the rebuff feels devalued, in some degree angry with the other, and typically withdraws.

The fourth kind of initiating event is *cumulative annoyance.* A first act may be unnoticed, a second, and even a third or fourth repetition may be ignored. But then some threshold is exceeded. "That does it!" says the respondent, and the fight is on. The example that opened this chapter illustrates the process:

> I asked him to move to the side. He told me to look out of the side mirror or turn around and look . . . . I said, "*Will* you move? I can't see through you." He just sat and glared. Twice more I asked him to move and finally he blew up and told me to pull over so he could drive.

Criticism, illegitimate demand, rebuff, and cumulative annoyance were identified through empirical analysis of events precipitating conflicts (Peterson, 1979). By a simple frequency count, those were the acts that most often initiated conflictual interactions. The conceptual significance of the events, however, can be most readily understood in reference to more general causal conditions. Both criticism and illegitimate demand imply perceived violations of norms. When one person criticizes another, the critical party claims that the other has failed somehow to measure up to his or her standards. The "illegitimacy" or "legitimacy" of any demand is defined by reference to normative expectations. When a demand exceeds the boundaries of reasonable expectation, it is considered illegitimate, and a reduction in conflict could presumably be brought about either by reducing the demands of one party or by extending the bounds of legitimacy as perceived by the other. Similarly, "rebuff" implies an expectation of response to appeal by one person and failure by the other to meet that expectation. "Cumulative annoyance" probably functions somewhat differently from the other three initiating events in regard to more general causal conditions. Implications of intentional norm violation are less clear. Instead, the abrasive but unwitting actions of one person appear to produce a progressively increasing state of irritation in the other. The irritation is tolerated up to a point, but then is expressed in some way designed to eliminate or reduce the annoying behavior.

Avoidance versus engagement

Once an initiating event has occurred, the interaction takes a decisive turn. The people involved may either engage each other in conflict or avoid it. In the following example, a perceived rebuff leads to some hurt feelings on the part of the husband, but he withdraws and avoids open conflict for a time. Later he expresses defiance and resentment toward his wife over another issue. She retaliates and, in his view, humiliates him, but he still avoids open dispute.

> We were in our car en route to a family gathering. The last few days had been going well, including this particular morning, until I asked her to sit next to me in the car, to which she sarcastically replied, "I'm fine." This brought up all sorts of old feelings of resentment within me but rather than try and settle a dispute at a family function I decided to can it for another time. Later I smoked a couple cigarettes (which I'm not supposed to do according to an agreement, but why adhere to an agreement like this when everything else is shitty between us). Upon discovering this, she promptly walked up and put out the cigarette plus announcing that I was not to smoke in front of her. (This was in the presence of other people.) Again I did nothing.

Still later, as the husband is trying to sleep during their return home, he becomes angry about the annoying behavior of his 3-year-old daughter, decides that the child was prompted by his wife, and conflict erupts.

> When it got to the point that she encouraged Susan to disturb me, *I got pissed!* I grabbed her by the hair and pulled her head a couple feet and informed her that I felt a little attitude rearrangement was in order.

If a conflict is engaged, it seems likely that at least one party considers the matter suficiently onerous to require action and believes that a favorable outcome can be attained. In general, powerful partners seem more likely than those with less power to join in battle rather than to avoid trouble. The dominant partners may stand to gain more by negotiation. By engaging in conflict, they may reestablish the generally superior position that the conflict threatened to upset. If successive violations become intolerable, however, even a dependent partner may be willing to endure whatever punishment the conflict entails and, by the act of engagement itself, gain greater influence over future interactions. Once the weaker partner "stands up and fights," the dominant one must thenceforth reckon with the costs of active conflict in pursuing personal aims at the expense of the other.

If the weaker person in an asymmetrical relationship sees conflict as a way of increasing power, conflict may be engaged. In experimental threat games, the more powerful participant usually begins by exploiting the weaker one, who often responds by initiating open dispute in which appeals for justice are made. These factors may explain why women are more likely than men to apply for marital counseling. Wives typically control fewer resources, have fewer alternatives, and are therefore less powerful than husbands (see Chapter 5, "Power"). In this position, they may see a chance to gain an ally in the therapist and enter treatment with prospects of greater equity in mind.

Avoidance must be mutual and probably occurs when the issues are seen either as insufficiently important to outweigh the distress conflict brings or as intractable, at least for the moment. Both parties perceive the risk of open conflict to be greater than the uncertain gains to be obtained through active dispute.

The consequences of avoidance and engagement are not well understood. It seems to be common lore among psychologists that engagement leads, all in all, to more satisfactory outcomes than avoidance. Yet Raush and his colleagues (1974) found some evidently happy couples who appeared to maintain their euphoria by systematic and successful denial of conflict. Whether denied conflicts would eventually find expression in displaced or misattributed form, as common belief and Deutsh's (1973) concept of latent conflict suggest, is an interesting but unanswered question. People in treatment to find better ways of managing conflict seem to need, sooner or later, to confront the issues that divide them if they are to develop effective ways of dealing with those issues. In a systematic study of avoidance and engagement of conflict among married couples, Knudson, Sommers, and Golding (1980) found that engagement of conflict was associated with increased understanding by each spouse of the perceptions of the other. Couples who avoided conflict tended to show decreases in consensual perceptions. For clients and clinicians concerned about conflict, however, the main question about avoidance and engagement is not whether people should become generally avoidant or confrontive, but how and under what conditions conflicts should be engaged and how and under what conditions they might better be avoided. The same question, in much the same form, might also interest social scientists.

Middle Stages of Conflict

Once conflict has been engaged, it appears to take either of two main turns, toward direct negotiation and resolution or toward escalation and intensification of conflict.

Direct negotiation

In most everyday dealings with one another, people settle their disagreements by direct negotiation. Each person states his or her position, seeks and obtains validation of the position by the other, and a straightforward problem-solving exchange ensues. Pertinent information is accurately expressed and received without distortion. Both parties work toward a solution until some acceptable outcome is attained.

Escalation

More interesting and memorable are conflicts that escalate from relatively placid disagreement to intensely angry fighting. Some knowledge about the process of escalation in close relationships is beginning to appear. Raush and his colleagues (1974) studied the behavior of 46 newlywed couples in four "quasiexperimental, quaisnaturalistic" situations which they called "improvisations." In each improvisation, separate instructions were given to the partners to create a conflict of interest. Two of the situations were "issue-oriented." One concerned a conflict over a choice of television programs and the other a conflict of plans for celebrating an anniversary. The other two improvisations were "relationship oriented." One partner was told that she or he had been feeling distant from the other and wanted to maintain the distance. The other partner was told to try to bridge the gap, to move closer psychologically to his or her partner. Roles were reversed for the final improvisation. The interactions were recorded and coded in accordance with a system developed inductively by the investigators. Results were analyzed sequentially and in other ways.

Among the many findings were that the relationship-oriented scenes were more threatening to the participants, aroused stronger emotions, and led to more use of coercion and threat than the issue-oriented scenes. A factor analysis based on interview data, questionnaires, interviewer ratings, and test responses allowed the identification of separate sets of "discordant" couples and "harmonious" couples within the sample. In general, the harmonious couples got along well together and were satisfied with their marriage. The discordant couples reported frequent conflicts and were dissatisfied with their marriages. In the improvisation task, the two groups behaved in different ways. Harmonious couples were likely to keep their discussions focused on cognitive issues and to keep the boundaries of conflict limited to specific differences. The discordant pairs were more likely to use coercion and personally rejecting statements. Wives in the discordant pairs tended to attack their husbands even in the issue-oriented scenes, while their husbands remained meek and submissive. But, in the relationship-oriented scenes, the discordant husbands changed, in the author's terms, from "lambs" to "tigers" and attacked their wives without mercy.

The coercion hypothesis of Patterson and his colleagues (e.g., Patterson & Reid, 1970), which has received some empirical support (Patterson, 1976b), provides an explanation of escalation in reinforcement terms. In a conflict, one or both members may apply aversive stimulation, such as criticism and threats, in order to get the other to comply. If compliance is not forthcoming, the aversive behavior increases in intensity and strength. Finally, one of the partners gives in, reinforcing the other for applying increasing levels of coercion. However, the compliant partner is also reinforced, by the cessation of aversive behavior, when he or she "gives in." The different events which were earlier described as initiating conflict could all be considered as coercive or as likely to elicit coercive behavior.

The long tradition of laboratory research on conflict has also produced reasonably clear knowledge about some of the conditions associated with the escalation of conflict. From this work, we know that biased perceptions and hostilities based on unvalidated beliefs about another are likely to intensify conflict and that, as conflicts escalate, strategies of power, along with tactics of threat, coercion, and deception, come increasingly into play (Deutsch, 1973, 1980). It seems reasonable to believe that conditions of these kinds would also hold for close relationships.

Perceived causal conditions in the escalation of conflict

The attributional process in escalating "natural" conflicts has not yet been examined, but retrospective studies of attributional conflict suggest some directions the process might take. Orvis, Kelley, and Butler (1976) asked members of 41 young couples to describe instances of behavior for which the two had different explanations. Nearly all of the examples the couples produced involved negative or unpleasant behavior by one person from which the attributional conflict developed. Their partners, who complained about the behavior, tended to explain it by referring to the personal characteristics or attitudes of the actors (ineptitude, poor judgment, selfishness, lack of concern). The actors who had in each case committed the objectionable act tended either to excuse the behavior as a consequence of extenuating circumstances, other people or objects, or the actor's temporary state, or to justify it by reference to higher values.

The tendency, in cases of conflict, for people to interpret another's behavior as an expression of the other's personal dispositions, is consistent with the general tendency to attribute one's own actions to situational influence, while another's actions are attributed to personal characteristics (Jones & Nisbett, 1971). In a conflictive interaction, the dominant situational influence *is* the opposing party. The same attributional theme is represented in the Kelley, Cunningham, and Stambul results (cited in Kelley, 1979), in which young couples were strongly disposed to describe problems by reference to the personal traits and attitudes of their partners, despite every encouragement to describe the problems in specific behavioral terms. It is also consistent with the tendency reported by clinicians of rather different persuasions (Bach & Wyden, 1968; R. L. Weiss, Hops, & Patterson, 1973) for married couples in conflict to express their disagreements in general personal terms, often by raising questions about the character of their opponents.

A related factor in the escalation of conflicts is the attribution of blame to the other. In one inquiry by Gottman (1979) and his colleagues, couples were interviewed to obtain play-by-play accounts of their most serious problems. Then they were asked to come to a mutually satisfactory resolution of the most troublesome problem. The interaction was recorded on videotape, coded, and analyzed both sequentially and nonsequentially. Each interaction was then divided into three phases: (1) an agenda-building stage, during which the participants expressed their feelings about the problems they were having; (2) an arguing phase dominated by disagreements; and (3) a negotiation phase, in which problem-solving, informational exchanges, and comments of agreement occurred most frequently. The agenda-building stage is of greatest interest for this discussion. One of the most consistent findings to emerge from several analyses was that couples in clinical treatment tended to move immediately into a pattern of cross-complaining during this stage. One of the partners, let us say a husband, would express a complaint, typically with negative affect and often in a sarcastic way. His wife would counter with her complaint, also with negative affect. Couples who got along better tended more frequently to display "validation loops" during the agenda-building phase. The husband would express his complaint. His wife would attempt to understand the complaint and to validate her perception of it.

All of this suggests that the escalation of conflicts in close relationships is accompanied by, and is possibly caused by, the attribution of blame to the other rather than to oneself or other circumstances (cf. Madden & Janoff-Bulman, 1979), by an attributional shift from specific characteristics to general ones, and particularly by a shift from behavioral description to personal blaming. Acts (you behaved thus and so) are interpreted as signs of traits (you always . . . ; you never . . .) or type characteristics (you are a . . .) on the part of the opponent. The features of causal conditions, especially durability and generality of effect, are thus attributed to events, and the conflict becomes more threatening than before. The redefinition of issues during the escalation of conflict is often quick and subtle. A wife may find herself talking heatedly about something entirely different from the issue that started a quarrel, without quite knowing how the discussion changed from one topic to another. The act of redefinition, even if it is not deliberate, may offer a short-run advantage to the person who does it. Reprehensible behavior may seem downright noble if it is phrased in terms of "principles" rather than described as it happened. Changing the grounds of discussion and personalizing issues are often used as tactics for winning arguments. In the long run, however, the generalization of issues is seldom helpful to either party. What began as a small, specific problem becomes a large, diffuse one. A manageable problem may become insoluble.

Conciliatory acts

If conflict has escalated to high levels of intensity, if insulting remarks have been exchanged, and especially if physical violence has occurred, it is difficult to move from open conflict to the rational problem-solving activity required for resolution. Before negotiation can begin under these conditions, an intermediate step is needed. This step usually takes the form of a conciliatory act, intended to reduce negative affect and to express a willingness to work toward resolution of the problem. In the most common conciliation pattern reported in the interaction records of married couples studied by Peterson (1979), two elements usually appeared. First, the problem was reframed as less important than maintaining the relationship. In one way or another, the conciliator said, "This problem has gotten out of hand; let's get it back in perspective." Second, the person making the first conciliatory move attributed some responsibility for the conflict to himself or herself, rather than blaming the other for the difficulties they were experiencing, and at the same time assumed some personal responsibility for working toward a resolution. Somehow or other, the conciliator said, "This problem is at least partly my fault. I will do what I can to solve it."

Conciliatory moves may be followed by reciprocal conciliation from the other. With reciprocal conciliation, anger is reduced on both sides and the combatants are able to move rationally and cooperatively toward resolution of the issue. In resolution of severe conflicts, the least that happens is a reversal of the process of escalation. Personal attacks are stopped. Coercion and threats of coercion are withdrawn. Concern that had been generalized across a range of issues is redirected toward more specific issues. With these changes, the partners can attend to solving the problem rather than injuring each other, and chances for a satisfactory outcome are improved.

If resolution is to be attempted following angry withdrawal, some kind of reconciliation seems necessary before any other negotiations can proceed. If feelings have run very high, the reconciliatory act will usually go beyond conciliation (reframing the issue to reduce it to manageable proportions, acknowledgment of personal responsibility, expression of interest in negotiation) to unusual expressions of affection and commitment to the relationship. If the reconciliatory move is reciprocated, further moves toward resolution, and possibly toward constructive change in the relationship, can proceed.

Termination of Conflict

In one way or another, all conflicts end. The forms of termination in natural settings are not well determined. Even the "quasinaturalistic" studies of Raush or Gottman tell us little in this regard because all such studies impose a demand on participants to "come to some kind of resolution." Attention is then focused on the ways people arrive at resolution, without admitting the substantial probability that some conflicts end with no resolution at all. Couples asked to write free narratives about significant interactions in their daily lives report a considerable number of conflicts ended by the angry withdrawal of one of the partners, an outcome few would regard as a resolution (Peterson, 1979). Without claiming that these are the only ways in which conflicts may come to a close, we will consider five kinds of endings, namely separation, domination, compromise, integrative agreement, and structural improvement. The endings are presented roughly in order from most destructive to most constructive, but it is well to remember that no one has ever demonstrated which kinds of outcomes are in fact constructive and which are destructive in close relationships. A constructive outcome is one that has a beneficial effect on later interactions. A destructive outcome is one that has a damaging effect on later interactions. No one so far has studied the linkage between early conflicts and later interactions in sufficient detail to allow assured statements about these kinds of consequent effects.

Separation

This ending is marked by the withdrawal of one or both parties without immediate resolution of the conflict. Under some conditions, separation may be a useful step in attaining later resolution. If two people are not making any progress in a dispute, they may realize that some time apart will give them a chance to arrive at more creative solutions to their problems. Even if only one party withdraws in the heat of battle, the act itself may be necessary to avoid violence or irreparable damage to the relationship. The time alone gives people a chance to "cool off," to reconsider their views of the situation, and possibly to think of better ways of dealing with it. Often, however, withdrawal may have a damaging effect. The withdrawal may be accompanied by an aggressive gesture, a "parting shot" of some kind, that leaves both parties feeling worse about each other and the relationship than before. In the reported interaction patterns of "disturbed," "average," and "satisfied" couples (Peterson, 1979), a cycle of aggression followed by withdrawal was most common among disturbed couples seeking professional help in dealing with marital problems. "Satisfied" couples, who claimed they were pleased with their marriages and who denied even having sought professional help, more commonly retaliated against aggression and fought conflicts through to some kind of resolution. Withdrawal itself offers no solution to the problem that started the conflict. Many people appear to let their angers subside and then come back together only to treat each other much as they did before the conflict, except, for a time, more gingerly. With no change in the conditions leading to conflict, further quarrels are likely.

Domination

Some conflicts end in conquest. One person continues to pursue the line of action leading to personal goals; the other gives way. The most obvious determinant of the outcome is differential power. The conditions related to power, such as control of resources and a readiness to move to available alternative relationships, should be correlated with attainment of the "winning" result. Except for any cases that might involve neurotic masochism, it seems that the "losers" will usually be characterized by the reverse, that is, few resources and alternatives. The effects of chronic domination are bound to be destructive for the loser, and even the winner, by ignoring the partner's wishes in order to gain victory at any cost, has failed to take advantage of the opportunities conflict can offer for constructive change.

In close relationships in which the participants are so strongly committed to maintaining the union that withdrawal is not a serious prospect, affective controls are likely to become predominant, and "victory" in battle may go to the participant who is willing to hold out longest in display of aversive behavior. In the coercive interactions described by Patterson and Reid (1970), children were frequently found to control their parents by escalating the intensity of commands, if necessary to the point of tantrums, until the parents gave in. The compliant behavior of the parent reinforced the coercive behavior of the child. The cessation of aversive behavior by the child reinforced the compliant behavior of the parent. A particularly unpleasant and durable pattern of interaction was maintained as a result.

Compromise

Following Pruitt and Carnevale (1980), we conceive of compromise as a solution to conflict in which both parties reduce their aspirations until a mutually acceptable alternative is found. Interests are diluted rather than reconciled. Compromise may be the best outcome available in unmixed competitive situations, in which one person's gain can occur only at the expense of the other. But, in most situations, it is likely that better solutions can be found.

Integrative agreements

An integrative agreement is one that simultaneously satisfies both parties' original goals and aspirations. Purely integrative agreements are rare because it is difficult to reconcile genuinely divergent interests. Most agreements fall somewhere between compromise and integration, with the creation of an alternative that satisfies somewhat modified goals and aspirations for one or both of the people involved. Pruitt and his collaborators (Pruitt & Carnevale, 1980; Pruitt & Lewis, 1977), who have written most instructively about the attainment of integrative agreements, distinguish several ways in which such agreements can be reached. Ways may be found to cut the costs of one or both parties. Novel ways may be found to compensate one or both parties for unavoidable losses. Ways may be found to bridge apparently divergent goals by inventing a new option that achieves both parties' major goals. When a conflict concerns two or more issues, a "logrolling" exchange of concessions may be arranged.

When integrative agreements are being developed, it is usually necessary for one or both of the participants to make concessions selectively. Some goals and demands must be modified while others are firmly maintained. Demands, goals, and aspirations, as Pruitt and Carnevale say, often "come in bundles," and, in order for concession to be seen as acceptable, it is usually necessary to "unlink" or separate the goals cognitively so that some may be attenuated while others are preserved. Integrative agreements are most likely to be sought at intermediate levels of conflict intensity. If the conflict is trivial, the people involved are likely to ignore it or choose an easier solution. If the conflict is too severe, they may be unable to stay with the difficult, cooperative problem-solving task long enough to work out a creative solution.

The keys to attaining integrative agreement, according to Pruitt and Carnevale (1980), are a mutually cooperative orientation and "flexible rigidity" in regard to the means and ends of conflict resolution. With a cooperative orientation, the situation is treated as a solvable problem for which a mutually acceptable solution is sought. This orientation is distinct from a competitive, "win-lose" orientation, in which one party tries to press the other toward his or her preferred outcome, or an "individualistic" orientation, in which one person is fundamentally unconcerned about the welfare of the other. In their "flexible rigidity" formulation, Pruitt and his colleagues have argued that attaining high joint benefit in integrative agreements requires both parties to be stubborn about their basic goals, but flexible about means for attaining goals. Lowering one's most valued goals and giving in on issues of true concern are merely compromise, unlikely to bring lasting, satisfactory results. Only by adhering firmly to one's central goals, but at the same time offering every readiness to try new ways of reaching the goals, can mutually beneficial, integrative agreements be reached.

Structural improvement

In close relationships, conflict and its most fortunate forms of resolution may lead to still more profound consequences than the simultaneous goal attainments of integrative agreement. The treatment of each person by the other may come to be guided by different rules. The affection of each partner for the other may grow stronger. The understanding of each partner for the other may be deepened. A severe conflict that has revealed previously unrecognized characteristics of the partners or of the relationship may be resolved through open communication and lead to a beneficial change in the causal conditions governing the relationship.

Unlike integrative agreement, which is most likely to occur at intermediate levels of conflict intensity, structural improvement is most likely to follow very serious conflict. In the heat of battle, new issues are exposed. The generalizing tendencies of escalation will lead to the inclusion of spurious issues as well as valid ones, but the combatants will find themselves with a longer list of problems than they had at the start of the fight, and some of these may be newly recognized issues of fundamental concern. The partners may also discover qualities in each other, and sometimes in themselves, that they had not known before, and, if the revelations are severely disturbing, separation may result. If affectional bonds are strong enough, however, the weary warriors will come back together through the process of reconciliation discussed above. They may then begin another process, of open communication, in which strongly experienced emotions are accurately expressed, spurious and valid issues are sorted out, new conceptions of one another developed, and, in the best of encounters, new levels of intimacy are attained.

Ideally, effective communication can occur only if several conditions can be met, though not every communicative exchange need involve full expression of all elements. For effective communication, each person must know what his or her individual interests are. Considering the ambivalence and denial that are so common in close relationships, it is not surprising that people often "do not know what they want." Most people in our society are not well taught in honest expressions of feeling or direct assertions of personal interest. When we feel sad, we are told to cheer up. When we feel angry, we are told to control ourselves. When we want something for our own sakes alone, we are told not to be selfish. Each person must also be able to express his or her interests to the other. This expression requires trust that the other will acknowledge the interests and be interested in helping fulfill them. Then the statement of one's own position must be articulated clearly enough to allow the other to understand it. If a clearly articulated message is delivered, it must be accurately received and understood by the other. The message must find a receptive ear and an open mind or it might as well never be stated. Receptivity and openness are difficult when one's most passionate wish is to express one's own view and, if possible, to have one's own way. Given accurate understanding, a constructive reply must follow, which demands that every condition governing assertion of the first message must hold for its reply. The second person, too, must know his or her own interests, trust the other enough to assert them, be capable of clear expression, and be ready in turn to receive without distortion the next message to come from the partner. There are many ways in which communicative failure may occur. The wonder is not that misstatements and misunderstandings appear in close relationships, but that valid expression and accurate understanding occur as often as they do.

If highly important but previously unrecognized issues are discussed and reconciled, it is likely that some change will take place in one or more of the causal conditions governing the relationship. Each person will know more about the other than before. Each person may attribute more highly valued qualities to the other than before. Having weathered the storm of previous conflict, each person may trust the other and their relationship more than before, and thus be willing to approach other previously avoided issues in a more hopeful and productive way. With these changes, the quality of the relationship will be improved over many situations and beyond the time of the immediate conflict with which the process began. For structural improvement of close relationships, a personalized form of "flexible rigidity" may be needed. Each person must be firm in asserting his or her own personal worth. But each person must also be allowing in regard to the qualities of the other, even those at variance with one's own values. In W. H. Auden's words, "the awareness of sameness is friendship; the awareness of difference is love."

The various outcomes of conflict discussed above, the principal stages in the process by which the outcomes are reached, and the main conditions that appear to determine the various turns conflictual interaction may take are summarized schematically in Figure 9.2. The figure is complex, but so are the possible courses of conflict.

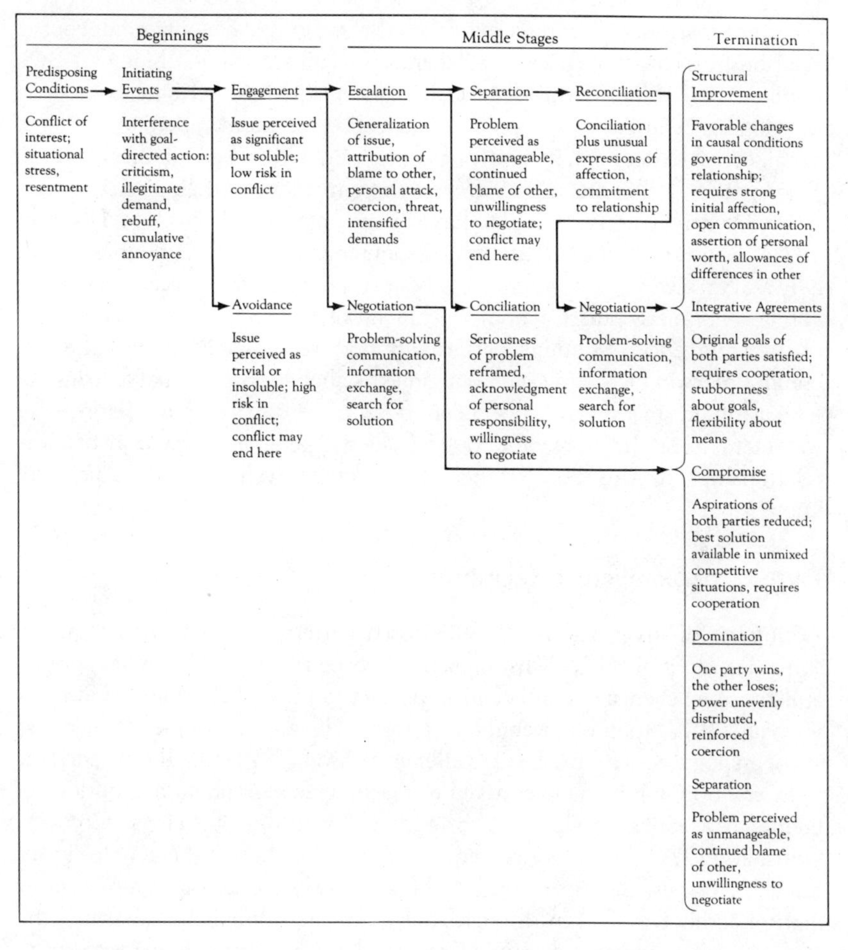

FIGURE 9.2
The possible courses of conflict from its beginnings, through its middle stages, to its termination. Arrows indicate the likely sequences, ending with avoidance or any of five possible terminations of the conflict.

CONFLICT IN THE DEVELOPMENT AND CHANGE OF CLOSE RELATIONSHIPS

In the preceding discussion, we have been concerned with conflict as a contemporaneous process. Events that take place in episodes and sequences of conflict have been described, and some causal conditions governing the events have been suggested. Now we shall consider the cumulative processes by which particular episodes of conflict, along with other experiences, lead to recurrent patterns of interaction and the various developmental courses relationships can take. Scarce as dependable data are in describing particular conflicts in natural settings, they are scarcer still in describing conflict in developing relationships. Some retrospective accounts of young couples (Braiker & Kelley, 1979) and of people who have been divorced (R. S. Weiss, 1975, 1976) are available. The summary views of therapists who have worked with divorcing couples have been obtained (Kressel & Deutsch, 1977). Raush et al. (1974) examined some members of their sample repeatedly and arrived at some inferences about developmental processes related to conflict. For many practical and ethical reasons, extended prospective studies of the emergence of conflict patterns in close relationships and of the role conflict plays in producing such outcomes as alienation, habitual fighting, or increasing intimacy have not been done. Even more than before, our statement must therefore be derived from the general conception of close relationships in this book rather than from a well-established body of knowledge.

Patterns of Conflictual Relationships

A number of investigators have discussed patterns of conflictual relationships. Raush et al. (1974) distinguished between couples who avoid conflict and those who engage conflict in improvisation tasks. Hawkins, Weisberg, and Ray (1977) found four communication styles among couples studied in a set of analogue situations. B. C. Miller and Olson (1976) found nine patterns of interaction among couples asked to reach agreement about attributions of blame in a series of vignettes from their Inventory of Marital Conflict. Gottman (1979) found four patterns of interaction among the couples in his research. Rands, Levinger, and Mellinger (1981) found four main types of marital conflict resolution through analysis of questionnaire data. When the studies are considered collectively, the methods by which the patterns were derived, as well as their contents, are seen to differ widely from one study to another. The findings are reminiscent of those reported previously in this chapter on the diversity of issues over which conflicts may arise. Diversity is also the rule, it appears, in describing patterns of conflict resolution. With multiple observations of conflict over a wide range of situations by a variety of methods, considerable diversity should be expected in the patterns of interaction that appear. As further investigations are completed and the results collated, it seems likely that a relatively small set of characterizations will accommodate most of the patterns, but the close descriptive studies required for clear topography have not yet been conducted.

The defining characteristics of close relationships suggest some of the differences that might be found in comparing different patterns of conflictual relationships. Intensity, frequency, diversity, and duration are not only conceptually distinct, but statistical correlations among them seem likely to be moderate. Some relationships may thus be marked by very intense conflicts that occur rarely over a single issue. Other relationships may involve frequent, low-intensity quarrels over many issues. Every other combination of the four characteristics might also occur. The relationships would all be "conflictual" in some sense, but they might also be different in important ways.

The developmental routes by which conflictual patterns come about differ from one relationship to another. It seems probable that some relationships would come to be ridden by conflict by slow incremental processes, while others would change more rapidly, in a stepwise, discontinuous way. With increasing experience in a relationship, one would expect increasing awareness by each party of the other's goals, including those whose attainment might interfere with one's own goal attainments. With increasing familiarity may come increasing willingness to express one's own position and assert one's own interests. Under these conditions, a gradual increase in the frequency and range of conflicts might be expected. With increases in conflict, other more satisfying kinds of interaction would likely decrease, partly by the simple displacement of one kind of interaction by another, and partly because unsatisfactorily resolved conflicts may leave residues of resentment and misunderstanding, which then predispose the partners to further conflict. In patterns of continuous deterioration, generally affectionate and enjoyable relationships would drift gradually into habits of dispute.

By clinical report and common observation, conflict can lead to sudden and radical change. In the film, *Going in Style,* Willie awakens from a dream about an experience he had as a young man. In words much as follows, he tells his friend about it.

> I had a dream about something that happened when my son was little. I came home from work and my wife said the kid had done something. I can't even remember what the hell it was. So I said, "Why did you do it?" and he said, "I didn't do it," so I cracked him on the behind. He said, "I didn't do it," so I hit him again. I was young. I didn't want him to get the best of me. Can you imagine? So he kept saying, "I didn't do it," and I kept hitting him on his little bottom, and finally he said, "OK, I did whatever the hell it was," and he ran over to his bed and put his face in the pillow. He wouldn't look at me. We never had any fun after that.

Discontinuous change could occur by transformation of any of the classes of causal conditions discussed throughout this book. Among personal conditions, a massive shift in the complex of beliefs and feelings one person in a relationship has about oneself, the other, or the relationship might transform the relationship in fundamental ways. For all the reasons we have mentioned, such changes have not been closely described by social scientists, though literary examples are plentiful and several theoretical formulations have been attempted.

To Raush and his colleagues (1974), significant shifts seem to occur in the *object relations schemata* of couples in conflict. Object relations schemata are "organized structures of images of the self and others, together with the needs and affects characterizing the relationships between the images; the schemata evolve out of contact with varying psychosocial contexts, and they influence the individual's actual and fantasied interpersonal interactions" (Raush et al., 1974, p. 43). Object relations schemata imply scripts for action. The images one has of oneself and another guide the general directions behavior will take. They determine, in part, the expectations each person has about the behavior of the other and imply rules of conduct that govern interactive behavior as it proceeds. If Herman thinks of Gladys as a despicable, untrustworthy, dangerous person, he will treat her with great caution and little fondness. If he regards her as lovable, loving, and the soul of integrity, he is more likely to open his heart to her and offer his affection. The concept of schemata, taken from Piaget's (1926) general theory of cognitive development, when applied to interpersonal object relations, is closely similar to Kelly's (1955) concept of personal and interpersonal constructs and to numerous other formulations in the field of person perception.

Object relations schemata develop from experience with others. New experiences are generally assimilated into previously formed schemata, in which case the schemata may become enriched or extended but remain structually the same. Some experiences, however, may require a basic change in the schemata themselves, to accommodate the new experience. Changes may occur in several ways. New evidence might demand a change in the concepts one has of oneself or another. A woman who sees her husband entering a motel with another woman will be forced then and there to change any ideas she may once have had about his fidelity. New interpretations may be made of previously unnoticed experience. A woman may receive interested attention from a man at a party, recall that she has received similar attentions in the past, and realize that other options besides staying with her mate might be available to her. Or one can reflect on a variety of past experiences and attribute new meanings to those experiences. Thus, a woman may ponder her career as a homemaker, decide her life has been more fundamentally dissatisfying then otherwise, and decide to pursue long submerged interests in a career as an artist. The cognitive–affective activities in each of these examples could lead to profound and sudden changes in marital relationships.

As with the development of any other interaction patterns, environmental, personal, and interactional changes may all contribute to the development of conflictual relationships. Major situational changes, such as moving away from friends and familiar surroundings to a different part of the country, or moving in with one's relatives, are likely to set new conditions for conflict both directly and indirectly. Any substantial change in the personal dispositions of one partner or another can contribute to conflict. If a husband becomes alcoholic or a wife loses self-esteem and "lets herself go," the relationship can hardly fail to be affected. The rise in conflict may be interactional. A wife may find that her husband is particularly sensitive about financial dependency on his mother and use this intelligence to attack his manhood when serious fights begin. The husband may know his wife is none too secure about the size of her hips and find ways to make her feel especially uncomely when the battle is raging. As each diminishes the security and fuels the resentment of the other, conditions for conflict are progressively intensified.

Principles Governing Changes in Conflict

Three very general principles appear to govern the shift from little to much conflict. One is a tendency toward reciprocity in social behavior. Withdrawal by one person is usually countered by the withdrawal of the other. A conciliatory move by one partner is usually followed by a conciliatory response from the other. Anger provokes anger. Affection begets affection.

A second principle has been stated as "Deutsch's crude law of social relations," which holds that "the characteristic processes and effects elicited by a given type of social relationship . . . . tend also to elicit that type of social relationship" (Deutsch, 1980, p. 365, italics omitted). Thus, cooperation both *induces* and *is induced by* perceived similarity in beliefs and attitudes, openness in communication, a readiness to be helpful, trusting and friendly attitudes, sensitivity to common interests, deemphasis of opposing interests and so on. Competition both *induces* and *is induced by* use of coercion, threat or deception, attempts to increase power differences in a personally favorable direction, poor communication, increased sensitivity to opposed interests, and minimization of similarities, and so on. The general definition of the relationship as perceived by participants and as defined by others thus influences the interactional processes that go on in the relationship. And as those processes occur, they enhance the very characteristics of the relationship that define it.

The third principle is the familiar one of reinforcement. Somehow the rewards from conflictual interaction must outweigh the costs if the behavior is maintained or increased. The gains may be obvious for at least one of the partners if he or she manages to achieve a valued goal by fighting for it. Except for some conflict-habituated couples who seem to enjoy fighting as one form of intimacy, however, the experience of conflict is apt to be aversive. The most difficult case to accommodate by a simple reinforcement principle is that in which the people involved fight constantly, do not seem to be held in their relationship by external constraints or lack of alternatives, but continue to do battle with one another frequently, sometimes viciously, to the evident gain of no one. A general principle of reinforcement can probably be sustained even in those cases, however, by recognizing that not all the gains to be had from conflict are directly apparent. Manifest issues may mask latent ones. The people in conflict may themselves be unclear about their own motives. Concealment of aims from an adversary may offer an advantage in competitive struggle. If the aim is socially reprehensible, public acknowledgment of it will rarely be straightforward. If the aim is personally repugnant to the person who holds it, some cognitive distortion about it is likely even if the distortion does not extend to outright repression. Instead of abandoning the well-established principle of reinforcement in cases of sustained, but evidently aversive interaction, it seems more reasonable to adopt the position that the participants are still "getting something" out of the interaction, but to recognize that rewards are often subtle and the identification of reinforcers is operationally difficult.

Developmental Patterns of Conflict Management

Although developmental courses remain to be empirically charted and principles governing change remain to be tested and refined, the conception we are proposing suggests some general developmental patterns related to conflict in close relationships. As a simplification, it may be useful to sketch these patterns and to consider briefly the causal conditions that appear to underlie them.

In several sections of this chapter, discussions are centered around an idealized distinction between *effective* and *ineffective* forms of conflict resolution. In general, the effective forms occur in supportive, often affectionate, relationships in which accurate communications occur, positive influences are employed, and conflicts are constructively resolved. Conversely, ineffective forms occur in competitive, often hostile, relationships in which communicative distortions are common, coercive influences are used, and more destructive outcomes of conflict are seen. In considerations of development, discussion can be centered about conditions conducive to *stability*, on the one hand, and to *instability* on the other. Drawing on conceptions of causal loops linking the events that occur in social interaction to causal conditions governing those interactions (see Chapter 2), two general conditions can be outlined. One is a condition of negative feedback, in which changes in the state of a system lead to some activity that tends to restore the prior state. The other condition is one of positive feedback, in which a change is followed by activity that intensifies and leads to progressive shift in the direction of the initial change. If these two gross dimensions of conflict in the development of close relationships are crossed in the usual way, four general developmental patterns appear, as shown in Table 9.1.

In patterns of *congeniality,* the basic goals of the people in the relationship are compatible, and means are developed for effective resolution of the minor differences that occur. For this kind of conflict resolution to occur, both P and O probably have to be reasonably secure, affectionately disposed toward one another, and have reasonably clear and well-validated views of themselves and each other. Their needs are acceptable, both privately and as disclosed to the other. When a need arises, P communicates a report of the need and an expectation of satisfactory response to O, who receives the message without undue distortion and responds in a way that fulfills P's needs. Since O has a genuine interest in P's welfare, the act that satisfies the partner is also personally gratifying, and, by the uncomplicated workings of reinforcement, the patterns of interaction will be repeated whenever the internal and external stimulus conditions for their occurrence are present. If any changes occur in the causal conditions governing the relationship, they are slight and incremental. The developmental result is a congenial stasis (see Chapter 4, "Emotion").

In patterns of *contention,* an expression of need by P is met by action from O that not only fails to satisfy P's need but interferes with P's course of action and initiates conflict. Some of the forms conflict may take in linear interaction sequences were discussed in the first part of this chapter and illustrated in Figure 9.2. Conflict begins when an action by one person interferes with the action of the other and then proceeds along any of several courses to any of several kinds of outcomes. If only by reinforcement derived from the cessation of aversive dispute, one pattern or another may stabilize as the relationship continues. The combatants may go through the same ritual time and again as their individual needs require exchange, but the mutual interests and flexibilities required for integrative solutions and creative changes do not appear.

TABLE 9.1
Developmental Patterns of Conflict Management

Effectiveness of Conflict Resolution	Stability of Relationship	
	Stable	Unstable
Effective	*Congeniality:* Static patterns of agreement	*Growth:* Progressive development
Ineffective	*Contention:* Static patterns of disagreement	*Alienation:* Metastatic deterioration

The outstanding feature of many repetitive conflict patterns is their rigidity. The interaction sequences are "self-perpetuating" (Wachtel, 1977a). For whatever motivational reasons, one of the people involved initiates a sequence that evokes complementary behavior from the partner, who then becomes an accomplice in playing out their familiar scenario. A relationship characterized by fixed interaction sequences of this kind does not grow. It congeals.

Classical experimental investigations (Maier, 1939; Masserman, 1943) have shown that very severe stress can lead to nearly intractable rigidities of behavior. Clinicians commonly assume that anxiety and insecurity contribute strongly to rigidities of thought as well as overt behavior. The new and uncertain is feared more than the unpleasant but familiar. The thought of change cannot be comfortably entertained. Under these conditions, the object relations schemata, and hence the feelings and actions of the adversaries, are unchanged from one fight to the next, except that their negative views of each other are repeatedly confirmed. As in congenial relationships, any changes that occur in causal conditions are slight and incremental. The developmental result is another form of stasis, only the content is combative rather than satisfying.

In *alienation,* conflicts bring about progressive changes in one or more of the causal conditions governing the relationship and lead eventually to mutual isolation. She nags; he withdraws. His withdrawal offers reason for further nagging, which leads to further withdrawal. The only solution that occurs to her is more nagging. The only solution that occurs to him is deeper withdrawal. As destructive conflicts continue, the evaluative views each person holds of the other become increasingly negative. Trust in the other as a satisfactory provider for one's needs is weakened, and with the loss of trust, clear and open disclosure is replaced by concealment and dissimulation. Positive influences are replaced by coercive controls. The developmental process follows a course of metastasis, and the outcome is separation, physical or psychological. Alienation seems likely when affectional bonds are weak, basic interests are divergent, and withdrawal occurs as a common way of terminating conflict. Seriously alienated pairs may stay together physically but have little involvement psychologically if external conditions favor maintaining the appearance of a continuing relationship. "Empty-shell" marriages and the relationships between many adolescent children and their parents appear to take this form.

In patterns of *growth,* the relationship may be disrupted by severe conflict from time to time, but the positive outcomes of effective conflict resolution discussed earlier in this chaper, that is, integrative agreements and structural improvements, build upon themselves to form the bases for greater intimacy and mutual enrichment of the experiences of those involved. A minimum condition for such a positive outcome seems to be the establishment of rational means for dealing with conflict. In one of the few prospective developmental studies of conflict to date, Raush et al. (1974) observed a general trend for newlywed couples to behave in less rejecting and more rational ways as their relationships continued, suggesting that they found more reasonable ways to deal with conflict as they stayed together in marriage. If affectional bonds are strong, and if cooperative attitudes, adherence to highly valued goals, and flexibility about means for attaining goals can be maintained, integrative agreements and occasional structural improvements will be seen. The relationship is then likely to become "closer," in every sense of the term, as the years go by.

In the positive development of relationships, one or more of the causal conditions influencing the relationship is changed in a constructive way. Conflict or other experience may lead to the revelation of new and disturbing issues, and with those revelations may come severe, though transient, failures in mutual satisfaction. If the affection of each person for the other is strong enough, if each person is secure enough to risk the creative explorations needed to find new ways of resolving differences, the ensuing developmental process may take the form of a dialectic, in which the action of one person is at first met by the contrary action of the other, but the actions, ideas, and emotions of both are eventually replaced by a new synthesis. With open disclosure, each person reveals a part of himself or herself the other had not known before. If the other accepts that and nourishes the special qualities of his or her partner, the developmental effect can be one of continually enriched understanding and progressively deepening intimacy.

The four developmental patterns just described are better regarded as conceptual simplications then as putative factual realities. The preceding chapter has shown how complex the process of development and change can be, how many conditions are involved in change, and how widely varied developmental outcomes may be. Still, the scheme suggested by our conception is quite consistent with the formulations of other investigators. Thus Cuber and Harroff (1965) describe five types of marriages: conflict-habituated, passive–congenial, vital, devitalized, and total. The parallels between conflict habituation and contention, passive congeniality and congeniality, vitality and growth, and devitalization and alienation seem fairly close. Cuber and Harroff's "total" marriages appear to combine the developmental processes we have proposed, as such processes probably would be combined in most complex, long-standing relationships. Developmental patterns and types of relationships are not the same, conceptually. Whether the general developmental patterns suggested by our conception will appear as typological categories in descriptive studies of marriages and other relationships is yet to be determined.

Reduction of Conflict

Knowledge about the development of conflictual relationships can be obtained not only by considering how conflicts increase but how they may decrease. Naturalistic longitudinal studies of conflict reduction have not yet appeared, but writings on clinical treatment suggest some ways in which relationships marked by frequent and severe conflict may become more harmonious. Since intervention in close relationships is the topic of Chapter 10, "Intervention," discussion at this point will be brief and focused on changes directly pertinent to conflict resolution.

Changes in behavior

Behaviorally inclined clinicians have recently extended the principles and practices of behavior therapy to interpersonal relationships. In behavioral approaches to treatment, the disturbing behavior itself is regarded as the problem, rather than as a sign of some underlying psychopathology, and principles of learning are applied more or less directly to change patterns of disturbed or disturbing behavior toward more acceptable forms. In the interpersonal case, the "symptom" is usually some form of dysfunctional interaction pattern. According to the research of Weiss, Patterson, and their colleagues at the University of Oregon, of Stuart and his colleagues, and of numerous other investigators (see Olson, 1976 for some reviews of pertinent work), it is exceedingly common for members of dysfunctional families to use aversive controls in their behavior toward one another. Each person in a close relationship must influence the behavior of others in some way if solutions are to be found for the inevitable conflicts of daily living. People who get along badly enough to find their way to clinics tend to inordinate uses of negative controls (criticism, coercion, threat, punishment). Treatment is frequently directed toward replacing those controls with more benign, positively reinforcing influences (encouragement, praise, appreciation). For people who have spent their lives hurting each other, redirection toward helping one another attain more effective and satisfying patterns of interaction can constitute a radical change.

Behavioral interventions require clear specifications of target behaviors. The generalizing tendencies discussed in the preceding section of this chapter as elements in the escalation of conflict are reversed in "pinpointing" techniques. A husband who complains that his wife is "too dependent" is led to specify in detail the kinds of activities his wife displays, to indicate the kinds of behavior he would consider more acceptable, and to work out, usually in some contractual form, a way by which he can help her attain the pattern of behavior they may both desire.

Changes in rules

Besides the kinds of contracts mentioned above, other changes may be made in the rules governing classes of conflictual behavior. Bach and Wyden (1968) have directed attention to the values of "fair fighting" in intimate relationships. Conflict, they say, is unavoidable. The difficulties people have come not so much from the conflict itself, nor even from the issues over which conflict arises, but from the failure to develop effective ways of resolving conflict. The subtle tricks people can use in their efforts to "win" fights (though at the same time they may be ruining the relationship) are legion. Grievances are hidden, only to be brought out later in "kitchen sink" attacks. The "Achilles heels" of participants are exposed, and pierced with exquisite skill when the enemy is most vulnerable. In clinical treatment, couples in conflict are taught to fight deliberately, to bring the conflictual patterns under voluntary control, but the therapist insists, and teaches clients to insist, on strict adherence to rules in the conduct of conflict. "Kitchen sink" generalizations are prohibited. The "beltline" is located for each party and "hitting below the belt" is forbidden by the referee. When people have been taught to "fight fair," they have at their disposal a general way of resolving conflicts that arise, and a means for reaching more constructive outcomes than they had found before.

Changes in communication

Many clinicians view the process of communication as central in the reduction of conflict as well as in its origins. Assuming that all people exhibit flawed communication at least some of the time, and that people in close relationships can all profit in some measure from training to improve communication, several standard programs have been designed to help people to learn to communicate more effectively than before (e.g., Gottman, Notarius, Gonso, & Markman, 1976; Guerney, 1977; S. Miller, Nunnaly, & Wackman, 1976; Stuart, 1976). The programs have several features in common. They all help people express themselves accurately. People are taught to "level" with one another. They are enjoined to state messages about beliefs and feelings in the first-person singular, that is, to use "I-messages" rather than "you" or "it" messages, because the feelings and beliefs they have are not "out there" in the other person or the situation, but within themselves, and the rules of communication training require correct attribution. By any of several means, the clients are taught to listen. People often have to be taught to keep still long enough to hear what another has to say. No interruptions are permitted. The receptive partner must attend and try to understand. Finally, communication training programs include some exercise in validation. Instead of unchecked "mind-reading," instead of assuming that each one knows what the other has attempted to convey, deliberate validation is required. A common technique has each person rephrasing the message his or her partner has just stated. If comprehension is accurate, dialogue can proceed. If not, necessary corrections can be made. By reducing whatever distortions may have characterized communications in the past, married couples and other closely related people can not only ameliorate one of the conditions that has contributed to conflict in the past, but acquire a general skill for avoiding needless conflict and establishing constructive relationships in the future.

Cognitive changes

Several forms of cognitive change may reduce conflict. In clinical practice, change is almost routinely brought about by reframing the problem that brought the clients in for treatment. As was noted above, a pattern of cross-complaining is very common in the interaction cycles of distressed married couples. At a more general level, it seems fair to say that most married couples seeking treatment for their problems will mutually find fault with their partners and, even if they do not express the wish directly, are inclined to hope that improvements will be brought about through some change in the other. A common clinical ploy in dealing with cross-complaining postures is to direct attention away from blaming and toward constructive changes that each person can initiate, quite independently of any return to be had from the partner.

Some clinicians will claim that clients are never fully accurate in defining the issues that bring them in for treatment. People who define issues correctly do not need treatment, and some reframing is therefore always required of people who seek treatment. At the very least, an issue defined in general terms requires specification. The manifest issues that clients present in a first interview may be seen on further study to be little more than a socially acceptable way for the clients to place themselves in a position to approach more painful and difficult problems. It is quite common for troubled married couples nearing divorce, for example, to seek "marriage counseling" so that they can assure friends, relatives, and themselves that they "did all they could" to save the marriage. It is common practice among many counselors not to confront clients immediately with the "truth" about their relationship, but instead to accept the clients' initial definition of the problem as operationally, but provisionally, valid and proceed with interventions appropriate to that definition. Conjoint discussions, communication training, behavior directives, paradoxical injunctions, and any of a wide range of procedures may be instituted. If, in fact, one or both of the partners covertly intends to leave the relationship, the meliorative procedures are likely to fail. At some point in that process, the treatment issue will have to be redefined, and therapy that started as "marriage counseling" becomes a consultation in divorce.

By therapeutic intervention or otherwise, cognitive changes seem very important for lasting reduction in the frequency and severity of conflict. The concepts and images that each partner has of the other are important causal conditions in the determination of conflict. Changes in those conditions may also determine the reduction of conflict.

Changes in family structure

Patterns of unproductive conflict, like all other interpersonal problems, are sometimes seen as expressions of dysfunctional family organization (e.g., Bowen, 1976; Minuchin, 1974). Family members may be enmeshed in apparently inextricable dependencies, or they may be psychologically isolated and disengaged from one another. Rules of authority may be confused. Power coalitions may shift from one issue to another across generational lines in ways that make rational decisions and constructive resolutions of conflict impossible to attain. In these cases, efforts to deal with conflict by changing the cognitive or behavioral dispositions of one person, or even the characteristic interactional patterns of any pair, are likely to fail. The full network of those involved in the production and maintenance of conflict must therefore be involved in reducing it. By joining the family, by observing and deciphering the subtle maneuvers through which various family members affect each other, by refusing to accept the role definitions clients attempt to cast on the professional, and by exercising his or her own influence to loosen enmeshments, embrace the alienated, and alter unsalutary coalitions, the family therapist attempts to change the structure of the family system in which conflict is seen to be rooted.

Proponents of the many forms of professional treatment for interpersonal conflict frequently write as if they were proposing a single therapeutic approach to all problems. This chapter suggests that different strategies and tactics will be required for different patterns of conflict. If couples are controlling each other aversively, a shift to positive controls may be helpful. If people fail to communicate effectively, the locus of failure needs to be identified and appropriate training directed toward clearer articulation, more attentive comprehension, or more careful validation, as the facts of each case may indicate. If the general features or particular details of each person's schemata are neither congruent with reality nor pragmatically useful as guides to effective behavior, appropriate cognitive changes may need to be brought about. If third-party influences impinge on dyadic conflict, changes in the larger network may be needed. Conflicts may arise in many ways. A developmental pattern marked by rigidly repeated episodes of contention may call for interventions to disturb the stasis. A pattern of progressive alienation may require a steadying influence to retard or reverse the metastatic process.

The reduction or constructive use of conflict requires accurate descriptive knowledge about the patterns of conflict that occur, reasonable inference about the causal conditions underlying the conflict, and only then the exercise of influence to bring about change.

NEEDS FOR KNOWLEDGE

Systematic knowledge about conflict in close relationships is severely limited. The resolution of conflict has been studied extensively in controlled situations, but study of the origins of conflict in natural settings has only begun. The role of conflict in the development of close relationships has been examined even less. The processes by which conflicts are precipitated, engaged, escalated, and resolved need closer descriptive study than they have received so far. The linkage of conflictual interactions to later events and causal conditions in changing relationships needs to be examined prospectively. A wider range of close relationships should be studied. Nearly all the ordered knowledge now available about conflict in close relationships comes from the study of young heterosexual couples. Older couples might have more to teach us. The conflicts of homosexuals should be at least as illuminating as of those who seek heterosexual partners. Good friends who drift apart might like to know why. Conflict among siblings must involve more than "sibling rivalry." The developing relationships of therapists and clients have never been studiously examined by third-party investigators. All we have are the reports of therapists. Systematic study of conflict must move beyond the pair to the larger and more complex networks of relationships in which pair relationships are embedded. Most contemporary family therapists assume that conflicts extend inherently beyond dyads to other family members, and over the three generations of which most families are composed. The assumption could use some systematic study, but the necessary research cannot begin until the scope of inquiry embraces larger networks. The associations of conflict viewed at an interactional level with more general cultural conditions, such as ethnic membership and social class, need to be examined. Linkage in the other direction, from the social psychology of conflictual interaction in intimate groups to the psychological processes going on within the individuals involved, is also imprtant for better understanding of interpersonal conflict. Many kinds of inquiry are needed but conceptually guided, descriptive, longitudinal research over a range of relationships seems needed most urgently if the processes and products of conflict in close relationships are to be more fully understood.

CHAPTER 10

Intervention

ANDREW CHRISTENSEN

Relationship problems often lead people to seek therapy. In a national survey (Gurin, Veroff, & Feld, 1960), more than half (59 percent) of the respondents who had sought professional help for personal problems said these problems concerned family relationships. Another survey found that more than half (54 percent) of the clients seen in short-term clinics specified marital or child–parent difficulties as the presenting problem (Parad & Parad, 1968).

People who seek help for interpersonal difficulties are met with a bewildering array of approaches to help solve relationship problems. Some therapists focus on historical factors in the person's life, while others insist on attention to the "here and now." Some therapists prefer to see the client alone, while others want to see all the members of the close relationship. Some therapists maintain a passive, interpretative stance, while others are directive and didactic. Therapists differ greatly in the attention they give to cognitive, emotional, or behavioral aspects of the problem.

The present chapter provides a comprehensive conceptualization of distress and treatment of close relationships based upon the ideas presented in

---

During the preparation of this chapter, the author was supported in part by NIMH Grant 32616. The author gratefully acknowledges the assistance of Richard Gilbert, Charles King, Ann Hazzard, and Louise Macbeth, who read and commented on a draft of this manuscript.

Chapter 2. Unlike a theory that states the important variables and specifies their interrelationships, this conception is meant to be a logical analysis of distress and treatment. It organizes the major classes of variables, without insisting on a particular allegiance to any one of them or arguing for a particular connection between any set of them. Rather than proposing new treatment approaches, the chapter aims at clarifying existing ones.

The chapter begins with a conceptual analysis of distress and treatment. Particular topics of analysis include dysfunction in close relationships, the client–therapist relationship, assessment, and intervention. As typically conceived, dysfunction refers to the deficits that account for distress; the client–therapist relationship serves as the interpersonal context for treatment; assessment provides the necessary information to guide treatment; and intervention refers to therapist actions that comprise the treatment. Following this conceptual section, a selective review of treatment strategies is provided, using a three-part organizational framework: individual, dyadic, and family approaches. In explicating treatment theories within each of these topic areas, we use distinctions made in Chapter 2 between events and causal conditions to highlight the general factors that each theory implicates in distress and the specific interaction patterns that the theory suggests will result from these factors. Brief explanations of the client–therapist relationship, assessment, and intervention, as seen by different treatment approaches within these three topic areas, provide a selective overview of the clinical–theoretical literature. Finally, a brief review of the empirical literature and a discussion of the implications of our conception for clinical work complete the chapter.

Our analysis is intended to apply to all close relationships, but the focus in this chapter is limited to the marital dyad and parent–child relations. The personal and societal significance of these particular relationships is obvious. Existing literature on psychological intervention in close relationships has focused on them almost exclusively. In addition, our analysis is limited to treatment whose purpose is relationship maintenance. Efforts to assist individuals in developing close relationships (e.g., Christensen, Arkowitz, & Anderson, 1975) and separation–divorce counseling to assist individuals in ending close relationships (Kaslow, 1981; Welch & Granvold, 1977) are excluded. Our analysis focuses on intervention strategies for relationships in which at least one of the partners wants to continue the relationship.

In this chapter, we attempt to provide some order to the confusing array of existing treatment ideas and practices. It would be both premature and presumptuous to suggest the "right" solutions to distress in close relationships. But we can clarify the alternatives and place them in context. What has been said of science can be said of clinical work: Truth arises more readily from error than from confusion.

CONCEPTUAL ANALYSIS OF DISTRESS AND TREATMENT

Dysfunction in Close Relationships

Dysfunctionality in close relationships can be defined by the property of facilitation versus interference, discussed in Chapter 2. Functional close relationships are characterized by high facilitation and low interference of individual and joint goals, needs, and desires. Dysfunctional close relationships are characterized by the reverse—high interference and/or low facilitation. The terms *needs, goals,* and *desires* are used more or less interchangeably to refer to those motives and corresponding outcomes that the individual values and puts effort into attaining. Conscious awareness of these needs is not necessarily implied. Joint goals as well as individual ones are included in the definition since participants in close relationships often share important goals but are unable to facilitate their realization. The husband and wife may both want children but may not be able to cooperate harmoniously in raising them.

Since each individual has a complex hierarchy of needs and goals, various combinations of facilitation and interference can define functional and dysfunctional relationships. Two lovers may facilitate such important desires in each other that significant interference with other goals is cheerfully tolerated. A marriage in which spouses have grown apart over many years may facilitate so few important needs in the other that any interference is a threat to the continuance of the relationship.

An important implication of this definition is that the criterion for dysfunction, the necessary level of interference and/or facilitation, must largely reflect individual standards. Research can reveal the kinds of needs typically fulfilled by close relationships. Clinicians and lay people alike can recognize extremes of asking too much or too little from a relationship. But, ultimately, what is enough for one may not be enough for another. On the basis of their own history and current alternatives, individuals determine a standard of acceptable need facilitation and interference. The definition of dysfunction must rely, to some degree, on that standard.

Dysfunction in close relationships often creates psychological distress. As discussed in Chapter 4, "Emotion," negative affect is a typical reaction to interpersonal interference with individual goals. Furthermore, couples often engage in conflict about these interferences. If the conflict is not resolved, dissatisfaction mounts about both the conflict and the unfulfilled needs. Chapter 9, "Conflict," describes these conflictual episodes; Chapter 1, "The Emerging Science of Relationships," mentions some of the serious consequences of conflict and isolation.

But dysfunction does not always lead to negative affect in close relationships. Some individuals, seeing easy substitutability for need satisfaction, can end close relationships and move on to others with apparent ease. Other persons can suffer repeated frustration with equanimity. Psychological processes, such as denial, may allow some to suffer, at least temporarily, interference with their most cherished goals and manifest little emotion. Furthermore, dysfunction is defined as much by a lack of facilitation as by an active interference. Rather than showing the usual signs of distress, like anger and depression, a couple may profess, quite calmly, a lack of love or feelings of distance or a sense that they are no longer "right for each other." Whatever the reasons, dysfunction should not be equated with distress in close relationships. Distress may indicate dysfunction, but dysfunction does not always produce psychological distress.

This conception of dysfunction presented here is descriptive. Dysfunction is a property of a relationship, a summary term capturing a consistency about the interconnections between P and O that facilitate or interfere with their individual and joint goals. Any implication about particular events is avoided. Members of close relationships may express their dysfunction in a myriad of events. Parent and child may argue repeatedly. Husband and wife may withdraw into resentful silence. The close relationship may move toward dissolution or remain stable and dysfunctional.

Implications about causal conditions are also avoided. In particular, this conception of dysfunction does not necessarily suggest pathology. Despite common views that only "sick" people or "crazy" environments lead to dysfunctional relationships, through normal processes of development, individuals in close relationships can change in ways that make them unable or unwilling to facilitate each other's goals. Their relationship can become dysfunctional even though they and their environments manifest no pathological deficiency. Some might argue that dysfunctional relationships that come to the attention of the clinician usually represent serious disturbances implying one or more significant deficits. Members of these relationships are not able to repair their differences or move on to other relationships. Such an orientation to pathology can, however, misconstrue normal developmental processes as "sick" ones.

Dysfunction in close relationships can result from any of the four major classes of causal conditions elaborated in Chapter 2. Characteristics of the members of the close relationship can create dysfunction. For example, one or both members may have unrealistic expectations about what needs can be facilitated in a close relationship. Deficits in the relationship itself, such as an implicit rule that prohibits discussion of dissatisfactions with the other, can create dysfunction. Conditions in the social environment, such as intrusive in-laws or demanding children, can prevent certain kinds of goal attainment. Conditions in the physical environment, such as intemperate climate or poor housing, can also have their effects.

As a property of the close relationship, dysfunction can be asymmetrical. The relationship may create a different pattern of facilitation and interference for each participant. The mother–infant relationship may fulfill virtually all the needs of the child but cause significant interference with some of the needs of the mother. A love affair may be fulfilling for one but limiting for the other. Thus, we could speak of a relationship as being functional for one and dysfunctional for the other. The pattern of therapy seeking by members of close relationships supports this notion of asymmetry. Husband and wife are often not in agreement about the need for help. Parent and child seldom share an equal enthusiasm about therapy. These differences can reflect a number of variables, but one important possibility is the difference in need fulfillment that each experiences in the relationship.

Asymmetry in dysfunction has important consequences for the relationship. When members of close relationships experience a lack of facilitation and/or an interference in important goals, they are unlikely to maintain the status quo. They may apply pressures to their partners to change the relationship, or they may seek fulfillment elsewhere. In almost all cases, P's reactions to a lack of need fulfillment will create interference for O. Whether P responds with nagging attempts to create change, with threats to break up, or with sullen depression, O's own need satisfaction will be threatened. Because of this interdependence between facilitation and interference in goals for both P and O, we can speak appropriately of a dysfunctional relationship when either P or O experiences significant lack of need fulfillment.

The Client–Therapist Relationship

The treatment of dysfunctional close relationships typically occurs in the context of another relationship, the one between client and therapist. Typically, it too qualifies as a "close relationship." The contact is usually quite frequent, at least once a week. The interconnections are usually limited to verbal interaction, but are diverse in their comprehensive coverage of the client's life. The contact is often durable, especially in long-term therapy. But, most importantly, the interconnections between client and therapist are strong. The most significant aspects of the client's life are discussed. Often, the client rethinks past sessions and rehearses future ones. As in any close relationship, these interconnections between the participants can alter causal conditions in their lives. The explicit goal of the therapeutic relationship is to change causal conditions in the client's life.

Although it has the defining characteristic of a close relationship, the client–therapist relationship is unique in several respects (Gilmore, 1973). More than most adult close relationships, it is distinguished by its asymmetry. The client discloses personal problems with little reciprocal disclosure by the therapist. The therapist provides assistance and support without any reciprocal aid from the client. Unlike most other close relationships, the therapeutic relationship is severely limited in time, place, and involvement. The intense interaction characteristic of therapy occurs under clearly specified circumstances—at a particular time and place each week. Few close relationships are so scheduled. Therapist ethics often prohibit involvement outside of the therapy context. For example, the ethical code of the American Psychological Association states that "psychologists make every effort to avoid dual relationships with clients . . . " (Ethical Standards of Psychologists, 1977, p. 4). Finally, the therapeutic relationship is unusual in that a desired outcome is termination. The implicit contract is that therapist and client will develop a close relationship as a means of furthering the client's autonomy until he or she no longer needs the therapeutic relationship.

The nature of the client–therapist relationship has important implications for treatment. Most importantly, the processes involved in close relationships are likely to occur in the client–therapist relationship and facilitate or hinder treatment progress. On the positive side, processes involved in establishing such conditions as trust and respect can further treatment goals in the same way that they facilitate the joint goals of members in other close relationships. On the negative side, processes like power struggles and misunderstandings can hinder the achievement of treatment goals in the same manner that they interfere with other close relationships.

A common assumption is that the dysfunctional patterns the client experiences in close relationships, and thus the focus of treatment, will appear in the client–therapist relationship. The husband may respond to the therapist as the demanding parent he sees in his wife. The wife may seek a coalition with the therapist against her husband as she does with her children. Dysfunctional patterns in the therapist's own close relationships may also appear in therapy. A hypersensitive therapist may overreact to the client's reasonable questions about the progress of therapy. An overworked therapist may identify with the workaholic husband, ignoring the wife's requests for more time together as a family.

Dysfunctional interaction between client and therapist need not have its roots in other close relationships but can result from present circumstances. The husband may use the therapist as an ally in his struggle with his wife, even though he does not typically employ others in marital conflict. The therapist may not have general difficulties with control in relationships, yet still get caught in a power struggle with a particular client. Whatever their origin, certain negative processes that happen in close relationships can occur in the client–therapist relationship and deflect work away from the stated goals.

Another property of the client–therapist relationship, its asymmetry, has important treatment implications. Progress in therapy rarely occurs in a simple straightforward fashion, with the client explaining the problem, the therapist providing recommendations or commentary, and the client following through. Often the client resists the treatment and struggles, however subtly, with the therapist. Certainly some of these problems result from the content of therapy, which is usually emotion packed for the client, but some issues result from or are accentuated by the asymmetry between therapist and client. The client must disclose. The client must admit, implicitly, an inability to handle close relationships and listen to one who, presumably, knows better. The "one down" position is obviously not a comfortable one for many.

Much of what we have said about the client–therapist relationship could be said about other professional relationships as well. Certainly, positive processes, such as trust, facilitate all these interpersonal contacts. Negative processes are hardly unique to the client–therapist relationship. Power struggles between teacher and student about the completion of assignments, for example, are common. Even the striking asymmetry is hardly unique, as anyone who has undressed before a physician knows.

Yet these processes may have much stronger effects in the client–therapist relationship because of the content involved. The issues are intimate, involving some of the most personal aspects of the client's life. The stakes are high, involving the client's major satisfactions in life. Feelings about self are crucially involved. Experiences of shame, inclination to avoid, or resentment at the therapeutic intrusion are common. A wife may experience discomfort in disrobing before her physician, but that can not equal her discomfort in hearing her husband complain about her lovemaking to the therapist. A husband may not relish the disclosure of his financial condition to his tax accountant, but that would be far easier than hearing his wife complain about his income and pressure him to face a long-avoided career decision.

Assessment of Dysfunctional Relationships

Goals of assessment

The first major task of treatment is assessment. To complete an assessment, the therapist must, in the terminology of this book, determine the patterns of events that together constitute the dysfunction in the relationship and determine the causal conditions that account for these events. How do relationship events facilitate and interfere with the needs and goals of the clients? What are the enduring conditions that account for this interference and lack of facilitation and how can they be changed?

Presenting complaints, the initial focus of assessment, provide the first indications of dysfunction. When a wife complains that her husband is always too tired for sex, the therapist gets a first indication of the frustrated goal and the form of the interference. However, clients often speak of their desires in very global terms and describe relationship events with difficulty. Their presenting complaint may simply be "I wished we loved each other more." Other clients focus on the interfering relationship events, with little awareness of the needs being frustrated: "We seem to argue all the time. Often I don't even know what we are really arguing about." Even when clients state the need and the interfering events clearly, they may not be giving the therapist an accurate assessment. If the husband in the above example becomes more sexually responsive, the wife may not view the therapy as successful. More fundamental dissatisfactions with the relationship might emerge. In any case, the task of the therapist in discovering the needs of the clients and the relationship events that facilitate and interfere with those needs remains difficult.

Therapists differ markedly in their assessment of goals and associated facilitating and interfering events. Some insist on detailed behavioral descriptions, the precise sequence of conflictual interaction. Others focus on the feelings that clients have about their needs and their partners' interference with these needs. Still others seek an account of the clients' cognitions. No matter what type of event therapists choose to focus on—behavior, affect, cognition—they also differ on the time frame within which these events are viewed. Some seek information about historical events in both the present close relationship and others preceding it, going back as far as the clients' relationship to their parents. Other therapists focus entirely on contemporary events or seek some information on both. Clearly, the number of relationship events that might be important for assessment is limitless.

The number of possible causal conditions that could determine these important relationship events is also extensive. There is little therapist consensus about crucial causal conditions. Does conflict indicate poor social skills, deep-rooted incompatabilities, or healthy involvement? Do thoughts of leaving imply immaturity, low commitment, or fear of closeness? Even simple evaluative decisions, like determining whether certain relationship events imply positive or negative causal conditions, can be problematic. As Dicks (1967) has noted about marital assessment, "If we are told they quarrel much more than before, it is difficult to assess if this is an advance or a deterioration. Nor, if they no longer fight, whether they have no bones of contention or whether they have ceased caring" (p. 308).

Certainly, therapists can and do ask clients their views of important events and causal conditions in their relationships. Even without direct questioning, clients often provide such information. Spouses, for example, may recount for their therapist specific events that provide clear evidence to them of unsavory causal conditions in the partner (e.g., selfishness, immaturity). Parents may provide detailed accounts of actions by their child that lead them to the conclusion that their child suffers from conditions that only a therapist can treat. For obvious reasons, therapists never treat these accounts as scientific conclusions drawn from objective data. In fact, some therapists treat client explanations as relationship events that themselves imply certain causal conditions. For example, psychodynamic therapists might view many spouses' conclusions about their mates as projections of unconscious impulses. Communication therapists might view some partners' conclusions about their mates as mislabeling due to communication disorders. To whatever extent they seek partners' explanations of their relationships, therapists must ultimately decide themselves what the critical events in clients' relationships are and what causal conditions these events imply.

The causal conditions that therapists derive from their examination of relationship events can be placed in one of four "locations" discussed in Chapter 2. (1) Some causal conditions, such as psychopathological disturbance, learning deficits, or physical illness, are located in the *individuals* who make up the relationship. (2) Incompatible needs, dysfunctional marital agreements, and incompatible role expectations are examples of *relational* (P × O) causal conditions that contribute to relationship distress. None of the latter conditions can be clearly attributed to either partner alone, but must be located in the interaction between participants. (3) Inadequate social support, diffuse social structure, and social pressures are examples of causal conditions in the *social environment* of the relationship. (4) Finally, crowding, noise, and weather are all causal conditions located in the *physical environment.* Any single set of events may suggest causal conditions of all four types. A couple's arguments about their partners, for example, may indicate cognitive distortions in one or both partners, inadequate rules about dealing with the extended family, intrusive social network influences (e.g., a demanding mother or father), and crowding (e.g., presence of one spouse's parents in the home). The difficult task of assessment is to locate and define the crucial causal conditions that account for the most variance in relationship distress.

Assessment procedures must especially identify the causal conditions that can be changed. Sometimes, crucial conditions are intractable, but other less important factors are susceptible to influence. Consider again the above example. The parent's presence in the home may be the crucial condition affecting the couple's relationship. However, social and economic conditions may prohibit any change in this living situation. The therapist must therefore determine what other conditions, which also contribute to the couple's distress, are subject to influence.

Methods of assessment

Four methods are typically used by therapists to assess relationship dysfunction: interviews, direct observation, questionnaires, and monitoring procedures. These same methods are used by researchers to study relationships (see Chapter 11, "Research Methods"). Although the goals of researchers are usually different from those of clinicians, their information gathering methods are similar.

Interviews are used most often since most clinicians believe that they provide the richest source of information, albeit through the eyes of the participants. Interviews can provide information about the individual, the dyad, the social environment, and the physical environment. Typically, the interview is used to get information about the presenting complaints in the relationship and about various historical aspects of the relationship.

Often, interviews centered around relationship complaints are conducted conjointly. This format gives therapists direct access to the clients' interpersonal behavior. Rather than hearing the clients' accounts of their arguments, therapists can witness them. Many therapists, viewing this direct observation of relationship interaction as crucial, attempt to structure such interaction apart from the conjoint interview. Therapists may ask family members to talk to one another during the session and may even leave the room and watch the discussion through observation mirrors. Sometimes therapists ask the clients to engage in specific communication tasks designed to elicit information about relationship processes (see Riskin & Faunce, 1972 for a list of interaction tasks). Thus, direct observation of relationship interaction provides information on both individual and dyadic events, but not on events in the social or physical environment. However, when the entire family is included in a therapy assessment session (e.g., Minuchin, 1974) or when home observation procedures are used (e.g., Christensen, 1979; Patterson, 1974b), information can be obtained directly on at least a small part of the social network.

A third method of assessment involves questionnaires. Like interviews, questionnaires are extremely versatile methods of gathering information. Although information on individual characteristics is most often gathered with questionnaires, data on most aspects of the dyad, the social environment, and the physical environment can be accessed with this method.

The last form of assessment requires clients to monitor their own and/or others' behavior. Like questionnaires, monitoring procedures are self-report instruments, but they require immediate recording of events rather than a retrospective report on them. Also, their focus is usually more specific. Clients may be asked to indicate on a daily checklist what behaviors have occurred in their relationship, as in the Spouse Observation Checklist (Patterson, 1976a; R. L. Weiss & Margolin, 1977). Or clients may be asked to provide a written account of an interaction, as in the Interaction Record (Peterson, 1979). Many therapists devise their own instruments relevant to the particular client problem. A therapist might request a couple, for example, to track their daily arguments, the circumstances surrounding them, and their feelings and thoughts about the interactions.

Assessment decisions about the nature of causal conditions and the best methods to ascertain these conditions are determined, to a great extent, by the theoretical leanings of the particular therapist. Therapists from different persuasions examine different events and infer different causal conditions. Certainly, the characteristic of the couple, their relationship, and their social and physical environment can be influential in assessment decisions. However, much of what therapists conclude about relationship distress reflects their theoretical training rather than the exigencies of the relationships assessed.

Intervention in Close Relationships

Initially, intervention follows both logically and temporally from assessment. Having determined the causal conditions that are responsible for the relationship distress, the therapist proceeds to alter those conditions. While the goal of assessment is to determine causal conditions by an examination of relationship events, the goal of treatment is to alter causal conditions by intervening in relationship events.

The therapist's assessment decision about the location of the responsible causal conditions usually determines who will be seen in therapy. Therapists work with the individual, the dyad, or parts of the social environment, depending in part on their assessment of who has the problem. Doing the therapeutic work "on location," as it were, is usually the most straightforward method of treatment. The clients can, however, interfere directly or indirectly with therapeutic plans. The therapist may conceptualize $P \times O$ factors as the ideal targets of treatment but discover that only the wife will come to therapy. The therapist may view marital difficulties as the source of child problems but refrain from intervention because of the sensitivity and defensiveness of the parents.

After issues about who will be seen in therapy are resolved, therapists must decide on particular intervention procedures. These procedures are aimed at changing causal conditions, but will only have direct effects on relationship events. Consider, for example, an overview of what occurs in psychotherapy. Therapists' contact with their clients is usually limited to a few hours of interaction per week or less. During this time, therapists ask questions, offer interpretations, reflect feelings, give advice, and disclose information, to name some of their more common activities (Goodman & Dooley, 1976). These therapist actions usually cause immediate responses in the clients.

Clients answer questions, comment on interpretations, elaborate on feelings, and so on. Thus, therapist behavior affects immediate client behavior, and vice versa, as in any dyadic interaction. However, both therapist and client intend that these simple conversations will have widespread impact on the client's life. In the terminology of this book, both therapist and client hope that events during the sessions will have impact on crucial causal conditions in the client's life.

Theories of intervention suggest explicitly or implicitly some connection between therapy events and causal conditions in the client's life. Psychoanalytic theory, for example, proposes that the interpretations of the therapist, if properly timed and stated, can gradually increase the self-awareness of the client. Interpretations are the events that ultimately change the causal condition of self-awareness, which is assumed to be responsible for diverse client behavior. Social learning theory suggests that certain kinds of training conducted by the therapist will alter the social skills of the client. The specific aspects of training, like role playing and reinforcement, are the events that alter the causal condition of social skill, which is held accountable for a variety of interpersonal behavior.

What was said earlier about assessment could also be said of intervention. The theoretical leanings of the therapist are a major determinant of what occurs. The results of the assessment, the client's response to intervention, plus the unique qualities of the client–therapist relationship will all influence intervention strategies. But the therapist's view of what should be done, as determined by his or her theoretical proclivities, will limit the range of possibilities considered and will suggest the most likely courses of action.

Because of their crucial importance in all phases of treatment, much of the remainder of this chapter will address some of the major clinical intervention theories. This material is organized according to three of the four types of causal conditions: individual, relational (dyadic), and social environmental (represented by family systems approaches). Since clinical intervention theories are rarely directed toward changes in the physical environment, this condition is excluded. However, we should note that some interventions are directed at the physical environment, as when a social worker assists a family in moving to better housing.

Under each of the three sections, we first consider an account of dysfunction in close relationships. Using a well-known theory, causal conditions that create dysfunctional interaction patterns of relationship events are described. Then, we consider treatment strategies that are based on the theory and that are designed to alter the postulated causal conditions. This discussion of treatment will cover the client–therapist relationship, assessment, and intervention. Lest the reader associate one theory with each major type of causal condition, we will briefly consider at least one other theoretical approach to dysfunction and treatment within each of the three sections. Treatment for child–parent, as well as marital, problems will be discussed in each of the sections.

INDIVIDUAL APPROACHES

Psychoanalytic Views

Dysfunction in close relationships

> This therapy is based on the recognition that unconscious ideas—or better, the unconsciousness of certain mental processes—are the direct cause of morbid symptoms. (S. Freud, 1904/1953, p. 266)

More than any other major theory, psychoanalytic psychology emphasizes causal conditions in the individual as the responsible factors that produce mental or behavioral disturbance. Sigmund Freud paid little attention to dyadic, social environmental, or physical conditions that might affect pathology but focused his attention almost exclusively on the individual client. Furthermore, Freud specified unconscious factors in the individual as the pathogenic causal conditions. He argued that qualities outside of the individual's normal awareness are crucial for understanding disturbed behavior. He suggested that unconscious conflicts related to instinctual impulses are the cause of psychological symptoms.

Although located in the individual, these causal conditions grow out of close relationships. Freud suggested that early child–parent interactions determine the child's unconscious conflicts and his or her defensive structure (manner of dealing with these conflicts). By their patterns of gratification and frustration of infantile needs, parents influence the child's entire psychological future. During the critical psychosexual stages, particularly during the oral, anal, and phallic periods, the child either achieves sufficient need satisfaction to advance to new levels of development or is fixated so that future activity is affected by unresolved needs from the past. In the context of this need gratification and frustration, the child participates in his or her first intimate relationships. During the oedipal period, for example, the child experiences both heterosexual attachment to the opposite-sex parent and interpersonal competition with the same-sex parent. If these first intimate contacts are resolved satisfactorily, the child identifies with the same-sex parent and releases his or her attachment to the opposite-sex parent. If the relationships are not resolved satisfactorily, the child's subsequent ability to relate to both sexes can be seriously damaged.

Consider, for example, the following case description from the psychoanalytic literature.

> Mrs. A, a 28-year-old teacher, married for three years, requested therapy because she felt increasingly less interested in her husband and sexually attracted to other men. She expressed ambivalence about ending the marriage because she objectively felt that her husband was a "good person" with whom she had much to share, and she was uncertain about the reasons for the change in her response to him. She requested individual therapy and did not want her husband included because she felt that he would be hurt and angered if he knew of a recent affair with a mutual friend. She was seen in intensive psychoanalytically oriented psychotherapy twice a week for six months. During this time the role of her unresolved oedipal conflicts in her search for another more sexually attractive man became increasingly apparent. She began psychoanalysis at this time, focusing on her intense desire to please her father and her competitiveness with her mother. She became aware that she viewed her husband as very different from her father and that this factor had played a significant part in her choice of him. The affair she'd had was with a man who more closely resembled her father in his manner and way of relating to her. The realization of these factors and her ability to work through the oedipal issues resulted in her decision to remain with Mr. A and to make a commitment to the marriage. (Nadelson, 1978, pp. 116–117)*

In this example, we see that historical events in the child–parent relationship, an unresolved oedipal complex, influenced contemporary interaction by means of unconscious conflicts. Mrs. A was sexually attracted to her father, but such desires were unacceptable to her, so the entire issue was repressed. The conscious manifestations of the conflict were her symptoms: a dissatisfaction with her husband and a desire for other men who happened to resemble her father. Therapy enabled her to resolve the conflict and thus eliminate the symptoms.

Later psychodynamic writers have made important changes in Freud's original formulations. Ego psychologists, for example, are less biologically and more socially oriented. They deemphasize infantile sexuality and focus more upon ego or reality functions. However, all psychodynamic formulations propose that individual causal conditions, formed during early child–parent interactions, determine the course of contemporary close relationships. For example, in a recent discussion of the psychoanalytic perspective on marriage, Meissner (1978) writes:

> For any given husband and wife, the marital dyadic relationship is so inextricably bound to the parent–child relationship in their families of origin . . . that no comprehensive psychoanalytic discussion of marital interaction can avoid the subject of parent–child interaction. The capacity to successfully function as a spouse is largely a consequence of the spouse's childhood relationships to his own parents. (p. 26)

---

*Reprinted by permission of Brunner/Mazel, Inc.

Most contemporary psychodynamic perspectives conceptualize individual causal conditions as unconscious conflicts. Clients have desires that are unacceptable to them and are thus repressed into their unconscious. Much of clients' behavior, including their symptoms, are reflections of this unconscious turmoil. The concept of projective identification, which has been discussed by a variety of psychodynamic thinkers (Dicks, 1967; Greenspan & Mannino, 1974; M. Klein, 1946; Zinner & Shapiro, 1972), describes how an individual's unconscious conflicts affect interaction in close relationships. Other psychodynamic theorists have used other labels to refer to somewhat similar phenomena, for example, "scapegoating" (Vogel & Bell, 1960), "family projection process" (Bowen, 1965), "irrational role assignments" (Framo, 1970), and "projective–introjective processes" (Meissner, 1978).

At the observable level, individuals reveal the process of projective identification by acting "as if" significant others are different from what they actually are. Parents may act *as if* a child is completely helpless and unable to learn to take care of itself. A spouse may act *as if* the partner is a vicious person, without any capacity for human feelings. These "as if" actions are the result of serious perceptual distortions. Family members ignore, select, exaggerate, and minimize aspects of the other to create a consistent view that may have little relation to reality. A spouse may ignore the partner's occupational skills, overemphasize the partner's difficulty in disciplining the children, and selectively recall instances of the partner's social failures in order to maintain the view that the partner is weak and inadequate.

At the intrapsychic level, the processes of projection and identification, motivated by unconscious conflict, account for these misdirected actions and distorted perceptions. First, family members project aspects of themselves onto each other. In conflict about the presence of a quality in themselves that they view as unacceptable, family members attribute that quality to another. The mother denies many of her own sexual desires but views her daughter as sexually promiscuous. The husband cannot accept his own weakness and inadequacy, so he views his wife as a "helpless female."

The use of projection satisfies one side of the conflict: It assures the person that someone else has the unacceptable quality. The second process, identification, satisfies the other side of the conflict: It provides an outlet for the expression of that quality. Family members identify with those on whom they project and thus receive, vicariously, gratification for the demands of the projected aspects of themselves. The mother experiences vicarious pleasure through her daughter's sexual escapades. The husband punishes his own inadequacy by berating his wife. As Dicks (1963) observed:

> [We were] able to understand the marital behavior of some of our patients when we interpreted it as persecution in the partner of traits, weaknesses, or faults rejected by the self; or again loving or seeking (often in vain fantasy) in the partner those parts of the self which were missing. (p. 125)

Unless family members are completely psychotic, they are somewhat constrained by the reality of the partner. The actual characteristics of the other limit the extent of projection possible. Therefore, family members unconsciously select a somewhat appropriate object for their projections. Assortive mating in neurotic individuals and the selection of a "symptom carrier" in disturbed families may be guided by the appropriateness of the particular other as the object of projections. In discussing marital choice, Gurman (1978) writes that "neurotic individuals seem to choose partners whose behavior will confirm their projections by acting upon them, thus protecting the projection from naturalistic modification by reality testing" (p. 456). Bowen (1976) suggests that mothers often choose a child born with a special feature (e.g., a defect) as a candidate for projection, since realistic problems serve as a convenient base for projected ones.

Selection processes are not the only reality aids for projective identification. Family members engage in conscious and unconscious collusion to support one another's projections. For example, Sager (1976) suggests that marital partners form an implicit contract to meet the unfulfilled needs of the other. With regard to adolescents, Zinner and Shapiro (1972) suggest that fear of losing the parents motivates the child to support the parental projections. Sometimes collusion can involve several family members, as when a wife colludes with her husband's projections onto the son.

Projection processes, because they involve inappropriate attributions, are usually frustrated, no matter how judicious the selection of the target or how wide ranging the collusion. At times, family members must act out their own identities rather than the identities ascribed to them. On these occasions, a variety of strategies may be used to force compliance with the projections. Dicks (1963) suggests that much marital disharmony is a result of coercive efforts to shape marital partners into patterns consistent with each other's projections.

> One or both partners fail to confirm the other's real personality or identity. Instead they require the other to conform to an inner role model and punish them if the expectation is disappointed. Much marital conflict can be shown to stem from strivings to coerce or mould the partner by very rigid and stereotyped tactics to these inner models. (p. 127)

In a similar manner, much child–parent conflict can be understood as attempts to force child compliance with parental projections.

Classical psychoanalytic treatment

In the classical psychoanalytic approach to treatment, a therapist sees no more than one member of a family (Abbate, 1964; Giovacchini, 1965; Gurman, 1978; Nadelson, 1978). However, in psychoanalysis of children, the therapist will often meet with the parents to learn something about the presenting complaints and to give them information about the treatment.

Psychoanalysis needs to be distinguished from psychoanalytically oriented or psychodynamic therapy. *Psychoanalytic* is the narrower term, referring specifically to Freudian theory and treatment techniques. *Psychodynamic* is the broader term, referring to theorists who owe some intellectual debt to Freud but who have developed their own concepts of the dynamic interplay between mental forces. Psychodynamic approaches emphasize the unconscious and the importance of early experience in shaping the mind of the individual, but usually focus on current, situational determinants of human functioning to a greater extent than traditional Freudian theory.

Many psychodynamic therapists have departed from the tradition of only individual therapy and will see spouses conjointly or concurrently (e.g., Ables & Brandsma, 1977; P. Martin, 1976; Sager, 1976) and will see families together (Ackerman, 1966). Later in this chapter, we will briefly discuss psychodynamic approaches. However, for purposes of illustrating a treatment that focuses on causal conditions in the individual, we will limit the present discussion to classical psychoanalysis. Three central issues, the client–therapist relationship, assessment, and intervention strategies, will be addressed.

A unique kind of client–therapist relationship is essential for psychoanalysis. Analysts strive to create a social situation of maximum intimacy and ambiguity. They ask their clients to "free associate"—to disclose anything that comes into their minds, no matter how personal or embarrassing. Clients are to minimize conscious control, giving immediate expression to thoughts and feelings. However, psychoanalysts are deliberately distant and aloof. In Anna Freud's words (1946/1976), "they remain impersonal and shadowy, a blank page" (p. 147). The client's disclosures are neither reciprocated nor supported. Clients must talk about their private thoughts and feelings without the usual social gratifications of mutuality ("I sometimes feel that way too") or comfort ("That must feel bad"). As Paolino (1978) notes, the interaction is characterized "by a combination of intimacy and deprivation unlike any other form of human relationship" (p. 90).

In this unique social situation, in which intimate emotions are stimulated but do not receive their expected consequences, clients cannot rely on social convention to guide their behavior. Instead, they use early experiences in intimate relationships to guide them in this new intimate relationship. They "transfer" experiences from the child–parent relationship to the therapeutic relationship. Clients respond to the therapist as if the therapist were a significant other from childhood. They project unconscious material onto the therapist, attributing qualities to the therapist that in reality represent the clients' own unresolved fears and needs. A client may act, for example, as if the therapist were a demanding father who must constantly be pleased. Thus, the client–therapist relationship recreates the child–parent relationship. As ontogeny recapitulates phylogeny, so the client–therapist relationship recapitulates the early family relationships of the client: ". . . the most significant pathogenic intrapsychic conflicts that were once reflected in a child–family relationship will invariably be reflected in the relationship between the adult patient and therapist if the therapist has the knowledge and technical skill to facilitate the process" (Paolino, 1978, p. 100).

This transference of infantile material to the therapist–client relationship is essential for resolving difficulties in the client's current close relationships. The unconscious conflicts that arose in early child–parent relations and contaminate the current client–therapist relationship are the precise problems that distort current marital and child–parent relationships. As Giovacchini (1965) writes: ". . . there comes a point in the analysis of a married person when he projects onto the analyst an image of the spouse. . . . During this phase of the transference neurosis, the analyst represents the spouse, and the spouse in turn represents an archaic infantile self-image" (p. 67).

Psychoanalysis with children requires some alterations in the client–therapist relationship. Since a child's verbal skills are limited, free association is not possible. However, one of the originators of child psychoanalysis, M. Klein (1932), made a crucial assumption that unstructured play could serve the same purpose for the child as free association did for the adult. In both play and free association, conscious control is minimized and the influence of the unconscious is more direct. In free association, adults describe fantasies and dreams whose purpose is often impulse gratification or resolution of painful experiences. Play serves similar functions.

In psychoanalytic play therapy, the analyst presents the child with various play objects, usually dolls that could represent figures in the child's life. During play, the analyst assumes that the child will reenact with the dolls some of the conflictual episodes in the child's relations with his or her family. However, transference to the analyst also occurs, whereby the child responds to the analyst in ways that reveal the child's common modes of responding to his or her parents.

The process of free association in adults or play in children provides psychoanalysts information for their assessment. By careful observation, they discern the nature of the unconscious conflicts that distort the client's current functioning in close relationships. This assessment is by no means straightforward. Analysts assume that clients cannot ultimately follow the directive to free associate. Unconscious forces will prevent a truly "free" association as clients block on some thoughts and disguise others. Children's play may likewise be contaminated by disguise and social expectation. Because of the indirect access to the client's unconscious material, psychoanalysts must consider their assessments tentative as they complete their exploration. Furthermore, the client's responses to the treatment interventions discussed below are used as additional information in the assessment.

Interpretation is the major intervention technique of psychoanalysis. As the adult client free associates or the child plays, the analyst offers interpretative commentary. These therapeutic explanations are carefully timed to lead clients gradually to insight and understanding of the conflicts that have interfered with their ability to deal objectively with the world and thus to relate to others effectively. The interpretations make the clients aware of repressed material so that they can ultimately accept it and integrate it as their own. The "bad" aspects of self, which were projected onto others, can now be recognized and reclaimed as one's own.

A major target of interpretations is resistance—the client's difficulty in following the therapeutic instructions. Since the unconscious material was repressed because of its potential for pain, the client will not expose it easily. Memory blocks, social conversation, avoidance of therapy, and difficulty in play are all events that can sabotage the therapeutic process. Interpretation of these events can lead the client to greater insight about the conflicts that cause them. A second target for interpretation is the transference relationship. Since the client's problems are recreated in the relationship with the therapist, interpretation can be directed to the immediate, contemporary manifestations of unconscious conflict. As Sigmund Freud (1912/1963) suggested, problems must be dealt with in vivo; they cannot be attacked "*in absentia* or *in effigie*" (p. 115).

Insight is not achieved quickly or by sudden realization, but rather by much "working through," whereby clients' needs and defenses are repeatedly brought into their consciousness by therapeutic interpretation. "Over and over, now here in one area and there in another, important defenses and their motivations are brought into the patients' consciousness" (Colby, 1951, p. 119). Furthermore, the insight achieved by clients is not just intellectual but involves emotional reexperiencing in the analytic situation. Clients must see and feel the irrationality of their reactions to the therapist and have "corrective emotional experiences" in the analytic relationship (F. Alexander & French, 1946). For example, a male client may gradually experience that the analyst is not demanding and critical like his mother and does not have to be placated. This emotional and intellectual relearning will extend beyond the therapy sessions so that the man will experience his wife differently also. He will not see her and react to her as if she were his demanding and critical mother. He will react to the reality of his wife rather than to his own projections.

Rational Emotive Therapy

A second example of an individual approach is Rational Emotive Therapy (RET), a popular treatment that has been influential in the recent cognitive movement in clinical psychology (Mahoney & Arnkoff, 1978). Albert Ellis, the founder of the movement, has applied his approach to a variety of human problems, including marriage (Ellis, 1964; Ellis & Harper, 1961/1977).

Like psychoanalysis, RET implicates causal conditions in the individual as the pathogenic agents leading to marital disturbance. In contrast to psychoanalysis, however, RET does not assume that these causal conditions are unconscious. Rather, the pathological causal conditions are beliefs and expectations that are available for scrutiny. Also in contrast to the classical psychoanalytic position discussed above, RET is conducted both conjointly and individually.

In their book on marriage, Ellis and Harper (1961/1977) define what is at "the very core of most serious marital disturbances: the emotional disturbance of the married partners" (p. 23). This disturbance "almost always results from the individual's mistaken, unrealistic, or illogical beliefs, or philosophies" (p. 24). Ellis and Harper suggest five major irrational beliefs that contribute most to disturbed marital interaction. The first belief, which they label "the dire need for love," occurs when partners assume that they *must* be loved and adored by their spouses, no matter what, and that it is horrible if such love is not forthcoming. Since human beings cannot provide complete, thorough, and unwavering love, anyone who expects such devotion from their partner will inevitably be frustrated and many engage in fruitless efforts to obtain an essentially unattainable response. According to Ellis and Harper, this unrealistic search for love accounts for the highest percentage of marital distress. A second irrational idea, labeled "perfectionism in achievement," is the belief that one should be competent and talented in all aspects of one's life. Husbands and wives who expect themselves to be perfect lovers, parents, providers, and housekeepers will face inevitable distress when they meet their own realistic limitations. If ideas of perfectionism are accompanied by the third irrational belief, "a philosophy of blame and punishment," then partners will spend fruitless time blaming themselves and their spouses for not meeting the unrealistic standards. Partners may assume that blame is a way of preventing future mistakes when, in fact, it often serves to increase their likelihood. The fourth irrational idea, labeled "catastrophizing frustrations," occurs when partners assume that it is horrible if their needs and desires are not met. Spouses assume that the world and, in particular, their partners should respond appropriately to their every need. When natural limitations in need fulfillment are met, these spouses then suffer major disappointments. The final belief, that "emotion is uncontrollable," occurs when spouses assume that human unhappiness is forced on them by events and people and is in no way self-caused. Spouses assume that they are the innocent victims of fate and that their emotional upset can only be changed if others, particularly their partners, change.

Marital discord results when spouses who have these irrational beliefs are confronted with experiences contrary to the beliefs. The spouse who expected undying love must deal with a partner whose affections fluctuate. The spouse who expected perfection must face the limitations of both self and partner. As spouses are frustrated, they engage in the negative interaction characteristic of distressed marriages. They blame each other for not meeting their expectations and demand future compliance. Rather than relieving their frustrations, this style of interaction further accentuates them.

> Disturbed marital interaction, in other words, arises when one mate reacts badly to the normal frustrations and the abnormal and unrealistic demands of the other mate, and in the process helps accentuate these frustrations and demands. Then the other mate, in his or her turn, also reacts poorly to the sensible requests and the unreasonable demands of the first mate; and increasing low frustration tolerance and outbursts of temper on the part of both spouses ensues. Disturbed individuals tend to respond anxiously or angrily even to relatively good life situations, since they have basically irrational or illogical attitudes or philosophic assumptions. (Ellis, 1964, pp. 1–2)

During treatment, RET therapists adopt a teacher–consultant role with clients. Acting as experts on human distress, these therapists question their clients about behavior, affect, and beliefs. They explain the causes of clients' troublesome behavior and point out more adaptive ways of thinking and responding. Like teachers, they give assignments to clients and challenge them to change.

Rational emotive therapists typically assess their clients' irrational beliefs by means of the clinical interview. Although several questionnaires have been constructed to tap such beliefs (e.g., Eisenberg & Zingle, 1975), these instruments are used primarily for research purposes. Information is gathered using an A-B-C paradigm. *A* refers to activating events, *B* to beliefs, and C to emotional consequences. Most clients can describe both activating conditions and emotional consequences and assume that the one causes the other. A spouse may describe the partner's flirtations (A), which arouse extreme anger and resentment (C). The task of the therapist is to discover what beliefs are intervening between the activating events and the emotional consequences and to convince the client that these beliefs are, in fact, the cause of the emotional upset. The spouse who is angered by the partner's flirtations may believe that such behavior implies a total absence of love and caring in the partner. This belief is the more significant cause of the anger and resentment.

Intervention in RET is oriented toward changing the client's irrational beliefs. Therapists label these ideas, explain their destructive effects, and challenge their validity. Often the therapy is somewhat disputational as therapists debate clients about the rationality of their beliefs. Clients are encouraged to engage in similar debates with themselves outside of the therapy session. As the irrational ideas are repeatedly attacked, therapists assist clients in replacing them with more rational ones. According to RET, these new ideas will lead to less emotional and more rational behavior. The spouse in the example above will not react so emotionally once the belief about the partner's flirtations is changed. Then, that spouse can engage in rational efforts to negotiate with the partner about conduct with opposite-sex friends.

Summary

In this section, two theoretical approaches to intervention in close relationships that focus on individual causal conditions have been examined. Psychoanalytic approaches implicate unconscious forces, particularly unconscious conflicts, as the pathogenic agent in close relationships. Cognitive approaches suggest beliefs and expectations as the primary determinants of distress.

Despite this individualized focus, both psychoanalytic theory and RET include some dyadic or P × O factors when explicating the specifics of disturbed interaction in close relationships. In the pattern of projective identification, family members collude with one another to support each other's projections. In the RET account of disturbed marital interaction, spouses' extreme reactions to the frustration of their unrealistic expectations accentuate both their demands and their frustrations. Neither approach views partners as totally free to "play out" their individual characteristics. The behavior of one member will support the characteristics of the other (e.g., collude with projections or attempt to meet unrealistic expectations) or will alter the characteristics of the other (e.g., disconfirm projections or frustrate expectations). Perhaps no theory can focus exclusively on individual causal conditions when describing close relationships unless it assumes that individuals "act out" their characteristics virtually oblivious to the other. In order to maintain a complete individual focus, intrachain connections must be postulated as the determining forces, with interchain connections merely serving as "releasers" of those forces.

DYADIC APPROACHES

Social Learning Views of Disturbance in Close Relationships

A major contribution of learning theory is to remove the causal focus from the individual and place it on the environment in which the individual functions. According to this approach, the important causal conditions for human behavior are not to be found by a search of the individual's traits, personality structure, or unconscious, but rather by a detailed examination of the environmental stimuli that impinge on the individual. The world, and not the person, is the culprit for disturbed functioning. In the early days of behaviorism, this environmentalism was extreme, as evidenced by J. B. Watson's (1924) famous challenge that he could select any healthy infant at random and train it to be a professional of his choosing, if he could but control the child's environment.

Social learning formulations, while drawing much from classical learning theory, emphasize the social environment as the major determinant of behavior. Furthermore, this approach incorporates cognitive variables in its analysis of environment–behavior relations. The human organism is no longer seen as a passive recipient of environmental events but rather as an active participant in those events. The organism raises its head between the stimulus and the response and mediates any stimulus–response connections. However, the role of the social environment has retained a primary, if not an exclusive, place in social learning formulations.

In interpersonal relationships, the most important environmental influence for P is the behavior of O. Likewise, the most important environmental influence for O is the behavior of P. These environmental influences have most often been conceptualized in social learning theory as reinforcement contingencies. According to the theory, these reinforcement contingencies are specific events that follow other events and thus increase the probability of the other events. P's laughter at O's jokes increases the probability that O will tell additional jokes in the future.

In social learning theory, the causal conditions often employed to explain both the reinforcing event and the reinforced event are habits, abilities, and skills. Because of their social skills, P and O generate jokes for the other and laugh at each other's jokes. Habits, abilities, and skills are not, however, static features of P and O. They are constantly being shaped by mutual reinforcement. By P's pattern of attention and enthusiastic response, P shapes O to be an excellent joke teller. By O's inattention and disinterest, O shapes P into being a poor joke teller. Thus, the habits, abilities, and skills that manifest themselves in close relationships are a reflection of the joint reinforcement history in the pair and are properly construed as dyadic, or $P \times O$, causal conditions.

According to social learning theory, distress in close relationships comes about largely as a result of the coercion process (Patterson & Hops, 1972; Patterson & Reid, 1970). This process, which applies equally well to marital relationships and parent–child relationships, illustrates the event level patterning of "reinforcement contingencies" as well as the $P \times O$ causal conditions of "skills."

Partners in a close relationship, no matter how compatible their needs and desires, must ultimately face situations in which their wishes do not mesh. Complete compatibility is impossible; interference between the events in P's and O's lives is inevitable. In these circumstances, P or O will desire the other to change and will employ a set of social skills for producing that change. For example, P may negotiate a compromise, invoke a norm like turn taking, or use charm and persuasion to handle the conflict. Most persons, however, also have repertoires of coercive skills to bring about desired changes. They have learned to use aversive stimulation, such as threats, complaints, and guilt induction, until the other complies.

The coercion hypothesis assumes that at some point partners resort to these aversive strategies to bring about desired changes. If the strategy is successful, then P has been reinforced for coercion and is likely to repeat this strategy on future occasions. O submits to the coercive demand in order to cease the aversive stimulation provided by P and is thus reinforced for compliance. P and O are, therefore, shaping each other to be coercive and compliant respectively.

Coercion may not lead to easy successes for P. Usually O does not comply with P's demand immediately, but resists until the aversiveness passes some threshold point. By so responding, O trains P not only to use coercion, but to use higher and higher levels of coercion in order to gain compliance. Thus, O shapes P's coercive habits. Consider the example provided by Patterson and Hops (1972) of marital coercion:

> Wife: "You still haven't fixed that screen door."
> Husband: (Makes no observable response, but sits surrounded by his newspaper.)
> Wife: (There is a decided rise in the decibel level of her voice.) "A lot of thanks I get for all I do. You said three weeks ago—"
> Husband: "Damn it, stop nagging at me. As soon as I walk in here and try to read the paper I get yelling and bitching."
>
> In this situation, the husband has the wife trained to increase the "volume" in order to get him to comply. She is more likely to resort to shouting next time she needs some change in his behavior. He, on the other hand, has learned that a vague promise will "turn off the pain." (pp. 424–425)

Or consider the example of child–parent coercive behavior provided by Patterson (1975) in a book for parents.

> When people think of punishment in a family, they ordinarily think of scolding, yelling, and spanking because these are the things which occur most often. However, a surprising number of parents use hitting as a means of punishing the child. They wait until the child's behavior is absolutely intolerable to them and then rush in, hitting to the right and left in a desperate attempt to get things calmed down. In one sense this "works." After the hitting, things are usually pretty quiet for a short time. The noise and clamor subside. This reinforces the parents for hitting. The next time the children get into a noisy hassle the parents are more likely to resort to hitting again. When the parents scold or hit, the children will stop being disruptive for a few minutes, which reinforces the parents for scolding or hitting. In this manner the children train parents to scold and yell a good part of

> the day. Parents who scold and hit are not "bad" people or "sick" people. They have simply allowed their children to train them to do things they really don't want to be doing. (pp. 20–21)

The coercion hypothesis does not assume that one person consistently takes the coercive role while the other takes the complaint role. Both members of the dyad may use coercion for their own purposes:

> Presumably, as one member of a system applies pain control techniques, the victims will eventually learn, via modeling and/or reinforcing contingencies, to initiate coercive interchanges. As the victims acquire coercive "skills," they will also be more likely to counter the coercive initiations of others by coercive measures of their own. This process produces extended interchanges in which both members of the dyad apply aversive stimuli. (Patterson, 1976b, p. 269)

As both partners become more skillful in the use of coercion, conflict will escalate. Partners resist the coercion of the other until coercion becomes more painful than compliance. In this way, partners train the other to be persistent in the application of coercion and to apply increasing intensities of aversive stimuli. Since both partners engage in these tactics, the entire interactional system is coercive. Partners get short-term benefits for these behaviors (immediate compliance from the other); however, the long-term effects are high levels of conflict.

Social Learning Approaches to Marital and Child–Parent Problems

Social learning or behavioral approaches to marital and child–parent problems developed somewhat independently. Nevertheless, they share certain broad similarities in the client–therapist relationship, assessment, and intervention. Whether conducting marital therapy, parent training, or family therapy, behavior therapists maintain a consultant–teacher role with their clients. In contrast to RET therapists, who maintain a similar role, behavior therapists tend to be more supportive and less challenging or disputational. They see an important aspect of their job as reinforcing client progress. During early stages of treatment, clients may engage in appropriate behavior primarily to please their supportive therapists. This dependency is seen as temporary until the client receives external reinforcement for new behaviors.

Assessment in behavioral marital and family therapy is typically comprehensive. During the clinical interview, clients are asked to specify their problems in detail and provide related history. Behavior checklists are often used as an aid in this specifying process. Subjective reactions are tapped by both clinical interview and questionnaire. In order to learn more about how family members discuss their problems, clinic observation of family interaction is often conducted. Sometimes home observation is done. The therapist may visit the home or request the clients to use a portable tape recorder to capture some relevant interaction. The heavily research-oriented programs employ sophisticated home observation procedures to aid in assessment (e.g., Christensen, 1979; Reid, 1978). Finally, members of the family may be asked to monitor the occurrence of certain problem behaviors. The purpose of this observation may simply be to track the frequency of occurrence of a behavior or to note the antecedent and consequent circumstances surrounding it.

As a result of this uniquely behavioral assessment, the therapist seeks to understand the specific problem behaviors, the client's feelings about them, their occurrence, the antecedent conditions that elicit them, and the consequent conditions that reinforce them. The assessment is also seen as an important first step for the client. The filtering of general dissatisfactions into specific concerns and the observation of one's own family behavior are thought necessary precursors to successful treatment. Furthermore, client observation and recording of relevant behaviors are often continued throughout treatment to maintain treatment focus and to provide some indication of progress.

An early step in intervention often involves some education about behavioral principles. Particularly in parent training approaches, clients are exposed to reading materials and didactic instruction. The book *Living with Children* by Patterson and Gullion (1971) is often used for training. Major principles taught include positive reinforcement, shaping, extinction, time out, and contracting.

Marital therapy relies less on didactic training than does parent training. Often, instruction is oriented more to principles of communication than to behavioral principles like reinforcement and shaping. Patterson's (1975) book *Families* is sometimes used in both marital and family therapy to explain behavioral principles to couples. Jacobson and Margolin (1979) and Gottman, Notarius, Gonso, and Markman (1976) have written communication training manuals to be used by couples. These manuals explain principles of expression, listening, and problem solving to be discussed below.

Both marital therapy and parent training programs maintain a dual focus during treatment: the increase of positive interaction and the reduction of problem behavior. During parent training, instruction and practice in the appropriate use of positive reinforcement is provided. The parents' repertoire of positive verbal and physical gestures is expanded; encouragement to administer these positive behaviors generously but contingently is provided. The parents are to "catch their child being good." As a means of decreasing problem behavior, parents are trained to ignore their child or provide "time outs" after the occurrence of inappropriate behavior. "Time out" is used for the more serious offenses and consists of brief periods of isolation in a nonreinforcing environment (e.g., sitting facing a wall, sitting in the bathroom). Sometimes parents are trained to set up contracts that specify consequences for both positive and negative behavior. Contracts are usually reserved for older children who can understand and participate in contingency agreements.

The procedures used in these programs to achieve these changes in parent behavior are simple instruction and supervised practice. The therapist advises the clients how to respond to child behavior, models these responses, and has the clients role play them. These training procedures may occur in the home or office and may involve the child or not. Training in the home with the child is usually considered the most powerful but most expensive intervention. After any role-playing exercise, the therapist provides feedback, sometimes with the aid of an audiotape or videotape replay. Sometimes parents are instructed to audio record some of their own efforts at home for further debriefing and feedback from the therapist.

The case example below, taken from W. H. Miller (1975), illustrates one part of the intervention stage for a mother and child in behavioral parent training undertaken because of the child's noncompliance and self-destructive behavior.

> The next week the therapist went to the home three times and used modeling, direct instruction, and feedback to demonstrate to Mrs. L that she could identify and increase Tracy's praisable behavior. During the initial visit the therapist first got Tracy's permission to play with her while building blocks on the floor. Then every few seconds the therapist gave Tracy a directive, like "please hand me that yellow block." With the first indication of compliance, the therapist praised Tracy. After a few minutes of pleasant play, also acknowledged by the therapist, Tracy was asked to build a small bridge as part of a larger construction. She resisted at first but, with the therapist praising approximations of her building attempts, she was able to complete the bridge. Next, the therapist named all the things Tracy had done that were "nice and showed how well she could play." Mrs. L was then asked to play briefly with Tracy in the same way she had just observed, and try to respond to Tracy only when she saw a behavior she liked and wanted to see more of. Mrs. L willingly tried to imitate the therapist and, although the therapist had to interrupt the play session a few times to remind Mrs. L to respond only to behavior she liked and wanted to see, she did get Tracy to comply with her simple requests. More importantly, Mrs. L began to praise Tracy, although in an unnatural, stilted fashion. Also, the therapist encouraged Mrs. L to name the behaviors she liked in the interaction.
>
> After the play session, Mrs. L acknowledged the improvement, but informed the therapist that she didn't like using praise when she didn't "feel" it. The therapist encouraged her to keep trying and indicated that she might feel better about the interaction later when she saw that Tracy could respond positively to her.
>
> The therapist repeated the procedure in Mrs. L's home on two other occasions; Tracy and her mother demonstrated increased positive interaction in each visit. After the third visit, the therapist asked Mrs. L to have a regular time each day

when she could practice improving Tracy's compliance with the shaping procedure.

After two weeks, Mrs. L reported that she was beginning to see things in Tracy's behavior that she genuinely liked for the first time since she was an infant. A clinic observation supported Mrs. L's report, and she not only showed considerably more praise for Tracy's increased compliance, but appeared to be much more natural in her wording and in her manner of praising Tracy. (p. 39)*

Like parent training, marital therapy requires a dual focus on increasing positive interaction and reducing problem behavior. Strategies to increase positive interaction require identification of positive behavior, assignments to increase the frequency of that behavior, and training in the expression of appreciation for its occurrence. As a first step, spouses are asked to list positive behaviors they would like to see their partner perform. The Spouse Observation Checklist (Patterson, 1976a; Weiss & Margolin, 1977), which contains a list of over 200 pleasing events, is often used to assist couples in this exercise. The only positive behaviors that are excluded from consideration are those that are an explicit source of conflict. If a main area of disagreement is how much the husband assists in housecleaning, the wife is not to request this behavior during this phase of therapy.

Once spouses are aware of the pleasing things they can do for one another, various interventions are employed to increase their frequency. Spouses may agree to provide "caring days" for each other by giving the other high levels of rewarding events (Stuart, 1976, 1980). The therapist may encourage the spouses to surprise each other with pleasing events or have them agree to do specific pleasing behaviors. An important focus is ensuring that spouses show appreciation for the pleasing events they do receive from their partners. Training in how to show appreciation is often provided. Throughout these interventions, the spouses often monitor the pleasing events exchanged and rate their daily marital satisfaction to determine the impact of the increase in positive behaviors.

The second major treatment stategy, problem-solving training, is perhaps the cornerstone of behavioral marriage therapy. The interventions to increase nonconflictual pleasing behaviors, while useful by themselves, are often a prelude to problem-solving training. If successful, these first interventions will increase the level of reinforcement in the relationship, reduce subjective levels of dissatisfaction, and prepare the couple for the demanding tasks of solving some of their relationship problems.

In problem-solving training, therapists teach clients to engage in a structured interaction consisting of two phases: problem definition and problem resolution. In the problem definition phase, couples are trained to state

---

*Reprinted by permission of Research Press Company.

problems in brief, specific, nonblaming terms. Partners are taught to express their feelings about the problem, to admit their role in the problem, and to paraphrase the other's view of the problem, even if that view is not consistent with their own. An important notion is that an issue is problematic and open for discussion even if only one partner is upset about it. Even though the other may not see it as a problem, he or she is trained to acknowledge the experience of the spouse and is shown that acknowledgment does not necessarily imply that he or she must change. When couples have completed the problem definition phase, they should summarize with a brief statement of the problem behavior and associated feelings. Jacobson and Margolin (1979) list the following as examples of well-defined problems:

> The problem is that, although we have been talking a lot more lately, we have not been communicating about things that are really important to me. You still can't talk to me about your feelings, and I still get little appreciation from you. I tend to turn off to you when I don't get this, and I feel closest to you when I get it . . . .
>
> When you let a week go by without initiating sex, I feel rejected. (p. 230)

Once an acceptable problem definition has been reached, the couple engages in the process of problem solution. First, they "brainstorm" to generate a large list of possible solutions, withholding judgment about the relative value of the solutions. Then, the couple negotiates and compromises until they reach a specific solution that they can implement. Considerable structuring and support by the therapist may be necessary to achieve an agreement. Once achieved, the agreement is recorded in writing. There is some controversy about the desirability of making the agreement a contingency contract, in which partners specify consequences for maintaining the terms of the agreement. Jacobson and Margolin (1979) argue that one should not use contingency contracts if couples are able to follow through with their agreements without them. Finding effective reinforcers for compliance with negotiated agreements is sometimes difficult; furthermore, partners may attribute change only to the existence of these external reinforcers and thus devalue the change.

An example of a problem definition and problem solution negotiated by a couple is reported in Jacobson and Margolin (1979).

> Problem: We don't spend enough quality time with the kids.
> Solution: 1) After supper while the kids are getting ready for bed (during baths, etc.) we'll clear table, etc.
>
> 2) Then, till the kids go to bed (7:30–8:00) we'll play or work as a group with no interruptions from TV or phone.
>
> 3) On Sunday, we'll seek some kind of church and then spend Sunday as a family day. (pp. 389–390)

At the beginning of problem-solving training, discussions are conducted during the therapy session and structured by the therapist, who sets up rules and coaches the communication. Later, the clients are encouraged to begin problem solving in their homes, but therapists recommend that they structure their own interaction in several ways. Clients should schedule their problem-solving discussions at a particular time and place where there are minimal distractions; often these discussions are best held late at night after children have gone to bed. Despite the temptation, clients should refrain from problem-solving discussions at other times, particularly those times when the problems occur. Discussions at "the scene of the crime," when emotions are high, are often doomed to failure. Clients should keep problem-solving sessions short (about 30 minutes to an hour) and have a preplanned agenda covering only one or two problems. All of these rules are recommended as a way of ensuring the clear discrimination of problem solving from other kinds of interaction. It is assumed that skill training will be more successful if focused on a specific, well-delineated interaction than if focused on comprehensive changes in the couple's natural way of relating. However, problem-solving training is assumed to have generalized effects, since agreements can be negotiated on any topic and thus have impact on any number of selected problems.

Rogerian Approaches to Marital and Child–Parent Problems

The theories of Carl Rogers (1942, 1951, 1959, 1961) have had enormous influence in the area of individual counseling. While Rogerian theory is primarily a theory of personality development and therapy, later writers (T. Gordon, 1970; Guerney, 1977) have taken some of Rogers' ideas and applied them to the treatment of close relationships.

Rogerian theory rests upon one single assumption about human motivation: that human beings have self-actualizing tendencies that can ensure their proper growth and development. Because of these tendencies, individuals avoid danger, satisfy biological drives, and fulfill various higher needs for mastery, social contact, and the like. Individuals learn from experience and can ultimately decide for themselves what is in their best interest. However, these actualizing forces can be blocked during childhood by the evaluations of significant others. Developing children need affection and approval from those close to them, but this positive regard is inevitably conditional upon behavior. Children learn when they are "good" and when they are "bad." These values are incorporated by them as their "conditions of worth." They feel worthwhile only when engaging in certain activities and avoiding others. If these conditions of worth are few and reasonable, children can be open to a variety of experiences and learn from them. If there are many conditions of worth and if these conditions have to do with inevitable human feelings, such as sexual arousal and anger, children may deny or distort large portions of their lives to maintain a sense of worth. This denial and distortion, meant to gain parental and ultimately self-acceptance, actually deprives them of necessary information to guide their behavior. Self-actualization is thus impeded. Consider, for example, children who learn that anger is wrong and cannot accept themselves unless they feel positive feelings toward others. When they are exposed to inevitable anger-producing situations, these children will deny those feelings and inhibit appropriate learning and responding to them.

The purpose of Rogerian therapy is to provide an atmosphere in which clients can achieve self-acceptance and be open to experience. When clients feel free to explore their feelings, rather than deny and distort them, self-actualizing tendencies will lead to growth and development. In a now classic paper, Rogers (1957) set forth the therapeutic conditions that would lead to client self-acceptance and growth: congruence, unconditional positive regard, and empathic understanding by the therapist for the client. Later writers have reformulated and relabeled the essential therapist conditions (Truax & Carkhuff, 1967; Tyler, 1969), but all emphasize such qualities as empathy and acceptance by the therapist leading to such reactions in the client as self-acceptance and exploration of feelings.

The startling aspect of Rogers' proposals is that these relationship conditions are necessary and sufficient for positive change. No special intellectual knowledge is necessary on the part of the therapist. Accurate psychological diagnosis is not essential and can be a waste of time or even destructive. And finally, these necessary and sufficient conditions of therapy are applicable to all or almost all types of clients and problems.

The application of Rogerian theory to the treatment of close relationships required one important assumption: that participants in these relationships could be taught some of the behaviors of both therapist and client. Similar to a therapist, they could provide empathic understanding to the other. Similar to a client, they could engage in an exploration of their feelings with the other.

Consider the approach developed by Guerney (1977), which he calls Relationship Enhancement (RE). According to Guerney, RE is applicable to all close relationships, although most of his attention is devoted to marital relationships. Consistent with his Rogerian emphasis, Guerney largely ignores assessment issues and assumes that the skills that RE teaches are useful for most couples, distressed or nondistressed. Psychotic individuals are, however, seen as inappropriate for RE.

Relationship enhancement is conducted in a group setting. A leader works with several couples together, training them in four sets of interpersonal skills: the expressive mode, the empathic responder mode, mode switching, and the facilitator mode. In the expressor mode, participants learn to state their feelings about their partner's specific behavior, without blaming their partner. For example, participants would be trained to say, "I feel annoyed when you come home late without calling," rather than, "You're so inconsiderate. You never keep your word."

> Expressor: Tonight I worked hard and long getting a good dinner ready and then you arrived a half hour later than I expected you, and everything was cold.
>
> Leader: Would it be accurate if you were to say, "I worked very hard preparing a good dinner tonight and when you came later than I expected you, it annoyed me very much"?
>
> Expressor: Yes.
>
> Leader: Try saying it like that—including that personal feeling in it. (Guerney, 1977, pp. 139–140)

In the empathic responder mode, participants learn to express accurate understanding to their partner. Nonverbal behavior like tone of voice, posture, and facial expression are emphasized. The verbal response taught in the empathic responder mode is essentially a restatement or reflection of the expressor's feeling. If the wife discusses how busy she feels with all her career and household responsibilities, the husband is taught to reflect this feeling (e.g., "You're feeling overwhelmed by all the things you're supposed to do") rather than defend his own work record ("But I always help with . . ."), offer a different interpretation of his wife's behavior ("You're not very organized"), or offer possible solutions ("If you would only . . .").

Participants are also taught how to switch from one mode to the other. Either responder or expressor can request a mode switch, but the switch should be agreed on and the responder should provide an acceptable response to the expressor's last statement before the switch occurs. Expressors are to request a switch when they have expressed their major feelings and suggestions about an issue or when they desire to know the other's views. Responders are to request a switch when they believe they have clarified the other's feelings, when their own feelings are impairing their ability to be empathic, or when they have something to say that might influence the other's perceptions favorably.

> Leader (to expressor): In what you're saying now, and in what you've said in your last comment as well, you're implying that you wonder whether your husband really cares whether or not you really succeed in your job. Is that something that you would like him to reply to now, or would you rather develop your own feelings about this further?
>
> Expressor: I'm a bit afraid to hear what he has to say, but yes, I guess I would like to know.
>
> Leader: Remember, you're free to ask for a reply whenever you want one. Ask him if he'd be willing to take the expressor mode now. (Guerney, 1977, p. 141)

Guerney (1977) assumes that problem resolution will proceed naturally and spontaneously after partners have expressed their feelings and experienced empathic responses from each other. If problem resolution does not follow, the leader may check with the partners to be sure all feelings have been expressed and understood. The leader may suggest that the couple begin to express solutions and may suggest that these same skills can be used to negotiate and compromise.

Despite the clear Rogerian focus of RE, behavioral principles are incorporated as a means of training these interpersonal skills. Reinforcement, modeling, graded practice, and shaping are used systematically by the leader. The final set of skills taught couples, the facilitator mode, involves the same set of skills the leader uses for training. Couples are trained to use social reinforcement and modeling to assist others in the development of RE skills. Over the course of the group therapy, the leader becomes more and more inactive as couples serve as facilitators for each other.

In the area of child–parent relations, the most popular Rogerian approach is probably Parent Effectiveness Training (PET) by T. Gordon (1970). Parent Effectiveness Training teaches parents how to listen to their children's feelings, how to express feelings to their children, and how to use these two skills in conflict resolution. Consistent with his Rogerian approach, Gordon assumes these skills are appropriate for most parent–child relations and thus obviates assessment issues. Like Guerney (1977), Gordon conducts most of the training in groups.

During PET, parents are taught a method of demonstrating acceptance to their child. The method, called active listening, requires that the parent listen to a child's statement and then feed back the child's message in the parent's own words, emphasizing the child's feelings. If the child says, "I don't like Jimmy. He always takes my toys," the parent might say "You're angry at Jimmy for taking your toys." Gordon contrasts active listening with a variety of other more common parental responses, such as probing and interrogating ("What did he take?"), advice ("Why didn't you take some of his toys?"), and reassurance ("Oh, Jimmy will give them back to you."). According to Gordon, the method of active listening is the appropriate response when the child expresses a concern or problem. The response communicates acceptance to the child, encourages the child to explore his or her feelings, and facilitates problem solving by the child. As Gordon (1970) states, "Active listening is a method of influencing children to find their own solutions to their own problems" (p. 66).

A second skill that PET teaches parents is the use of "I-messages." When parents have concerns they wish to communicate to their children, Gordon suggests they state their feelings about the specific behavior or situation, for example, "I feel real tired and don't feel like playing with you now." These messages are contrasted with "you-messages" that may command ("stop bothering me"), blame ("you are a pest"), instruct ("you shouldn't bother me when I'm tired"), or interpret ("you're just trying to get my attention"). Gordon suggests that I-messages are less likely to provoke resistance or rebellion in the child since they do not communicate blame. The I-messages leave the responsibility for behavior change with the child and communicate a parental trust in the child's own ability to correct difficult situations. Furthermore, these messages model more appropriate communication for the child.

Another major skill that PET teaches parents is a method of conflict resolution. Gordon discusses the disadvantage of problem resolution in which either the parent or the child wins and the other gives in. Gordon proposes a process whereby both parent and child engage in a joint attempt to find a solution acceptable to both. This process requires a generous use of both active listening and I-messages so parent and child understand clearly what each wants. The search for solutions can be based both on accurate information about each person's feelings and upon the positive feelings of good will that the use of these two skills create.

Summary

In this section, we have considered approaches to the treatment of distressed relationships that focus on the dyad as the unit of analysis. The problem does not lie within the individual but between the individuals. Their interaction influences each one to behave in ways that further exaccerbate their tensions rather than relieve them. In the social learning view, individuals reinforce each other for increasing levels of coercive behavior. In the Rogerian view, the lack of open expression and mutual understanding inhibits the growth of each individual and the relationship. T. Gordon (1970) has expressed this dyadic view of relationship problems succinctly:

> These conflicts between needs of the parent and needs of the child are not only *inevitable in every family* but are bound to *occur frequently*. They run all the way from rather unimportant differences to critical fights. They are problems in the *relationship*—not owned solely by the child nor solely by the parent. Both parent and child are involved in the problem—the needs of both are at stake. So *the relationship owns the problem*. (p. 149, emphasis in original)

Intervention from a dyadic perspective naturally involves both members of the relationship. As was indicated above, behavioral and Rogerian approaches to therapy are conducted conjointly. The sole exception concerns parent–child relations. Because of the child's obvious social and intellectual limitations, therapy for parent–child problems often becomes parent training. However, involvement of the child in therapy is often considered beneficial (Patterson, Reid, Jones, & Conger, 1975), dependent on the child's age. Despite the frequent exclusion of the child, the focus is on alteration of the child–parent relationship by altering the parent's behavior.

FAMILY SYSTEMS APPROACHES

> The therapist . . . regards the identified patient merely as the family member who is expressing, in the most visible way, a problem affecting the entire system. This does not mean that the identified patient is irrelevant to therapy. He will need special attention. But the whole family must be the target of therapeutic interventions. (Minuchin, 1974, pp. 129–130)

Family therapists assume that the causal conditions for psychological problems lie not in the individual or in isolated dyads but in the social environment. As their name suggests, these therapists focus on the family system as the most important social environment for most individuals. They use the term "family" loosely, referring to members of the nuclear family, important members of the extended family (e.g., a live-in grandmother), and even important nonfamily members (e.g., a live-in baby sitter). Family therapists focus on the relationships among family members as both the breeding ground for psychological problems and the terrain on which any healing must take place. The common use of the term "identified patient" reflects their limited concession to what is often the family's individual conceptualization of the problem.

Although the family system may be the most important aspect of the social environment, it is one limited part of the individual's social life. Friends, neighbors, and fellow workers undoubtedly have their effects on close relationships. However, only one systematic attempt has been made to go beyond the family to the larger social environment. In social network therapy (Speck, 1967; Speck & Attneave, 1973), important members of the social environment, as well as family members, are included in therapy. This approach is conceptually interesting, but presents practical problems of considerable magnitude.

Consideration of family systems therapies will allow us to integrate the two repeated themes of this chapter: child–parent problems and marital problems. Family therapists examine both and believe that an important relationship often exists between the two. Parents have difficulty with their children in part because of their spousal conflicts. And spouses have difficulty with each other in part because of the way they deal with their children. Framo's (1975) claim that "whenever you have a disturbed child, you have a disturbed marriage" expresses this view in extreme form.

In the analysis of any living system, one can distinguish between the structure of the system and the process or functioning of the system. In the former, one examines the static features of the system, such as the organization of its parts. In the latter, one examines the dynamic interplay between parts, such as feedback and homeostatic mechanisms. In this section, we first discuss Minuchin's (1974) structural family theory as an example of systems thinking directed at family structure. Then we discuss communication approaches as an example of systems thinking directed at family process and functioning.

Minuchin's Structural Family Theory

Dysfunction in close relationships

Minuchin implicates family structure as the primary determinant of family interaction. By structure, Minuchin means the hierarchical organization and differentiation of the family into subunits. The most common type of family organization, for example, is along generational lines, with parents in hierarchical relation to their children. Minuchin uses the structural concept of boundary to define the types of organization and differentiation. Boundaries "are the rules defining who participates and how" (Minuchin, 1974, p. 53). They apply to each individual and each combination of individuals in the family as well as to the family as a whole. Boundaries exist on a continuum from rigid, in which family members have little significant interaction and are disengaged from one another, to diffuse, in which family members are enmeshed in each other's lives and have little autonomy. Between these extremes are clear boundaries, in which families have considerable room for interaction with one another but in which clear rules for this interaction exist.

Minuchin explains dysfunctional behavior in terms of boundaries that exist within the family. If families have rigid boundaries that lead to disengagement or diffuse boundaries that lead to enmeshment, problems are liable to result.

> Members of enmeshed subsystems or families may be handicapped in that the heightened sense of belonging requires a major yielding of autonomy. The lack of subsystem differentiation discourages autonomous exploration and mastery of problems. In children particularly, cognitive–affective skills are thereby inhibited. Members of disengaged subsystems or families may function autonomously but have a skewed sense of independence and lack feelings of loyalty and belonging and the capacity for interdependence and for requesting support when needed. (Minuchin, 1974, p. 55)

Both diffuse and rigid boundaries may be present simultaneously in the family. If a mother has an overly attached relationship with a son that excludes the father, Minuchin would describe the family structure as one in which mother and son are separated by a diffuse boundary, while the father and mother–son dyad have a rigid boundary between them. The mother is enmeshed with the son, while the father is disengaged from both parental and spousal functions. In this situation, both the marital dyad and the parent–child dyad could exhibit various symptoms. Minuchin would argue that one cannot understand these symptoms by examining either the individuals or the particular dyad. The crucial causal conditions lie in the structure of the family of which the individuals and dyads are a part.

A clear example of family dysfunction that results from boundary problems is provided by Aponte and Hoffman (1973) in their case study of the R family, which had an anorectic child. The R's consisted of a mother, father, and three children: Laura, age 14; Jill, age 12; and Steven, age 10. The family sought help because of Laura's excessive dieting and refusal to eat the amount and kind of foods her parent prescribed. At the beginning of therapy she weighed 62 pounds.

Initially, the family defined their only problem as revolving around Laura—her weight loss and apparent unhappiness. When asked about disagreements, the family insisted that there were none. In fact, the family defined itself as very close, enjoying much contact between members. As an example of their closeness, the family described their policy of open doors throughout the house. Parents and children closed their bedroom doors only while changing clothes; otherwise doors were kept open. Another example of closeness concerned evening activity. Parents and children would sit in the parent's bedroom watching television. Mr. R and the children massaged and cuddled on the bed during these evening periods while Mrs. R did needlepoint or attended to various chores in the house. Recently, Laura had shown less inclination to engage in this contact with Mr. R, but he continued it with the other two children.

After some probing of the family about their typical interactions, the structure became apparent. The father was enmeshed in the children's lives, while the mother was disengaged from the father. The father in particular had an intrusive relationship with the children. While his contact with them was not overtly sexual, the closeness he had with them was at the expense of closeness with his wife. The mother was left out but did not acknowledge her feelings. Laura was rebelling against this closeness and intrusiveness by refusing to eat and by withdrawing from the evening sessions in front of the television. "Not eating has been a way of trying to keep some kind of territory for herself; some area she can control" (p. 32).

Most of the therapy focused on the husband–wife relationship. The therapist supported the wife's feeble efforts to express dissatisfaction with the marital relationship and attempted to extricate the father from his over-involvement with the children.

> As the conflict between the couple came to the fore, the involvement between Laura and her father got less. The father and mother began to have intense

> quarrels, and the father went to the hospital for two anxiety attacks, which seemed to be caused by a sense of isolation from both his children and his wife. In therapy sessions, he began complaining that he had too much responsibility and his wife was not carrying her share of the load. As a result, the wife became more active. Once the couple had established a better partnership, Barragan (the therapist) gave them the task of making their daughter gain ten pounds in three weeks. The father, now that he had the support of his wife, was able to make the girl eat. For the first two weeks, they didn't get anywhere. But in the third week, Laura gained eight pounds and two weeks later had gained twenty-four. The family had not sought futher help, but a follow-up six months later showed that Laura was not only in good physical shape but had a boyfriend. (p. 44)*

This particular case study is an apt illustration of Minuchin's structural views in that the family's rule about open doors is a metaphor for their enmeshed family structure. The case study concludes with the comments that Laura "began to lead an independent existence. . . . She could finally shut her door" (p. 44).

Minuchin's structural family therapy

Minuchin's structural approach requires therapists to be active participants in their client families. They must become accepted members of the family in order to provide leadership and direction. They educate, give assignments, challenge, and support the family. They form shifting coalitions, allying with first one individual or subsystem, then another. They are active group leaders.

Minuchin divides therapeutic activity into two kinds of operations: joining–accommodation and restructuring. The therapist must first "join" the family, that is, relate directly to the family in such a way as to establish a position of leadership in the family. Joining operations require that the therapist initially confirm and support the existing family structure. The therapist may, for example, respect temporarily the family's definition of the problem and their choice of spokesperson about the problem. The therapist may use many of the time-honored means of gaining rapport with clients—reflections of feelings, requests for elaboration, and various verbal and nonverbal indications of interest. The therapist may accommodate his or her style to the family's style. If the family has a slow, depressed interactional tone or if the family has an energetic tempo of conversation, the therapist may adjust his or her tempo correspondingly. The therapist may also use self-disclosure to create a sense of unity with the family.

During the process of joining–accommodation, the therapist observes the family interaction and forms a diagnosis of the family. The diagnosis is a

*Reprinted by permission of Family Process, Inc.

characterization of the family structure. Minuchin has a series of symbols to represent the hierarchical arrangement, the type of boundaries, and the location of conflict and coalition in families. A map of the family structure, using these symbols, is the best representation of a structural family diagnosis.

Restructuring operations are the active agents of family change. Minuchin discusses a host of these interventions. For example, one kind of intervention involves the manipulation of space in the therapy sessions. The structural therapist pays close attention to how the family members seat themselves, since such seating arrangements are often used as clues to patterns of coalition and isolation in the family. During an intervention, the therapist may alter these seating patterns, asking family members who rarely interact with one another to face each other and discuss some problem without interference by others. The therapist may include only certain participants in a session as a way of supporting one subsystem and weakening another.

> The therapist can join with an isolated father and his children, excluding the mother, to allow new functions to develop. A dominating parent can be sent behind a oneway mirror, so that he can participate without being able to control. Young children can be brought into a session to provide a lessening of conflict. The therapist always works with his map of the total family in mind. Even when he is working intensively with a subgroup, his goal is the total restructuring of the family. (Minuchin, 1974, p. 147)

Another type of intervention involves the therapist's active alliance with family subsystems. The therapist may ally with a weak member of the family, to the point of criticizing others for their actions against that weak member. The therapist may praise a subsystem in the family as a means of strengthening that subsystem. The therapist may side with one member against another and prohibit other family members from participation so that a direct conflict, without interference from others, may occur. The following extract provides an example of these interventions and an important caution about them.

> For example, in one family, conflict in the spouse subsystem is avoided by scapegoating a son. When the wife challenges her husband for not making more money, he directs his attention to the boy, correcting his behavior and thereby re-establishing his own threatened sense of competence. When the husband challenges the wife for being a sloppy housekeeper, she begins to talk about the child's misbehavior in school. The therapist joins the husband in a coalition against the wife, strongly supporting his demand for more order in the house. This technique results immediately in the emergence of spouse conflict which, with the therapist' help, is negotiated within the spouse subsystem, freeing the boy.
>
> When the therapist joins one family member, he must be acutely aware of his ally's threshold of endurance and of the other family members' threshold of endurance. He runs a strong risk of alienating the target of the coalition and,

> frequently, the whole family. Even in the midst of an attack, it is important to convey some support to the target. If the therapist is working with a cotherapist, the cotherapist must support the target of the coalition. If the therapist is working alone, he must convey the sense that he recognizes the target's side to the story. For example, even as the therapist is helping the husband attack his wife for slovenliness, he interprets the wife's behavior as a resistance to arbitrary control. (Minuchin, 1974, p. 149)

Another set of interventions involve task assignments. The therapist may ask family members to enact certain behaviors in the session or at home. To increase the father's executive functions, the therapist may insist that only the father bring up child problems in the session. To increase the boundaries around the marital dyad, the therapist may request that the couple spend more time together without the children. In the case of the R family, discussed above, the therapist made assignments about keeping doors closed and knocking before entering. In making task assignments, as in other interventions, the therapist's goal is to alter the structure of the family.

Family Systems Communication Therapy

Systems theories that focus on family process and functioning rather than on family structure include the strategic therapy of Erickson and Haley (Haley, 1963, 1973, 1976) and a variety of communication therapies (e.g., Jackson & Weakland, 1961; Sluzki, 1978; Watzlawick, 1976; Weakland, 1976). Most of these theorists had some contact with the Mental Research Institute of Palo Alto and contributed intellectually to the systems thinking that emerged there in the 1960s and 1970s; all of the theorists focus on human communication as the key to understanding family behavior. Differences exist between various theorists, but for our short exposition here we will consider them together.

These theorists see all interpersonal behavior as communication, as attempts by partners to define the relationship and to influence the other. It is obvious that communication is not limited to the verbal message given but includes the tone and posture that serve to qualify that message. However, these theorists also include so-called involuntary behavior as interpersonal communication. The husband's depression, the wife's "uncontrollable temper," and the child's crying spells are seen as interpersonal messages.

Another important tenet of this approach is that interpersonal behavior is maintained by current communication processes. While interpersonal behavior may have its origins in past experiences, its persistence is due to contemporary interaction. Thus, the communication approaches view psychological symptoms as interpersonal behavior maintained by current interactional processes. The husband's depression may be an attempt to assert his influence and counterbalance the wife's domineering. The depression continues because of these power dynamics between them. A child's deviance may be in response to a pattern of parental attention to another sibling. The pattern may continue because of the parent's continued positive response to the sibling and negative attention to the child.

Assessment in the communication approaches requires the therapist to determine the specific presenting complaints (symptoms) and the current interactions maintaining these complaints. Through the clinical interview, the therapist inquires about the family's problems and their attempts to solve them. The therapist may structure a direct conversation between family members about a particular problem to observe these attempted solutions. This information is crucial because a family's attempts to deal with a problem are often a major factor in maintaining it. The wife's attempts to convince the husband of the irrationality of his depression or the parent's attempts to lecture the child on his deviance may perpetuate rather than ease the troubling symptoms.

Communication theorists endorse the homeostasis principle of system functioning—the notion that a system will resist efforts to change and will return to its steady state after change. The therapist, therefore, should not offer a system's interpretation of the problem or direct appropriate behavior change. Such efforts would be met with resistance. Were the therapist to suggest that the husband's depression is a controllable response to the wife's domineering and that they should openly discuss power issues in the relationship, the couple might well protest and defend their initial view of the problem more strongly. Were the therapist to suggest to the teenager that his outbursts are attempts to control his parents' behavior and that he should engage in more positive influence attempts, the teenager would be conspicuously unresponsive.

The approach to change must therefore be indirect. The therapist must act paradoxically, bringing about major changes without appearing to do so. The treatment must undermine the family's own view of the problem and create changes in the regular patterns of their responses to one another while seeming to support the current equilibrium. Then changes will take place surprisingly, with little resistance.

This view of family treatment requires a unique client-therapist relationship. As in Minuchin's approach, therapists take an active leadership role in the family. They educate, challenge, and direct, but in a more wily and wizardly fashion. They disclose little of their elaborate strategy for change to the family. Often, they surprise the family with unexpected interpretations and unusual assignments. They are active but crafty leaders.

Two major types of intervention are symptom prescription and relabeling. In symptom prescription, the therapist asks the family members to continue having the symptoms they complain about. Various rationales may be offered for this prescription, such as the need to gather information on exactly how bad the symptom is. Although symptom prescription requires the family to continue their problem behavior, the therapist may request certain "minor" changes in the symptom. The therapist may request that the symptoms be particularly severe during the following week. The therapist may request the symptom-afflicted member to have the problem only in certain locations or to move to those locations when the symptom occurs. The therapist may request that partners temporarily deceive each other about a symptom, telling the other that it is bad when it is not and that it is okay when it is severe.

Symptom prescriptions are seen as a "no lose" treatment strategy for the therapist. If the family resists the prescription and has fewer symptoms, then they are automatically getting better. If the family has the symptoms, a pattern of compliance to the therapist has been established that may be useful in further therapeutic directives. In either case, the notion of client control over symptoms has been promoted. If the family follows the therapist directive to have the symptom but alter it in the ways suggested, further evidence of the symptom's controllability is established, plus the pattern of behavior around the symptom can be broken.

> Through prescribing a symptom to a symptomatic member in the presence of the mate, the therapist aims at shattering the very pattern that perpetuates the symptom. That effect occurs through two mechanisms: (a) When the patient is told "fake the symptom, and fake it well," the other member of the dyad is implicitly being told that the symptomatic behavior to which he/she may be exposed to *may* be false, therefore inhibiting "spontaneous" responses that in turn may reinforce and perpetuate the symptom; and (b) it subtly increases the consensus about the patient's control over the symptom, and decreases the chances of his/her claiming spontaneity. At the same time, it evokes its counterpart, that if the subject can *produce* a symptom through a prescription, he/she may also be able to *reduce* it. One of the key interactional attributes of symptoms, namely, the fact that they are considered spontaneous by the participants, is drastically questioned by a well-placed symptom description. (Sluzki, 1978, p. 382)

Several examples of the successful use of symptom prescription are reported in the literature. Hare-Mustin (1975) worked with the family of a 4-year-old boy who threw frequent, spectacular temper tantrums. The therapist helped the family decide on special locations and times for the tantrums, giving the child some "control" over these decisions. The tantrums dropped dramatically, despite the therapist's expressed concern over the rapid change and encouragement of the child to continue having the tantrums at the appointed times and places. Jacobson and Margolin (1979) report the successful use of symptom prescription with a couple who had frequent fights between therapy sessions. These fights were very destructive, often including physical violence, and threatened to destroy the relationship before therapy could reverse the pattern. The therapist expressed a great need to understand the fights since they seemed so destructive and yet so different from the couple's polite and subdued behavior in the session. The therapist requested that the couple fight in the session while the therapist observed from behind a one-way mirror. The couple was unable to generate a fight. Therefore, the therapist gave the couple a portable cassette recorder to take home with the instruction that they were to turn on the tape recorder whenever they began a fight.

> The therapist convinced the couple to comply with the plan. When they returned the following week reporting that, for the first time in months, they had not engaged in one argument, the therapist appeared bewildered and his response was punitive. He insisted that they fight at least once the following week; otherwise, progress would be retarded. Alas, the fighting was over between this couple, and therapy had to proceed without this invaluable assessment information. Therapy proceeded smoothly and successfully subsequent to the ceasefire induced by the instructions to fight "on the air." (Jacobson and Margolin, 1979, p. 153)

A second therapeutic strategy is *relabeling* (Haley, 1963; Soper and L'Abate, 1977) or reframing (Watzlawick, 1976; Soper & L'Abate, 1977). The therapist offers a different interpretation of the family's behavior in order to facilitate change in that behavior. Relabeling or reframing is often used to redefine the role relations among family members, highlighting the positive aspects of one member's behavior that had been seen as negative and highlighting the negative aspects of another member's behavior that had been seen as positive. These reinterpretations can also be used to remove the pressure on one family member by focusing on another. The therapist, for example, may express more concern about a sibling than about the child labeled as problematic.

Often, relabeling and reframing are used together with symptom prescription. The therapist redefines the problem in order to make a symptom prescription more understandable, as in the following example.

> The present author recently saw a couple who were locked in a relationship in which the wife's nagging, complaining negativism contrasted sharply with the husband's appeasing optimism. Each of them was expressing mounting dissatisfaction and resentment, and both concurred in defining her as the victimizer and him as the victim. I praised her for having decided to take the heavy load of the villain . . . , though to help to make him look good while she remained the "mean one" must have had some kicks for her. But, I continued, there are no kicks like the kicks of being the "good guy," and he was selfishly keeping most of the assets to himself. They looked quite surprised, but went along with the reasoning and the metaphors. I prescribed that he should respond to any statement of hers by being as pessimistic as possible—regardless of how really pessimistic he might be at that moment, without leaking to her the true nature of his feelings. I prescribed her to remain as pessimistic as before, "in order not to make the task more difficult for

> him". . . To analyze these prescriptions: (1) The presciption acknowledged that he *could* be genuinely pessimistic; (2) her habitual behavior was being relabeled as a gesture of good will; (3) if they followed instructions, they would quite quickly reach a deadlock that would force her to extricate the positive side of the matter; (4) nobody—but the therapist, perhaps—would be blamed for any escalation into pessimism between them, totally eliminating any label of victim and victimizer; and (5) they were being granted a humorous way out of confrontations, used by them thereafter on occasions. In fact, both expressed a remarkable reduction of the conflicts and a general improvement in the relationship from there on, while I correspondingly, expressed concern about their unexplainable and too dramatic change. . . . (Sluzki, 1978, pp. 377–378)*

Summary

In this section, we have examined two therapies that go beyond the individual and the dyad and focus on the family system. Minuchin's approach emphasizes the organization of family systems as the crucial determinant of pathology. Communication approaches emphasize feedback and control mechanisms as the causal agents for family symptoms. The critical concern with the social environment can be seen in Sluzki's (1978) advice to therapists having difficulty with a marital case: "If even then you find yourself unable to detect meaningful regularities and/or to produce change, *then* increase the number of participants in the session (introduce offspring or parents or co-therapist)" (p. 391).

RELATIONS AMONG THERAPY APPROACHES

The differences among the preceding clinical theories are striking. Variation is apparent on every crucial aspect of treatment: the nature of dysfunction, the client–therapist relationship, assessment, and intervention. Nevertheless, important similarities exist among all or almost all of these approaches. Two similarities are especially relevant for a discussion of close relationships. Despite their different foci, each approach promotes individual autonomy by clarifying the needs and desires of the client. A sorting through of expectations for self and others is undertaken, directly or indirectly. In the conjoint approaches, this autonomy is further enhanced by training in direct communication. Family members are often asked to speak for themselves rather than for each other. If a mother talks about what "we think" and what "we feel" in reference to her daughter, the therapist may direct her to

---

*Reprinted by permission of Brunner/Mazel, Inc.

rephrase her comments as "I think" and "I feel" as a means of differentiating her from her daughter. If family members make impersonal statements about personal issues, the therapist may paraphrase the statement in personal terms or request the family member to talk in personal terms. The husband's statement that "women don't need to work when their husbands are making good money" may be met with a therapist response that "you don't want Mary to work since you feel you are making sufficient money."

Another important similarity is a reinterpretation of client problems emphasizing the client's part in these problems. Despite enormous differences in their construal of events, therapists from different approaches move the client away from blame and accusation toward individual responsibility. Psychoanalytic approaches make the client aware of projections onto others. Behavioral approaches demonstrate to clients their unwitting reinforcement of aversive behavior in others. Rogerian approaches force partners to hear how their behavior leads to negative reactions in the other. Family communications theorists offer deliberate, strategic reinterpretations to clients to alter their current construals of victims and villains.

Until now, we have discussed relatively pure models of clinical treatment of close relationships. As might be expected, several theorists have attempted a *conceptual* integration of one or more of the viewpoints discussed. The influential work of Ackerman (1956, 1958, 1966) and Bowen (1965, 1966, 1976, 1978) merged psychoanalytic and systems approaches to family therapy. Recently, behaviorists have attempted an integration of social learning concepts with systems approaches (Barton & Alexander, 1981; Birchler & Spinks, 1980; R. L. Weiss, 1980). Though not explicitly addressed to close relationships, Wachtel (1977b) attempted to combine behavioral and psychodynamic viewpoints. Gurman (1980) recently suggested that the major challenge for the field of marital and family therapy in the 1980s is a rapprochement and integration among the competing methods and theories.

More common than a theoretical integration is a *clinical* integration of various approaches. Strict adherence to the models discussed above may be more the exception than the rule. While being predominantly oriented toward one or another model, many clinicians attempt other kinds of interventions as the exigencies of a particular case or their own predilections suggest. The behaviorist may provide interpretative commentary of individual dysfunction; the structural therapist may engage in some operant reinforcement strategies.

Perhaps the most common form of treatment of close relationships is an eclectic approach within a psychodynamic framework. The majority of therapists who treat marital and family problems have been influenced by psychodynamic formulations (Group for the Advancement of Psychiatry, 1970) and accept the notion of unconscious conflicts and motivation as a major determinant of human behavior. However, in their actual therapy sessions, they borrow unabashedly from other approaches. In case formulations, they entertain other than intrapsychic concepts of human functioning.

Psychodynamically oriented therapists move freely between the various levels of analysis discussed in this chapter: the individual, the dyad, and the social environment. For example, Ables and Brandsma (1977) argue that "a couple's behavior can be understood both from an interactional and intrapsychic point of view" (p. 340). The therapist must "be able to recognize present behaviors that are rooted in unconscious assumptions, wishes and fears that will not yield to rational persuasion or 'reasonable' solutions" (p. 31). But the therapist must also "recognize that couples influence each other both in eliciting and in maintaining behaviors. An interlocking interaction often exists because it serves certain important functions, healthy or otherwise, for each individual, thus contributing to a stable equilibrium" (p. 340).

Psychodynamically oriented therapists are also eclectic in the techniques they employ. Ables and Brandsma (1977), for example, use contracting and negotiation for couples, train the Rogerian skills of listening and expression, and employ a collage of other techniques besides the psychoanalytic strategy of interpretation. The purpose of these techniques and their results are often interpreted psychodynamically. For example, Gurman (1980) describes a series of active negotiations between a couple to satisfy the wife's desires for more intimacy with the husband. The wife's failure to meet the negotiated agreements led to an exploration of her fear of intimacy, which was rooted in a childhood rejection by her father. Gurman suggests that the behavioral negotiation approach assisted in the rapid discovery of the psychodynamic problem.

Any particular therapeutic approach or integration of approaches must be evaluated however, not by the extent of its usage, but by its demonstrated effectiveness in altering dysfunctional relationships. Only empirical research on the outcome of therapy can provide such an evaluation.

OUTCOME RESEARCH ON INTERVENTION IN CLOSE RELATIONSHIPS

The research on the outcome of marital and family therapy is extensive and has generated numerous reviews (Beck, 1975; Gurman & Kniskern, 1978; Jacobson & Martin, 1976; Wells & Dezen, 1978). The quantity, if not always the quality, of these studies is impressive. In perhaps the most comprehensive review of the literature, Gurman and Kniskern (1978) cited over 200 relevant studies that examined such variables as type of treatment, length of therapy, and use of cotherapists.

The data are quite supportive of the overall efficacy of marital and family therapy. For example, a number of studies have compared marital or family interventions with control conditions in an attempt to determine if treatment has any effect at all. Gurman and Kniskern (1978) reviewed these studies under the classifications of marital versus family intervention and behavioral versus nonbehavioral methods. The distinction between marital versus family intervention depended on whether a child was a focus of treatment; the distinction between behavioral versus nonbehavioral methods depended on whether behavioral–social learning approaches or other methods (e.g., psychodynamic and systems approaches) were employed. In the marital area, Gurman and Kniskern classified 10 out of 15 studies as demonstrating the superiority of nonbehavioral methods over control conditions, while 7 out of 11 studies demonstrated the superiority of behavioral methods over control conditions. In the family area, they classified 8 out of 13 studies as demonstrating the superiority of nonbehavioral methods over control conditions, while 8 out of 8 studies demonstrated the superiority of behavioral methods over control conditions. These studies are heterogeneous, employing different populations, procedures, therapists, and dependent measures. The quality of the studies also varies considerably. Therefore, one should pay little attention to the slight differences in "hit rate" between the various classifications. Taken as a whole, however, these data provide marked support for the efficacy of marital and family intervention.

Outcome studies have also compared the relative effectiveness of different kinds of treatment. Many of these investigations have compared marital or family therapy with some form of individual treatment. In these comparisons, the preponderance of evidence suggests that marital or family therapy is superior to individual therapy when the problems concern family living (Gurman & Kniskern, 1978).

A smaller number of studies have examined the relative efficacy of different theoretical approaches to marital or family therapy. Not enough research has been done to indicate which type of marital or family therapy is superior. However, an examination of one exemplary study illustrates the attempt and points to some problems.

Alexander and his associates at the University of Utah (J. Alexander & Parsons, 1973; N. C. Klein, Alexander, & Parsons, 1977; B. V. Parsons & Alexander, 1973) conducted one of the most impressive comparisons of family therapies. In their study, 99 families with children referred by juvenile court for "soft" delinquency offenses were randomly assigned to a no-treatment control condition or to one of three therapy conditions: a behaviorally oriented family systems approach, a client-centered (Rogerian) family approach, and an eclectic–dynamic family therapy approach. Outcome data, available on 86 families, included family interaction measures at posttreatment assessment, recidivism rates for the target child at 6–18 months follow-up, and court referrals at 2½–3½ years after the end of treatment for siblings of the target child. All measures indicated the superiority of the behavioral systems approach over both control and alternate treatment conditions. The recidivism measure was further bolstered by comparison with county base rates, which roughly equaled the no-treatment control group but greatly exceeded the behavioral group. Perhaps most striking were the positive effects of the behavioral group on court referrals for the siblings of the referred child. These findings indicate the system-wide nature of the changes induced by therapy.

In commentary on this research, Gurman and Kniskern (1978) argue that the study should not be viewed as a test of behavioral family therapy since the investigators did not use a strict behavioral approach but incorporated strategies used by nonbehavioral therapists as well. Alexander and his associates, like many therapists, are somewhat eclectic and do combine strategies. While useful clinically, these combinations prevent careful tests of theoretically derived treatment.

This commentary suggests an important problem in outcome research—the specification of the independent variable. Although laboratory research usually involves the comparison of carefully defined conditions, which can be easily described in methods sections, the conditions in outcome research refer to months of interpersonal contact between therapist and client. Besides all the problems of experimental design and measurement of change that plague outcome research, the investigator is faced with the elementary question of what is actually being compared and how it relates to some theory of intervention.

Because of these problems, some writers (e.g., Bergin, 1971; Gottman & Markman, 1978) have suggested that researchers not ask which therapies work and which ones work best. The question needs to be focused on particular problems, procedures, and therapist–client relationships. Paul's (1969) well-known question is apt: "What treatment, by whom, is most effective for this individual, with that specific problem, under which set of circumstances, and how does it come about?" (p. 62).

CONCLUSIONS AND IMPLICATIONS

In this chapter, we have used the concepts from Chapter 2 to analyze four important aspects of the treatment of close relationships: dysfunction, the client–therapist relationship, assessment, and intervention. We have also used our conceptual framework to organize some of the major theoretical conceptions about treatment. Our discussion of theories was necessarily brief and illustrative; the purpose was to demonstrate the analytic potential of our framework rather than to comprehensively review existing theory. At this point, we consider the broader implications of our framework for clinical work with distressed relationships.

Our conceptual framework suggests that the most basic unit of analysis of close relationships is the event—any specific change in P or O. P's thought that O may not love him any more, O's angry words, P's feelings of anticipation—all of these "happenings" are the basic material of a relationship. Clinicians must involve themselves in these details. Despite the temptation to remain at a broad level of analysis, a failure to focus on events will often lead to a misunderstanding of the relationship and difficulty in changing it. Clinicians who rely on clients' general descriptions of their relationships rather than seeking the specific events that led to those descriptions may well hear highly idiosyncratic and distorted accounts. Clinicians who intervene without attention to relationship events are prevented from knowing if their efforts achieved changes in the daily behavior, thoughts, and feelings of their clients.

Relationship events are important for understanding the client–therapist relationship, assessment, and intervention. Therapists must decide what events will develop facilitative relationships with their clients. During assessment, therapists and clients must determine what are the important events in the clients' close relationships that are troublesome and need changing. During intervention, therapists must determine what events will lead to growth and development. It is easy to remain at a broad level in discussions of clinical work, to talk about self-awareness, growth, personality integration, and social skill learning. What is most difficult is to describe the kinds of events that will achieve these lofty goals.

Our framework highlights several kinds of events—affect, cognition, and overt behavior—that occur concurrently and sequentially in the behavior chains of both P and O. Clinical theories have typically focused more strongly on one or another of these types of activity. For example, Rogerian approaches focus on affect, Rational Emotive Therapy emphasizes cognition, and behavioral approaches focus on overt behavior. An important question for close relationship research concerns the patterning and relative importance of these various intrachain events. Do thought and affect follow behavior, as the behaviorist has suggested, or do beliefs and expectations determine affect and behavior, as Rational Emotive Therapists have suggested? Another important question concerns the connections between intrachain and interchain events. Are P's complaints and criticism of O primarily a function of P's excessively high expectations about marriage or is their major function to control O's behavior and keep the focus off P's own inadequacies?

Despite its importance, a focus on events carries with it certain dangers. Clinician and client both can become lost in detail. Perhaps every therapist has experienced the unproductive situation in which therapy becomes a forum for endless stories of the client's functioning. Even the most simple of relationships contains a multitude of complex events. Comprehensive assessment of these events can mire therapy in an obsessive–compulsive search. What must always be remembered is that the purpose of an examination of events is to discover the properties of the relationship—the themes or patterns that characterize it. Events are examined to discern the regularities that constitute distress in the relationship. P's accusations and O's resentful reassurances, P's demand for more time and attention and O's withdrawal and avoidance, P and O's retaliatory aggressiveness—these are the kinds of properties that the clinician looks for amidst the multitude of events. These are the kinds of properties that must change if therapy is to be successful.

The examples so far have been of negative events and properties. Their positive counterparts must also be examined in assessment and strengthened in intervention. The couple's playful style of physical affection, the recreational time they spend together, their cooperative efforts in response to external stress—these positive characteristics of the relationship must be addressed in treatment, if only to acknowledge their existence and importance to the relationship.

Upon examination of properties, causal conditions must be hypothesized and efforts made to change them. Our framework highlights the diverse array of causal conditions that can affect close relationships. From a clinical perspective, the framework suggests that a variety of things can go wrong in a close relationship. Causal conditions in P, O, their interaction, their social environment, or their physical environment can bring about distress. In one relationship, the paranoid suspicions of a spouse may be a crucial disruptor. In another, the appearance of a third party may alter the relationship irreversibly. Perhaps more commonly, a number of factors act together in complex ways to create problems in relationships.

With such heterogeneity in the causal factors creating dysfunction, a flexible array of treatment strategies is necessary. One can hardly endorse our analytic framework without also endorsing a broad eclecticism in clinical work. Therapists who focus exclusively on individual, dyadic, or family conditions limit their potential impact. For example, the practice of seeing only individuals or the practice of always seeing couples together—both of which are common stances in therapy—are contraindicated by our framework. From this framework, it is not surprising that the field of marital and family therapy is opening up, is experimenting with a variety of treatment approaches, and is moving toward diversity rather than a gradual focus on the single best treatment.

A. A. Lazarus (1971) distinguished between theoretical eclecticism and technical eclecticism. In the first case, theoretical ideas are mixed in unsystematic ways; in the second case, techniques are used based on empirical evidence, independent of theoretical grounding. Our framework suggests a third kind of eclecticism, based on a broad conception of the phenomena creating dysfunction. Such a conception would not be inconsistent with technical eclecticism but would imply that different theories may be necessary for different aspects of the phenomena of close relationships.

Although our conception of close relationships (Chapter 2) covers a variety of causal conditions relevant to close relationships, P × O causal conditions have been highlighted. For the lay person, factors in P *or* O are likely to be most salient. In their discussion and analysis of relationship dissatisfaction, family members are liable to attribute their distress to characteristics of one person. Clinicians have also centered many of their causal analyses on P or O. As we saw, classical Freudian theory puts its emphasis on aspects of the individual partners. In contrast, our conception highlights factors involving the mesh between P and O and factors emerging from P's and O's interactions as being crucial to an understanding of relationships. The agreements that develop over time, the coordinated habits, and the unique mesh of characteristics are factors that, although profound and enduring, may not easily capture the participants' or the scientists' attentions.

It is interesting that the outcome data on marital and family therapy versus individual therapy come out in favor of the approaches that include all participants in the relationship (Gurman & Kniskern, 1978). We can speculate that attention to P × O causal conditions, which is more easily possible when all members of the relationship are seen, may account for the superiority of the marital and family approaches. Of course, we cannot exclude other reasons, such as the greater number of causal conditions available to the marital or family therapist than to the individual therapist. Furthermore, a more direct assessment of these causal conditions is possible. When therapists see only a single member of the relationship, they are at "arms-length" from many of the relevant causal conditions. This distance can lead to some inaccuracies in the assessment process as well as limitations in intervention. As noted in Chapter 11, "Research Methods," discrepancies between partners in their accounts of important activities are common in marriages and families.

From the heterogeneous array of potential causal conditions, the clinician needs to consider not only those that account for the most variance in relationship distress but also those that are most susceptible to change. Clinicians must focus on causal conditions that are both important and modifiable. The husband's physical disability, no matter how instrumental in the marital discord, may be unchangeable. The wife's paranoia may be a crucial disruptor of the family, but she refuses to participate in treatment. Poverty may be a consistent stressor for the family but essentially unmodifiable during the course of therapy. Therefore, the therapist in each of these cases must focus where change is possible.

Therapists' intervention in causal conditions takes place through events. Through simple conversation, therapists alter their clients' immediate behavior with the hopes of changing more general factors in their lives. Though many clients expect immediate and dramatic change, the experience of therapists and the view of most clinical theories is that change in causal conditions is gradual and cumulative rather than sudden and disjunctive. The "insight" that is the goal of psychoanalytic therapy is achieved by much "working through"; the change in interpersonal competence that is the goal of much behavior therapy takes place through "shaping"; the change in family structure that is the goal of some systems approaches takes place through many directives and reinterpretations.

Changes in our understanding of the ways to improve relationships have also been gradual and will require much additional "shaping," direction, and reinterpretation. While current theories provide insight and guidance, they are often a source of division, controversy, and confusion. Empirical efforts have offered promising signs of the potential of marital and family therapy, but much more work needs to be done to test the various theories and methods. In the current state of affairs, clinical work is as much art as science.

In this complex and changing field, we assert the value of a simple but comprehensive conception of close relationships. This conception identifies the essential matter of close relationships and suggests the features of which any analysis must take account. Although it does not endorse any particular causal view of relationships or their treatment, it does set forth the probable alternatives and indicates their complex and reciprocal interrelationships. This conception can guide clinicians and researchers alike.

CHAPTER 11

Research Methods

JOHN H. HARVEY, ANDREW CHRISTENSEN,
and EVIE McCLINTOCK

Dora and Mark Jones are participating in a research study. When they arrive at the laboratory, they are separated and each is briefed about role playing a scene with the other. Dora is told that it is their anniversary and she has prepared a special dinner to surprise him. Mark is told that to celebrate their anniversary, he has made reservations at an expensive restaurant. Each is unaware of the other's plans. Then they are asked to role play what will happen when he returns home from work. After they have finished this scene, they are asked to interact in some additional situations. All scenes are videotaped. The researcher, who is interested in how couples deal with conflict, analyzes the interactions of Mr. and Mrs. Jones along with those of a number of other couples and identifies patterns of communication and conflict resolution.

Jane Far and Don Weldon, who have lived together for two years, are also participating in a research study. The investigator requests that they each complete a long checklist of daily couple activities for several consecutive evenings. The researcher conducting this study is interested in comparing the responses of Jane and Don and other cohabiting couples to the responses of a sample of married couples.

Ella and Ben Wilson are participating in a study that involves long interviews with the investigator. The partners are questioned individually and jointly about their feelings and experiences during their courtship years, their early years of marriage, and the previous year of marriage. The investigator is interested in studying stages in relationship development.

These examples illustrate some of the methods available to the researcher for describing and explaining relationship phenomena. The choice of an appropriate method or methods is determined by several factors. First, the type of question the researcher is trying to answer will suggest certain methods. Obviously, a study of subjective feelings of love will differ from a study of children's aggressive behavior in the family. Second, the availability of appropriate measuring instruments will affect the methodology. If valid and reliable measures of "subjective love" exist, these will likely be employed by the investigator interested in this phenomenon. Third, the possibility of manipulation will affect methodology. If children's aggression can be altered by therapeutic intervention, the researcher may employ that manipulation to investigate the phenomenon. Finally, ethical considerations may dictate or rule out particular methods. We could not, for example, study the effects of extramarital sex on feelings of subjective love by informing one partner that the other had had an affair.

In many of the substantive chapters of this book, attention has been directed toward methodological issues. Chapter 3, "Interaction," described operations for analyzing interactional data. Chapter 4, "Emotion," emphasized the problems associated with interpreting global, self-report ratings of feelings. A salient feature of Chapter 9, "Conflict," was the technique for obtaining husbands' and wives' interaction records to study episodes of marital conflict.

Since the phenomena addressed in these chapters differ in significant ways, the methodological issues also differ. Nevertheless, there are some issues that are common across the diverse domains of research on close relationships. In this chapter, we will consider some of the most important of these. First, we will examine methods of description—the ways researchers can collect data on close relationships. Next, we will examine the methods of causal analysis—the strategies the researcher can employ to detect causal patterns. Finally, we will examine some ethical considerations that limit both types of analyses.

The complexities of close relationships pose a great challenge to investigators interested in describing and understanding the thoughts, feelings, and behaviors of members of these relationships. Despite this complexity and the problems associated with it, we agree with Hinde's (1978) assertion that a science of interpersonal relationships is at this time both necessary and possible. He writes that

> some may regard the term "science" as simply inappropriate in this context because of the complexity and intangible nature of human interpersonal relationships. But a science need not underestimate the complexity of its subject matter, and it is just

> because the issues are sometimes too close to us to be seen clearly that such an approach is necessary. (p. 373)

An important assumption of this entire chapter is that an adequate methodology exists for fruitful, scientific investigation of close relationships. Though there is plenty of room for methodological advancement, current strategies provide powerful analytic tools.

DESCRIPTION

The initial step in any study of close relationships is description. This is true whether one's inquiry is motivated by a desire to understand the dynamics, explain the patterns, or predict the course of relationships. Systematic descriptions involve the assessment of events and properties of the dyad. They provide the basis for taxonomic work that organizes our knowledge of relationships and serve as the foundation for causal analyses whose aim is to explain the sources of observed relational regularities.

There are three major sources of descriptive information about relationships and the context within which they occur: the members of the relationship, third parties who have access and can observe the relationship directly, and records and traces of the partners' behavior. Information can be obtained directly from the members of the relationship in the form of participant reports. Participants can describe their own or their partner's feelings, thoughts and behavior, the properties of their relationship, and its perceived causal context. A second source of information about relationships is researchers or other third parties who, although not directly a part of the relationship, have the opportunity to observe it. These outsiders to the relationship can provide information about the behavior and personal attributes of the participants and their relationship. Such reports take the form of detailed accounts of observed interactional events, ratings of the observed behavior, or ratings of the properties/attributes of the partners and their relationship. The collection of participant or observer reports by a researcher usually involves the establishment of a direct or indirect relationship with the partners. Whether the data are collected through interviews, questionnaires, or observations, some contact with one or both members of the dyad has to be established to obtain the desired information, and the characteristics of this third-party affiliation may influence the primary relationship under investigation.

Information about relationships can also be collected through examination of records and traces of the partners' behavior. Public records, such as marriage and divorce documents, physical traces, such as patterns of carpet wear in the home, and personal papers, such as letters and diaries, can reveal information about close relationships. Since these records and traces are not normally generated with the researcher in mind, they provided a limited source of information. Nevertheless, the information is often free of the reactive effects of more obtrustive procedures that involve the clear presence of the researcher.

In discussing the descriptive analysis of relationships, we will first consider the two most frequently employed data collection strategies—participant and observer reports—and then examine their reliability and validity.

Participant Reports

Participant reports have been the most common source of information about close relationships. A survey of 12 American journals (Ruano, Bruce & McDermott, 1969) found that, between 1962 and 1967, 65 percent of the empirical studies of marriage and the family employed participant-report methods. More recent reviews of the marital interaction literature (Gottman, 1979) suggest that the popularity of participant reports remained unchanged in the 1970s. This method has several unique advantages. First, participant verbalizations about subjective experiences are the only currently available means for investigators to tap directly such covert activities as perceptions, feelings, thoughts, expectations, and memories. Second, participants can describe events that, though overt, are usually private to the relationship, such as sexual behavior or conflict. Finally, participant reports are relatively easy to obtain and involve much less inconvenience and expense than observer reports. These advantages derive primarily from the special vantage point from which members can view their own relationship (Levinger, 1977). As insiders, they possess a wealth of highly detailed information that is unavailable to the incidental observer, such as a researcher, or even to more permanent observers like friends or other family members. And such information can be transmitted to a researcher at relatively little expense in terms of time and money.

The three most typical formats for obtaining participant report data are self-administered questionnaires, behavioral records kept by the participants, and interviews. All three formats usually involve the participants' responding to questions structured by the researcher. Hence, participants' subjective memories of events necessarily are filtered through the researcher's questions, which constrain what is reported as well as how it is reported. The resulting information is inevitably a joint product of the structure imposed by the researcher and the information contributed by the participant.

Several distinctions among self-administered questionnaires, behavioral records kept by participants, and interviews are noteworthy. Questionnaires are generally designed to obtain summary information. Participants are typically required to generalize over large classes of events in order to report molar properties (e.g., level of conflict) or to infer causal conditions (e.g, personality traits). Because of the complexity and/or ambiguity of this cognitive task, this instrument *by itself* may not represent a highly valid approach to assessing proerties or understanding causal conditions. Nevertheless, questionnaires can provide an accurate account of the participants' impressions, which themselves may be important causal conditions.

In contrast to questionnaires, behavioral records typically produce more specific and molecular evidence about some event or events. They are usually accounts of interaction. Records are most often obtained immediately after events have occurred and consist of tallies or written descriptions of these events. With proper aggregation, however, these records can provide information on properties of the relationship. For example, P's records of arguments with O, if obtained over a long enough period of time, can provide good estimates of the level of conflict in their relationship.

Finally, interviews are unique in that they involve direct participation by third parties (i.e., the investigator), who can influence the nature of the evidence as it is being obtained. In all methods of participant report, the investigator structures the research questions, but only in the interview can this structuring take place interactively. The researcher can probe further; the participant can seek clarification and elaboration. This special feature of interviews can enable a more complete and accurate assessment of the topic of interest, but it may also adversely affect the obtained information, as when participants say what they think the investigator wants to hear.

Researcher-imposed structure

In all three methods of participant report—questionnaires, behavioral records, and interviews—the researcher elicits information with specific questions. These questions structure the task for the participants, requiring them to exclude and include particular kinds of information.

When the investigators ask questions about relationship events, they usually do so in an attempt to assess a particular property of interaction. The events are considered instances of the property under consideration. For example, the Spouse Observation Checklist, developed by Patterson and Weiss and their co-workers (Patterson, 1976a; R. L. Weiss & Margolin, 1977), inquires about the occurrence of 400 different behavioral events categorized a priori as pleasing, e.g., "we took a shower or bath together," or displeasing, e.g., "spouse said something unkind to me." The occurrence of specific events indicates the degree of pleasingness or displeasingness of the interaction.

To develop such instruments as the Spouse Observation Checklist, investigators need to conceptualize and define a "universe of admissible observations" (Cronbach, Glesar, Nanda, & Rajaratnam, 1972) that includes all the events or patterns that reflect a particular property. In some instances, researchers may approximate the universe of admissible observations by eliciting a large sample of relatively unconstrained events from one group of respondents and then using them in a more structured way with another group. Investigators then may sample the conceptually or empirically defined domain of events and present participants with a subset that they consider representative (see Triandis, Vassiliou, & Nassiakou, 1968).

The questions or items in structured instruments differ on a dimension of molecularity versus molarity. The items in the Spouse Observation Checklist refer to molecular phenomena—events. Many of the questions that investigators pose to respondents deal with more molar characteristics—such properties of the relationship as the distribution of power in a marriage or the frequency and intensity of certain experiences, such as satisfaction, trust, or commitment. Although information about relationship properties is often the goal of the investigator, the accuracy with which properties can be directly assessed by global reports is questionable. Christensen, Sullaway, and King (1983) found, for example, that spouses were less likely to agree when reporting on molar than on molecular items.

When researchers are uncertain what kinds of events or properties occur in relationships, they often employ unstructured instruments that allow respondents to provide their own descriptions of events. For example, time-budget studies that assess the use of time and division of labor in households (e.g., R. A. Berk & Berk, 1979; Robinson, 1977) require respondents to describe their daily activities in their own words and to indicate how long each lasted. Peterson's (1979) "interaction record" requires individuals to describe in detail a "significant interaction" with their spouses, indicating who initiated it, when and where it took place, and how it developed and ended.

Unstructured formats are seldom used in relationship research. One reason is that they elicit considerable variation in responses, and their reduction to a limited set of categories through coding is time consuming and expensive. However, unstructured formats can capture the unique flavor of each relationship and can often generate information that contributes insight about relationship dynamics.

Investigators often wish to ascertain whether reported patterns of events recur across situations. For example, if interested in disagreement as a possible indicator of marital distress, investigators will want to determine whether disagreements occur across a wide range of issues. They will try to think of a large number of contexts in which disagreements may occur and select a subset considered representative of the whole. Thus, Spanier's (1976) Dyadic Adjustment Scale inquires about agreement–disagreement over religious matters, sexual relations, handling family finances, household tasks, and so on.

The time interval between the occurrence of events and the reporting of these events also varies across different forms of participant reports. Concurrent reports are made simultaneously or nearly simultaneously with the events. An example is found in a study of marital interactions by Wills, Weiss, and Patterson (1974), who asked spouses to record the incidence of pleasing and displeasing spouse behaviors at the time that they occurred, using a portable event recorder. Retrospective reports are made after the relevant events have occurred. Most participant reports are retrospective; people are asked about events that transpired anywhere from a few hours to months or years before. Time considerations are also involved in the structuring of the questions. Sometimes questions are anchored to a particular time interval, such as the last 24 hours, the last week, or the last 4 years. Alternatively, time may remain unspecified, allowing respondents to summarize information over any time interval they desire. It is generally thought that, the more specific and proximal the time interval designated for the response is, the more accurate the report will be.

Cognitive processes of respondents

The constraints that investigators impose through the structure of their questions require different types of cognitive work on the part of the respondent. Consider one type of question commonly used in questionnaires and interviews: "How often do you and your spouse engage in activity X?" (e.g., quarrel, make love, visit relatives). How does a respondent go about answering this question?

First, a respondent must understand the activity in question, that is to say, translate the researcher's question into an instance or instances of behavior that occur in the relationship. To the extent that the researcher and the respondent share the same categories of meaning and the question is specific enough, the respondent's translation will be fairly compatible with the researcher's definition. Even specific terms may not, however, have the same meaning for researcher and participant. In a study of college students, D. G. Berger and Wenger (1973) found considerable confusion about the meaning of "loss of virginity." For example, 41 percent of the students agreed that female virginity was lost if the woman brought herself to an orgasm.

Once the nature of the event pattern has been understood by the respondent, there is a need to retrieve from memory one or more instance of occurrence and to describe them to the investigator. Frequently, researchers ask respondents to aggregate the information they possess so as to make global judgments about relationship events. Currently, little is known about the mental mathematics that respondents use to estimate the frequency of occurrence of particular events or about the information processing involved in answering global questions, such as "how satisfied are you with your relationship?"

There is ample evidence, however, that various forms of error and bias can occur as participants answer such questions. One form of error (Ericsson & Simon, 1980) stems from the nature of the questions; general and long-term retrospective questions tend to elicit less accurate reports than specific and short-term questions. It has been repeatedly demonstrated that, the longer the time interval between the occurrence of an event and its recall, the more distorted or inaccurate the recall is (Messé, Buldain, & Watts, 1982). Distortions tend to occur because of memory decay, because the events were not given proper attention when they occurred, and for a variety of other reasons. In such cases, respondents may use inferential processes to fill in memory gaps and end up misrepresenting reality (Ericsson & Simon, 1980).

Bias may arise because individuals tend to recall some types of information more readily than others. The often noted divergence of participants' reports about their relationship has been attributed to differential recall. Egocentric bias, the tendency of some participants to report more of their own than of their partner's contribution and behaviors, is one form of bias underlying divergent perceptions of relationships. This phenomenon, reported by M. Ross and Sicoly (1979), has also been observed by Thompson and Kelley (1981) and by Christensen, Sullaway, and King (1983). Ross and Sicoly conjecture that egocentric bias is based on differential availability of self information and other information. When queried about differential contributions of self and partner, respondents are more likely to recall instances of their own rather than other's behavior. Thompson and Kelley also show that respondents have more self information available but suggest that, when making judgments, respondents use information stored in dispositional or typical form rather than single instances.

Problems with egocentric bias and recall raise serious questions about the use of participants as accurate reporters of their own behavior. To the extent that we can understand the nature and direction of these biases through methodological research, we might be able to develop techniques and instruments to counteract them. Questions about relationships should be based upon an informed understanding of the ways people collect, store, retrieve, and interpret relevant information from their memory. Thus, methodological research on verbal report data and substantive research on cognitive processes are overlapping enterprises.

Observer Reports

The second most commonly used source of information about close relationships are observer reports. Observers may view live action or recordings produced by mechanical devices, such as videotapes or audiotapes. Observers may record interaction events in a number of alternative ways. In some cases they keep narrative records of the events, which are then analyzed by the researcher. This approach has been used extensively by Barker and his colleagues (1963) for analyzing the stream of behavior; by Blurton Jones and others (1972) in ethological studies of parent–child interactions; and by sociologists and anthropologists in ethnographies and case studies (e.g., Lofland, 1971). Alternatively, observers can be trained to transform the observed events immediately into one of a number of mutually exclusive categories provided by the researcher. This type of recording, called behavioral coding, may be done by hand or with the help of mechanical devices, such as the Datamyte, a device that facilitates the storage and analysis of a large body of observational data (Sidowski, 1977). Furthermore, observers sometimes evaluate the observed behaviors on rating scales developed by the investigator to tap some underlying property of interest. For example, observers may rate the warmth, symmetry, cooperation, or conflict in relationships on seven-point scales.

Observer reports have been employed less frequently than participant reports for several reasons. First, this form of data collection tends to be more time consuming and expensive than participant reports. For example, observation of mechanically obtained records often requires many hours for transcribing and coding the data. Gottman (1979) estimates that, in his observational system, every hour of videotaped interaction requires 28 hours of transcribing and coding. An additional drawback is that observational records provide information about only the "objective" reality of relationships and very little about the "subjective" reality as perceived by the participants. The latter has often been the primary interest of many investigators.

On the other hand, observer reports have some unique advantages over participant reports. Observational records make available to the investigator samples of interactive behavior that have not been filtered through the perceptions of the participants. Levinger (1977) observes that "the members of a relationship are blind to some of its peculiarities" (p. 146); observer reports often allow the researcher to identify patterns that the members may not recognize or be able to describe.

Observer reports vary in terms of the setting in which the behavior is observed and in terms of the extent to which the investigator elicits or structures the observed behavioral interaction. The setting or locale of observation is important because the ultimate goal of relationship researchers is to be able to generalize from the behavior samples that they obtain to behaviors that occur in the everyday life of the relationship. Whenever behavior is observed in artificial settings, such as the laboratory, the question arises whether it is representative of daily interactions or is the outcome of the unique stimulus characteristics of the setting. Similar generalizability concerns are also associated with the extent to which the investigator elicits the observed behaviors. In some cases, investigators observe naturally occurring interactions; in other instances, they elicit particular behaviors by requesting the members of the close relationship to respond to structured tasks. In the latter case, similar issues of representativeness arise, namely, whether the observed behavior reflects reactions to the elicitation procedures or stable patterns of the relationship. These two parameters—the type of setting (natural versus laboratory) and the degree of structure imposed by the investigator—provide two dimensions for categorizing various methods employed in collecting observational data (Gilbert & Christensen, in press).

Unstructured interaction in natural settings

Ethological investigations of parent–child interaction are typical of studies that have employed observers to record interactions in natural settings, such as playgrounds or nursery schools (Blurton Jones, 1972). Similarly, psychological studies of parent–infant behavior have observed unelicited interactions during feeding (e.g., Bakeman & Brown, 1977) or when mother and infant are in the home (Ainsworth, 1972). Dreyer and Dreyer (1973) had observers monitor family dinnertime behavior. To obtain data about interaction across a variety of home situations, Steinglass (1979) assigned to each spouse an observer who moved throughout the home following the person observed.

Behaviors observed in the home and other nonlaboratory settings are only approximations of naturally occurring interactions. The naturalness of behavior is inevitably compromised to some degree by the presence of the observer, though this influence can be minimized in several ways (e.g., by the placement of the observer). In some cases investigators alter the naturalistic situation further by requiring families to remain in one room and to refrain from watching television, answering phones, or receiving visitors during the observation (S. M. Johnson & Bolstad, 1975; Patterson, 1979).

Because observer presence has been found to alter the behavior of the observed, some investigators have tried to record naturally occurring interaction using mechanical devices. An early example of mechanical recording of unelicited interaction is reported by Soskin and John (1963), who attached portable transmitters on the members of a couple and recorded their interaction. Their apparatus was bulky and obtrusive. Couples were constantly aware that they were being recorded. Christensen (1979) has recently developed an automated recording system capable of sampling family interaction in multiple situations. Two small microphones are placed in two high-interaction areas of the home (e.g., the kitchen and the family room) and wired to a footlocker that contains a stereo recorder and timing equipment. The equipment is located in an inconspicuous area, such as a closet. Family members specify 3 hours during the day when interaction is most likely to occur. Timers in the footlocker activate the equipment at random 15-minute intervals during these high-interaction periods. Thus, family members do not know exactly when they are being recorded, and the technique has been shown to be relatively nonreactive (Christensen & Hazzard, in press).

Unstructured interaction in laboratory settings

This data collection strategy has been rarely employed. J. J. Snyder (1977) examined reinforcement contingencies across distressed and nondistressed families by videotaping unelicited interactions through a one-way mirror in a room set up to be a living room and playroom. Birchler, Weiss, and Vincent (1975) videotaped 5-minute segments of unstructured marital interaction that took place while the investigators were preparing the subsequent structured tasks. Unstructured laboratory observations can be a useful adjunct to structured ones, as well as a useful method on their own. To the degree that they allow participants to act as they normally would in public places, they reduce the artificiality introduced by the structured tasks that investigators often employ to elicit interaction.

Structured interaction in natural settings

There are few examples of structured interaction observed in natural settings. This method has primarily been used to assess the impact of setting familiarity–artificiality on dyadic interaction. Gottman (1979) used structured laboratory tasks in the subjects' home to compare laboratory and home-based interactions. S. Martin, Johnson, Johansson, and Wahl (1976) also used home observations to demonstrate inconsistencies between parent–child behavior at home and in the laboratory.

Structured interaction in laboratory settings

Researchers often elicit interaction by presenting participants with structured laboratory tasks that require them to react in particular ways. A number of different types of structured tasks have been used in relationship research: game playing, decision making, conflict resolution, and stimulus interpretation (Gilbert & Christensen, in press).

Game playing has been widely used in studies of parent–child interaction because it provides the researcher an opportunity to observe the reciprocal information and behavior control strategies of parents and children (Hess & Shipman, 1965). With couples, game-playing tasks afford investigators the chance to observe the couple's ability to establish and comply with rules, as well as the cooperative or competitive mode of coordination they adopt (see Straus & Tallman, 1971, for a description of SIMFAM, and Speer, 1972, for a description of the Prisoner's Dilemma game as used with couples). Decision-making tasks provide information about the processes that dyads or families employ to arrive at more or less mutually satisfactory solutions or plans of action (Gottman, 1979; Riskin & Faunce, 1972). Conflict-resolution tasks enable an observer to evaluate the manner in which dyads cope with differences of opinion or conflicts of interest, as well as the processes they employ to restore equilibrium (e.g., Raush, Barry, Hertel, & Swain's, 1974, improvisation scenes and Strodtbeck's, 1951, revealed difference technique). Finally, stimulus-interpretation tasks aim to disclose modes of information transmission and strengths or deficits in communication when partners are trying to reach a consensus about an ambiguous stimulus like the Thematic Apperception Test or Rorschach cards (Singer & Wynne, 1966).

These structured tasks offer two main advantages. First, they provide data about processes that might be difficult to observe in natural situations, either because the events tend to be private and inaccessible to outside observers, or because the events occur with such low frequency that it would be very expensive to attempt to observe them *in vivo.* Second, structured tasks constrain behavioral variability and yield data that are comparable across families. The structure permits analysis of all the obtained interactions along similar dimensions, an analysis that would not be possible if each couple were engaged in a different activity (e.g., talking versus watching television). The major disadvantage is that the investigator cannot know whether the interaction observed is characteristic or whether it is greatly influenced by the setting and the task. In order to overcome this disadvantage, investigators try to obtain data on several tasks (e.g., several decision-making or conflict-resolution issues) in order to assess how much of the behavioral variance is accounted for by the specific situations to which the partners are responding and how much is common across tasks. An alternative validity check suggested by Christensen and Hazzard (in press) is to ask respondents to rate their interaction, indicating whether they feel it is representative of the way they typically interact.

The Causal Context of Participant and Observer Reports

A major concern of researchers is that the data they obtain through participant or observer reports accurately reflect properties and events of relationships and not other extraneous factors, such as biases of the observers or respondents and reactivity to the research situation. The many factors that threaten the reliability and validity of data can be thought of as constituting part of the causal context of participant and observer reports.

In this section, we consider four categories of contextual factors that can reduce the reliability and validity of relationship measurement: (1) the relationship between the investigator and the participants, (2) the immediate stimulus situation in which the data are elicited, (3) characteristics of the respondents, and (4) characteristics of their relationship. These factors do not operate independently of each other; rather, they tend to overlap and interact. For example, characteristics of the individuals, such as literacy, interact with characteristics of the questions presented to them, such as sentence complexity. The data-collection situation is also part of the relationship between the researcher and the participants. Separation of the four factors only helps simplify the discussion.

The relationship between researcher and participant

Most research on relationships involves some interaction between the researcher and one or more members of the relationship. Even when information is obtained through anonymous questionnaires, the data collection involves a symbolic relationship between the respondent and the investigator. When respondents are observed or interviewed, the relationship becomes closer and more intimate. How does the relationship between the researcher and the participants affect what the researcher observes, how the dyad behaves, or what the participants report about their relationship (see Levinger, 1977)? The investigator–participant relationship can create biases in both the researcher and the respondents.

RESPONDENT BIASES ELICITED BY THE RESEARCH RELATIONSHIP. Studies of human behavior share an important drawback not generally experienced in most other fields of scientific investigation, namely, that the objects of investigation tend to alter their characteristics when being observed (Christensen & Hazzard, in press). In questionnaire studies, this effect has been discussed as a "social desirability" bias, in interviews as "impression management," and in observational studies as "reactivity to observation." These response sets can result from the conscious or unconscious desire of people to present themselves to others in ways that will benefit their personal interests (Weary-Bradley, 1978). Most often, participants want to present themselves in a "good light." This desire is triggered when the respondent enters the research relationship, and its effect on behavior raises serious questions about the accuracy of the obtained information.

Edwards (1957) was one of the first investigators to examine the tendency of respondents to report good rather than bad things about themselves. He found that agreement with questionnaire items was directly related to the social desirability of the items. Since then, several scales have been developed to assess individual differences in social-desirability tendencies. Scores on these scales are often correlated with scores on instruments measuring other characteristics of the individuals or their relationships in an attempt to estimate how much of the variation in the data of interest is accounted for by social-desirability bias. Alternatively, some researchers obtain a priori ratings of the social desirability of individual items and then try to minimize the effect by eliminating highly socially desirable items. A more general strategy recommended by several writers (e.g., Crano & Brewer, 1973; Selltiz, Wrightsman, & Cook, 1976) is to reassure respondents of the anonymity of their responses so as to minimize their attempt to impress the researcher positively. To answer some questions, however, the identity of the respondent cannot be concealed; so the "anonymity strategy" has its own limitations.

In interview studies, the relationship between researcher and respondent is closer than in questionnaire studies and may thus create greater reactivity. Respondents may try to impress interviewers by misrepresenting information in such a way as to create a positive image of themselves. Exaggerations and inaccuracies aimed at presenting a respectable front have often been documented in survey studies (Selltiz, Wrightsman, & Cook, 1976). Under some conditions, interviewees may exaggerate the severity of their symptoms, as when an interviewer hints to insecure chronic mental patients that they are under consideration for discharge (Braginsky, Braginsky, & Ring, 1969). The respondent's report can also be affected by the demand characteristics of the interview situation and by the interviewer's expectations (Hyman, 1954). To counteract reactive effects in interviews, researchers try to put the respondent at ease, to begin the interview with innocuous questions and then move to more personal issues, and to minimize their influence of the respondent during the course of the interview.

In observational studies, reactivity has also been a major concern (Johnson & Bolstad, 1973). Several methodological studies have documented that interaction in parent–child pairs changes when the individuals become aware that they are being observed (Zegiob, Arnold, & Forehand, 1975; Zegiob & Forehand, 1978). Other studies have examined whether observed behavior changes as a function of repeated observation sessions. Since reactive effects are assumed to diminish over time, the fact that behavior does not change as a function of repeated observation has been used as indirect proof of the absence of initial reactivity (Patterson & Cobb, 1971). A different approach to observational reactivity is illustrated by "fakeability" studies, in which participants are instructed to behave in particular ways that could foster a positive or negative impression of themselves. These studies assume that the processes through which people alter their behavior voluntarily at the request of the investigator are analogous to those used when they want to impress the investigator. Vincent, Friedman, Nugent, and Messerly (1979) showed that the discussion of hypothetical conflicts by couples was altered by giving instructions to present themselves in one condition as "blissful" and in another condition as "conflicted." However, when R. S. Cohen and Christensen (1980) instructed couples to discuss their *own* problems while trying to exhibit "normal," "good," or "bad" behavior, they found no difference among these three conditions. Thus, the use of less extreme instructions and the couple's own conflicts apparently eliminated the effects of the different instructional sets.

Two seemingly contradictory strategies can be used to solve reactivity in questionnaires, interviews, and observations. One stresses complete unobtrusiveness of the data-collection procedure, maximizing distance in the researcher–respondent relationship. The other advocates forming a close and trusting relationship with the respondents to allow them to present themselves to the researcher in as "natural" a way as possible. Participant-observer studies that involve repeated interactions between researcher and respondents have used the latter strategy to obtain intimate views of relationships (Komarovsky, 1967; L. Rubin, 1976). Ultimately, an investigator's research goals determine which solution is more appropriate for any particular study. If one attempts to approach the objective reality of relationships as reflected in unbiased samples of behavior, a relatively unobtrusive data-collection approach would recommend itself. To the extent that one is interested in the subjective reality of the participants and wants to document not only how they act but also what they think and feel about their relationship, the formation of a relatively close relationship between researcher and respondents appears more appropriate. In many studies, a combination of both approaches would yield the most useful data.

INVESTIGATOR BIASES ELICITED BY THE RESEARCH RELATIONSHIP. Investigators enter the researcher–respondent relationship with expectations that can bias the information they obtain. Interviewers may anticipate a certain distribution of responses and try to fulfill this expectation by means of leading questions, failure to probe, errors in recording responses, and even more subtle cuing of particular responses in the respondent through voice intonation or postural cues (Blair, 1980; Hyman, 1954). To date, empirical support of interviewer bias has not been demonstrated definitively (DeLamater & MacCorquodale, 1975; Sudman, Bradburn, Blair, & Stocking, 1977), but it seems very likely that such biases do occur.

All these sources of bias can operate in observational studies. In addition, observer bias may arise from idiosyncratic perceptions or interpretations of the observed behavior. Observers, as well as interviewers, can become emotionally involved with one or both members of the dyad and misinterpret or misreport behaviors. Extensive observer training prior to the actual data collection and constant checks of interobserver reliability have been used as a means to control for observer bias. But even these procedures do not necessarily guard against observers' differential response to particular individuals. Yarrow and Waxler (1979) reported several instances in which observer accuracy, as indicated by the degree of agreement between observers, was dependent on such characteristics of the person being observed as sex or social class. They suggest that observer accuracy might be greater with couples or individuals whose behavior is aligned with the expectations and norms of the observer than those whose behavior is not.

Finally, observer behavior may change as a function of the time spent observing. O'Leary and Kent (1972) discuss the phenomenon of "reliability decay," in which interobserver agreement declines during periods when observers think they are not being checked. In addition, they report "observer drift," in which pairs of observers modify the coding system although they continue to exhibit high agreement with each other. Gottman (1979) suggests that, when coding from videotapes, these problems can be attenuated by a written code manual, by training sessions on coded materials, and by using one reliability checker who independently codes samples of every tape.

The stimulus environment of data collection

The immediate stimulus environment of observed or reported behaviors includes the setting in which the data collection is taking place and the structure of the task presented to the respondent.

The possible impact of the data-collection situation on participant reports has not attracted much attention. It is assumed that, since participant reports often refer to behaviors already enacted, the immediate context in which reporting takes place will not strongly affect the recall or evaluation of these events. Although this assumption may sometimes be reasonable, aspects of the immediate data-collection situation can definitely influence responses. Interviews or questionnaires administered in the home have the asset of keeping respondents in their natural setting, but, at the same time, they can be plagued by distracting factors, such as the presence of third parties, interruptions by other family members, unexpected visitors, or telephone calls. Distractions are particularly acute when interviews take place in low-income households in which lack of space precludes privacy. Additional factors that can affect responses are associated with the structure of the instrument. These factors include length, difficulty, and order effects, which can create interviewee confusion and fatigue.

Characteristics of the data-collection setting are likely to influence those behaviors that are the subject matter of observer reports. For example, the artificiality inherent in laboratory settings raises questions about the generalizability of the observed behaviors to natural settings. How much are observations obtained in the laboratory the product of the setting and the procedures used there, and how much can they be generalized to other settings? It is clear to most researchers that the processes observed in the laboratory do not necessarily correspond to routine interactions in relationships. Many of the interactive processes that are of research interest, such as conflict, have low base rates in the natural setting. To the extent, however, that laboratory assessments can be considered representative of those infrequent occasions in which couples or parent–child pairs self initiate low-base-rate interactions, they can be a very valuable source of information. Some studies have even used the artificiality and strangeness of the laboratory as part of the stimulus configuration eliciting behaviors of interest. For example, Ainsworth (1972) assessed child attachment to the mother in the laboratory, where the laboratory itself becomes the "strange situation" in which the child is left alone by the mother.

Characteristics of the participants (P and O)

There are several characteristics of the respondents that can affect their responses to questionnaires or interviews. These characteristics of P and O interact with characteristics of the stimulus situation and result in biased reports. One often overlooked source of bias in participant report data is the inability of a respondent to read or understand the language of the questions. With very few exceptions (e.g., Bienvenu, 1970), authors do not specify the degree of literacy required for responding to questionnaires. In addition, respondents may not conceptualize their interactions or relationships in the terms employed by the investigator. Many questions require considerable translation on the part of a respondent to approximate the meaning categories of the researcher. For example, questions concerned with power or equity may not be understood by respondents who have never thought of their relationship in those terms.

Properties of the close relationship

Several studies have demonstrated that participant reports of relationships are affected by unintended features of those relationships. For example, the satisfaction that individuals experience in the relationship might cause them to overreport or underreport particular events. Christensen and his colleagues (Christensen & Nies, 1980; Christensen et al., 1983) found that distressed couples agreed less than nondistressed couples in reporting their own and their partner's behavior. The length of the relationship has also been linked to systematic variations in particular reports. Christensen et al. (1983) found a greater tendency for participants to attribute responsibility for negative events to the partner and/or deny them in self as the length of the relationship increased.

Reporting biases associated with particular properties of relationships can generate problems for the investigator. For example, differences in reported behavior patterns that the investigator may attribute to socioeconomic or other factors may simply reflect differential reporting due to differences in marital satisfaction or length of relationship between the groups that are being compared. As researchers become more aware of the systematic biases that can affect participant reports, they should be better able to control these factors or assess their effects. Furthermore, to the extent that characteristics of the individual participant or of their relationship can bias their reports, it might be necessary for investigators either to obtain reports from both partners or use multiple methods to study relationships. This approach might be particularly important when the researcher is using distressed or unhappy couples as reporters of their behavior.

The Integration of Data from Different Sources

In the foregoing section of the chapter, we discussed the tools available to researchers for describing relationships. In particular, we examined participant and observer reports and their causal contexts. Because of the unique advantages and disadvantages of each of these methods, several researchers (e.g., Levinger, 1977; Olson, 1977) have advocated the use of both observer and participant reports within the same study.

The use of multiple methods or sources of information originated in personality and attitude research with the objective of validating the measurement of particular constructs. Such validation efforts are clearly needed in close relationship research, since different measures of the same construct are often in disagreement (e.g., Olson & Rabunsky, 1972; Sullaway & Christensen, in press). Multimethod or multiscore approaches have additional important functions besides validation (Olson, 1977). Central among them is the identification of the divergence between subjective and objective reports and between the reports of two partners.

The identification of both types of divergence can be very significant both for theoretical and empirical reasons. Discrepancies between participants' reports and their observed behaviors can illuminate the dynamics of relationships and at the same time suggest areas for therapeutic intervention. For example, O'Dawd's research (reported in Olson, 1977) identified a discrepancy between reported and observed behaviors of mothers of addicted adolescents. Mothers of addicted and nonaddicted youths reported equal levels of support for their children. However, when the two groups of mothers were observed interacting with their children, mothers of addicted adolescents were found to give significantly less behavioral support to their child than did mothers of nonaddicted youths.

Partner disagreements in the interpretation of events may reflect important communication deficits. For example, Gottman, Notarius, Markman, Bank, Yoppi, and Rubin (1976) contrasted the intent and the impact of messages exchanged between spouses. They identified a significant discrepancy between the intent of the sender and the perception of the receiver in distressed, but not in nondistressed, couples and concluded that distressed couples had problems reading accurately each other's behavior. The findings in both the above studies would not have been possible if only one data source were used.

A multimethod or multisource approach cannot be recommended indiscriminately for all close relationship research. Its utilization should depend on its appropriateness for the goals of the study and on the resources available to the investigator. However, ultimately, a comprehensive description of close relationships will need to incorporate both subjective and objective accounts and integrate the divergent data obtained from multiple sources via alternative methodologies.

This section has focused on the description of close relationships. While most studies gather descriptive information only as part of a more general causal analysis, studies that simply describe important features of relationships serve an important function in the field. In order to answer questions about *why* people do what they do in close relationships, we must first have sufficient information about *what* they do in these relationships. If our interest is in conflict, for example, we must first know what kinds of conflict typically occur before we can accurately answer why it occurs. Because much of what happens in close relationships is private, because relationships show great variability, and because "common knowledge" includes much misleading information, such as media stereotypes, we need systematic efforts to describe what actually occurs in close relationships. This work can then provide a firm basis for theoretical efforts and causal analyses.

CAUSAL ANALYSIS

Neither the social scientist nor the casual observer can doubt that the causal dynamics of close relationships are complex. Analysis of the phenomena, as in Chapter 2, or simple reflection on personal experience often reveals the multiple strands of cause and effect in operation. Prior to our discussion of the methodology of causal analysis, it may be useful to enumerate some reasons for this complexity. First, few causes in close relationships have simple, one-to-one, determinant effects. We know, for example, that conflict is often initiated by a rebuff (see Chapter 9, "Conflict"). However, a rebuff does not always lead to conflict. The partner may shrug it off, inquire sympathetically about the other's mood, or engage in a variety of nonconflictual responses. Second, causes in close relationships are usually multiple. The likelihood of a rebuff creating an argument is affected by a variety of variables, such as the mood of the participants and the social situation. Third, causes in close relationships exist reciprocally in causal loops. Whether a rebuff leads to a conflictual episode depends on the mutual interaction surrounding that rebuff. The spouse who received the rebuff can attack the other, and an escalating argument may occur. On the other hand, the spouse may meet the rebuff with concern about the partner's mood, the partner may then apologize for the rebuff, and a positive interaction may ensue. Finally, many of the important causal factors in close relationships result from the separate and joint histories of the participants. These conditions, stored in the memories and jointly constructed environments of the participants, are not immediately available to the investigator. The sensitivities of the partner to rebuff, the unique interpersonal context of the rebuff, and the couple's skills at handling interpersonal tension are all important conditions affecting the likelihood of conflict.

Science has provided tools for handling this complexity. Methodological design and statistical analysis can assist in sorting out the causal dynamics in close relationships. While some (e.g., Meehl, 1978) have questioned the power or applicability of these tools for understanding social behavior, we argue that science has already shed considerable light on close relationships (as indicated by the research and theory summarized in this book) and offers potential for further knowledge about this area of human behavior. In the following pages, we will explain and illustrate how scientific methodology can be employed to achieve a causal understanding of close relationships.

Research Designs

To assert a causal relation between an independent variable A and a dependent variable B, one must establish covariation between A and B, temporal precedence of A before B, and the absence of competing explanations for the mutual variation. In the *classic experimental design,* variable A is manipulated across groups of subjects created by random assignment, and variable B is subsequently measured. If values of B are associated with the manipulations of A, both covariation and temporal precedence are established. Since subjects are randomly assigned to groups, equivalence on all other variables besides A can generally be assumed, and competing hypotheses can therefore be discarded. Thus, the classic experimental design can establish causality.

Consider, for example, the income maintenance experiments conducted in Seattle and Denver (Hannan, Tuma, & Groeneveld, 1977). Approximately 5,000 low-income families in these two areas were assigned to either a control condition or one of several income-maintenance conditions. Assignment was random within categories of family income, race–ethnicity, marital status, and site. Income guarantees existed for all members of the family, whether they remained together or not. Families were interviewed three times a year during the 3- to 5-year experiment and for a 2-year period after the experiment. Results indicated that income maintenance produced a large and statistically significant increase in the rate of marital dissolutions.

From this study, we can infer a causal relation between income maintenance and marital disruption. The two variables covary, and the first precedes the second. Because of the random assignment of this large group of subjects to the various conditions, we can be relatively confident in discounting the possibility that other variables accounted for the effect. We do not know the extent to which these results are generalizable. For example, it is probably unlikely that they would replicate with families from higher income levels. We do not know the proximal causal factors that accounted for these results. Perhaps the couples who terminated their relationship had long been dissatisfied with each other and simply needed assurances of financial support before making a decision. Perhaps the income-maintenance intervention added to the dissatisfaction by increasing arguments about money. Whatever the generality of these results or their detailed explanation, we can conclude that income maintenance increased marital disruptions in these participants.

The major alternative to the classic experiment has been the *correlational design,* in which two or more variables are observed without the manipulation of either. In many cases, only covariation can be established; the investigator is uncertain if A caused B, if B caused A, or if some other variables caused both A and B. Consider, for example, a recent study by Hendrick (1981), which found a significant positive correlation between marital satisfaction and self-disclosure. While providing evidence for covariation, these data alone do not permit a determination of temporal precedence or rejection of alternative hypotheses. Self-disclosure may increase marital satisfaction, marital satisfaction may increase self-disclosure, or other variables, such as attitude similarity, may account for both. As we shall see below, Hendrick's study was more complex than indicated so far and, in fact, provided additional evidence on the causal factors accounting for marital satisfaction.

While evidence from correlational research cannot definitely confirm a causal hypothesis, it is relevant to these hypotheses. If correlational analyses reveal significant covariation in the expected direction, the data can be interpreted as being "consistent with" or "supportive of" the hypothesis. If correlational analyses do not reveal the expected covariation, the data can be interpreted, within the sampling and measurement constraints of the study, as disproving the hypothesis.

While the classic experimental and correlational designs serve as the models for research on close relationships, several variations of these designs deserve attention. Sometimes these variations can illuminate causal relations more clearly than can the classic designs, or sometimes they can be employed when more powerful experimental designs would be impossible.

Variations of the Classic Experiment

First, we consider three variations of the classic experimental design: the single-subject design, the natural experiment, and the analogue experiment. In *single-subject* designs, experimental manipulations are employed on a single subject. Observations after each manipulation establish covariation and temporal precedence. Alternative hypotheses are ruled out, not through random assignment of subjects to manipulated conditions, but through repeated manipulations over time on the same subject. Thus, each subject serves as its own control.

Two major variations of single-subject research are the reversal or withdrawal design and the multiple baseline design (Hersen & Barlow, 1976). In the first case, repeated manipulations are employed on the same behavior of the same subject. For example, during a baseline time period, a mother makes requests of her son but provides no feedback for his subsequent compliance or noncompliance. During time period 2, the mother verbally reinforces compliance and ignores noncompliance. During time period 3, she returns to the conditions of baseline. During time period 4, she again reinforces compliance and ignores noncompliance. If the investigator observes systematic variation in the child's compliance in response to these conditions, we can assert a causal influence of reinforcing feedback on the child's compliance. It is unlikely that the child's compliance changes coincidentally with the mother's feedback.

In the second type of single-subject design, several behaviors are observed continuously ("multiple baselines"), and an independent variable is manipulated sequentially across these behaviors. Assume, for example, that the frequencies of three behaviors—fights with siblings, swearing, and destructiveness—are observed continuously in a young girl. After a baseline observation period, her mother applies "time out" each time the child starts a fight with siblings. In this procedure, the girl is removed to a solitary, nonreinforcing location, such as a bathroom, and required to stay there alone for a brief period. After this condition has been in effect, the mother applies "time out" to each incident of swearing as well. Finally, the mother applies "time out" to destructive behavior also. If each behavior decreases in frequency when the "time out" procedure is applied to it, we can assert that "time out" had a causal effect on this child's behavior.

Both of these designs require repeated introductions of a treatment condition on a single subject. If only one manipulation were employed, it could be argued that other conditions, which coincidentally occurred with the manipulation, were responsible for the observed effects. The girl in the above example may have fought less frequently because her sibling became preoccupied and unavailable for such interaction. However, when manipulations are repeated and when their timing is determined a priori, coincidental effects can be ruled out with relative assurance.

While single-subject designs allow for causal interpretations of results, these interpretations are limited to the single subject investigated. Therefore, investigators typically repeat the experiment several times with different subjects to demonstrate generalization across subjects (Hersen & Barlow, 1976).

A second variation of the classic experimental design is the "*natural experiment*" (Bronfenbrenner, 1977; D. T. Campbell & Stanley, 1966). Sometimes naturally occurring changes mimic the classic experiment and can be exploited by the alert investigator. Ideally, the natural experiment takes advantage of situations in which certain people are exposed to a variable of interest and in which the exposure is "lawless, arbitrary, uncorrelated with prior conditions" (Campbell & Stanley, 1966, p. 65). Assume for example, that a large company has to "lay off" a certain percent of its employees. In one plant, at which a large number of employees have been hired at approximately the same time, termination cannot be decided on the basis of seniority. Thus, a lottery system is established. If we compare rates of marital stress (e.g., as indicated by marital separation) between the groups that kept and lost their jobs, we could establish a causal relationship between unemployment and marital stress.

A third variation of the classic experiment design is the *analogue study*. The term *analogue* is used because the experimental manipulation is meant to mimic some natural process. If a couple is asked to act "as if" they experienced some state defined by the experimenter, or if a mood induction technique like hypnosis is used to create some emotion, or if drugs are used to create a certain physiological state, then the investigator is contriving an experimental analogue of a natural process. It can, however, be an experiment in the sense that manipulations are employed across ramdomly assigned groups or within single subjects. The central issue for analogue research is external validity (Campbell & Stanley, 1966). To what extent do the manipulations create the kinds of conditions that occur naturally? Because analogue studies raise major questions regarding external validity, we can distinguish them as a special class of experimental studies. In making this point, we should recognize that there are particular and sometimes imposing generalization issues associated with all forms of scientific inquiry, including the classic experimental design and even nonexperimental studies conducted in natural settings.

Variations of the Classic Correlational Design

In the simplest correlational design, the investigator is only able to establish covariation between A and B. Direction of causal effects or the presence of alternative hypotheses cannot be ruled out. However, it may be possible to go beyond mere covariation in correlational research to establish temporal precedence of one variable over another or to provide evidence relevant to plausible alternative hypotheses.

Sequential correlation designs can establish temporal precedence of one variable over the other as well as covariation. In the simplest and most obvious case, the investigator relates two variables that occur at different time periods. If marital arguments regularly follow visits by in-laws, but visits by in-laws do not regularly follow marital arguments, the direction of any possible causal relation between the two conditions is established. However, other factors, such as family stresses, may precipitate both the visits and the arguments.

Correlational analyses between sequential variables can be conducted within subjects as well as across subjects. If we observe that a father often criticizes his daughter after the mother criticizes him, we can establish covariation between mother-criticizes-father and father-criticizes-daughter, and we can establish temporal precedence of the former over the latter. Other variables may be causing both, but if there is a causal relation between the two, we know the direction. Obviously, it may be difficult for the investigator to obtain the detailed observations necessary to develop such a conclusion.

These kinds of studies, which observe the temporal covariation of behaviors, employ sequential analyses (Gottman, 1979; Patterson, 1979; Sackett, 1979) and describe covariation in terms of probability rather than correlational statistics. Typically, the unconditional probability of some class of behavior is compared to the conditional probability of that behavior, given the probability of a specific alternative behavior. For example, father-criticizes-daughter may occur in only 5 percent of the total observational intervals, but it may occur in 25 percent of the intervals following mother-criticizes-father. Statistical comparisons of these probabilities can establish a significant relation between the two behaviors.

Variations of the classic correlational design can address the problem of alternative causal hypotheses as well as the problem of temporal precedence of variables. If correlational designs are conducted on multiple variables, the investigator can gather evidence regarding these alternative hypotheses. In fact, the investigator rarely measures only two variables and their correlation with one another. Additional measures, which can illuminate the causal analysis, often can be assessed with little extra expense. Consider Hendrick's (1981) study of self-disclosure and marital satisfaction, which was discussed above. She also measured attitude similarity and found that it was significantly correlated with marital satisfaction but not with self-disclosure. Given these data, the hypothesis suggested earlier, that attitude similarity accounts for the relationship between self-disclosure and marital satisfaction, seems unlikely.

Many statistical techniques are available for examining the relationships among correlated variables. Although explication of these procedures is beyond the scope of this chapter, we will mention some of the more common techniques and refer the reader to detailed sources for more thorough descriptions (Bentler, 1980; Kenny, 1979; Kerlinger & Pedhazur, 1973; Li, 1975). Through partial correlation techniques, the investigator can compute the relation between two variables with the effects of a third statistically removed. In multiple regression or discriminant analysis, a variety of independent variables are related to one criterion variable. The investigator can determine the strength of relationship between any particular independent variable and the criterion in the context of other variables. Factor analysis enables the investigator to determine statistically which variables cluster together such that a set of dimensions may account for their variability. Finally, path analysis allows the investigator to test empirically whether a set of variables intercorrelates as predicted by theory.

Another correlational design, which typically involves multiple variables, is the *ex post facto design.* Two or more groups of subjects who differ "after the fact" on the independent variable but who are matched or equated on several other relevant variables are contrasted. Differences between or among groups in the independent variable are created, not by experimental manipulation, but by the way the groups are constituted. For example, distressed and nondistressed couples who are demographically similar are compared on an interactional variable. The investigator may then assume that a difference between the groups on the dependent variable can be attributed to the independent variable (i.e., the level of distress), rather than to the variable on which the groups were equated (i.e., demographic characteristics).

The ex post facto design has wide appeal. If experimental manipulations are impossible, at least the investigator can locate subjects who differ on the independent variable. Equating on one or two alternative causal variables is usually not difficult. However, results from these designs are often inappropriately treated as experimental data. Because some control of alternative hypotheses is achieved and because the design has the appearance of an experiment, including a similarity in statistical analyses, researchers may not be sufficiently cautious in interpreting their results.

The limitations of the ex post facto design are several. First, if investigators are to control for alternative hypotheses, they must have knowledge in advance of the set of possible causal variables. Rarely is this the case. Second, researchers must equate groups on all these alternative variables. Pragmatically, it is difficult to equate groups on more than a few variables. Furthermore, such efforts can generate highly unrepresentative samples of subjects (Meehl, 1971). If severely distressed families are typically from low-income homes and happy familes are typically from middle- and upper-income homes, a comparative study that equates on income will include an unrepresentative sample of distressed families, an unrepresentative sample of nondistressed families, or both. Finally, equating groups on alternative variables can sometimes prevent the discovery of a true relation between the independent and dependent variable (Meehl, 1971). Assume, for example, that an investigator hypothesizes that social skill deficits in conflict resolution are responsible for higher levels of conflict in distressed couples. Further assume that socioeconomic status affects social skills by affecting exposure to socially appropriate models. If the researcher investigated the social skill hypothesis, but matched distressed and nondistressed groups on socioeconomic status, as common practice would suggest, the researcher might fail to discover a true effect. Thus, decisions about equating groups on alternative variables must be based on the researcher's causal model of proximal (e.g., social skills deficit) and distal (e.g., socioeconomic status) variables that affect the phenomenon of interest (e.g., marital distress).

This discussion highlights what almost all social scientists know, but sometimes forget—that definite causal conclusions cannot be made from correlational data. Even when temporal precedence can be established, even when some control over alternative causal variables can be maintained, even when the design has the appearance of an experiment, a causal relationship cannot be conclusively inferred.

With this important qualification in mind, we can nevertheless assert that correlational research is relevant to causal hypotheses. In the absence of experimental data, correlational data can advance knowledge. Even in the simplest case of the correlation of two currently measured variables, a causal hypothesis is tested. If a zero correlation is obtained, or if the correlation is opposite in direction to that predicted, the credibility of the hypothesis has been sharply diminished. If a significant correlation in the expected direction is obtained, the credibility of the hypothesis is enhanced. The more elaborate correlational designs can offer additional evidence on the plausibility of alternative hypotheses. By establishing the temporal precedence of one of the variables or by disconfirming any effect of an alternative variable, these designs can narrow the range of possible causal explanations.

Time and Causal Designs

In all of the designs discussed earlier, time is an essential feature. In classic experimental designs, a variable is manipulated at time 1 and effects on the dependent variable are measured at time 2. In single-subject experimental designs, time is even more crucial since experimental manipulations are conducted at two or more occasions in a single subject or group of subjects. While time is not an important feature of simple correlational designs, it is crucial for sequential correlational designs.

Therefore, most of the designs we have discussed above are technically "longitudinal" since the independent and dependent variables are manipulated or measured at different points in time. However, the term *longitudinal design* usually refers to cases in which measurements are made at much longer intervals than in typical experiments—usually months or years instead of hours or days. Thus, longitudinal designs are useful for measuring long-term effects rather than immediate effects. They are appropriate for examining historical as opposed to contemporaneous effects (Chapter 2).

The logic of longitudinal designs is no different from the logic of designs used to measure immediate effects. The experimenter simply repeats over time the measurement of the dependent variables. Most of the designs discussed above may be used in a longitudinal format. The investigator may employ, for example, a longitudinal, experimental design for establishing long-term causal effects. Clinical studies that assess the long-term effects of treatment across randomly assigned experimental and control groups represent the case in point. Correlational designs may also be used in a longitudinal format to assess the relations between variables at different time lags. For example, cross-lagged panel correlation designs (D. T. Campbell & Stanley, 1966; Crano & Mellon, 1978; Kenny, 1975) assess variables at two different points in time. By computing concurrent and lagged correlations (time 1 with time 2), the investigator may establish covariation, temporal precedence, and evidence relevant to competing hypotheses (but see Rogosa, 1980, for a critique).

Longitudinal research is often proposed but rarely employed in research on close relationships. Since much of our interest is in the long-term effects of causal conditions, longitudinal designs are an obvious choice. However, the time and expense of such studies, combined with problems of subject attrition, are often prohibitive. Therefore, cross-sectional and retrospective designs have been employed as approximations to longitudinal designs.

In *cross-sectional research,* the investigator samples participants who are at different stages of development rather than following them through these stages. It can be considered a special case of the ex post facto design. The investigator, for example, examines two groups of married couples who are matched on various demographic criteria but who differ in number of years married. Cross-sectional research can establish covariation but, unlike a true longitudinal design, cannot establish temporal precedence of variables or developmental sequence. If an investigator found that older couples were more alike in their values and interests than younger couples, the investigator could not conclude that a mutual movement toward similarity developed over time, since the older couples might have been more similar when they first met or they may have experienced special historical events (e.g., the Great Depression) that promoted similar values and interests.

Retrospective designs also examine the long-term effects of causal variables but depend on the subject's memory to measure these variables. Couples might be asked, for example, to report on conditions at the beginning of their marriage and late in their marriage. Comparisons between these two reports suggest temporal covariations between variables. Because of the possible distortions in subjects' memories, as noted in our earlier discussion of participant reports, retrospective designs are often subject to the criticism that they tell more about subjects' constructions of events than about the actual events themselves.

Causal Analysis of the Close Relationship

Chapter 2 distinguished two kinds of causal effects: the momentary event-to-event causal connections and the effects of broad causal conditions. The logic of causal analysis, discussed above, applies equally well to both of these causal phenomena. For example, if we were interested in the event-to-event connections between a maternal behavior, such as ignoring, and a child behavior, such as whining, we might instruct mothers in one group to ignore their child's whining and instruct mothers in another group to respond to the whining. A comparison of whining in the two groups would indicate the nature of the event-to-event connections. However, if we were interested in the effects of a broad causal condition, such as financial security on marriage, we might conduct an income maintenance experiment like that discussed earlier in this chapter. In both cases, we would use a similar design, but, in the one case, we would investigate event-to-event connections, and, in the other, we would investigate the effects of a broad causal condition.

The social scientist rarely tries to understand the causal network of a single, specific event. Explanations of specific events are of interest to the lay person and the clinician and are of major importance in some legal cases. When the social scientist is invited to consult or testify in a legal case, this kind of explanation is required. However, in these settings, the event of concern is historical (e.g., a murder), and standard causal designs are impossible. Thus, in the legal setting, the social scientist is limited to post hoc analyses, and the limitations of such analyses are severe. In their typical work, social scientists are not concerned with a single, historical event but with broad causal conditions and recurrent event-to-event connections.

The following sections illustrate the major kinds of causal analysis the researcher of close relationships is likely to encounter. Specifically, we consider the effects of the physical environment, the social environment, and individual P and O characteristics on the dyadic relationship. Consistent with this book's emphasis on causal loops, we also consider the effects of the dyad on itself, on the individual members, and on the physical and social environment. As we discuss these causal analyses, we provide illustrative research sampled from the possible designs discussed above.

Effects of external factors on the dyad

The analysis of causal conditions external to the dyad, such as the effects of the physical and social environment, is more straightforward than the analysis of internal causal conditions. These external factors can be clearly distinguished from the dyad members and their relationship; thus, the measurement of the independent variable is unlikely to be confounded or contaminated by the measurement of the dependent variable.

Research on the effects of the physical environment has been primarily correlational in nature. For example, Rosenblatt and Cunningham (1976) investigated the hypothesis that television watching leads to family tension. The assumption was that noise, distraction, and discrepant preferences for programs and volume would generate family tension. A sample of 64 respondents answered questions about television watching in their home and about family tensions. Indices of the two variables were significantly correlated ($r = .42$), a finding that would support their hypothesis. On the basis of additional analyses, however, the investigators concluded that family tensions were leading to television watching, rather than the reverse. We will return to this evidence below.

Perhaps the most common study of the effects of the social environment on the dyad has been the investigation of marital or family therapy. Many of these investigations have been experimental, involving treatment manipulations across randomly assigned groups or within individual dyads. A study by Jacobson (1977) is especially interesting since it involves a combination of these two experimental approaches. Ten married couples who were dissatisfied with their marital relationship were randomly divided into an experimental treatment group and a waiting list control group. All couples were assessed with both self-report and observational measures. In addition, most spouses in the treatment condition monitored the daily occurrence of two problem behaviors for a 2-week period prior to treatment and throughout the course of treatment. Eight sessions of behaviorally oriented marital therapy, consisting of training in problem solving and contingency contracting, were provided the couples in the experimental condition. These couples negotiated contingency contracts successively on the two problem behaviors that they monitored. Results showed superior performance of the experimental group over the control group on measures of problem solving and marital satisfaction. In addition, multiple baseline comparisons conducted within the treatment couples showed that changes in the problem behaviors generally occurred when the contracts regarding these behaviors were implemented.

Effects of internal factors

Here we refer to the effects of P and O characteristics on dyadic interaction and to the effects of dyad characteristics on subsequent dyadic interaction. Distinguishing between these two kinds of effects can be difficult. If we observe, for example, an increasing asymmetry in the interaction of P and O, can we attribute it to the dominance of P, the submissiveness of O, or to a mutual accomodation between them?

INDIVIDUAL FACTORS. The experimental ideal for determining the effects of individual characteristics on dyadic interaction would require the random assignment of characteristics to members of the dyad or the random assignment of subjects to close relationships. While such experiments are ethically and practically impossible, a variety of experimental approximations can be conducted. Consider, for example, a series of studies by Patterson and his associates (cited in Patterson, 1979) that were able to demonstrate the effects of certain mother behaviors like ignoring on the probability of whining by the child. Using sequential analysis, these investigators first established what behaviors of the mothers regularly preceeded child whining. Then they employed a series of single-subject, reversal designs to experimentally establish the causal relation between these behaviors by the mothers and the whining of their children. Mothers were instructed to increase the frequency of these previously identified behaviors during the middle phase of an experimental period. Observations of the occurrence of child whining preceeding, during, and following that manipulation established the causal effects of the mothers' behaviors on the probability of child whining. Furthermore, these effects were demonstrated both in the laboratory and home settings.

Manipulations of dyad membership is another strategy for investigating the effects of P and O characteristics. Investigators have paired subjects for brief laboratory interactions (e.g., M. Snyder, Tanke, & Berscheid, 1977) or naturalistic dates (Walster, Aronson, Abrahams, & Rottman, 1966). By randomly pairing individuals or by experimentally manipulating the information participants have about each other, investigators have shown the effects on dyadic interaction of individual characteristics, such as physical attractiveness.

Correlational methodology represents the most common approach to assessing the effects of the individual on the dyad. The investigator examines the association between individual characteristics and dyad properties. At times, the temporal precedence of one over the other can be established. The classic studies of Newcomb (1961) on the acquaintanceship process, for example, assessed college students before they met each other and periodically while they lived together. Newcomb was able to show how certain individual attitudes were predictive of later friendship pairings.

Most correlational work on the effects of the individual on the dyad is only able to establish covariation, not temporal precedence. The ex post facto design is, for example, one common methodology. The investigator compares a group of dyads in which P has a particular characteristic (e.g., marriages with an employed wife) with a group of dyads in which P does not have that characteristic (e.g., marriages in which the wife is a full-time homemaker).

An interesting and elaborate way of analyzing the effects of the individual on the dyad involves a combination of the ex post facto design and an experimental manipulation of dyad membership. Consider, for example, a study by Liem (1974) that was based on an earlier methodology employed by Waxler (1974) and Haley (1968). The investigator had parents of schizophrenic sons and parents of normal sons interact with their own sons, with normal-stranger sons, and with schizophrenic-stranger sons. Through this design, Liem was able to show that communication problems in families with a schizophrenic son are more likely the result of the son than of the parents. Birchler, Weiss, and Vincent (1975) used a similar methodology with distressed and nondistressed married couples. Each member of a married couple interacted with his or her own spouse, with an opposite-sex, nondistressed-stranger spouse, and with an opposite-sex, distressed-stranger spouse. Their results cast doubt on individual effects on distress and suggested instead that "distress appears to be a function of a dyadic interaction. . . . Individuals learn how to relate in an ineffective manner specifically with their spouses while retaining social competency in interaction with others" (p. 359).

P × O FACTORS. Two classes of effects of P × O on their subsequent interaction can be distinguished. The first kind of effect, similar to a statistical interaction, occurs because of the combination of P and O characteristics. If we are interested, for example, in the interactional effects of various combinations of dominant and submissive persons or in the interactional effects of certain mood combinations between P and O, our independent variable is the hypothesized match of P and O attributes.

To analyze the effects of a P × O combination, the researcher can employ the kinds of manipulations of dyad membership discussed above. Here the researcher is not concerned with the main effects of some type of P characteristic (e.g., schizophrenia) but with the interaction of P and O characteristics (e.g., schizophrenia in combination with an intrusive parent figure or schizophrenia in combination with a distant parent figure).

Another method for discovering the effects of P × O combinations is analogue experimentation. Through role playing, hypnosis, or instructions, the investigator can attempt to create in existing dyads the P × O combinations of interest. Consider, for example, a recent investigation of demand characteristics in observations of marital interactions by Vincent et al. (1979). Samples of nondistressed and distressed couples first completed a problem-solving interaction task, which was videotaped. Next, half of the couples were told to complete a similar task "as if they were trying to present themselves as the most happy, blissful, and contented couple that they could imagine" (p. 560). The remaining couples were told to complete a similar task "as if they were trying to present themselves as the most unhappy, conflicted, and distressed couple that they could imagine" (p. 560). The effect of these "fake good" and "fake bad" instructions were apparent on a number of behavioral categories. However, very little evidence for differential responsiveness by distressed and nondistressed couples to these instructions was apparent. This study illustrates the use of instruction to create a dyadic combination (a mutual intent to fake good or bad) and measures its effects on behavior.

A second kind of dyadic effect occurs because of the joint activities of P and O. Not only their individual characteristics, whether examined separately or in combination, but also their shared history can affect the interaction of P and O. If P and O make an agreement with one another, if they shape each other's behavior through mutual reinforcement, if they come to have similar attitudes by reading the same books and going to the same movies—all of these joint activities will have impact of their future interaction.

This second kind of dyadic effect can be examined through longitudinal research that correlates early dyadic characteristics with later dyadic effects. C. T. Hill, Rubin, and Peplau (1976) showed that dating couples who initially reported more closeness, love, and exclusiveness were more likely to be together at a 2-year follow-up. King and Christensen (in press) demonstrated that dating couples who had engaged in a number of important behaviors indicative of increasing intimacy were more likely to be together 5 months later. These behaviors included such activities as expressions of affection (e.g., saying "I love you" to the partner) and sharing of resources (e.g., loaning each other money). While not as compelling as longitudinal research, cross-sectional research and retrospective studies can also be employed to examine the effects of the dyad. For example, Braiker and Kelley (1979) had young married couples report on the history of their relationships. From these data, the investigators were able to identify certain stages in the development of close relationships and to show the course of important conditions, such as love and conflict–negativity, during these stages.

The ex post facto design can also be employed to examine dyadic effects resulting from shared experiences. Consider, for example, the common comparison of distressed and nondistressed relationships. The experimenter can use such a comparison to elucidate the interactional effects of repeated experiences of unsatisfactory relating. However, since most investigators are more interested in factors that lead to distress than in factors that result from distress, they typically employ this design to explore possible causal factors in distressed relations. Consider, for example, Patterson's (1976b, 1979) work on the coercion phenomenon. He has provided evidence that negative reinforcement arrangements, in which family members use aversive responses to terminate the aversive responses of others, are more likely in problem families than in normal families. His hypothesis is that this process of coercion is causally related to the relationship characteristics of dissatisfaction and distress.

Effects of the dyad on close relationship members and the external world

Researchers have treated the close relationship more often as a dependent variable than as an independent variable. They have used many of the strategies discussed above to examine the impact of various conditions on the relationship. But they have not used similar strategies to study the impact of the close relationship on its members and on the social and physical surroundings. A more complete understanding of close relationships requires examination of the full circle of causal effects. In this section, we illustrate some of the strategies for exploring the effects of the close relationship.

EFFECTS ON THE INDIVIDUAL. Some of the most important effects of the dyad concern its impact on individual participants. One of the most interesting general questions is "What types of dyadic experiences and characteristics lead to individual distress, and what types of such experiences lead to individual fulfillment?" Using a primarily correlational methodology, research is beginning to address these issues. For example, a large body of correlational literature has documented an association between marital disruption and a variety of physical and emotional disorders. Experiences as diverse as psychopathology, automobile accidents, suicide, homicide, alcoholism, and high blood pressure have been linked to marital disruption (see Bloom, Asher, & White, 1978, for a review).

Because of the limitations of correlational methodology in exposing causal effects, some investigators have used longitudinal designs to establish temporal order as well as covariation. In a 2-year longitudinal investigation of life events and psychiatric symptomology, Myers, Lindenthal, Pepper, and Ostrander (1972) demonstrated that major changes in interpersonal functioning, whether positive (e.g., marriage) or negative (e.g., separation from spouse, trouble with boss) were associated with increased psychiatric symptoms.

Brim (1958) was able to exploit a natural experiment to demonstrate some effects of the dyad on the individual. Using data collected by Koch (1954, 1955), Brim showed that children with opposite-sex siblings had more traits of the opposite sex than children with same-sex siblings. The mechanisms that brought about these changes were not clear (e.g., parental attitudes, modeling between opposite-sex sibs); however, since sex of children is determined by processes that approximate randomness, one can be confident of the causal status of sexual composition of sibling pairs in determining the behavior of the individual children.

Classic experimental designs have also been used to examine the effects of the dyad upon the individual. McLean, Ogston, and Grauer (1973) randomly assigned 20 depressed outpatients to either a behaviorally oriented marital therapy condition or a comparison group that received treatment from other available agencies. Treatment in the comparison group consisted of either medication, group therapy, individual psychotherapy, or some combination of these, but in no case was the spouse involved. Posttreatment comparison on questionnaire and observation measures favored the marital therapy condition over the comparison group. Thus, efforts to alter interaction in the marital dyad had clear effects on the individual depression, beyond what was achieved by alternative treatments that did not involve the spouse.

EFFECTS ON THE SOCIAL AND PHYSICAL WORLD. The effects of the dyad are not limited to the individual members, but extend to their social and physical surroundings. Questions about effects of the marital dyad upon family and child functioning are of great social importance. Using correlational methodologies, a number of investigators have shown a relationship between marital problems and child disturbance (e.g., Christensen & Margolin, 1981; S. M. Johnson & Lobitz, 1974; Love & Kaswan, 1974; Oltmanns, Broderick, & O'Leary, 1977). Furthermore, marital disruption (separation and/or divorce) has been associated with child disturbances (Hetherington, 1979; Hetherington, Cox, & Cox, 1977).

Relatively few investigations have examined the effect of close relationships on the physical world. The previously mentioned study by Rosenblatt and Cunningham (1976) on television watching and family tensions is an interesting exception. They suggested that family members may use television as a way of escaping or avoiding conflict. "Operation of the television set could be used by family members to avoid tense interaction and the expression of anger and aggression" (p. 105). These investigators found a significant positive correlation between the two variables, but this finding was consistent with the hypothesis above as well as its reverse, that television watching creates tension. Faced with this interpretative difficulty, the authors used additional variables in an attempt to differentiate the hypotheses and concluded that the relationship between family tension and television watching is "due more to the use of television watching in order to avoid tense interaction than to television set operation as a source of frustration" (p. 109). Thus, they used a correlational design with multiple variables to show the effect of interaction (tension) on the physical environment (television set operation). Of course, members of close relationships modify their physical environment in many more important ways than this example suggests. Families clear fields, build homes, and litter their environment. But studies of how close relations affect these kinds of changes are yet to be done.

Some Recommendations for Causal Analysis

We have discussed the variety of designs that are possible for investigating causal effects in close relationships, and we have shown how these designs can be used to investigate specific causal questions about close relationships.

Our exposition has been illustrative rather than comprehensive. The field currently provides many more questions than answers. What we have tried to show is that the tools for seeking these answers are available.

Methodological expertise is not sufficient by itself for answering causal questions. Clear conceptualization is necessary to guide methodological decisions. In this spirit, we offer several methodological recommendations based on the conception of close relationships detailed in this book.

1. Investigators can profitably rely more often on longitudinal designs. Our definition of close relationship includes a criterion of longevity. In order to be described as close, a relationship must endure over time. Some of the major questions about close relationships relate to their durability: e.g., what leads to long-term relationships and what stages occur in close relationships? A full understanding of these questions, and of close relationships themselves, is not possible without longitudinal research.
2. Investigators can usefully employ designs that investigate the intricate causal "loops" frequently involved in relationship processes. Many of the important conditions in close relationships function both as cause and effect. The characteristics of the members, the dyad, and the social and physical environment affect the relationship and are affected by the relationship. Investigators must focus on both sides of the reciprocal effects in order to come to a full understanding of close relationships.
3. Investigators can usefully employ single-subject designs. Because close relationships are durable social systems that are constantly subjected to interacting effects, they often become more and more unique over time. In order to understand the range of possibilities for close relationships, we must focus increasingly on individual dyads. In order to understand the effects of causal conditions on unique dyads, we must examine those unique dyads. Single-subject designs measure individual dyads repeatedly and are thus the strategies that enable the researcher to comment about the response of particular dyads.
4. Investigators should employ experimental designs whenever possible. Since we desire to understand causal effects, experimental designs offer an inherent advantage over correlational designs. When we cannot employ experimental designs, we should try to use sequential correlational designs and multiple variable correlational designs rather than simple correlational designs. With the former designs, we can restrict the range of possible causes by establishing temporal order and/or negating alternative plausible hypotheses.

Of course, there are limits to the kind of manipulations we can employ. We cannot subject relationships to harm. We must have the informed consent of our participants. However, we can engage in manipulations designed to better the lives of our participants. Such manipulations need not involve formal therapy. They can include information and exposure to new experiences that may affect the lives of participants in positive ways. If we want to understand family communication, we can expose families to positive models of open discussion and problem solving. If we want to understand sibling relations, we can experiment with games that encourage cooperation play. The opportunities for enhancing close relationships, and at the same time increasing our knowledge of them, are great in number and scope.

ETHICAL ISSUES IN RESEARCH ON CLOSE RELATIONSHIPS

The activities of description and causal analysis often carry risks for research subjects. Both report and observational methodologies may invade their privacy. During an interview, participants may reveal more than they had intended about themselves. During an observational session, they may get so involved in the task at hand that they do or say things that embarrass them later. If the observation or report procedures are conducted conjointly, participants may reveal information to their partners that they later regret. In each case, the researcher's data collection procedures have instigated revelations that the participants did not intend. Furthermore, these procedures may provoke subsequent thought and discussion by the participants that have unintended relationship consequences (Z. Rubin & Mitchell, 1976).

The experimental manipulations required for causal analyses can also have negative consequences for research subjects. Nowhere is this clearer than in the income maintenance studies reported above. Although the marital disruptions that ensued from the experimental manipulations may have been positive outcomes in many cases, such dramatic effects of research must give the investigator pause. Even less powerful manipulations may have unintended, negative consequences. Asking subjects to engage in problem-solving tasks or instructing them to act "distant" from their spouse can create distress by encouraging them to behave in ways that are unfamiliar or emotionally volatile for them. If deception is involved in these manipulations, however temporary, the ethical problems are compounded.

Research subjects often seek information about the results of the project or their own performance in it. The investigator can hardly refuse these reasonable requests for information, and, in most cases, no ethical problems will be involved. However, if the information reflects poorly on the participant or the relationship, some distress can be generated. Should the investigator inform a married couple that they performed poorly on a conflict resolution task and that performance on this task predicts breakup?

Certain commonly accepted and, in some cases, required procedures are necessary to handle the ethical issues discussed above. Research subjects should be informed in writing of all procedures, benefits, and risks of the research prior to participation. They must freely give their consent prior to participation and be able to terminate their involvement freely at any time. Debriefing should be conducted at the end of the study whether or not any of the procedures may have generated negative reactions. Certain resources should be available to accommodate any negative effects on the research. These will depend on the nature of the study (e.g., referrals for counseling in studies of distressed couples, medical backup for drug studies). These and other issues are thoroughly discussed in the American Psychological Association manual on research ethics (APA, 1973).

Beyond these standard procedures, certain qualities of the researcher will go far to ensure ethical conduct. These qualities include sensitivity to the participant's reactions to the study; common sense in distinguishing important reactions from unimportant ones; and a generous supply of courtesy, respect, and skill in dealing with people. Only careful selection, training, and supervision of research staff can ensure that these qualities will be regularly exhibited in the research context.

CONCLUSIONS

We have examined a number of methodological issues of interest to investigators of close relationships. Clearly, sophisticated methodologies are of major importance in producing reliable information about relationships. Nevertheless, methodology is not the most important component of the research enterprise. Without well-conceived theory to guide the researcher's inquiries, analyses, and interpretations, methodological sophistication can be useless. Methods are tools to be used in collecting systematic information about relationships. But theory identifies the empirical phenomena that should be examined and provides a context within which the obtained information can be understood.

Theory and method are not independent. Theoretical development often generates new methodological strategies for studying relationships. And methodologies often produce new information that stimulates theoretical development. Even research that seems clearly designed to examine only methodological or theoretical issues often provides valuable information about the other domain. Studies designed to test aspects of human cognitive processing can inform our efforts to construct reliable self-report questionnaires. Studies of reactivity to observation can enrich our understanding of situational factors that influence behavior.

Thus, theory and methods are linked in a complex, dynamic, and interactive relationship. Advances in one will affect advances in the other. The causal loops that pervade close relationships are also apparent in the ways we study those relationships.

CHAPTER 12

Epilogue: An Essential Science

HAROLD H. KELLEY

At the beginning of this book, in Chapter 1, "The Emerging Science of Relationships," we observed that basic knowledge within the social, behavioral, and biological sciences is essential to an understanding of human relationships. Here we wish to make a case for the opposite point, that basic knowledge of close relationships is *essential* to the other disciplines, especially psychology and sociology. This view, too, was expressed in Chapter 1: ". . . since many human characteristics are determined by the nature of social relationships, the knowledge contributed by a science of relationships is ultimately critical to the full development of psychology as well as many other of the behavioral and biological sciences" (pp. 9–10). "No attempt to understand human behavior, in the individual case or in the collective, will be wholly successful until we understand the close relationships that form the foundation and theme of the human condition" (p. 19). In this final chapter, now that we have developed our framework and applied it to various problems, we can use it further to explain why the science of close relationships is indispensable for the full development of both psychological science and social science.

The general validity of our point will, perhaps, not be questioned. It is widely accepted that close relationships affect both their individual members and the social collections within which they exist. Furthermore, it has often been noted that these effects make it important to know the basic facts about the functioning of such relationships. For example, from the side of psychology, we might note Sears' (1951) comments:

> . . . the fact is that a large proportion of the properties of a person that compose his personality are originally formed in diadic situations and are measurable only by reference to diadic situations or symbolic representations of them. (p. 479)
> Individual and group behavior are so inextricably intertwined, both as to cause and effect, that an adequate behavior theory must combine both in a single internally congruent system. (p. 476)

From the side of sociology, there are analogous observations about the part that close relationships play in the structuring and continuity of society. For example, R. H. Turner (1970) describes the family as a basic unit in social stratification, as follows:

> Parents . . . are agents of various groups in carrying out their socialization responsibilities. In an important sense they are agents of the class system, and the most crucial imprint they place on the child is his projection into one class or another as he approaches adulthood. (p. 452)

Turner also observes that the interaction in the family affects how it moves up or down within the social system, as in this example:

> The behavior of the husband and wife in entertaining friends or acquaintances in their home is suffused with possibilities for consolidating, improving, or worsening the social standing of the family. (p. 447)

Another sociologist, Hagestad (1981), describes the family as a mediator between society and the individual:

> . . . the life course is shaped not only through a set of distal societal factors, such as age-related norms and historical events. A powerful proximal social force is found in the family. (p. 14)
> . . . the personal meaning of societal change often depends on family meanings: concrete effects on family resources and interpretations of what took place. (p. 28)

Our framework for the analysis of close relationships, as schematized, for example, in Figure 2.5, enables us systematically to examine the relations between the science of close relationships and other disciplines. The interaction in a close relationship is viewed as existing in a causal loop with its causal conditions. Thus, it is affected by and, in turn, affects its personal, relational, and environmental conditions. The framework also emphasizes that the nature of these effects depends on the interaction process, constituted by the interchain connections within the relationship. For example, the relationship is affected by its E_{soc} and, in turn, affects that environment. It is also affected by the dispositions and traits of its members, the P and O conditions, and in turn, affects them. In each case, the nature of the effects is determined by the causal connections between the two persons' chains of events. The close relationship's interaction also provides possible causal links between different types of causal condition. For example, it may link E_{soc} with E_{phys}, as when a young couple furnishes, arranges, and maintains their home in ways that are influenced by their friends and relatives.

Here, we wish to consider the particular relations between a science of close relationships and the neighboring sciences concerned, respectively, with the social environment (the social sciences, especially sociology) and with the personal conditions (the individual sciences, especially psychology). For this purpose, a useful reconfiguration of our schema is that shown in Figure 12.1. The close relationship (i.e., its interaction) is located between E_{soc} and P and O, the relative vertical positions defining a scale of increasing size of unit of analysis, with the close relationship being intermediate between the individual units and the social units. The figure shows separately the causal loops between the relationship and its social environment, on the one hand, and between the relationship and its personal causal conditions, on the other hand.

Figure 12.1 serves to identify a number of aspects of the relation between a science of relationships and such related sciences as psychology and sociology. First, it calls our attention to the possibility that close relationships must enter into the generalizations of the neighboring disciplines. The reason is that the causal conditions that are the subject matter of those disciplines may be affected in important ways by close relationship interactions. Psychology is concerned with characteristics of P and O, such as their motives, skills, habits, values, attitudes, and so on. Various close relationships in which P and O have experience may be causally linked to those characteristics. Sociology deals with characteristics of E_{soc}, such as social structure, norms, organization, and so on. Close relationships, individually and collectively, have causal links to their social environments and, therefore, may exercise influence on their stability, change, and so on.

Second, Figure 12.1 calls our attention to the possibility that close relationships may form part of the causal linkage between E_{soc} and P or O. That is to say, close relationships may form a part of the interface between society and the individual. They may mediate influence "downward" from society to the person, through such processes as socialization, family mobility, interpretation of social events, and control of social contacts. And close relationships may also mediate influence "upward" from individual to society, for example, when, through shaping individuals' reproductive activities, they collectively change the demographic features of a nation, or when they provide support for the deviant individual who becomes an effective agent of social change.

The comments above outline some of the *possible* ways that various relationships are linked causally to P, O, and E_{soc}, and provide *possible*

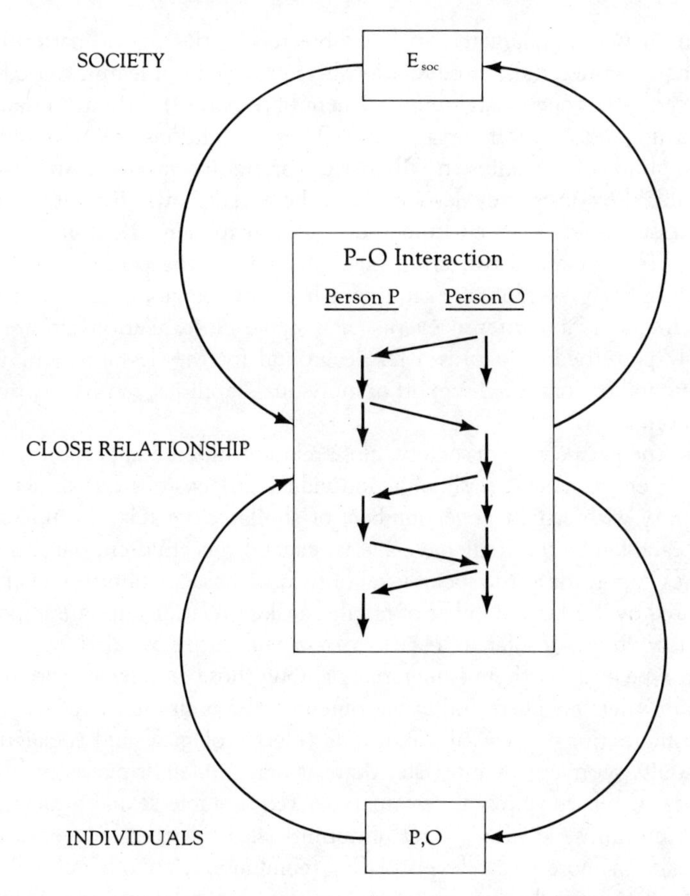

FIGURE 12.1
Schematic outline of a close relationship's causal links with its social environment and personal causal conditions. The interaction within the close relationship both (1) affects and is affected by the E_{soc} *and P and O causal conditions and (2) mediates effects from society to individuals and from individuals to society.*

linkages between the individual and social levels. A consideration of the specific properties of *close* relationships suggests that these linkages are not merely possible but are *probable* and, indeed, are likely to be *very important.* Our definition of "close" identifies a class of relationship in which individuals are exposed to strong interactional influences of diverse types and over a considerable period of time. This definition refers to relationships, such as those between lovers, spouses, close friends, and longtime co-workers, that, by their very nature, have strong impact on both individuals and society.

From the perspective of the individual, close relationships are the sites of frequent and intense events that occur over a long time period. To the extent the interaction is characterized by stable regularities in the patterning of interchain connections, close relationships provide ideal learning conditions (e.g., consistent cueing and reinforcement of responses)—though not always for learning "ideal" skills or attitudes. These conditions are epitomized by parent–child relationships in which the parents interact consistently with the child, the consistency deriving from the parents' own dispositions (e.g., interaction habits acquired from their earlier experiences); from social influences (e.g., manuals on child raising); or from the parent–child interaction itself (e.g., a mutually acquired pattern of coercion and compliance). If the frequent and intense events of a stable close relationship are consistently patterned, it affords a fertile ground for the development, modification, and/or continued support of individuals' abilities, motives, attitudes, and so on.

From the perspective of society, close relationships are important not only for their effectiveness in shaping individual propensities but also for the uniformity with which large numbers of similar close relationships do so. Thus, each family unit may have a great effect on its children, but this effect becomes important at the social level only insofar as a certain type of effect is replicated over a large number of families. Close relationships are important to society, then, insofar as (1) there is a *multiplicity* of relationships of a certain type and (2) there is *uniformity* among those of a given type in their effects on their members and environments. The multiplicity reflects recurrent combinations of causal conditions (e.g., biological and social needs, propinquity, convergent interests) that, at many different places and times, give rise to relationships of certain types (e.g., stable sexual attachments, close friendships, kinship). The uniformity also reflects homogenizing influences from more molar levels of E_{soc} (community, church, school, work group). In their multiplicity, certain types of close relationship provide the building blocks for these social organizations, inasmuch as individuals usually belong to them as members of various close relationships rather than as separate persons. And the organizational membership of close relationships acts to heighten the uniformity among them. Thus, the many family groups in a parish are potential elements for the formation of a church's congregation. Once formed, the congregation acts to create and maintain uniformities among its member families, for example, in their stability, rules for spousal interaction, and moral principles taught their children.

In their multiplicity and uniformity, and through the strong, enduring effects they have on their members, close relationships collectively have marked effects on the society. For example, being affected in similar ways by a change in climate, such as an extended drought, large numbers of families migrate from one place to another, with enormous consequences for social organization, even at the national level. In such instances, we observe that close relationships collectively constitute part of the linkage between E_{phys}

and E_{soc}. Similarly, changing economic conditions may create the necessity for wives, even those with small children, to enter the labor market. As many couples find it necessary to respond to common economic circumstances in this particular way, they collectively create the need for expanding the resources available for child day care (e.g., facilities, qualified personnel). In such instances, we see that close relationships collectively form part of the linkage between one feature of E_{soc} (economic conditions) and another (organizations and occupations).

Another fact highlighted by our scheme is that the nature of the causal linkage provided by the close relationship depends on the specific interaction processes. In the examples above, whether and for where a family leaves a drought stricken area and whether and where the wife seeks employment will be determined by decision processes that include some sort of discussion between husband and wife (and possibly other family members as well) and a resolution of their differing preferences through mutual influence (see Chapters 5, "Power," and 9, "Conflict"). Only if these various processes, as they occur in thousands of couples, are similar in their outcomes, will the environmental influences have uniform effects over the collective mass of the affected families.

Our schematic drawing of the interchain connections between the intrachain strands of activities and events, as in Figure 12.1, provides a general description of possibly important consequences of interaction process. From a psychological perspective, the interchain connections are important insofar as they give rise (1) to stable patterns of intrachain connections (e.g., characteristic ways of overtly expressing one's internal feelings, ways of encoding one's thoughts into language) and (2) to changes in the connections between intrachain events, on the one hand, and events in the external environment or the activities of other people, on the other hand (e.g., certain affective responses to another person's distress, tendencies to approach challenging tasks with confidence, stereotyped beliefs elicited by hearing certain speech dialects). These are interactional accounts of the origin of such psychological features as thoughts, motives, habits, skills, attitudes, and social dispositions.

From the perspective of society, the interaction processes are important insofar as they (1) form the basis for stable relationships, and (2) comprise ways in which the members are, individually and jointly, responsive to certain social cues, on the one hand, and effective in producing environmental events of social significance, on the other hand. Thus, the interchain connections are proximally responsible for the degree of stability of each relationship (see Chapter 7, "Love and Commitment"), for the extent of a couple's adherence to the moral teachings of their respective families of origin, and for the effectiveness with which they are able to accumulate economic resources, purchase a home, and so on.

The foregoing implies that, to attain a full understanding of the part close relationships play in relation to the causal conditions of their particular concerns, scientists in neighboring disciplines will find it necessary to turn to the results of interaction process analysis from close relationship researchers. In the absence of such results, our scientific neighbors will be compelled either to make untested assumptions about interaction processes or to investigate the processes for themselves. The latter has often been the case in the past, reflecting the absence of researchers who regard themselves as specialists in the study of close interaction. Throughout the preceding chapters, we have seen ways in which the "emerging" science of close relationships is being founded on the studies of interaction conducted by investigators with roots in individual psychology (e.g., clinical, developmental) or in sociology (e.g., family, small groups). These studies of interaction reflect the "essential" nature of knowledge about close relationships being promulgated here, but also reveal a niche in the network of the behavioral and social sciences in which a new, distinctive discipline needs to be established and developed.

From both the psychological and sociological perspectives, close relationships have both a historical and a contemporaneous relevance. On the one hand, our schematic drawing, as in Figure 12.1, shows the interlinkage that exists, at any given time, between a relationship and its causal conditions—the contemporaneous aspect. On the other hand, the causal loops constituted by the upward and downward links imply that a relationship and its conditions are in a dynamic interplay that extends over time. As noted in Chapter 2, in certain cases, that interplay may be characterized in terms of positive or negative feedback, with initial changes being either amplified over time by the causal loops, or counteracted and dampened out by them. In view of the causal loops, the states of the relationship and of its causal conditions at any given point in time can, in principle, be traced back to earlier states and processes—the historical aspect of the close relationship analysis.

In the earlier quotation, Sears (1951) highlights the dual historical and contemporaneous relevance of close relationships for psychology: as the situations in which personality characteristics "are originally formed" (historical) and as the situations with reference to which these characteristics are presently manifested or "measurable" (contemporaneous). In relation to the latter, we may note that every psychological assessment and experiment conducted with humans is inevitably a social interaction. As a consequence, the level and variability of the results must be accounted for, in part, by social interaction phenomena. These phenomena are important not only in measuring "social" variables, such as attitudes and conformity, where the social situations greatly affect the observations (see Chapter 11, "Research Methods"), but also in more "basic" experimental procedures. For example, Kimble (1967) reports that the performance of subjects in eyelid-conditioning experiments is affected by their attitudes toward the social aspects of the situation. Those subjects who are hostile and resistant to being manipulated by the procedure show less conditioning of their responses.

To illustrate the interplay between close relationship research and other levels of scientific activity, we will now consider several specific areas of research. To suggest the broad relevance of our assertion that a science of close relationships is essential, these examples are drawn from diverse domains: psychosomatic medicine, economic behavior, and the social development of children. These examples are used to make a number of different points, but we will see two recurrent themes: (1) growing recognition of the necessity of examining causal links in which close relationships are implicated and (2) insights that these links are well understood only through close analysis of the interaction events internal to relationships.

SOCIAL SUPPORT

We noted in Chapter 1, "The Emerging Science of Relationships," that close relationships can influence mental and physical health. From the perspective of psychology or medicine, the close relationship is a causal condition linked to various symptoms of the person. This condition is usually described in terms of the social support provided the individual by interaction in various relationships. The most basic evidence here is provided by the correlation between people's relationships and their health. Traupmann and Hatfield (1981) provide a recent summary of such evidence. People who are married or have close intimates are happier and have less mental illness. They are also less vulnerable to physical illness, have fewer psychosomatic symptoms, and have lower mortality rates.

These results are usually interpreted as reflecting a contribution that close interaction makes to the individual's well-being. However, there are obvious problems in inferring causality from such correlational evidence. For example, one alternative interpretation might be that a person's propensity to unhappiness or illness may, in some way, contribute to lack of intimate relationships. Another interpretation might be that both the illness and the lack of social support stem from some third factor, such as a general incompetence in coping with life's problems.

The case for a support-to-health direction of causation is somewhat strengthened by more complex correlational data. There is evidence that the degree of social support modulates the correlation between external stress and health. For example, the "stress" a person experiences at work (e.g., urgent demands, work interruptions, heavy work load, role ambiguity, many responsibilities) is more closely correlated with indicators of "strain" (e.g., blood pressure, ulcers, neurotic symptoms) for workers with little or no social support than for those with much social support (French, 1974; House,

1980). Other evidence indicates that people endure stressful life changes better if they have social support (Cobb, 1976). If we can assume that work stress or traumatic life events impinge on people pretty much at random, then these results suggest that social support plays a causal role in attenuating the impact of E_{soc} and E_{phys} on the individual. Of course, that assumption cannot be confidently made in many cases, inasmuch as people often contribute in complex ways to the quality of their environment and the catastrophes that befall them. The causal mediation is most convincingly indicated by studies such as that of Cobb and Kasl (summarized by Kahn & Antonucci, 1981), in which workers who lost their jobs through a plant closing (for which, as individuals, they could hardly have been responsible) were contrasted with comparable workers in similar plants that did not shut down. Job loss produced severe mental and physical health effects, but these effects were less for persons with social support (i.e., having spouses, friends, and relatives).

If we may conclude, as most authors in this area do, that social support affects the impact on the individual of external stresses, there remains the question of what exactly constitutes "social support." It has been measured by individuals' reports of the number and quality of their interpersonal relations, for example, workers' ratings of their relations on the job, with supervisors and co-workers, and off the job, with spouses, relatives, and friends. What has been left unclear are the specific interactional events required for these relations to be "supportive." As Rook and Peplau describe in their review (1982), researchers in the area have proposed many different social exchanges and provisions as constituting social support. For example, Kahn and Antonucci (1981) suggest that social support be defined as "interpersonal transactions that include one or more of the following key elements: affect, affirmation, and aid" (p. 392). These elements include, respectively, expressions of liking, admiration, respect, or love; expressions of agreement or acknowledgment of the appropriateness or rightness of something another person says or does; and giving direct aid or assistance, including things, money, information, and time. Dunkel-Schetter and Wortman (1981) provide a quite similar list of the important components of social support, also suggestive of interaction events and scenarios that, if the authors' hunches are correct, should be observed in supportive interactions. These components are (1) emotional support: communication of love, concern, or respect; (2) tangible support: material aid or help with tasks; and (3) appraisal support: providing information enabling the other person to evaluate self and experiences.

Dunkel-Schetter and Wortman are especially explicit in calling for an analysis of the interaction between persons who are victims of life crises and the others with whom they have close contacts. Working from the perspective of their studies of victims of cancer and rape, these authors thoughtfully analyze the particular interaction scenarios that commonly occur between victims, who need to express their feelings and find out that these are normal, and their confidantes, who are made to feel uncomfortably vulnerable by the victim's plight and who wrongly assume that the interaction should avoid negative feelings and should be directed to cheering up the victim.

The last two decades of research on social support make an impressive case for the relation between social support and various aspects of health. The research seems now to have reached a point at which more precise understanding of the processes and mechanisms involved will require longitudinal studies of the details of people's social interactions. Longitudinal work is necessary to pin down the exact causal role played by social support. A specification of interaction events is necessary to identify the essential components of social support. From the evidence presently at hand, there is little reason to doubt that social support will prove to be a complex phenomenon consisting of several different effective ingredients. It is also likely that these various ingredients have their effect on health by quite different mechanisms. For example, some may have their effects through direct psychophysiological pathways, in much the way that petting and handling rabbits fed a high cholesterol diet reduced their heart disease (Nerem, Levesque, & Cornhill, 1980). (These investigators comment that different laboratories working with variations in animal diets may obtain divergent results because of differences among them in "sociopsychological environment." This observation provides a dramatic example of our earlier point that, whether recognized or not, much basic research involves social relationships.) In contrast, other components of social support may have their effects through exposing the person to influences that induce him or her to follow good health practices or to comply fully with a medical regimen (Cobb, 1976). It is apparent that description of the relevant interaction features will be an essential part of any investigation that proposes to identify the specific mechanisms that mediate the support-to-health linkage.

ECONOMIC BEHAVIOR

We consider here the intersection between economic analysis of consumer behavior and sociological analysis of family decision making.

Most of the attention of economists is directed at understanding the workings of the entire economy or of large sectors of it. Often this macroanalysis is supplemented with microanalysis, in which investigators study the behavior of separate units of the economy and attempt to understand how the outputs from these units collectively determine the aggregate phenomena of the economy. The question of the optimal unit for the microanalysis has been answered in various ways. One reviewer, Ferber (1973), observes that an early approach treated the individual as the economic actor. The individual was initially regarded as "economic man," who made decisions only on the basis of prices, income, and financial resources. This view was subsequently updated by recognition of the noneconomic considerations that enter individual "economic" decisions. Only later have close relationships, specifically, the family, been brought under study as the relevant economic unit.

It has often been sociologists who have found it necessary to emphasize that the family is the primary decision-making unit in such activities as working, saving, and buying. For example, R. Hill (1961) asserted:

> The nuclear family is the decision-making unit in asset accumulation. Many choices and decisions are made by family heads, to be sure, but they are made *for* the family and often involve some participation by the children. (p. 58)

More recently, Ferber (1973) has expressed concern about the lack of attention given the consumer activities of the family and the failure to realize that these activities would provide a more realistic account of economic behavior than the prevailing individualistic emphasis. In a similar vein, in comments about marketing research, H. L. Davis (1976) observed that

> the view of consumers as individual decision makers is still very much alive despite commonsense observation that the family is the relevant decision-making unit and a growing research interest in the field. (p. 242)

While there are these continuing concerns about the matter, it is now common practice in economic research for family units to figure prominently in the analysis. And families often play a similarly important role in the thought and research that underlies the marketing of various products. For example, one textbook author in the field of consumer behavior writes:

> The process by which families make decisions concerning their purchases of products and services is of major concern to the marketing executive. . . . Family influence studies have consistently shown that a number of family members may be involved in the decision to purchase a wide range of products and brands. (Runyon, 1980, p. 186)

In sum, it is now assumed in economic research that close relationships play an important part in the causal linkage between the individual's needs, goals, and efforts, on the one hand, and the overall economy, on the other. Sellers of goods assume the same linkage, recognizing that although their advertisements are read or seen by individuals, purchase decisions are usually made in and influenced by the family.

The research on family behavior, including marketing research, has been greatly influenced by the studies of Blood and Wolfe (1960) and subsequent investigators of family power (see Chapter 5, "Power"). As a consequence, the research has focused, in H. L. Davis' (1976) words, "on the outcomes of decision making rather than on the process that has led to those outcomes" (p. 252). Decision outcomes have usually been described in terms of who made a given decision or who had the greatest influence on it, whether the husband, the wife, or both. From this work, a number of generalizations emerge, for example, that individuals play a greater part in decisions about products they themselves use, and that joint decisions are more likely the greater the outlay required relative to family income (Ferber, 1973). Outcome research has undoubtedly been useful for marketing practitioners, for example, in indicating the particular types of purchase decision that are most influenced by husbands versus wives (H. L. Davis, 1976). The research has also been fruitful for sociological theory, for example, in establishing links between the husband's power in the family and such causal conditions as the spouses' relative resources and cultural norms (see Chapter 5).

On the other hand, there is much dissatisfaction with the current lack of knowledge about the relevant interaction processes. As H. L. Davis (1976) observes, the result of the focus on outcomes is

> that little has been learned about how families actually reach decisions. . . . It is as though one has tried to understand the game of chess by looking only at the outcomes of each game, ignoring entirely the strategies used by each player. (p. 252)

There are a number of illustrations of how knowledge about interaction process can enhance the understanding of the collective effects of family economic behavior. Two examples can be mentioned here.

First, failure to study interaction process has probably led to underestimates of the importance in family economic decisions of women and children. Evidence on this point is provided by Turk and Bell's (1972) findings that questionnaire measures show wives and children to have far less influence than they are observed to have in actual family discussions. There are a number of indications that the widely used simple "power" questions (e.g., about who is "boss" or who "wins out" in case of disagreement) elicit answers heavily influenced by social norms about the husband's authority. Biases of this sort can be reduced through more focused and thorough questioning, and, employing such methods, researchers have found clear evidence, for example, of children's influence on purchase decisions. High school and college-age children seem to serve to interpret social change to their parents and, as a consequence, influence decisions about movies, fashions, and such (Hagestad, 1981; Katz & Lazarsfeld, 1955). The simple questions probably underestimate the influence of wives in traditional marriages by failing to detect the subtle methods by which they impress their preferences on family decisions while publicly maintaining their husbands' authority. The nature and scope of the influence of family members given less authority by cultural norms can only be identified in the details of interaction, derived from observations of actual discussions or detailed accounts by the participants of their recent transactions.

Second, in the absence of information about interaction process, researchers create unrealistic models of "decision making." Overly simple notions of "economic man" find their parallels in current ideas about stages of the family decision process. These consist of idealized scenarios of the successive steps that lead to an economic action, for example, recognition of a problem or need, gathering information, deliberation and discussion, decision, and, finally, action. In contrast to this neat scenario is the view provided by evidence that many purchases are made with little deliberation and that many well-developed plans are never acted on (Foote, 1974; Katona & Mueller, 1954). Foote's brief summary of families' explanations for their unplanned actions and unfulfilled plans provides insight into why simple stage models are inappropriate. Important facts are the long duration of the family, the changes in its environmental conditions that occur over time, and the interdependence among its various actions, due to limited resources. For example, an important household appliance finally (though unexpectedly) breaks down, an immediate replacement is required, and money planned for another use is diverted to the emergency need. An unplanned purchase is made under great time pressure, and an existing decision is postponed or modified. Such examples suggest that models of family decision making must take account of the interplay between environmental changes and the overlapping, long-lasting, and often ill-formulated plans of the family. More realistic models must be based on evidence about the succession of transactions around many different problems as they occur over lengthy periods of time and as they are given a temporal patterning by events external to the family.

Idealized stage models of decision making are also generally unrealistic in their implicit assumptions about the immediate circumstances surrounding decisions. Weick (1971) points out that families usually conduct their decision-making discussions under abominable conditions, for example, in the evening when the members are tired, in the midst of distractions, and with each new problem having to compete with other unfinished business. Here again, it seems that detailed description of the specific occasions and content of decision transactions, obtained from observation or detailed reports, is necessary for a realistic modeling of the process.

As students of the close relationship, we, of course, are interested in all the details of a family's inner workings. However, economists and marketers are likely to feel that, for their purposes, they need to know only the "bottom line" of the decision-making enterprise—the action taken. In contrast to that view, it can be argued that even they must often give attention to the details of the process leading to the final action. The reason is that the details of interaction process often provide clues about the conditions under which a family's actions will change. The point is a variant of one we have repeatedly emphasized: A family's economic behavior is linked to its P, O, P × O, E_{soc}, and E_{phys} conditions *through its internal processes.* Knowing that teenage children interpret certain social changes to their parents enables us better to predict, for certain types of family and types of purchase, the probable shifts in buying behavior with changing times. Knowing how plans are modified and reprioritized in situations requiring expedient action will enable us to anticipate how the family's economic behavior will respond to the wear and tear of their E_{phys} and to unanticipated shifts in their E_{soc}.

SOCIAL RECIPROCITY

The preceding two examples of social support and economic behavior illustrate some of the many ways in which close relationships mediate contemporaneous effects from society to person (e.g., the effect of stress at work place on the person's well-being) or from person to society (e.g., the effect of people's needs and actions on the economy). Our next example deals with historical (developmental) effects, illustrating some of the many ways in which close relationships prepare individuals for social life. In a sense, society depends on close relationships to prepare individuals in these various ways.

From the perspective of collective life, among the most important of people's propensities are their tendencies to reciprocate the actions of other persons and to realize that action they direct toward others is likely to be reciprocated. Sociologists have argued that social cohesion and equilibrium depend on positive reciprocity—on what Gouldner (1960), in an important statement on the subject, describes as "a generalized moral norm of reciprocity which defines certain actions and obligations as repayments for benefits received" (p. 170, italics deleted). Here, we briefly consider some of the evidence from recent studies of children's early social interactions suggesting the origins in close relationships of dispositions relevant to generalized reciprocity.

The term *reciprocity* has been used to refer to many different phenomena, but here we adopt Cairns' (1979) usage: ". . . reciprocity can be said to occur when the acts of two or more persons support each other in a relationship and their actions become similar to each other" (p. 298, italics deleted). Cognitively, reciprocity involves an understanding "that one's acts may eventually breed counteracts of a similar sort" (p. 298) from one's partner. This conception encompasses both positive reciprocity, the repayment of benefits received, and negative reciprocity, *lex talionis:* "an eye for an eye. . . ." The conception also ties reciprocity closely to expectations about the consequences of one's actions. These expectations provide a basis for recognizing the interplay between one's own effects on other persons and their effects on oneself, this recognition being necessary for internalizing the "norm of reciprocity."

From the perspective of our schematic framework, we may say that early interaction is the source of much of the child's intrachain organization. Although such organization does not absolutely depend on social stimuli, it does play an important part, as Cairns (1979) observes:

> From birth onward, social acts tend to play a key role in behavior organization because they are (a) more readily enmeshed with ongoing activities of the child, and (b) are more compelling (salient, intrusive) than are nonsocial events. (p. 316)

Our framework further suggests that reciprocity requires a special tuning of the intrachain sequences and their interchain connections so that P's and O's activities are mutually facilitative. Early interaction in close relationships must somehow affect the meshing between P's and O's behavior and, through cumulative experiences, affect their reciprocity-relevant interaction dispositions, that is, P and O causal conditions relating to attentiveness, responsiveness, communication, expectations, and so on.

We can mention here only a small part of the rich literature from developmental psychology that is relevant to social reciprocity. For more details, the reader is referred to Cairns' (1979) overview of the area. Some of the early origins of reciprocity in mother–child relationships are identified by Brazelton, Koslowski, and Main (1974), who conducted a microscopic analysis of intense mother–infant interaction during the first 20 months of the child's life. From the earliest weeks on, these pairs were observed to exhibit cycles of attention and withdrawal from attention. The infant typically initiates a cycle by orienting toward the mother, and she then responds in ways (smiles, vocalizations, touching) that accelerate the infant's looking at and responding to her. The infant soon reaches a peak of excitement followed by a deceleration of responsive activity and a withdrawal of gaze or turning away. A sensitive mother allows this withdrawal or even encourages it by looking away first. Her "reward" for doing so is that she receives a longer period of attention when the child next turns back to her. In this sort of interaction, a basic social skill is learned: paying attention to other persons. Through the interaction framework provided by the mother's cues, the child learns

> how to contain *himself,* how to control motor responses, and how to attend for longer and longer periods. (Brazelton et al., 1974, p. 70)

Other investigators identify additional skills that are learned in early interaction and that are relevant to social reciprocity. These include, for example, how to anticipate another person's action, how to interpret affect, how to read intentions, and how to give and take turns (e.g., Kaye, 1977; Schaffer, 1979). Mother–infant interaction is also seen as providing concepts necessary for later language learning, as Bruner (1975) argues here:

> The facts of language acquisition could not be as they are unless fundamental concepts about action and attention are available to children at the beginning of learning. . . . These concepts must be ones that are developed in mutuality with a speaker of the language. (p. 6)

As Bruner further explains, the necessary knowledge grows out of acting jointly with the mother, paying attention with her to a common focus, distinguishing segments of an organized action sequence, and learning to trade positions with her from being the recipient of an action to being the agent. These preverbal experiences create a child who is strongly oriented in its initial language use to pursuing and commenting on action being undertaken jointly with another person.

Current developmental research emphasizes the bidirectional nature of the influence between mother and infant. This emphasis is the result of careful studies of the details of interaction sequences that reveal ways in which the infant affects the behavior of its caretaker (e.g., M. Lewis & Rosenblum, 1974). Thus, infants are themselves modified by early interaction, but they also modify their mothers, for instance, affecting how the mothers feed them (Ainsworth & Bell, 1969; Kaye, 1977). As in the example above from Brazelton et al. (1974), sensitive mothers permit the infant to play an active part in shaping their interaction patterns. The bidirectionality of influence in mother–infant relationships is of basic importance for the development of social reciprocity inasmuch as it enables the child to learn that its actions are effective in controlling its social environment.

Socialization experiences in close relationships do not end after the first few years of life, and they extend well beyond the mother–infant dyad. Subsequent relations of significance include those with age peers—siblings and playmates. Recent observations of interaction between age-mates reveal that its extent during infancy is far greater than had previously been thought (Eckerman, 1979). Whereas encounters at 1 year of age are brief and often negative (reflecting competition over objects or space), the next 2 years see marked increases in the positiveness and complexity of interaction. From the age of 2, and with increasing frequency in the following years, pairs show strong tendencies to join each other in common activities, with one child's shift to a new activity being quickly followed by the other's joining in. As Cairns, Green, and MacCombie (1980) summarize:

> Perhaps the most striking outcome of these analyses has been the finding that children tend to entrap each other in common activities, common patterns of communication, and common feelings and behavior. . . . There is, thus, an

> *interpersonal bias* whereby children attend to the actions of others and participate with them in a common "plan" for behavior. (p. 94)

These studies of interaction in close relationships reveal some of the experiential bases of social attentiveness and responsiveness, the organization of behavior necessary for pursuing joint plans, and the basis of social give and take. In brief, we see here the origins of some of the many components that are expressed in cooperative adult activities requiring communication and coordination.

If Gouldner's (1960) sociological analysis of reciprocity were to be rewritten today, in the light of the recent knowledge gained from developmental studies of close interaction, would it be written any differently? We think so. Admittedly, there are huge unexplored gaps between the social tendencies observed in developmental research, on the one hand, and the generalized norms of reciprocal expectations and obligations that Gouldner imputed to adults, on the other. However, the elements of conformity to the norm—the knowledge of the effects of one's own actions on others, the expectations that "one's acts may eventually breed counteracts of a similar sort" (Cairns, 1979)—are found to be early products of close interaction. This observation suggests that the adult norm has a firm basis in skills and concepts that are well established and widely distributed. From this perspective, the generalized reciprocity norm need not be seen primarily as a means of social control necessary to suppress egocentric and egoistic individuals if there is to be social order. Rather, there is a good fit between society's needs for reciprocity-prone members and the tendencies that individuals are likely to bring to adult social life from experiences in their early close relationships. The origins of adult reciprocity behavior are to be found in the complex interplay between individual tendencies derived developmentally from close relationships and contemporaneous social and group processes that elicit and shape those tendencies.

CONCLUSION

Our thoughts about these and similar examples lead us to several conclusions. There are many ways in which psychological and social sciences can benefit from, and even find necessary, a science of close relationships. Because close relationships have such strong effects on their members and are so pervasive in any society, the dynamics of change in psychological and societal causal conditions cannot be fully understood without taking relationships into account.

As our examples suggest, these benefits of close relationship research will be greatest if investigators provide fine-grained, longitudinal descriptions of interaction. Of course, it is difficult to decide what to select from the complex, extended flow of interaction for inclusion in one's description. In any given period of research, investigators' selections are likely to be guided by prevailing theories about what is important. As a consequence, the selection is likely to impose somewhat artificial categories on interaction events, to fail to make distinctions within categories that later appear to be obvious, and to fail to record important aspects of interaction. Thus, much time and effort can be spent producing descriptions of interaction that are of little use to later researchers. With technological developments that permit extensive recording and storing of information, we hope that the degree of selectivity necessary in describing close interaction can be greatly reduced. We envisage here a system of minimally intrusive sound and video recording of natural interaction, supplemented with various postinteraction records (e.g., of the cognitive and affective threads of the intrachain strands) obtained as soon as possible after the interaction and possibly cued by the interaction recordings themselves. This area is of great promise for the creative invention and study of descriptive devices. We do not pretend to see clearly what will be required, but wish here simply to emphasize the need. If this need can be met, we might anticipate a future in which there exist libraries of comprehensive and detailed records of close interaction, records that lend themselves to descriptive analysis and reanalysis and that enable the development of causal theories to proceed inductively as well as deductively.

Having concluded that a science of close relationships is essential, we must finally return to a recognition of the interdependence between different levels of knowedge and the mutual interrelatedness among scientific efforts at the individual, close relationship, and societal levels. Close relationship research cannot afford, any more than can individual or social research, to be isolated from the efforts and results in neighboring disciplines. Many of its questions as well as many of its answers must be drawn from these associated areas. Thus, our framework, with its outline of the interrelations among causal conditions and between different levels of phenomena, applies in a general way to the process of scientific work in the behavioral and social sciences. This opinion does not reflect a delusional fixation of the present authors on interconnections and feedback loops—a *folie à neuf.* It is the very existence of causal links among our variables, configured in circular patterns, that requires all of us in the neighboring disciplines to be interdependent in our work.

References

Abbate, G. M. Child analysis at different developmental stages. *Journal of the American Psychoanalytic Association,* 1964, *12*(1), 135–150.

Ables, B. S., & Brandsma, J. M. *Therapy for couples.* San Francisco: Jossey-Bass, 1977.

Ackerman, N. W. Interlocking pathology in family relationships. In S. Rado & G. Daniels (Eds.), *Changing concepts of psychoanalytic medicine.* New York: Grune & Stratton, 1956.

Ackerman, N. W. *The psychodynamics of family life.* New York: Basic Books, 1958.

Ackerman, N. W. *Treating the troubled family.* New York: Basic Books, 1966.

Ainsworth, M. D. S. Object relations, dependency and attachment: A theoretical review of the infant–mother relationship. *Child Development,* 1969, *40,* 969–1025.

Ainsworth, M. D. S. Attachment and dependency: A comparison. In J. L. Gewirtz (Ed.), *Attachment and dependency.* Washington, D.C.: V. H. Winston, 1972.

Ainsworth, M. D. S., & Bell, S. M. Some contemporary patterns of mother–infant interaction in the feeding situation. In A. Ambrose (Ed.), *Stimulation in early infancy.* London: Academic Press, 1969.

Albert, S., & Kessler, S. Ending social encounters. *Journal of Experimental Social Psychology,* 1978, *14,* 541–553.

Aldous, J., Osmond, M. W., & Hicks, M. W. Men's work and men's families. In W. R. Burr, R. Hill, F. I. Nye, & I. L. Reiss (Eds.), *Contemporary theories about the family* (Vol. 1). New York: Free Press, 1979.

Alexander, F., & French, T. M. *Psychoanalytic therapy.* New York: Ronald Press, 1946.

Alexander, J., & Parsons, B. Short-term behavioral intervention with delinquent families: Impact on family process and recidivism. *Journal of Abnormal Psychology,* 1973, *81,* 219–225.

Almquist, E. M. Women in the labor force. *Signs,* 1977, *2,* 843–855.

Altman, I., & Haythorn, W. W. The effects of social isolation and group composition on performance. *Human Relations,* 1967, *20,* 313–340.

Altman, I., & Taylor, D. A. *Social penetration: The development of interpersonal relationships.* New York: Holt, Rinehart & Winston, 1973.

Altman, I., Vinsel, A., & Brown, B. A. Dialectic conceptions in social psychology: An application to social penetration and privacy regulation. *Advances in Experimental Social Psychology,* 1981, *14,* 108–160.

American Council of Life Insurance. *The Family Economist,* February 15, 1978.

American Psychological Association. Ethical principles in the conduct of research with human participants. Washington, D.C.: APA, 1973.

Andelin, H. B. *Fascinating womanhood.* New York: Bantam, 1963.

Anderson, J. R. *Cognitive psychology and its implications.* San Francisco: W. H. Freeman and Company, 1980.

Aponte, H., & Hoffman, L. The open door: A structural approach to a family with an anorectic child. *Family Process,* 1973, *12*(1), 1–44.

Arnold, M. B. *Emotion and personality* (2 vols.). New York: Columbia University Press, 1960.

Aronson, E. The effect of effort on the attractiveness of rewarded and unrewarded stimuli. *Journal of Abnormal and Social Psychology,* 1961, *63,* 375–380.

Aronson, E., & Mills, J. The effect of severity of initiation on liking for a group. *Journal of Abnormal and Social Psychology,* 1959, *59,* 177–181.

Asher, S., & Gottman, J. (Eds.), *The development of children's friendships.* Cambridge, England: Cambridge University Press, 1981.

Bach, G. R., & Wyden, P. *The intimate enemy: How to fight fair in love and marriage.* New York: Avon, 1968.

Bahr, S. J. Effects on power and division of labor in the family. In L. N. W. Hoffman & F. I. Nye (Eds.), *Working mothers: An evaluative review of the consequences for wife, husband, and child.* San Francisco: Jossey-Bass, 1975.

Bahr, S. J., & Rollins, B. C. Crisis and conjugal power. *Journal of Marriage and the Family,* 1971, *33,* 360–367.

Bakeman, R., & Brown, J. V. Behavioral dialogues: An approach to the assessment of mother–infant interaction. *Child Development,* 1977, *48,* 195–201.

Bakeman, R., & Dabbs, J. M., Jr. Social interaction observed: Some approaches to the analysis of behavior streams. *Personality and Social Psychology Bulletin,* 1976, *2,* 335–345.

Bales, R. F. *Interaction process analysis: A method for the study of small groups.* Reading, Mass.: Addison-Wesley, 1950.

Bales, R. F. *Personality and interpersonal behavior.* New York: Holt, Rinehart & Winston, 1970.

Bandura, A. Social learning through imitation. In M. R. Jones (Ed.), *Nebraska symposium on motivation, 1962*. Lincoln: University of Nebraska Press, 1962.

Bandura, A. *Social learning theory*. Englewood Cliffs, N.J.: Prentice-Hall, 1977.

Bandura, A., & Huston, A. C. Identification as a process of incidental learning. *Journal of Abnormal and Social Psychology*, 1961, *63*, 311–318.

Bandura, A., Ross, D., & Ross, S. A. A comparative test of the status envy, social power, and secondary reinforcement theories of identificatory learning. *Journal of Abnormal and Social Psychology*, 1963, *67*, 527–534. (a)

Bandura, A., Ross, D., & Ross, S. A. Vicarious reinforcement and imitative learning. *Journal of Abnormal and Social Psychology*, 1963, *67*, 601–607. (b)

Bandura, A., & Walters, R. H. *Social learning and personality development*. New York: Holt, Rinehart & Winston, 1963.

Bane, M. J. *Here to stay: American families in the twentieth century*. New York: Basic Books, 1976.

Banton, M. *Roles: An introduction to the study of social relations*. New York: Basic Books, 1965.

Barker, R. G. (Ed.). *The stream of behavior*. New York: Appleton-Century-Crofts, 1963.

Barnett, L. R., & Nietzel, M. T. Relationship of instrumental and affectional behaviors and self-esteem to marital satisfaction in distressed and nondistressed couples. *Journal of Consulting and Clinical Psychology*, 1979, *47*, 946–957.

Barton, C., & Alexander, J. Functional family therapy. In A. S. Gurman & D. P. Kniskern (Eds.), *Handbook of family therapy*. New York: Brunner/Mazel, 1981.

Basow, S. A. *Sex-role stereotypes: Traditions and alternatives*. Monterey, Calif.: Brooks/Cole, 1980.

Bates, F. L., & Harvey, C. C. *The structure of social systems*. New York: Gardner Press, 1975.

Bateson, G., & Jackson, D. D. Some varieties of pathogenic organization. In D. McRioch (Ed.), *Disorders of communication*. Association for Research in Nervous and Mental Disease, Vol. 42. Research Publications, Baltimore, 1964.

Beck, D. F. Research findings on the outcome of marital counseling. *Social Casework*, 1975, *56*, 153–181.

Becker, J. D. A model of encoding of experiential information. In R. C. Schank & R. M. Colby (Eds.), *Computer models of thought and language*. San Francisco: W. H. Freeman and Company, 1973.

Becker, H. S. Notes on the concept of commitment. *American Journal of Sociology*, 1960, *66*, 32–40.

Becker, H. S. Personal changes in adult life. *Sociometry*, 1964, *27*, 40–53.

Beckman, L. J., & Houser, B. B. The more you have, the more you do: The relationship between wife's employment, sex-role attitudes and household behavior. *Psychology of Women Quarterly*, 1979, *4*, 160–174.

Bell, J. E. Family group therapy as a treatment method. Paper presented at the Eastern Psychological Association Convention, Boston, 1953.

Bell, R. Q. A reinterpretation of the direction of effects in studies of socialization. *Psychological Review*, 1968, *75*, 81–95.

Bell, R. Q. Stimulus control of parent or caretaker behavior by offspring. *Developmental Psychology*, 1971, *4*, 63–72.

Bem, D. J., & Allen, A. On predicting some of the people some of the time: The search for cross-situational consistencies in behavior. *Psychological Review*, 1974, *81*, 506–520.

Bem, S. L. Gender schema theory: A cognitive account of sex typing. *Psychological Review*, 1981, *88*, 354–364.

Benne, K. D., & Sheats, P. Functional roles of group members. *Journal of Social Issues*, 1948, *4*(2), 41–49.

Bentler, P. M. Multivariate analysis with latent variables: Causal modeling. In M. R. Rosenzweig & L. W. Porter (Eds.), *Annual Review of Psychology* (Vol. 31). Palo Alto, Calif.: Annual Reviews, 1980.

Bentler, P. M., & Huba, G. J. Simple minitheories of love. *Journal of Personality and Social Psychology*, 1979, *37*, 124–130.

Berardo, F. M. Special issue: Decade review. *Journal of Marriage and the Family*, 1980, *42* (Whole No. 4).

Berger, D. G., & Wenger, M. G. The ideology of virginity. *Journal of Marriage and the Family*, 1973, *35*, 666–676.

Berger, J., Cohen, B. P., & Zelditch, M., Jr. Status characteristics and social interaction. *American Sociological Review*, 1972, *37*, 241–255.

Berger, J., Rosenholtz, S. J., & Zelditch, M. Status organizing processes. In A. Inkeles, N. S. Smelser, & R. H. Turner (Eds.), *Annual Review of Sociology* (Vol. 6). Palo Alto, Calif.: Annual Reviews, 1980.

Bergin, A. E. The evaluation of therapeutic outcomes. In A. E. Bergin & S. L. Garfield (Eds.), *Handbook of psychotherapy and behavior change*. New York: Wiley, 1971.

Bergler, E. *Divorce won't help*. New York: Harper, 1948.

Berk, R. A., & Berk, S. F. *Labor and leisure at home: Content and organization of the household day*. Beverly Hills, Calif.: Sage, 1979.

Berk, S. F. (Ed.). *Women and household labor*. Beverly Hills, Calif.: Sage, 1980.

Berkman, L. S., & Syme, S. L. Social networks, host resistance, and mortality: A nine-year follow-up study of Alameda County residents. *American Journal of Epidemiology*, 1979, *109*, 186–204.

Berlyne, D. *Conflict, arousal and curiosity*. New York: Academic Press, 1960.

Berlyne, D. Attention. In E. Carterette & M. Friedman (Eds.), *Handbook of perception* (Vol. 1). New York: Academic Press, 1974.

Berman, S., & Weiss, V. *Relationships*. New York: Hawthorn, 1978.

Bernal, G., & Baker, J. Towards a metacommunicational framework of couple interactions. *Family Processes*, 1979, *18*, 293–302.

Bernard, J. *The future of marriage*. New York: World, 1972.

Bernard, J. *The female world*. New York: Free Press, 1981. (a)

Bernard, J. The good-provider role: Its rise and fall. *American Psychologist*, 1981, *36*, 1–12. (b)

Berscheid, E. Commentary. *Medical Aspects of Human Sexuality*, 1980, *14*(9), 41–42.

Berscheid, E. Attraction and emotion in interpersonal relationships. In M. S. Clark & S. T. Fiske (Eds.), *Affect and cognition*. Hillsdale, N.J.: Erlbaum, 1982.

Berscheid, E. Interpersonal attraction. In G. Lindzey & E. Aronson (Eds.), *Handbook of social psychology* (3rd ed.). Reading, Mass.: Addison-Wesley, in press-a.

Berscheid, E. *The problem of emotion in close relationships.* New York: Plenum, in press-b.

Berscheid, E., & Campbell, B. The changing longevity of heterosexual close relationships: A commentary and forecast. In M. J. Lerner & S. C. Lerner (Eds.), *The justice motive in social behavior.* New York: Plenum, 1981.

Berscheid, E., & Fei, J. Romantic love and sexual jealousy. In G. Clanton & L. G. Smith (Eds.), *Jealousy.* Englewood Cliffs, N.J.: Prentice-Hall, 1977.

Berscheid, E., Gangestad, S., & Kulakowski, D. Emotion in close relationships: An overview of theory with implications for counseling. In S. D. Brown & R. W. Lent (Eds.), *Handbook of counseling psychology.* New York: Wiley, in press.

Berscheid, E., & Graziano, W. The initiation of social relationships and social attraction. In R. L. Burgess & T. L. Huston (Eds.), *Social exchange in developing relationships.* New York: Academic Press, 1979.

Berscheid, E., Graziano, W., Monson, T., & Dermer, M. Outcome dependency: Attention, attribution, and attraction. *Journal of Personality and Social Psychology,* 1976, *34,* 978–989.

Berscheid, E., & Walster, E. A little bit about love. In T. L. Huston (Ed.), *Foundations of interpersonal attraction.* New York: Academic Press, 1974.

Berscheid, E., & Walster, E. *Interpersonal attraction* (2nd ed.). Reading, Mass.: Addison-Wesley, 1978.

Biddle, B. J. *Role theory: Expectations, identities and behaviors.* New York: Academic Press, 1979.

Biddle, B. J., & Thomas, E. J. (Eds.), *Role theory: Concepts and research.* New York: Wiley, 1966.

Bienvenu, M. J. Measurement of marital communication. *Family Coordinator,* 1970, *19,* 26–31.

Bierstedt, R. An analysis of social power. *American Sociological Review,* 1950, 6, 7–30.

Bijou, S. W., & Baer, D. M. *Child development,* Vol. 1, *A systematic and empirical theory.* New York: Appleton-Century-Crofts, 1961.

Bindra, D. Emotion and behavior theory: Current research in historical perspective. In P. Beach (Ed.), *Physiological correlates of emotion.* New York: Academic Press, 1970.

Birchler, G. R., & Spinks, S. H. Behavioral-systems marital and family therapy: Integration and clinical application. *American Journal of Family Therapy,* 1980, 8(2), 6–28.

Birchler, G. R., Weiss, R. L., & Vincent, J. P. Multimethod analysis of social reinforcement exchange between maritally distressed and nondistressed spouse and stranger dyads. *Journal of Personality and Social Psychology,* 1975, *31,* 349–360.

Blair, E. Using practice interviews to predict interviewer behavior. *Public Opinion Quarterly,* 1980, *44,* 257–259.

Blau, P. M. *The dynamics of bureaucracy.* Chicago: University of Chicago Press, 1955.

Blau, P. M. *Exchange and power in social life.* New York: Wiley, 1964.

Block, J. Studies in the phenomenology of emotions. *Journal of Abnormal and Social Psychology,* 1957, *54,* 358–363.

Blood, R. O., Jr. *Love match and arranged marriage.* New York: Free Press, 1967.

Blood, R. O., Jr., & Blood, M. *Marriage* (3rd ed.). New York: Free Press, 1978.

Blood, R. O., Jr., & Hamblin, R. L. The effects of the wife's employment on the family power structure. *Social Forces,* 1958, *26,* 347–352.

Blood, R. O., Jr., & Wolfe, D. M. *Husbands and wives: The dynamics of married living.* Glencoe, Ill.: Free Press, 1960.

Bloom, B. L., Asher, S. J., & White, S. W. Marital disruption as a stressor: A review and analysis. *Psychological Bulletin,* 1978, *85,* 867–894.

Bloom, B. L., White, S. W., & Asher, S. J. Marital disruption as a stressful life event. In G. Levinger & O. C. Moles (Eds.), *Divorce and separation.* New York: Basic Books, 1979.

Blumer, H. *Symbolic interactionism.* Englewood Cliffs, N.J.: Prentice-Hall, 1969.

Blurton Jones, N. (Ed.). *Ethological studies of child behavior.* London: Cambridge University Press, 1972.

Blurton Jones, N., & Leach, G. M. Behavior of children and their mothers at separation and greeting. In N. Blurton Jones (Ed.), *Ethological studies of child behavior.* London: Cambridge University Press, 1972.

Bohannan, P. The six stations of divorce. In P. Bohannan (Ed.), *Divorce and after.* New York: Doubleday, 1970.

Bolton, C. D. Mate selection as the development of a relationship. *Marriage and Family Living,* 1961, *23,* 234–240.

Bott, E. *Family and social network.* London: Tavistock, 1957.

Bott, E. *Family and social network* (2nd ed.). New York: Free Press, 1971.

Boulding, K. *Conflict and defense: A general theory.* New York: Harper, 1962.

Bowen, M. Family psychotherapy with schizophrenia in the hospital and in private practice. In I. Boszormenyi-Nagy & S. L. Framo (Eds.), *Intensive family therapy.* New York: Hoeber, 1965.

Bowen, M. The use of family theory in clinical practice. *Comprehensive Psychiatry,* 1966, *7,* 345–374.

Bowen, M. Family therapy and family group therapy. In D. H. L. Olson (Ed.), *Treating relationships.* Lake Mills, Iowa: Graphic Press, 1976.

Bowen, M. *Family therapy in clinical practice.* New York: Aronson, 1978.

Bowers, K. S. Situationism in psychology: An analysis and a critique. *Psychological Review,* 1973, *80,* 307–336.

Bowlby, J. *Attachment and loss,* Vol. 1, *Attachment.* New York: Basic Books, 1969.

Bowlby, J. *Attachment and loss,* Vol. 2, *Separation: Anxiety and anger.* New York: Basic Books, 1973.

Braginsky, B. M., Braginsky, D. D., & Ring, K. *Methods of madness: The mental hospital as a last resort.* New York: Holt, Rinehart & Winston, 1969.

Braiker, H. B., & Kelley, H. H. Conflict in the development of close relationships. In R. L. Burgess & T. L. Huston (Eds.), *Social exchange in developing relationships.* New York: Academic Press, 1979.

Brain, R. *Friends and lovers.* New York: Basic Books, 1976.

Brazelton, T. B., Koslowski, B., & Main, M. The origins of reciprocity: The early mother–infant interaction. In M. Lewis & L. A. Rosenblum (Eds.), *The effects of the infant on its caregiver.* New York: Wiley, 1974.

Brickman, P., & Campbell, D. T. Hedonic relativism and planning the good society. In M. H. Appley (Ed.), *Adaptation-level theory.* New York: Academic Press, 1971.

Brim, O. G. Family structure and sex role learning by children: A further analysis of Helen Koch's data. *Sociometry,* 1958, *21,* 1–16.

Broderick, C. B., & Schrader, S. S. The history of professional marriage and family therapy. In A. S. Gurman & D. P. Kniskern (Eds.), *Handbook of family therapy.* New York: Brunner/Mazel, 1981.

Bronfenbrenner, U. Toward an experimental ecology of human development. *American Psychologist,* 1977, *32,* 513–531.

Bronfenbrenner, U. *The ecology of human development: Experiments by nature and design.* Cambridge, Mass.: Harvard University Press, 1979.

Brooks, B. Mother is a freshperson: A study of power in the family. Unpublished master's thesis, University of Massachusetts, Amherst, 1978.

Bruner, J. S. The ontogenesis of speech acts. *Journal of Child Language,* 1975, *2,* 1–19.

Bryson, R. B., Bryson, J. B., Licht, M. H., & Licht, B. G. The professional pair: Husband and wife psychologists. *American Psychologist,* 1976, *31,* 10–16.

Burgess, E. W., & Cottrell, L. S., Jr. *Predicting success or failure in marriage.* Englewood Cliffs, N.J.: Prentice-Hall, 1939.

Burgess, E. W., & Locke, H. J. *The family: From institution to companionship* (2nd ed.). New York: American, 1960.

Burgess, E. W., & Wallin, P. *Engagement and marriage.* Philadelphia: Lippincott, 1953.

Burgess, E. W., Wallin, P., & Shultz, G. D. *Courtship, engagement, and marriage.* Philadelphia: Lippincott, 1954.

Burgess, R. L., & Huston, T. L. (Eds.). *Social exchange in developing relationships.* New York: Academic Press, 1979.

Burić, O., & Zečević, A. Family authority, marital satisfaction, and the social network in Yugoslavia. *Journal of Marriage and the Family,* 1967, *29,* 325–336.

Burke, P. J. The development of task and social–emotional role differentiation. *Sociometry,* 1967, *30,* 379–392.

Burke, P. J. Role differentiation and the legitimation of task activity. *Sociometry,* 1968, *31,* 404–411.

Burke, P. J. Participation and leadership in small groups. *American Sociological Review,* 1974, *39,* 832–843.

Byrne, D. *The attraction paradigm.* New York: Academic Press, 1971.

Byrne, D., & Blaylock, D. Similarity and assumed similarity of attitudes between husbands and wives. *Journal of Abnormal and Social Psychology,* 1963, *67,* 636–640.

Cairns, R. B. *Social development: The origins and plasticity of interchanges.* San Francisco: W. H. Freeman and Company, 1979.

Cairns, R. B., & Green, J. A. How to assess personality and social patterns: Observations or ratings? In R. B. Cairns (Ed.), *The analysis of social interactions.* Hillsdale, N.J.: Erlbaum, 1979.

Cairns, R. B., Green, J. A., & MacCombie, D. J. The dynamics of social development. In E. C. Simmel (Ed.), *Early experiences and early behavior: Implications for social development.* New York: Academic Press, 1980.

Caldwell, M. A. Negotiation in sexual encounters. Unpublished doctoral dissertation, University of California, Santa Barbara, 1979.

Campbell, A., Converse, P. E., & Rodgers, W. L. *The quality of American life.* New York: Russell Sage Foundation, 1976.

Campbell, D. T., & Stanley, J. *Experimental and quasi-experimental designs for research.* Skokie, Ill.: Rand McNally, 1966.

Candland, D. K. The persistent problems of emotion. In D. K. Candland, J. P. Fell, E. Keen, A. I. Leshner, R. Plutchik, & R. M. Tarpy, *Emotion.* Monterey, Calif.: Brooks/Cole, 1977.

Candland, D. K., Fell, J. P., Keen, E., Leshner, A. I., Plutchik, R., & Tarpy, R. M. *Emotion.* Monterey, Calif.: Brooks/Cole, 1977.

Cannon, W. B. The James–Lange theory of emotions: A critical examination and an alternative theory. *American Journal of Psychology,* 1927, *39,* 106–124.

Cannon, W. B. *Bodily changes in pain, hunger, fear, and rage* (2nd ed.). New York: Appleton, 1929.

Caputo, D. V. The parents of the schizophrenic. *Family Process,* 1963, *2,* 339–356.

Carr, H. A. *Psychology, a study of mental activity.* New York: Longmans, Green, 1929.

Carson, R. C. *Interaction concepts of personality.* Chicago: Aldine, 1969.

Cartwright, D. P. A field theoretical conception of power. In D. P. Cartwright (Ed.), *Studies in social power.* Ann Arbor, Mich.: Institute for Social Research, 1959.

Cartwright, D. P. The nature of group cohesiveness. In D. P. Cartwright & A. Zander (Eds.), *Group dynamics: Research and theory* (3rd ed.). New York: Harper & Row, 1968.

Cartwright, D. P., & Zander, A. *Group dynamics: Research and theory* (3rd ed.). New York: Harper & Row, 1968.

Castellan, N. J., Jr. The analysis of behavior sequences. In R. B. Cairns (Ed.), *The analysis of social interactions.* Hillsdale, N.J.: Erlbaum, 1979.

Centers, R. *Sexual attraction and love: An instrumental theory.* Springfield, Ill.: Charles C. Thomas, 1975.

Centers, R., Raven, B. H., & Rodrigues, A. Conjugal power structure: A re-examination. *American Sociological Review,* 1971, *36,* 264–278.

Chapple, E. D. Quantitative analysis of complex organizational systems. *Human Organization,* 1962, *21,* 67–80.

Chase, S. *Roads to agreement.* New York: Harper & Bros., 1951.

Cherlin, A. Work life and marital dissolution. In G. Levinger & O. C. Moles (Eds.), *Divorce and separation.* New York: Basic Books, 1979.

Cherry, C. *On human communication.* Cambridge: MIT Press, 1978.

Christensen, A. Naturalistic observation of families: A system for random audio recordings. *Behavior Therapy,* 1979, *10,* 418–422.

Christensen, A., Arkowitz, H., & Anderson, J. Practice dating as treatment for college dating inhibitions. *Behavior Research and Therapy,* 1975, *13,* 321–331.

Christensen, A., & Hazzard, A. Reactive effects during naturalistic observation of families. *Behavioral Assessment,* in press.

Christensen, A., & Margolin, G. Correlational and sequential analyses of marital and child problems. Paper presented at the American Psychological Association Convention, Los Angeles, 1981.

Christensen, A., & Nies, D. C. The spouse observation checklist: Empirical analysis and critique. *The American Journal of Family Therapy,* 1980, *8,* 69–79.

Christensen, A., Sullaway, M., & King, C. Systematic error in behavioral reports of dyadic interaction: Egocentric bias and content effects. *Behavioral Assessment*, 1983, *5*, 131–142.

Clanton, G., & Smith, L. G. (Eds.). *Jealousy*. Englewood Cliffs, N.J.: Prentice-Hall, 1977.

Clark, M. S., & Mills, J. Interpersonal attraction in exchange and communal relationships. *Journal of Personality and Social Psychology*, 1979, *37*, 12–24.

Cobb, S. Social support as a moderator of life stress. *Psychosomatic Medicine*, 1976, *38*, 300–314.

Cochran, S. D., & Peplau, L. A. Values of attachment and autonomy in heterosexual relationships. Unpublished manuscript, University of California, Los Angeles, 1983.

Cohen, L. J. The operational definition of human attachment. *Psychological Bulletin*, 1974, *81*, 207–217.

Cohen, R. S., & Christensen, A. Further examination of demand characteristics in marital interaction. *Journal of Clinical and Consulting Psychology*, 1980, *48*, 121–123.

Colby, K. M. *A primer for psychotherapists*. New York: Ronald Press, 1951.

Cole, R. A., & Jakimik, J. Understanding speech: How words are heard. In G. Underwood (Ed.), *Strategies of information processing*. London: Academic Press, 1978.

Condon, W. S., & Ogston, W. D. Speech and body motion synchrony of the speaker–hearer. In D. L. Horton & J. J. Jenkins (Eds.), *Perception of language*. Columbus, Ohio: Merrill, 1971.

Cook, T. D., & Flay, B. R. The persistence of experimentally induced attitude change. In L. Berkowitz (Ed.), *Advances in experimental social psychology* (Vol. 11). New York: Academic Press, 1978.

Cooley, C. H. *Social organization*. New York: Schocken, 1962. (Originally published, 1909.)

Coser, L. A. *The functions of social conflict*. Glencoe, Ill.: Free Press, 1954.

Coser, L. A. *Continuities in the study of social conflict*. New York: Free Press, 1967.

Cowan, C. P., & Cowan, P. A. Couple role arrangements and satisfaction during family formation. Paper presented at meetings of the Society for Research in Child Development, April, 1981.

Crano, W. D., & Aronoff, J. A cross-cultural study of expressive and instrumental role complementarity in the family. *American Sociological Review*, 1978, *43*, 463–471.

Crano, W. D., & Brewer, M. B. *Principles of research in social psychology*. New York: McGraw-Hill, 1973.

Crano, W. D., & Mellon, P. M. Causal influence of teachers' expectations on children's academic performance: A cross-lagged panel analysis. *Journal of Educational Psychology*, 1978, *70*, 39–49.

Cronbach, L. J., Gleser, G. C., Nanda, H., & Rajaratnam, N. *The dependability of behavioral measures*. New York: Wiley, 1972.

Cuber, J. F., & Harroff, P. B. *The significant Americans: A study of sexual behavior among the affluent*. New York: Appleton-Century-Crofts, 1965.

Cuber, J. F., & Harroff, P. B. *Sex and the significant Americans*. New York: Appleton-Century-Crofts, 1966.

Dahrendorf, R. *Class and class conflict in industrial society.* Stanford, Calif.: Stanford University Press, 1959.

Darley, J. M., & Fazio, R. H. Expectancy confirmation processes arising in the social interaction sequence. *American Psychologist,* 1980, *35,* 867–881.

Darwin, C. *Expression of the emotions in man and animals.* New York: Appleton, 1899.

Davis, H. L. Decision making within the household. *Journal of Consumer Research,* 1976, *2*(4), 241–260.

Davis, H. L., & Rigaux, B. D. Perception of marital roles in decision processes. *Journal of Consumer Research,* 1974, *1*(1), 51–61.

Davis, K. *Human society.* New York: Macmillan, 1949.

Davitz, J. R. *The language of emotion.* New York: Academic Press, 1969.

Dawkins, R. *The selfish gene.* New York: Oxford University Press, 1976.

Deaux, K. *The behavior of women and men.* Monterey, Calif.: Brooks/Cole, 1976.

DeBurger, J. E. Marital problems, help-seeking, and emotional orientation as revealed in help-request letters. *Journal of Marriage and the Family,* 1967, *29,* 712–721.

Degler, C. N. *At odds: Women and the family in America from the Revolution to the present.* New York: Oxford University Press, 1980.

DeLamater, J., & MacCorquodale, P. The effects of interview schedule variations on reported sexual behaviors. *Sociologial Methods and Research,* 1975, *4,* 215–236.

DeLora, J. S., Warren, C. A. B., & Ellison, C. R. *Understanding sexual interaction* (2nd ed.). Boston: Houghton Mifflin, 1981.

Deutsch, M. Field theory in social psychology. In G. Lindzey (Ed.), *Handbook of social psychology.* Reading, Mass.: Addison-Wesley, 1954.

Deutsch, M. *The resolution of conflict: Constructive and destructive processes.* New Haven: Yale University Press, 1973.

Deutsch, M. Equity, equality, and need: What determines which value will be used as a basis for distributive justice? *Journal of Social Issues,* 1975, *31*(3), 137–149.

Deutsch, M. Fifty years of conflict. In L. Festinger (Ed.), *Four decades of social psychology.* New York: Oxford University Press, 1980.

Dickman, H. R. The perception of behavioral units. In R. G. Barker (Ed.), *The stream of behavior.* New York: Appleton-Century-Crofts, 1963.

Dicks, H. V. Object relations theory and marital studies. *British Journal of Medical Psychology,* 1963, *36,* 125–129.

Dicks, H. V. *Marital tensions.* New York: Basic Books, 1967.

Diebold, A. R., Jr. Anthropological perspectives. In T. A. Sebeok (Ed.), *Animal communication.* Bloomington: Indiana University Press, 1968.

Dion, K. K., & Dion, K. L. Self-esteem and romantic love. *Journal of Personality,* 1975, *43,* 39–57.

Dion, K. L., & Dion, K. K. Correlates of romantic love. *Journal of Consulting and Clinical Psychology,* 1973, *41,* 51–56.

Dizard, J. *Social change in the family.* Chicago: Community and Family Study Center, University of Chicago, 1968.

Doehrman, S. R. Psychological aspects of recovery from coronary heart disease: A review. *Social Science and Medicine,* 1977, *11,* 199–218.

Doob, L. *Social psychology.* New York: Holt, 1952.

Dreyer, C. A., & Dreyer, A. S. Family dinner-time as a unique behavior habitat. *Family Process*, 1973, *12*, 291–301.

Driscoll, R., Davis, K. E., & Lipitz, M. E. Parental interference and romantic love: The Romeo and Juliet effect. *Journal of Personality and Social Psychology*, 1972, *24*, 1–10.

Duck, S. W., & Gilmour, R. (Eds.), *Personal relationships. 2: Developing personal relationships*. London: Academic Press, 1981.

Duncan, S., Jr., & Fiske, D. W. *Face-to-face interaction: Research, methods, and theory*. Hillsdale, N.J.: Erlbaum Associates, 1977.

Dunkel-Schetter, C. & Wortman, C. B. Dilemmas of social support: Parallels between victimization and aging. In S. B. Kiesler, J. N. Morgan, & V. K. Oppenheimer (Eds.), *Aging: Social change*. New York: Academic Press, 1981.

Durkheim, E. [*Suicide*.] New York: Free Press, 1951. (Originally published, 1897.)

Dutton, D. G., & Aron, A. P. Some evidence for heightened sexual attraction under conditions of high anxiety. *Journal of Personality and Social Psychology, 1974, 30,* 510–517.

Dyck, A. J. Criteria for marking the social contact. In R. G. Barker (Ed.), *The stream of behavior*. New York: Appleton-Century-Crofts, 1963.

Eckerman, C. O. The human infant in social interaction. In R. B. Cairns (Ed.), *The analysis of social interactions: Methods, issues, and illustrations*. Hillsdale, N.J.: Erlbaum, 1979.

Edwards, A. L. *The social desirability variable in personality assessment and research*. New York: Dryden, 1957.

Eibl-Eibesfeldt, I. Similarities and differences between cultures in expressive movements. In R. Hinde (Ed.), *Non-verbal communication*. Cambridge: Cambridge University Press, 1972.

Eidelson, R. J. Interpersonal satisfaction and level of involvement: A curvilinear relationship. *Journal of Personality and Social Psychology*, 1980, *39*, 460–470.

Eisenberg, J. M., & Zingle, H. W. Marital adjustments and irrational ideas. *Journal of Marriage and Family Counseling*, 1975, *1*, 81–91.

Ellis, A. The nature of disturbed marital interaction. Paper presented at the American Psychological Association Convention, 1964.

Ellis, A. *Humanistic psychotherapy: The rational-emotive approach*. New York: Julian Press & McGraw-Hill Paperbacks, 1974.

Ellis, A., & Harper, R. A *guide to successful marriage*. Los Angeles: Wilshire, 1977. (Originally published, 1961.)

Emerson, R. Power-dependence relations. *American Sociological Review*, 1962, *27*, 31–41.

Emerson, R. Power-dependence relations: Two experiments. *Sociometry*, 1964, *27*, 282–294.

Ericsson, K. A., & Simon, H. A. Verbal reports as data. *Psychological Review*, 1980, *87*, 215–251.

Ethical Standards of Psychologists. Washington, D.C.: American Psychological Association, 1977.

Falbo, T., & Peplau, L. A. Power strategies in intimate relationships. *Journal of Personality and Social Psychology*, 1980, *38*, 618–628.

Ferber, R. Family decision making and economic behavior: A review. In E. B. Sheldon (Ed.), *Family economic behavior: Problems and prospects.* Philadelphia: Lippincott, 1973.

Ferreira, A. J. Family myths. In P. Watzlawick & J. H. Weakland (Eds.), *The interactional view.* New York: Norton, 1977.

Festinger, L. *A theory of cognitive dissonance.* Stanford, Calif.: Stanford University Press, 1957.

Fink, C. F. Some conceptual difficulties in the theory of social conflict. *Journal of Conflict Resolution,* 1968, *12,* 412–460.

Fishman, P. M. Interaction: The work women do. *Social Problems,* 1978, *25,* 397–406.

Fitz, D., & Gerstenzang, S. *Anger in everyday life: When, where, and with whom?* St. Louis: University of Missouri, 1978. (ERIC Document Reproduction Service No. ED 160 966.)

Fletcher, G. J. O. Causal attributions for marital separation. Unpublished doctoral dissertation, University of Waikato, Hamilton, New Zealand, 1981.

Foa, U. G., & Foa, E. B. *Societal structures of the mind.* Springfield, Ill.: Charles C. Thomas, 1974.

Foa, U. G., & Foa, E. B. *Resource theory of social exchange.* Morristown, N.J.: General Learning Press, 1975.

Foote, N. Unfulfilled plans and unplanned actions. *Advances in Consumer Research,* 1974, *1,* 529–531.

Forgas, J. P. The perception of social episodes: Categorical and dimensional representations in two different social milieux. *Journal of Personality and Social Psychology,* 1976, *33,* 199–209.

Forgas, J. P., & Dobosz, B. Dimensions of romantic involvement: Towards a taxonomy of heterosexual relationships. *Social Psychology Quarterly,* 1980, *43,* 290–300.

Fraiberg, S. Blind infants and their mothers: An examination of the sign system. In M. Bullowa (Ed.), *Before speech.* London: Cambridge University Press, 1979.

Framo, J. L. Symptoms from a transactional viewpoint. In N. Ackerman (Ed.), *Family therapy in transition.* Boston: Little, Brown, 1970.

Framo, J. L. Personal reflections of a family therapist. *Journal of Marriage and Family Counseling,* 1975, *1,* 1–22.

Freedman, J. *Happy people: What happiness is, who has it, and why.* New York: Harcourt Brace Jovanovich, 1978.

French, J. R. P., Jr. Person role fit. In A. McLean (Ed.), *Occupational stress.* Springfield, Ill.: Charles C. Thomas, 1974.

French, J. R. P., Jr., & Raven, B. The bases of social power. In D. Cartwright (Ed.), *Studies in social power.* Ann Arbor, Mich.: Institute for Social Research, 1959.

Freud, A. The psycho-analytic treatment of children. Excerpted in C. E. Shaefer (Ed.), *Therapeutic use of child's play.* New York: Aronson, 1976. (Originally published, 1946.)

Freud, S. [*On psychotherapy.*] *Standard edition* (Vol. 7). (J. Strachey, Ed. and trans.) London: Hogarth Press, 1953. (Originally published, 1904.)

Freud, S. [*Civilization and its discontents.*] (J. Strachey, Ed. and trans.) New York: Norton, 1961. (Originally published, 1930.)

Freud, S. [*The dynamics of the tranference.*] (J. Riviere, Trans.) New York: Collier Books, 1963. (Originally published, 1912).

Friedrich, C. J. *Man and his government.* New York: McGraw-Hill, 1963.

Furstenberg, F. F., Jr. Premarital pregnancy and marital instability. *Journal of Social Issues,* 1976, *32*(1), 67–86.

Gadlin, H. Private lives and public order: A critical view of the history of intimate relations in the United States. In G. Levinger & H. L. Raush (Eds.), *Close relationships: Perspectives on the meaning of intimacy.* Amherst: University of Massachusetts Press, 1977.

Gamson, W. *Power and discontent.* Homewood, Ill.: Dorsey Press, 1968.

Gans, H. *The urban villagers.* New York: Free Press, 1962.

Gelles, R. J., & Straus, M. A. Violence in the American family. *Journal of Social Issues,* 1979, *35*(2), 15–39.

Gewirtz, J. L., & Baer, D. M. Deprivation and satiation of social reinforcers as drive conditions. *Journal of Abnormal and Social Psychology,* 1958, *57,* 165–172.

Gerard, H. B., & Matthewson, G. C. The effects of severity of initiation on liking for a group: A replication. *Journal of Experimental Social Psychology,* 1966, *2,* 278–287.

Gilbert, R., & Christensen, A. Observational assessment of marital and family interaction: Methodological considerations. In L. L'Abate (Ed.), *Handbook of family psychology.* Homewood, Ill.: Dorsey, in press.

Gillespie, D. L. Who has the power? The marital struggle. *Journal of Marriage and the Family,* 1971, *33,* 445–458.

Gilmore, S. K. *The counselor-in-training.* Englewood Cliffs, N.J.: Prentice-Hall, 1973.

Giovacchini, P. L. Treatment of marital disharmonies: The classical approach. In B. L. Greene (Ed.), *The psychotherapies of marital disharmony.* New York: Free Press, 1965.

Glenn, N. D. Psychological well-being in the post-parental stage: Some evidence from national surveys. *Journal of Marriage and the Family,* 1975, *37,* 105–110.

Glick, P. C., & Norton, A. J. Marrying, divorcing, and living together in the U.S. today. *Population Bulletin,* 1977, *32,* 1–39.

Glick, P. C., & Spanier, G. B. Cohabitation in the United States. In P. J. Stein (Ed.), *Single life: Unmarried adults in social context.* New York: St. Martin's Press, 1981.

Goffman, E. Replies and responses. *Language in Society,* 1976, *5,* 257–313.

Goldhammer, H., & Shils, E. Types of power and status. *American Journal of Sociology,* 1939, *45,* 171–182.

Goode, W. J. *After divorce.* Glencoe, Ill.: Free Press, 1956.

Goode, W. J., Hopkins, E., & McClure, H. M. *Social systems and family patterns: A propositional inventory.* Indianapolis: Bobbs-Merrill, 1971.

Goodman, G., & Dooley, D. A framework for help-intended communication. *Psychotherapy: Theory, Research, and Practice,* 1976, *13,* 106–117.

Gordon, S. L. Sentiments: The social context of emotion. Unpublished doctoral dissertation, University of California, Los Angeles, 1979.

Gordon, S. L. The sociology of sentiments and emotion. In M. Rosenberg & R. H. Turner (Eds.), *Social psychology: Sociological perspectives.* New York: Basic Books, 1981.

Gordon, T. *Parent effectiveness training.* New York: Wyden, 1970.

Gottman, J. M. *Marital interaction: Experimental investigations.* New York: Academic Press, 1979.

Gottman, J. M., & Bakeman, R. The sequential analysis of observational data. In M. Lamb, S. Suomi, & G. Stephenson (Eds.), *The study of social interaction.* Madison: University of Wisconsin Press, 1979.

Gottman, J. M., & Markman, H. Experimental designs in psychotherapy research. In S. L. Garfield & A. E. Bergin (Eds.), *Handbook of psychotherapy and behavior change: An empirical analysis* (2nd ed.). New York: Wiley, 1978.

Gottman, J. M., & Notarius, C. Sequential analysis of observational data using Markov chains. In T. Kratochwill (Ed.), *Strategies to evaluate change in single-subject research.* New York: Academic Press, 1978.

Gottman, J. M., Notarius, C., Gonso, J., & Markman, H. *A couple's guide to communication.* Champaign, Ill.: Research Press, 1976.

Gottman, J. M., Notarius, C., Markman, H., Bank, S., Yoppi, B., & Rubin, M. E. Behavior exchange theory and marital decision making. *Journal of Personality and Social Psychology,* 1976, *34,* 14–23.

Gottman, J. M., & Parkhurst, J. T. A developmental theory of friendship and acquaintanceship processes. In W. A. Collins (Ed.), *Minnesota symposia on child psychology* (Vol. 13). Hillsdale, N.J.: Erlbaum, 1980.

Gouldner, A. W. The norm of reciprocity: A preliminary statement. *American Sociological Review,* 1960, *25,* 161–178.

Greenspan, S. I., & Mannino, F. V. A model for brief intervention with couples based on projective identification. *American Journal of Psychiatry,* 1974, *131,* 1103–1106.

Greenwald, A. G. The totalitarian ego: Fabrication and revision of personal history. *American Psychologist,* 1980, *35,* 603–618.

Gross, H. E. Considering "A biosocial perspective on parenting." *Signs,* 1979, *4,* 695–717.

Gross, N., Mason, W. S., & McEachern, A. W. *Explorations in role analysis.* New York: Wiley, 1966.

Group for the Advancement of Psychiatry. *Treatment of families in conflict: The clinical study of family process.* New York: Aronson, 1970.

Guerin, P. J. Family therapy: The first twenty-five years. In P. J. Guerin (Ed.), *Family therapy: Theory and practice.* New York: Gardner, 1976.

Guerney, B. G., Jr. *Relationship enhancement: Skill training programs for therapy, problem prevention, and enrichment.* San Francisco: Jossey-Bass, 1977.

Gurin, G., Veroff, J., & Feld, S. *Americans view their mental health: A nationwide interview survey.* New York: Basic Books, 1960.

Gurman, A. S. Contemporary marital therapies: A critique and comparative analysis of psychoanalytic, behavioral, and system theory approaches. In T. J. Paolino & B. S. McCrady (Eds.), *Marriage and marital therapy: Psychoanalytic, behavioral, and systems theory perspectives.* New York: Brunner/Mazel, 1978.

Gurman, A. S. Behavioral marriage therapy in the 1980's: The challenge of integration. *The American Journal of Family Therapy,* 1980, 8(2), 86–96.

Gurman, A. S., & Kniskern, D. P. Research on marital and family therapy: Progress, perspective, and prospect. In S. L. Garfield & A. E. Bergin (Eds.), *Handbook of psychotherapy and behavior change: An empirical analysis* (2nd ed.). New York: Wiley, 1978.

Gurman, A. S., & Kniskern, D. P. Family therapy outcome research: Knowns and unknowns. In A. S. Gurman & D. P. Kniskern (Eds.), *Handbook of family therapy*. New York: Brunner/Mazel, 1981.

Haas, L. Benefits and problems of egalitarian marriage: A study of role-sharing couples. Unpublished manuscript, Indiana University–Purdue University at Indianapolis, 1980.

Hacker, H. M. Blabbermouths and clams: Sex differences in self-disclosure in same-sex and cross-sex friendship dyads. *Psychology of Women Quarterly*, 1981, *5*, 385–401.

Hadley, T., & Jacob, T. Relationships among measures of family power. *Journal of Personality and Social Psychology*, 1973, *27*, 6–12.

Hadley, T., & Jacob, T. The measurement of family power. *Sociometry*, 1976, *39*, 384–395.

Hagestad, G. O. Problems and promises in the social psychology of intergenerational relations. In R. W. Fogel, E. Hatfield, S. B. Kiesler, & E. Shanas (Eds.), *Aging: Stability and change in the family*. New York: Academic Press, 1981.

Hagestad, G. O., & Smyer, M. A. Divorce in mid-life: Implications for parent caring. Paper presented at meetings of the American Orthopsychiatric Association, April, 1980.

Haley, J. *Strategies of psychotherapy*. New York: Grune & Stratton, 1963.

Haley, J. Testing parental instructions to schizophrenic and normal children: A pilot study. *Journal of Abnormal Psychology*, 1968, *73*, 559–565.

Haley, J. *Uncommon therapy: The psychiatric techniques of Milton H. Erickson, M.D.* New York: Norton, 1973.

Haley, J. *Problem solving therapy*. San Francisco: Jossey-Bass, 1976.

Hannan, M. T., Tuma, N. B., & Groeneveld, L. P. Income and marital events: Evidence from an income-maintenance experiment. *American Journal of Sociology*, 1977, *82*, 1186–1211.

Hare-Mustin, R. Treatment of temper tantrums by paradoxical intervention. *Family Process*, 1975, *14*, 481–486.

Harré, R., & Secord, P. F. *The explanation of social behavior*. Totowa, N.J.: Rowman & Littlefield, 1972.

Harris, C. C. *The family*. London: Allen & Unwin, 1969.

Harris, L., & Associates. *The Harris survey yearbook of public opinion 1970*. New York: Harris, 1971.

Hartup, W. W. The peer system. In P. H. Mussen & E. M. Hetherington (Eds.), *Handbook of child psychology: Socialization, personality, and social development* (Vol. 4). New York: Wiley, 1983.

Harvey, J. H., Wells, G. L., & Alvarez, M. D. Attribution in the context of conflict and separation in close relationships. In J. H. Harvey, W. Ickes, & R. F. Kidd (Eds.), *New directions in attribution research* (Vol. 2). Hillsdale, N.J.: Erlbaum, 1978.

Hawkins, J. L., Weisberg, C., & Ray, D. L. Perception of behavioral conformity, imputation of consensus, and marital satisfaction. *Journal of Marriage and the Family*, 1977, *39*, 479–490.

Hazzard, S. *The transit of Venus*. New York: Viking, 1980.

Hebb, D. O. *The organization of behavior.* New York: Wiley, 1949.

Heer, D. M. Dominance and the working wife. *Social Forces,* 1958, *36,* 341–347.

Heer, D. M. Husband and wife perceptions of family power structure. *Marriage and Family Living,* 1962, *24,* 65–67.

Heider, F. *The psychology of interpersonal relations.* New York: Wiley, 1958.

Heise, D. R. *Understanding events: Affect and the construction of social action.* Cambridge: Cambridge University Press, 1979.

Heiss, J. (Ed.). *Family roles and interaction* (2nd ed.). Chicago: Rand McNally, 1976.

Heiss, J. Social roles. In M. Rosenberg & R. H. Turner (Eds.), *Social psychology: Sociological perspectives.* New York: Basic Books, 1981.

Hendrick, S. S. Self-disclosure and marital satisfaction. *Journal of Personality and Social Psychology,* 1981, *40,* 1150–1159.

Henley, N. M. *Body politics: Power, sex and nonverbal communication.* Englewood Cliffs, N.J.: Prentice-Hall, 1977.

Hersen, M., & Barlow, D. H. *Single-case experimental designs: Strategies for studying behavior change.* New York: Pergamon Press, 1976.

Hess, R. D., & Shipman, V. C. Early experience and the socialization of cognitive modes in children. *Child Development,* 1965, *36,* 869–886.

Hetherington, E. M. Divorce: A child's perspective. *American Psychologist,* 1979, *34,* 851–858.

Hetherington, E. M., Cox, M., & Cox, R. The aftermath of divorce. In J. H. Stevens, Jr., & E. M. Matthews (Eds.), *Mother–child, father–child relations.* Washington, D.C.: National Association for the Education of Young Children, 1977.

Hicks, M. W., & Platt, M. Marital happiness and stability: A review of the research in the sixties. *Journal of Marriage and the Family,* 1970, *32,* 553–574.

Hilbert, R. A. Toward an improved understanding of "role." *Theory and Society,* 1981, *10,* 207–225.

Hill, C. T., Rubin, Z., & Peplau, L. A. Breakups before marriage: The end of 103 affairs. *Journal of Social Issues,* 1976, *32*(1), 147–167.

Hill, R. Patterns of decision-making and the accumulation of family assets. In N. N. Foote (Ed.), *Household decision-making: Consumer behavior* (Vol. 4). New York: New York University Press, 1961.

Hinde, R. A. Interpersonal relationships: In quest of a science. *Psychological Medicine,* 1978, *8,* 373–386.

Hinde, R. A. *Towards understanding relationships.* London: Academic Press, 1979.

Hochschild, A. R. Emotion work, feeling rules, and social structure. *American Journal of Sociology,* 1979, *85,* 551–575.

Hoffman, L. W., & Manis, J. D. Influences of children on marital interaction and parental satisfactions and dissatisfactions. In R. M. Lerner & G. B. Spanier (Eds.), *Child influences on marital and family interaction.* New York: Academic Press, 1978.

Holman, T. B., & Burr, W. R. Beyond the beyond: The growth of family theories in the 1970s. *Journal of Marriage and the Family,* 1980, *42,* 729–742.

Holmes, D. S. Differential change in affective intensity and the forgetting of unpleasant personal experiences. *Journal of Personality and Social Psychology,* 1970, *15,* 234–239.

Holmes, J. G. The exchange process in close relationships: Microbehavior and macromotives. In M. J. Lerner & S. C. Lerner (Eds.), *The justice motive in social behavior.* New York: Plenum, 1981.

Holmstrom, L. L. *The two-career family.* Cambridge, Mass.: Schenkman, 1972.

Homans, G. C. *Social behavior: Its elementary forms.* New York: Harcourt, Brace & World, 1961.

Hopkins, J. R. Sexual behavior in adolescence. *Journal of Social Issues,* 1977, *33*(2), 67–85.

Hotchner, T. *Pregnancy and childbirth.* New York: Avon, 1979.

House, J. S. *Occupational stress and the mental and physical health of factory workers.* Ann Arbor: University of Michigan Institute for Social Research, Survey Research Center, 1980.

Hunt, J. McV., Cole, M. W., & Reis, E. S. Situational cues distinguishing anger, fear and sorrow. *American Journal of Psychology,* 1958, *71,* 136–151.

Hunt, M. *The affair.* New York: World, 1969.

Hunt, M. *Sexual behavior in the 1970s.* New York: Dell, 1974.

Hunt, M., & Hunt, B. *The divorce experience.* New York: McGraw-Hill, 1977.

Huston, T. L., & Burgess, R. L. Social exchange in developing relationships: An overview. In R. L. Burgess & T. L. Huston (Eds.), *Social exchange in developing relationships.* New York: Academic Press, 1979.

Huston, T. L., & Levinger, G. Interpersonal attraction and relationships. *Annual Review of Psychology,* 1978, *29,* 115–156.

Huston, T. L., Surra, C. A., Fitzgerald, N. M., & Cate, R. M. From courtship to marriage: Mate selection as an interpersonal process. In S. W. Duck & R. Gilmour (Eds.), *Personal relationships. 2: Developing personal relationships.* London: Academic Press, 1981.

Hyman, H. *Interviewing in Social Research.* Chicago: University of Chicago Press, 1954.

Izard, C. E. *Human emotions.* New York: Plenum Press, 1977.

Jackson, D. D. Family interaction, family homeostasis and some implications for conjoint family psychotherapy. In J. H. Masserman (Ed.), *Individual and family dynamics.* New York: Grune & Stratton, 1959.

Jackson, D. D. Family rules: Marital quid pro quo. In P. Watzlawick & J. H. Weakland (Eds.), *The interactional view.* New York: Norton, 1977.

Jackson, D. D., & Weakland, J. Conjoint family therapy: Some consideration on theory, technique, and results. *Psychiatry,* 1961, *24,* 30–45.

Jacobson, N. S. Problem solving and contingency contracting in the treatment of marital discord. *Journal of Consulting and Clinical Psychology,* 1977, *45,* 92–100.

Jacobson, N. S. and Margolin, G. *Marital therapy: strategies based on social learning and behavior exchange principles.* New York: Brunner/Mazel, 1979.

Jacobson, N. S., & Martin, B. Behavioral marriage therapy: Current status. *Psychological Bulletin,* 1976, *83,* 540–556.

Jacobson, N. S., Waldron, H., & Moore, D. Toward a behavioral profile of marital distress. *Journal of Consulting and Clinical Psychology,* 1980, *48,* 696–703.

Jaffe, D. T., & Kanter, R. M. Couple strains in communal households: A four-factor model of the separation process. *Journal of Social Issues,* 1976, *32*(1), 169–191.

Jahoda, M. Conformity and independence: A psychological analysis. *Human Relations,* 1959, *12,* 99–120.

James, W. What is emotion? *Mind,* 1884, 9, 188–204.

Janoff-Bulman, R. Characterological versus behavioral self-blame: Inquiries into depression and rape. *Journal of Personality and Social Psychology,* 1979, *37,* 1798–1809.

Jefferson, G. Side sequences. In D. Sudnow (Ed.), *Studies in social interaction.* New York: Free Press, 1972.

Jessor, R., & Jessor, S. L. The perceived environment in behavioral science: Some conceptual issues and some illustrative data. *American Behavioral Scientist,* 1973, *16,* 801–828.

Johnson, M. P. Personal and structural commitment: Sources of consistency in the development of relationships. Unpublished paper, Pennsylvania State University, Department of Sociology, 1978.

Johnson, S. M., & Bolstad, O. D. Methodological issues in naturalistic observation: Some problems and solutions for field research. In L. A. Hamerlynck, L. C. Handy, & E. J. Mash (Eds.), *Behavior change: Methodology, concepts, and practice.* Champaign, Ill.: Research Press, 1973.

Johnson, S. M., & Bolstad, O. D. Reactivity to home observation: A comparison of audio recorded behavior with observers present or absent. *Journal of Applied Behavior Analysis,* 1975, *8,* 181–185.

Johnson, S. M., & Lobitz, G. K. The personal and marital adjustment of parents as related to observed child deviance and parenting behaviors. *Journal of Abnormal Child Psychology,* 1974, *2,* 193–207.

Jones, E. E., & Gerard, H. B. *Foundations of social psychology.* New York: Wiley, 1967.

Jones, E. E., & Nisbett, R. E. *The actor and the observer: Divergent perceptions of the causes of behavior.* Morristown, N.J.: General Learning Press, 1971.

Jones, E. E., & Pittman, T. S. Toward a general theory of strategic self-presentation. In J. Suls (Ed.), *Psychological perspectives on the self* (Vol. 1). Hillsdale, N.J.: Erlbaum, 1982.

Jones, E. E., & Wortman, C. *Ingratiation: An attributional approach.* Morristown, N.J.: General Learning Press, 1973.

Jung, C. G. *Analytical psychology: Its theory and practice.* New York: Random House, 1968.

Kahn, R. L., & Antonucci, T. C. Convoys of social support: A life-course approach. In S. B. Kiesler, J. N. Morgan, & V. K. Oppenheimer (Eds.), *Aging: Social change.* New York: Academic Press, 1981.

Kandel, D. B., & Lesser, G. S. Marital decision-making in American and Danish urban families: A research note. *Journal of Marriage and the Family,* 1972, *34,* 134–138.

Kanter, R. M. Commitment and social organization: A study of commitment mechanisms in utopian communities. *American Sociological Review,* 1968, *33,* 499–517.

Kanter, R. M. *Commitment and community: Communes and utopias in sociological perspective.* Cambridge: Harvard University Press, 1972.

Kantor, D., & Lehr, W. *Inside the family: Toward a theory of family process.* New York: Harper Colophon, 1975.

Kaslow, F. W. The history of family therapy in the United States: A kaleidoscopic overview. *Marriage and Family Review,* 1980, 3(1/2), 77–111.

Kaslow, F. W. Divorce and divorce therapy. In A. S. Gurman & D. P. Kniskern (Eds.), *Handbook of family therapy.* New York: Brunner/Mazel, 1981.

Katona, G., & Mueller, E. A study of purchase decisions. *Consumer Behavior,* 1954, *1,* 30–87.

Katz, E., & Lazarsfeld, P. F. *Personal influence.* Glencoe, Ill.: Free Press, 1955.

Kaye, K. Toward the origin of dialogue. In H. R. Schaffer (Ed.), *Studies in mother–infant interaction.* New York: Academic Press, 1977.

Keele, S. W. Movement control in skilled motor performance. *Psychological Bulletin,* 1968, *70,* 387–403.

Kelley, H. H. Causal schemata and the attribution process. In E. E. Jones, D. E. Kanouse, H. H. Kelley, R. E. Nisbett, S. Valins, & B. Weiner (Eds.), *Attribution: Perceiving the causes of behavior.* Morristown, N.J.: General Learning Press, 1971.

Kelley, H. H. An application of attribution theory to research methodology for close relationships. In G. Levinger & H. L. Raush (Eds.), *Close relationships: Perspectives on the meaning of intimacy.* Amherst, Mass.: University of Massachusetts Press, 1977.

Kelley, H. H. *Personal relationships: Their structures and processes.* Hillsdale, N.J.: Erlbaum, 1979.

Kelley, H. H. The causes of behavior: Their perception and regulation. In L. Festinger (Ed.), *Retrospections on social psychology.* New York: Oxford University Press, 1980.

Kelley, H. H. Marriage relationships and aging. In R. W. Fogel, E. Hatfield, S. B. Kiesler, & E. Shanas (Eds.), *Aging: Stability and change in the family.* New York: Academic Press, 1981.

Kelley, H. H., Cunningham, J. D., Grisham, J. A., Lefebvre, L. M., Sink, C. R., & Yablon, G. Sex differences in comments made during conflict in close heterosexual pairs. *Sex Roles,* 1978, *4,* 473–491.

Kelley, H. H., & Michela, J. Attribution theory and research. In M. R. Rosenzweig & L. Porter (Eds.), *Annual Review of Psychology* (Vol. 31). Palo Alto, Calif.: Annual Reviews, 1980.

Kelley, H. H., & Ring, K. Some effects of "suspicious" versus "trusting" training schedules. *Journal of Abnormal and Social Psychology,* 1961, *63,* 294–301.

Kelley, H. H., & Thibaut, J. W. *Interpersonal relations: A theory of interdependence.* New York: Wiley-Interscience, 1978.

Kelly, G. A. *The psychology of personal constructs.* New York: Norton, 1955.

Kelman, H. C. Processes of opinion change. *Public Opinion Quarterly,* 1961, *25,* 57–78.

Kelvin, P. Predictability, power and vulnerability in interpersonal attraction. In S. W. Duck (Ed.), *Theory and practice in interpersonal attraction.* London: Academic Press, 1977.

Kendon, A. Movement coordination in social interaction: Some examples described. *Acta Psychologica,* 1970, *32,* 110–125.

Kenkel, W. F. Influence differentiation in family decision making. *Sociology and Social Research,* 1957, *42,* 18–25.

Kenkel, W. F. Sex of observer and spousal roles in decision-making. *Marriage and Family Living,* 1961, *23,* 185–186.

Kenny, D. A. Cross-lagged panel correlation: A test for spuriousness. *Psychological Bulletin,* 1975, *82,* 887–903.

Kenny, D. A. *Correlation and causality.* New York: Wiley-Interscience, 1979.

Kerckhoff, A. C., & Davis, K. E. Value consensus and need complementarity in mate selection. *American Sociological Review,* 1962, *27,* 295–303.

Kerlinger, F. N., & Pedhazur, E. J. *Multiple regression in behavioral research.* New York: Holt, Rinehart & Winston, 1973.

Kiesler, C. A. *The psychology of commitment.* New York: Academic Press, 1971.

Kimble, G. A. Attitudinal factors in eyelid conditioning. In G. A. Kimble (Ed.), *Foundations of conditioning and learning.* New York: Appleton-Century-Crofts, 1967.

King, C. E., & Christensen, A. The Relationship Events Scale: A Guttman scaling of progress in courtship. *Journal of Marriage and the Family,* in press.

Kinsey, A. C., Pomeroy, W. B., & Martin, C. E. *Sexual behavior in the human male.* Philadelphia: Saunders, 1948.

Kinsey, A. C., Pomeroy, W. B., Martin, C. E., & Gebhard, P. H. *Sexual behavior in the human female.* Philadelphia: Saunders, 1953.

Kintsch, W., & van Dijk, T. A. Text comprehension and production. *Psychological Review,* 1978, *85,* 363–394.

Kirch, A. M. (Ed.). *The anatomy of love.* New York: Dell, 1960.

Klein, M. *The psycho-analysis of children.* London: Hogarth Press, 1932.

Klein, M. Notes on some schizoid mechanisms. *International Journal of Psychoanalysis,* 1946, *27,* 99–110.

Klein, N. C., Alexander, J. F., & Parsons, B. V. Impact of family systems intervention on recidivism and sibling delinquency: A model of primary prevention and program evaluation. *Journal of Consulting and Clinical Psychology,* 1977, *45,* 469–474.

Klinger, E. *Structure and functions of fantasy.* New York: Wiley-Interscience, 1971.

Klinger, E. *Meaning and void: Inner experience and the incentives in people's lives.* Minneapolis: University of Minnesota Press, 1977.

Knudson, R. M., Sommers, A. A., & Golding, S. L. Interpersonal perception and mode of resolution in marital conflict. *Journal of Personality and Social Psychology,* 1980, *38,* 751–763.

Koch, H. L. The relation of primary mental abilities in five- and six-year-olds to sex of child and characteristics of his sibling. *Child Development,* 1954, *25,* 210–223.

Koch, H. L. Some personality correlates of sex, sibling position, and sex of sibling among five- and six-year-old children. *Genetic Psychology Monographs,* 1955, *52,* 3–50.

Kohen, J. A., Brown, C. A., & Feldberg, R. Divorced mothers: The costs and benefits of female family control. In G. Levinger & O. C. Moles (Eds.), *Divorce and separation.* New York: Basic Books, 1979.

Komarovsky, M. *Blue-collar marriage.* New York: Random House, 1967.

Komarovsky, M. *Dilemmas of masculinity.* New York: Norton, 1976.

Koren, P., Carlton, K., & Shaw, D. Marital conflict: Relations among behaviors, outcomes, and distress. *Journal of Consulting and Clinical Psychology,* 1980, *48,* 460–468.

Kressel, K., & Deutsch, M. Divorce therapy: An in-depth survey of therapists' views. *Family Process,* 1977, *16,* 413–443.

Lakoff, R. Language in context. *Language,* 1972, *48,* 907–924.

Lamb, M. E. Influence of the child on marital quality and interaction during the prenatal, perinatal, and infancy periods. In R. M. Lerner & G. B. Spanier (Eds.), *Child influences on marital and family interaction.* New York: Academic Press, 1978.

Lange, C. G. *The emotions.* Baltimore: Williams & Wilkins, 1922. (Originally published, 1885.)

Laws, J. L. A feminist review of the marital adjustment literature: The rape of the Locke. *Journal of Marriage and the Family,* 1971, *33,* 483–516.

Layton, B. D., & Moehle, D. Attributed influence: The importance of observing change. *Journal of Experimental Social Psychology,* 1980, *16,* 243–252.

Lazarus, A. A. *Behavior therapy and beyond.* New York: McGraw-Hill, 1971.

Lazarus, R. S. Emotions and adaptation: Conceptual and empirical relations. In W. J. Arnold (Ed.), *Nebraska Symposium on Motivation.* Lincoln: University of Nebraska Press, 1968.

Lederer, W. J., & Jackson, D. D. *The mirages of marriage.* New York: Norton, 1968.

Lee, J. A. A typology of styles of loving. *Personality and Social Psychology Bulletin,* 1977, *3,* 173–182.

Leeper, R. W. A motivational theory of emotion to replace "emotion as disorganized response." *Psychological Review,* 1948, *55,* 5–21.

Leik, R. K., & Leik, S. A. Transition to interpersonal commitment. In R. L. Hamblin & J. H. Kunkel (Eds.), *Behavioral therapy in sociology.* New Brunswick, N.J.: Transaction Books, 1977.

Lein, L. Male participation in home life: Impact of social supports and breadwinner responsibility on the allocation of tasks. *The Family Coordinator,* 1979, *28,* 489–496.

LeMasters, E. E. Parenthood as crisis. *Marriage and Family Living,* 1957, *19,* 352–355.

LeMasters, E. E. *Blue-collar aristocrats: Life-styles at a working-class tavern.* Madison: University of Wisconsin Press, 1975.

Lerner, M. J., Miller, D. T., & Holmes, J. G. Deserving and the emergence of forms of justice. *Advances in Experimental Social Psychology,* 1976, *9,* 133–162.

Levenson, S. *In one era and out the other.* New York: Hall, 1973.

LeVine, R. A. Anthropology and the study of conflict: Introduction. *Journal of Conflict Resolution,* 1961, *5,* 3–15.

Levinger, G. The development of perceptions and behavior in newly formed social power relationships. In D. Cartwright (Ed.), *Studies in social power.* Ann Arbor, Mich.: Institute for Social Research, 1959.

Levinger, G. Task and social behavior in marriage. *Sociometry,* 1964, *27,* 433–448.

Levinger, G. Marital cohesiveness and dissolution: An integrative review. *Journal of Marriage and the Family,* 1965, *27,* 19–28.

Levinger, G. Sources of marital dissatisfaction among applicants for divorce. *American Journal of Orthopsychiatry,* 1966, *36,* 803–807.

Levinger, G. Supplementary methods in family research. In W. D. Winter & A. J. Ferreira (Eds.), *Research in Family Interaction.* Palo Alto: Science and Behavior Books, 1969.

Levinger, G. A social psychological perspective on marital dissolution. *Journal of Social Issues,* 1976, *32*(1), 21–47.

Levinger, G. Reviewing the close relationship. In G. Levinger & H. L. Raush (Eds.), *Close relationships.* Amherst, Mass.: University of Massachusetts Press, 1977.

Levinger, G. Marital cohesiveness at the brink: The fate of applications for divorce. In G. Levinger & O. C. Moles (Eds.), *Divorce and separation.* New York: Basic Books, 1979. (a)

Levinger, G. A social exchange view on the dissolution of pair relationships. In R. L. Burgess & T. L. Huston (Eds.), *Social exchange in developing relationships.* New York: Academic Press, 1979. (b)

Levinger, G. Toward the analysis of close relationships. *Journal of Experimental Social Psychology,* 1980, *16,* 510–544.

Levinger, G., & Breedlove, J. Interpersonal attraction and agreement: A study of marriage partners. *Journal of Personality and Social Psychology,* 1966, *3,* 367–372.

Levinger, G., & Moles, O. C. (Eds.), *Divorce and separation: Context, causes, and consequences.* New York: Basic Books, 1979.

Levinger, G., & Raush, H. (Eds.), *Close relationships: Perspectives on the meaning of intimacy.* Amherst, Mass.: University of Massachusetts Press, 1977.

Levinger, G., Senn, D. J., & Jorgensen, B. W. Progress toward permanence in courtship: A test of the Kerckhoff–Davis hypotheses. *Sociometry,* 1970, *33,* 427–443.

Levinger, G., & Snoek, J. D. *Attraction in relationship: A new look at interpersonal attraction.* Morristown, N.J.: General Learning Press, 1972.

Lewin, K. Defining the "field at a given time." *Psychological Review,* 1943, *50,* 292–310.

Lewin, K. *Resolving social conflicts.* New York: Harper & Bros., 1948.

Lewin, K. *Field theory in social science.* New York: Harper & Bros., 1951.(a)

Lewin, K. Intention, will and need. In D. Rapaport (Ed.), *Organization and pathology of thought.* New York: Columbia University Press, 1951. (Originally published, 1926.)(b)

Lewis, G. H. Role differentiation. *American Sociological Review,* 1972, *37,* 424–434.

Lewis, M., & Rosenblum, L. A. *The effect of the infant on its caregiver.* New York: Wiley, 1974.

Lewis, R. A. A developmental framework for premarital dyadic formation. *Family Process,* 1972, *11,* 17–48.

Lewis, R. A., & Spanier, G. B. Theorizing about the quality and stability of marriage. In W. R. Burr, R. Hill, F. I. Nye, & I. L. Reiss (Eds.), *Contemporary theories about the family* (Vol. 1). New York: Free Press, 1979.

Li, C. C. *Path-analysis: A primer.* Pacific Grove, Calif.: Boxwood Press, 1975.

Liem, J. H. Effects of verbal communications of parents and children: A comparison of normal and schizophrenic families. *Journal of Consulting and Clinical Psychology,* 1974, *42,* 438–450.

Linton, R. *The study of man.* New York: Appleton-Century, 1936.

Lippitt, R., Polansky, N., Redl, F., & Rosen, S. The dynamics of power. *Human Relations,* 1952, *5,* 37–64.

Lofland, J. *Analyzing social settings.* Belmont, Calif.: Wadsworth, 1971.

London, H., & Nisbett, R. E. (Eds.). *Thought and feeling: Cognitive alteration of feeling states.* Chicago: Aldine, 1974.

Lott, A. J., & Lott, B. E. A learning theory approach to interpersonal attitudes. In A. G. Greenwald & T. M. Ostrom (Eds.), *Psychological foundations of attitudes.* New York: Academic Press, 1968.

Love, L. R., & Kaswan, J. W. *Troubled children: Their families, schools, and treatments.* New York: Wiley, 1974.

Lupri, E. Contemporary authority patterns in the West German family: A study in cross-national validation. *Journal of Marriage and the Family,* 1969, *31,* 134–144.

Lynch, J. J. *The broken heart: The medical consequences of loneliness.* New York: Basic Books, 1977.

McCall, G. J. The social organization of relationships. In G. J. McCall, M. M. McCall, N. K. Denzin, G. D. Suttles, & S. B. Kurth (Eds.), *Social relationships.* Chicago: Aldine, 1970.

McCall, G. J., & Simmons, J. L. *Identities and interactions: An examination of human associations in everyday life* (rev. ed.). New York: Free Press, 1978.

McClintock, E., & Baron, J. N. SCIP: Social contact interactive parser. Unpublished manuscript, Family Care Center, Santa Barbara, 1979.

McClintock, M. Comment on "A biosocial perspective on parenting." *Signs,* 1979, *4,* 703–710.

Maccoby, E. The choice of variables in the study of socialization. *Sociometry,* 1961, *24,* 357–371.

Mack, R. W., & Snyder, R. C. The analysis of social conflict: Toward an overview and synthesis. *Journal of Conflict Resolution,* 1957, *1,* 212–248.

McLean, P. D., Ogston, K., & Grauer, L. A behavioral approach to the treatment of depression. *Journal of Behavioral Therapy and Experimental Psychiatry,* 1973, *4,* 323–330.

Madden, J. E., & Janoff-Bulman, R. Blaming and marital satisfaction: Wives' attributions for conflict in marriage. Unpublished manuscript. Department of Psychology, University of Massachusetts, Amherst, 1979.

Mahoney, M., & Arnkoff, D. Cognitive and self-control therapies. In S. L. Garfield & A. E. Bergin (Eds.), *Handbook of psychotherapy and behavior change* (2nd ed.). New York: Wiley, 1978.

Maier, N. R. F. *Studies of abnormal behavior in the rat: The neurotic pattern and an analysis of the situation which produces it.* New York: Harper & Bros., 1939.

Mandler, G. *Mind and emotion.* New York: Wiley, 1975.

Martin, B. Parent–child relations. In E. M. Hetherington (Ed.), *Review of child development research* (Vol. 5). Chicago: University of Chicago Press, 1976.

Martin, P. *A marital therapy manual.* New York: Brunner/Mazel, 1976.

Martin, S., Johnson, S. M., Johansson, S., & Wahl, G. The comparability of behavioral data in laboratory and natural settings. In E. J. Mash, L. A. Hamerlynck, & L. C. Hardy (Eds.), *Behavior modification and families.* New York: Brunner/Mazel, 1976.

Mason, K. O., Czajka, J. L., & Arber, S. Change in U.S. women's sex-role attitudes, 1964–1974. *American Sociological Review,* 1976, *41,* 573–596.

Masserman, J. H. *Behavior and neurosis: An experimental psychoanalytic approach to psychobiologic principles.* Chicago: University of Chicago Press, 1943.

Masters, J. C., & Wellman, H. M. The study of human infant attachment: A procedural critique. *Psychological Bulletin,* 1974, *81,* 218–237.

Matlin, M. W., & Stang, D. J. *The Pollyana principle: Selectivity in language, memory, and thought.* Cambridge, Mass.: Schenkman, 1978.

May, R. *Love and will.* New York: Norton, 1969.

Mayer, N. *The male mid-life crisis: Fresh starts after 40.* New York: Doubleday, 1978.

Mayhew, B. H., Gray, L. N., & Richardson, J. T. Behavioral measurement of operating power structures: Characterization of asymmetrical interactions. *Sociometry,* 1969, 474–489.

Mazur, A. Status interactions in dyads. Paper presented at the Annual Meeting of the Society for General Systems Research, New York City, January 1975.

Mead, G. H. *Mind, self, and society.* Chicago: University of Chicago Press, 1934.

Meehl, P. E. High school yearbooks: A reply to Schwarz. *Journal of Abnormal Psychology,* 1971, *77,* 143–148.

Meehl, P. E. Theoretical risks and tabular asterisks: Sir Karl, Sir Ronald, and the slow progress of soft psychology. *Journal of Consulting and Clinical Psychology,* 1978, *46,* 806–834.

Meeker, B. F., & Weitzel-O'Neill, P. A. Sex roles and interpersonal behavior in task-oriented groups. *American Sociological Review,* 1977, *42,* 91–105.

Mehrabian, A. *Nonverbal communication.* Chicago: Aldine, 1972.

Meissner, W. W. The conceptualization of marriage and family dynamics from a psychoanalytic perspective. In T. J. Paolino & B. S. McCrady (Eds.), *Marriage and marital therapy: Psychoanalytic, behavioral, and systems theory perspectives.* New York: Brunner/Mazel, 1978.

Mellen, S. L. W. *The evolution of love.* San Francisco: W. H. Freeman and Company, 1981.

Merton, R. K. *Social theory and social structure* (rev. ed.). New York: Free Press, 1968.

Messé, L. A., Buldain, R. W., & Watts, B. Recall of social events with the passage of time. *Personality and Social Psychology Bulletin,* 1981, *7,* 33–38.

Mettee, D., & Aronson, E. Affective reactions to appraisal from others. In T. Huston (Ed.), *Foundations of interpersonal attraction.* New York: Academic Press, 1974.

Michel, A. Comparative data concerning the interaction in French and American families. *Journal of Marriage and the Family,* 1967, *29,* 337–344.

Michela, J. L. Perceived changes in marital relationships following myocardial infarction. Unpublished doctoral dissertation, University of California, Los Angeles, 1981.

Miller, B. C., & Olson, D. H. Cluster analysis as a method for defining types of marriage interaction. Paper presented at the preconference at the methodology workshop of the National Council on Family Relations, October 1976.

Miller, E. D., Hintz, R. A., & Couch, C. J. The elements and structure of openings. *The Sociological Quarterly,* 1975, *16,* 479–499.

Miller, G. A. The magical number seven, plus or minus two: Some limits in our capacity for processing information. *Psychological Review,* 1956, *63,* 81–97.

Miller, G. A., Gallanter, E., & Pribram, K. H. *Plans and the structure of behavior.* New York: Holt, 1960.

Miller, R. L. Marital dissolution: Paths to breakup. Unpublished doctoral dissertation, University of Massachusetts, Amherst, 1982.

Miller, S., Nunnaly, E. W., & Wackman, D. Minnesota Couples Communication Program (MCCP): Premarital and marital groups. In D. H. Olson (Ed.), *Treating relationships.* Lake Mills, Iowa: Graphic, 1976.

Miller, W. H. *Systematic parent training: Procedures, cases and issues.* Champaign, Ill.: Research Press, 1975.

Minsky, A. A framework for representing knowledge. In P. H. Winston (Ed.), *The psychology of computer vision.* New York: McGraw-Hill, 1975.

Minuchin, S. *Families and family therapy.* Cambridge: Harvard University Press, 1974.

Mishler, E. G., & Waxler, N. E. *Interaction in families: An experimental study of family processes and schizophrenia.* New York: Wiley, 1968.

Mitchell, G. *Human sex differences: A primatologist's perspective.* New York: Van Nostrand Reinhold, 1981.

Mitchell, R. E. Some social implications of high density housing. *American Sociological Review,* 1971, *36,* 18–29.

Murray, R. The influence of crowding on children's behavior. In D. Canter & T. Lee (Eds.), *Psychology and the built environment.* London: Architectural Press, 1974.

Murstein, B. I. Stimulus-value-role: A theory of marital choice. *Journal of Marriage and the Family,* 1970, *32,* 465–481.

Murstein, B. I., Cerreto, M., & MacDonald, M. G. A theory and investigation of the effect of exchange-orientation on marriage and friendship. *Journal of Marriage and the Family,* 1977, *39,* 543–548.

Myers, J. K., Lindenthal, J. J., Pepper, M., & Ostrander, D. R. Life events and mental status: A longitudinal study. *Journal of Health and Social Behavior,* 1972, *13,* 398–406.

Nadelson, C. C. Marital therapy. In T. J. Paolino & B. S. McCrady (Eds.), *Marriage and marital therapy: Psychoanalytic, behavioral, and systems theory perspectives.* New York: Brunner/Mazel, 1978.

Nagel, J. H. *The descriptive analysis of power.* New Haven: Yale University Press, 1975.

Napier, A. Y. The rejection–intrusion pattern: A central family dynamic. *Journal of Marriage and Family Counseling,* 1978, *4,* 5–12.

Neisser, U. *Cognitive psychology.* New York: Appleton-Century-Crofts, 1967.

Neisser, U. *Cognition and reality: Principles and implications of cognitive psychology.* San Francisco: W. H. Freeman and Company, 1976.

Nerem, R. M., Levesque, M. J., & Cornhill, J. F. Social environment as a factor in diet-induced atherosclerosis. *Science,* 1980, *208,* 1475–1476.

Neubeck, G. (Ed.). *Extra-marital relations.* Englewood Cliffs, N.J.: Prentice-Hall, 1969.

Newcomb, T. M. *The acquaintance process.* New York: Holt, Rinehart & Winston, 1961.

Newman, H. M., & Langer, E. J. A cognitive arousal model of intimate relationship formation, stabilization, and disintegration. Unpublished manuscript, Graduate Center of City University of New York, 1977.

Newman, H. M., & Langer, E. J. Post-divorce adaptation and the attribution of responsibility. *Sex Roles,* 1981, *7,* 223–232.

Newtson, D. Attribution and the unit of perception of ongoing behavior. *Journal of Personality and Social Psychology,* 1973, *28,* 28–38.

Newtson, D. Foundations of attribution: The perception of ongoing behavior. In J. H. Harvey, W. J. Ickes, & R. F. Kidd (Eds.), *New directions in attribution research* (Vol. 1). Hillsdale, N.J.: Erlbaum, 1976.

Newtson, D., & Engquist, G. The perceptual organization of ongoing behavior. *Journal of Experimental Social Psychology,* 1976, *12,* 436–450.

Newtson, D., Engquist, G., & Bois, J. The objective basis of behavior units. *Journal of Personality and Social Psychology,* 1977, *35,* 847–862.

Noller, P. Misunderstandings in marital communication: A study of couples' nonverbal communication. *Journal of Personality and Social Psychology,* 1980, *39,* 1135–1148.

Norman, D. A., & Rumelhart, D. E. *Explorations in Cognition.* San Francisco: W. H. Freeman and Company, 1975.

Nowlis, V., & Nowlis, H. H. The description and analysis of moods. *Annals of the New York Academy of Science,* 1956, *65,* 345–355.

O'Leary, K. D., & Kent, R. Behavior modification for social action: Research tactics and problems. In L. A. Hamerlynck, L. C. Handy, & E. J. Mash (Eds.), *Behavior change: Methodology, concepts and practice.* Champaign, Ill.: Research Press, 1972.

Olsen, T. *Tell me a riddle.* New York: Dell, 1956.

Olson, D. H. The measurement of family power by self-report and behavioral methods. *Journal of Marriage and the Family,* 1969, *31,* 545–550.

Olson, D. H. Inventory of marital conflicts (IMC): An experimental interaction procedure. *Journal of Marriage and the Family,* 1970, *32,* 443–448.

Olson, D. H. Marital and family therapy: Integrative review and critique. In G. B. Broderick (Ed.), *A decade of family research and action 1960–1969.* Minneapolis: National Council on Family Relations, 1971.

Olson, D. H. *Treating relationships.* Lake Mills, Iowa: Graphic, 1976.

Olson, D. H. Insiders' and outsiders' views of relationships: Research studies. In G. Levinger & H. Raush (Eds.), *Close relationships: Perspectives on the meaning of intimacy.* Amherst: University of Massachusetts Press, 1977.

Olson, D. H., & Rabunsky, C. Validity of four measures of family power. *Journal of Marriage and the Family,* 1972, *34,* 224–234.

Oltmanns, T. F., Broderick, J. E., & O'Leary, K. D. Marital adjustment and the efficacy of behavior therapy with children. *Journal of Consulting and Clinical Psychology,* 1977, *45,* 724–729.

O'Neill, W. L. *Divorce in the progressive era.* New Haven: Yale University Press, 1973.

Oppenheim, F. E. *Dimensions of freedom.* New York: St. Martin's Press, 1961.

Oppenheimer, V. K. The sociology of women's economic role in the family. *American Sociological Review,* 1977, *42,* 387–406.

Orden, S. R., & Bradburn, N. M. Dimensions of marriage happiness. *American Journal of Sociology,* 1968, *73,* 715–731.

O'Rourke, J. F. Field and laboratory: The decision-making behavior of family groups in two experimental conditions. *Sociometry,* 1963, *26,* 422–435.

Orvis, B. R., Kelley, H. H., & Butler, D. Attributional conflict in young couples. In J. H. Harvey, W. J. Ickes, & R. E. Kidd (Eds.), *New directions in attribution research* (Vol. 1). Hillsdale, N.J.: Erlbaum, 1976.

Osgood, C. E., May, W. H., & Miron, M. S. *Cross-cultural universals of affective meaning.* Urbana: University of Illinois Press, 1975.

Paolino, T. J. Introduction: Some basic concepts of psychoanalytic psychotherapy. In T. J. Paolino & B. S. McCrady (Eds.), *Marriage and marital therapy: Psychoanalytic, behavioral, and systems theory perspectives.* New York: Brunner/Mazel, 1978.

Parad, L. G., & Parad, H. J. A study of crisis-oriented planned short-term treatment: Part II. *Social Casework,* 1968, *49,* 418–426.

Parke, R. D. Interactional designs. In R. B. Cairns (Ed.), *The analysis of social interactions.* Hillsdale, N.J.: Erlbaum, 1979. (a)

Parke, R. D. Perspectives on father–infant interaction. In J. D. Osofsky (Ed.), *Handbook of infant development.* New York: Wiley, 1979. (b)

Parsons, B. V., & Alexander, J. F. Short-term family intervention: A therapy outcome study. *Journal of Consulting and Clinical Psychology,* 1973, *41,* 195–201.

Parsons, T. *The social system.* New York: Free Press, 1951.

Parsons, T., & Bales, R. F. *Family, socialization and interaction process.* Glencoe, Ill.: Free Press, 1955.

Patterson, G. R. Changes in status of family members as controlling stimuli: A basis for describing treatment process. In L. Hamerlynck, L. C. Handy, & E. J. Mash (Eds.), *Behavior change: Methodology, concepts, and practice.* Champaign, Ill.: Research Press, 1973.

Patterson, G. R. A basis for identifying stimuli which control behavior in natural settings. *Child Development,* 1974, *45,* 900–911. (a)

Patterson, G. R. Interventions for boys with conduct problems, multiple settings, treatments and criteria. *Journal of Consulting and Clinical Psychology,* 1974, *42,* 471–481. (b)

Patterson, G. R. *Families: Applications of social learning to family life* (rev. ed.). Champaign, Ill.: Research Press, 1975.

Patterson, G. R. Some procedures for assessing changes in marital interaction patterns. *Oregon Research Institute Research Bulletin,* 1976, *16*(7). (a)

Patterson, G. R. The aggressive child: Victim and architect of a coercive system. In E. Mash, L. Hamerlynck, & L. Handy (Eds.), *Behavior modification and families.* New York: Brunner/Mazel, 1976. (b)

Patterson, G. R. A three-stage functional analysis for children's coercive behaviors: A tactic for developing a performance theory. In D. Baer, B. C. Etzel, & J. M. LeBlanc (Eds.), *New developments in behavioral research: Theory, methods, and applications. In honor of Sidney W. Bijou.* Hillsdale, N.J.: Erlbaum, 1977.

Patterson, G. R. A performance theory for coercive family interaction. In R. B. Cairns (Ed.), *The analysis of social interactions: Methods, issues, and illustrations.* Hillsdale, N.J.: Erlbaum, 1979.

Patterson, G. R. Mothers: The unacknowledged victims. *Monographs of the Society for Research in Child Development,* 1981, *45*(5, Serial No. 185).

Patterson, G. R., & Cobb, J. A. A dyadic analysis of "aggressive" behaviors. In J. P. Hill (Ed.), *Proceedings of the Fifth Annual Minnesota Symposia on Child Psychology* (Vol. 5). Minneapolis: University of Minnesota, 1971.

Patterson, G. R., & Gullion, M. E. *Living with children: New methods for parents and teachers* (rev. ed.). Champaign, Ill.: Research Press, 1971.

Patterson, G. R., & Hops, H. Coercion, A game for two: Intervention techniques for marital conflict. In R. Ulrich & P. Mountjoy (Eds.), *The experimental analysis of social behavior.* New York: Appleton-Century-Crofts, 1972.

Patterson, G. R., & Moore, D. R. Interactive patterns as units. In M. E. Lamb, S. J. Suomi, & G. R. Stevenson (Eds.), *The study of social interaction: Methodological issues.* Madison: University of Wisconsin Press, 1979.

Patterson, G. R., & Reid, J. B. Reciprocity and coercion: Two facets of social systems. In C. Neuringer & J. L. Michael (Eds.), *Behavior modification in clinical psychology.* New York: Appleton-Century-Crofts, 1970.

Patterson, G. R., Reid, J. B., Jones, R. R., & Conger, R. E. *A social learning approach to family intervention.* Vol. 1: *Families with aggressive children.* Eugene, Ore.: Castalia, 1975.

Patterson, G. R., & Whalen, K. Establishing causal status for controlling stimuli found in the natural environment. Unpublished manuscript, 1978.

Paul, G. L. Behavior modification research: Design and tactics. In C. M. Franks (Ed.), *Behavior therapy: Appraisal and status.* New York: McGraw-Hill, 1969.

Payne, E. E., & Mussen, P. H. Parent–child relations and father identification among adolescent boys. *Journal of Abnormal and Social Psychology,* 1956, *52,* 358–362.

Pendse, S. G. Category perception, language and brain hemispheres: An information transmission approach. *Behavioral Science,* 1978, *23,* 421–428.

Peplau, L. A. Power in dating relationships. In J. Freeman (Ed.), *Women: A feminist perspective* (2nd ed.). Palo Alto, Calif.: Mayfield, 1979.

Peplau, L. A., & Gordon, S. L. The intimate relationships of lesbians and gay men. In E. R. Allgeier & N. B. McCormick (Eds.), *The changing boundaries: Gender roles and sexual behavior.* Palo Alto, Calif.: Mayfield, 1983.

Peplau, L. A., & Gordon, S. L. Women and men in love: Sex differences in close heterosexual relationships. In V. E. O'Leary, R. K. Unger, & B. S. Wallston (Eds.), *Women, gender and social psychology.* Hillsdale, N.J.: Erlbaum, in press.

Peplau, L. A., & Perlman, D. (Eds.). *Loneliness: A sourcebook of current theory, research and therapy.* New York: Wiley-Interscience, 1982.

Peplau, L. A., & Rook, K. Dual-career relationships: The college couple perspective. Paper presented at the annual meeting of the Western Psychological Association, San Francisco, April 1978.

Perrucci, C. C., Potter, H. R., & Rhoads, D. L. Determinants of male family-role performance. *Psychology of Women Quarterly,* 1978, 3, 53–66.

Peterson, D. R. Assessing interpersonal relationships by means of interaction records. *Behavioral Assessment,* 1979, *1,* 221–236.

Phillips, S. U. Some sources of cultural variability in the regulation of talk. *Language in Society,* 1976, *5,* 81–95.

Piaget, J. *The language and thought of the child.* New York: Harcourt, Brace, 1926.

Pineo, P. C. Disenchantment in the later years of marriage. *Marriage and Family Living,* 1961, *23,* 3–11.

Pines, A., & Aronson, E. Antecedents, correlates, and consequences of sexual jealousy. *Journal of Personality,* 1982, in press.

Pleck, J. H. The male sex role: Definitions, problems and sources of change. *Journal of Social Issues,* 1976, *32*(3), 155–164.

Pleck, J. H. Changing patterns of work and family roles. Paper presented at the annual meeting of the American Psychological Association, Los Angeles, August 1981. (a)

Pleck, J. H. *The myth of masculinity.* Cambridge, Mass.: MIT Press, 1981. (b)

Pleck, J. H., & Rustad, M. *Husbands' and wives' time in family work and paid employment in the 1975–1976 Study of Time Use.* Wellesley, Mass.: Wellesley College Center for Research on Women, 1980.

Plutchik, R. *The emotions: Facts, theories, and a new model.* New York: Random House, 1962.

Plutchik, R. *Emotion: A psychoevolutionary synthesis.* New York: Harper & Row, 1980.

Pollner, M., & Wikler, L. "Cognitive enterprise" in einem Fall von Folié à Famille. In H.-G. Soeffner (Ed.), *Interpretative Verfahrèn in den Sozial- und Textwissenschaften.* Stuttgart: J. B. Metzlersche Verlagsbuchhandlung, 1979.

Poloma, M. M., & Garland, T. N. The myth of the egalitarian family: Family roles and the professionally employed wife. In A. Theodore (Ed.), *The professional woman.* Cambridge, Mass.: Schenkman, 1971.

Porter, L. W., Steers, R. M., Mowday, R. T., & Boulian, P. V. Organizational commitment, job satisfaction, and turnover among psychiatric technicians. *Journal of Applied Psychology,* 1974, *59,* 603–608.

Powers, W. T. *Behavior: The control of perception.* Chicago: Aldine, 1973.

Pribram, K. H. Emotion: Steps toward a neuropsychological theory. In D. C. Glass (Ed.), *Neurophysiology and emotion.* New York: Rockefeller University Press & Russell Sage Foundation, 1967.

Pruitt, D. G., & Carnevale, P. J. D. The development of integrative agreements in social conflict. In V. J. Derlega & J. Grzelak (Eds.), *Living with other people.* New York: Academic Press, 1980.

Pruitt, D. G., & Kimmel, M. J. Twenty years of experimental gaming: Critique, synthesis, and suggestions for the future. *Annual Review of Psychology,* 1977, *28,* 363–392.

Pruitt, D. G., & Lewis, S. A. The psychology of integrative bargaining. In D. Druckman (Ed.), *Negotiations: A social psychological perspective.* Beverly Hills, Calif.: Sage-Halstead, 1977.

Purdy, T. F. Perceptions of involvement in close relationships. Senior honors thesis, University of Massachusetts, Amherst, 1978.

Rainwater, L. *Family design.* Chicago: Aldine, 1965.

Rands, M. Social networks before and after marital separation. Unpublished doctoral dissertation, University of Massachusetts, Amherst, 1980.

Rands, M., & Levinger, G. Implicit theories of relationship: An intergenerational study. *Journal of Personality and Social Psychology,* 1979, *37,* 645–661.

Rands, M., Levinger, G., & Mellinger, G. Patterns of conflict resolution and marital satisfaction. *Journal of Family Issues,* 1981, *2,* 297–321.

Rapoport, A. *Fights, games, and debates.* Ann Arbor: University of Michigan Press, 1960.

Rapoport, R., & Rapoport, R. *Dual-career families re-examined.* New York: Harper Colophon Books, 1976.

Rapoport, R., Rapoport, R., & Thiessen, V. Couple symmetry and enjoyment. *Journal of Marriage and the Family,* 1974, *36,* 588–591.

Raush, H. L. Interaction sequences. *Journal of Personality and Social Psychology,* 1965, *2,* 487–499.

Raush, H. L., Barry, W. A., Hertel, R. K., Swain, M. A. *Communication, conflict and marriage.* San Francisco: Jossey-Bass, 1974.

Raven, B. H. Social influence and power. In I. D. Steiner & M. Fishbein (Eds.), *Current studies in social psychology*. New York: Holt, 1965.

Raven, B. H., Centers, R., & Rodrigues, A. The bases of conjugal power. In R. E. Cromwell & D. H. Olson (Eds.), *Power in families*. New York: Wiley, 1975.

Reagan, R. W. An interview with President Reagan. *First Monday*, 1981, *11*(6), 4.

Reid, J. B. (Ed.). *A social learning approach to family intervention*. Vol. 2. *Observation in home settings*. Eugene, Ore.: Castalia, 1978.

Reik, T. *A psychologist looks at love*. New York: Lancer, 1972.

Reiss, I. L. *Family systems in America* (3rd ed.). New York: Holt, Rinehart & Winston, 1980.

Ridley, C. A., & Avery, A. W. Social network influence on the dyadic relationship. In R. L. Burgess & T. L. Huston (Eds.), *Social exchange in developing relationships*. New York: Academic Press, 1979.

Ring, K., & Kelley, H. H. A comparison of augmentation and reduction as modes of influence. *Journal of Abnormal and Social Psychology*, 1963, *66*, 95–102.

Riskin, J., & Faunce, E. E. An evaluative review of family interaction research. *Family Process*, 1972, *11*, 365–455.

Robinson, J. P. *How Americans use time: A social–psychological analysis*. New York: Praeger, 1977.

Robinson, J. P., Yerby, J., Fieweger, M., & Somerick, N. Sex-role differences in time use. *Sex Roles*, 1977, *3*, 443–458.

Rodman, H. Marital power in France, Greece, Yugoslavia, and the United States: A cross-national discussion. *Journal of Marriage and the Family*, 1967, *29*, 320–324.

Rodman, H. Marital power and the theory of resources in cultural context. *Journal of Comparative Family Studies*, 1972, *3*, 50–69.

Rogers, C. R. *Counseling and psychotherapy*. Boston: Houghton Mifflin, 1942.

Rogers, C. R. *Client-centered therapy: Its current practice implications and theory*. Boston: Houghton Mifflin, 1951.

Rogers, C. R. The necessary and sufficient conditions of therapeutic personality change. *Journal of Consulting Psychology*, 1957, *21*, 95–103.

Rogers, C. R. A theory of therapy, personality, and interpersonal relationships, as developed in the client-centered framework. In S. Koch (Ed.), *Psychology: A study of a science* (Vol. 3). New York: Basic Books, 1959.

Rogers, C. R. *On becoming a person*. Boston: Houghton Mifflin, 1961.

Rogosa, D. A critique of cross-lagged correlation. *Psychological Bulletin*, 1980, *88*, 245–258.

Rollins, B. C., & Bahr, S. J. A theory of power relationships in marriage. *Journal of Marriage and the Family*, 1976, *38*, 619–627.

Rollins, B. C., & Cannon, K. L. Marital satisfaction over the life cycle: A re-evaluation. *Journal of Marriage and the Family*, 1974, *36*, 271–283.

Rook, K. S., & Peplau, L. A. Perspectives on helping the lonely. In L. A. Peplau & D. Perlman (Eds.), *Loneliness: A sourcebook of current theory, research, and therapy*. New York: Wiley-Interscience, 1982.

Roper Organization. *The Virginia Slims American opinion poll* (Vol. 3). New York: Roper, 1974.

Rosch, E., & Mervis, C. B. Family resemblances: Studies in the internal structure of categories. *Cognitive Psychology*, 1975, *7*, 573–605.

Rosenblatt, P. C. Needed research on commitment in marriage. In G. Levinger & H. L. Raush (Eds.), *Close relationships: Perspectives on the meaning of intimacy.* Amherst: University of Massachusetts Press, 1977.

Rosenblatt, P. C., & Cunningham, M. R. Television watching and family tensions. *Journal of Marriage and the Family,* 1976, *38,* 105–111.

Ross, H. S., & Goldman, B. D. Establishing new social relations in infancy. In T. Alloway, P. Pliner, & L. Krames (Eds.), *Attachment behavior* (Vol. 3). New York: Plenum, 1977.

Ross, H. S., & Sawhill, I. V. *Time of transition: The growth of families headed by women.* Washington, D.C.: Urban Institute, 1975.

Ross, M., & Sicoly, F. Egocentric biases in availability and attribution. *Journal of Personality and Social Psychology,* 1979, *37,* 322–336.

Rossi, A. A biosocial perspective on parenting. *Daedalus,* 1977, *106,* 1–31.

Ruano, B. J., Bruce, J. D., & McDermott, M. M. Pilgrim's Progress II: Recent trends and prospects in family research. *Journal of Marriage and the Family,* 1969, *31,* 688–698.

Rubin, L. *Worlds of pain: Life in the working class family.* New York: Basic Books, 1976.

Rubin, Z. Measurement of romantic love. *Journal of Personality and Social Psychology,* 1970, *16,* 265–273.

Rubin, Z. *Liking and loving: An invitation to social psychology.* New York: Holt, Rinehart & Winston, 1973.

Rubin, Z. *Children's friendships.* Cambridge, Mass.: Harvard University Press, 1980.

Rubin, Z., & Hartup, W. W. (Eds.). *Relationships: Their role in children's development.* New York: Cambridge University Press, in press.

Rubin, Z., Hill, C. T., Peplau, L. A., & Dunkel-Schetter, C. Self-disclosure in dating couples: Sex roles and the ethic of openness. *Journal of Marriage and the Family,* 1980, *42,* 305–318.

Rubin, Z., & Levinger, G. Theory and data badly mated: A critique of Murstein's SVR theory and Lewis's PDF model of mate selection. *Journal of Marriage and the Family,* 1974, *36,* 226–231.

Rubin, Z., & Mitchell, C. Couples research as couples counseling: Some unintended effects of studying close relationships. *American Psychologist,* 1976, *31,* 17–25.

Rubin, Z., Peplau, L. A., & Hill, C. T. Loving and leaving: Sex differences in romantic attachments. *Sex Roles,* 1981, *7,* 821–835.

Runyon, K. E. *Consumer behavior and the practice of marketing* (2nd ed.). Columbus, Ohio: Merrill, 1980.

Rusbult, C. E. Commitment and satisfaction in romantic associations: A test of the investment model. *Journal of Experimental Social Psychology,* 1980, *16,* 172–186.

Russell, B. *Power: A new social analysis.* New York: Norton, 1937.

Russell, J. A. Affective space is bipolar. *Journal of Personality and Social Psychology,* 1979, *37,* 345–356.

Ryan, M. P. *Womanhood in America: From Colonial times to the present* (2nd ed.). New York: Watts, 1979.

Sackett, G. P. The lag sequential analysis of contingency and cyclicity in behavioral interaction research. In J. Osofsky (Ed.), *Handbook of infant development.* New York: Wiley-Interscience, 1979.

Safilios-Rothschild, C. A comparison of power structure and marital satisfaction in urban Greek and French families. *Journal of Marriage and the Family*, 1967, *29*, 345–352.

Safilios-Rothschild, C. Family sociology or wives' family sociology: A cross-cultural examination of decision-making. *Journal of Marriage and the Family*, 1969, *31*, 290–301.

Safilios-Rothschild, C. The study of family power structure: A review 1960-1969. *Journal of Marriage and the Family*, 1970, *32*, 539–552.

Safilios-Rothschild, C. A macro- and micro-examination of family power and love: An exchange model. *Journal of Marriage and the Family*, 1976, *38*, 355–362.

Sager, C. *Marriage contracts and couple therapy: Hidden forces in intimate relationships.* New York: Brunner/Mazel, 1976.

Scanzoni, J. *Opportunity and the family.* New York: Free Press, 1970.

Scanzoni, J. A historical perspective on husband-wife bargaining power and marital dissolution. In G. Levinger & O. C. Moles (Eds.), *Divorce and separation: Context, causes, and consequences.* New York: Basic Books, 1979. (a)

Scanzoni, J. Social exchange and behavioral interdependence. In R. L. Burgess & T. L. Huston (Eds.), *Social exchange in developing relationships.* New York: Academic Press, 1979. (b)

Scanzoni, J. Social processes and power in families. In W. R. Burr, R. Hill, F. I. Nye, & I. L. Reiss (Eds.), *Contemporary theories about the family* (Vol. 1). New York: Free Press, 1979. (c)

Scanzoni, J., & Fox, G. L. Sex roles, family, and society: The seventies and beyond. *Journal of Marriage and the Family*, 1980, *42*, 743–756.

Scanzoni, L., & Scanzoni, J. *Men, women and change: A sociology of marriage and the family.* New York: McGraw-Hill, 1976.

Schachter, S. *The psychology of affiliation.* Stanford, Calif.: Stanford University Press, 1959.

Schachter, S. The interaction of cognitive and physiological determinants of emotional state. In L. Berkowitz (Ed.), *Advances in experimental social psychology* (Vol. 1). New York: Academic Press, 1964.

Schachter, S., & Singer, J. E. Cognitive, social and physiological determinants of emotional state. *Psychological Review*, 1962, *69*, 379–399.

Schaffer, H. R. Acquiring the concept of dialogue. In M. M. Bornstein & W. Kessen (Eds.), *Psychological development from infancy.* Hillsdale, N.J.: Erlbaum, 1979.

Schank, R., & Abelson, R. *Scripts, plans, goals and understanding.* Hillsdale, N.J.: Erlbaum, 1977.

Scheflen, A. E. *Communicational structure: Analysis of a psychotherapy transaction.* Bloomington: Indiana University Press, 1973.

Schlesinger, I. M. *Production and comprehension of utterances.* Hillsdale, N.J.: Erlbaum, 1977.

Schmidt, R. A. Anticipation and timing in human motor performance. *Psychological Bulletin*, 1968, *70*, 631–646.

Schwartz, G., & Merten, D. *Love and commitment.* Beverly Hills, Calif.: Sage, 1980.

Schweder, R. A., & D'Andrade, R. G. The systematic distortion hypothesis. In R. Schweder (Ed.), *Fallible judgment in behavioral research.* San Francisco: Jossey-Bass, 1980.

Sears, R. R. A theoretical framework for personality and social behavior. *American Psychologist,* 1951, *6,* 476–483.

Sears, R. R., Maccoby, E., & Levin, H. *Patterns of child-rearing.* Evanston, Ill.: Row, Peterson, 1957.

Selltiz, C., Wrightsman, L. S., & Cook, S. W. *Research methods in social relations* (3rd ed.). New York: Holt, Rinehart & Winston, 1976.

Selye, H. *Stress without distress.* New York: Lippincott, 1974.

Seyfried, B. A. Complementarity in interpersonal attraction. In S. Duck (Ed.), *Theory and practice in interpersonal attraction.* London: Academic Press, 1977.

Shibutani, T. *Society and personality: An interactionist approach to social psychology.* Englewood Cliffs, N.J.: Prentice-Hall, 1961.

Sidowski, J. B. Observational research: Some instrumented systems for scoring and storing behavioral data. *Behavior Research Methods and Instrumentation,* 1977, *9,* 403–404.

Simmel, G. *Conflict.* Glencoe, Ill.: Free Press, 1955.

Singer, M. T., & Wynne, L. C. Principles of scoring communication defects and deviances in parents of schizophrenics: Rorschach and TAT scoring manuals. *Psychiatry,* 1966, *29,* 260–288.

Skinner, B. F. *Science and human behavior.* New York: Macmillan, 1953.

Skolnick, A. *The intimate environment: Exploring marriage and the family* (2nd ed.). Boston: Little, Brown, 1978.

Skolnick, A. Married lives: Longitudinal perspectives on marriage. In D. Eichorn, J. Clausen, N. Haan, M. Honzik, & P. Mussen (Eds.), *Present and past in middle life.* New York: Academic Press, 1981.

Sluzki, C. E. Marital therapy from a systems theory perspective. In T. J. Paolino and B. S. McCrady (Eds.), *Marriage and marital therapy: Psychoanalytic, behavioral, and systems theory perspectives.* New York: Brunner/Mazel, 1978.

Smelser, W. T. Personality influences in social situations. *Journal of Abnormal and Social Psychology,* 1961, *62,* 535–542.

Snyder, C. R., & Fromkin, H. L. *Uniqueness: The human pursuit of difference.* New York: Plenum, 1980.

Snyder, J. J. Reinforcement analysis of interaction in problem and nonproblem families. *Journal of Abnormal Psychology,* 1977, *86,* 528–535.

Snyder, M., & Swann, W. B. Behavioral confirmation in social interaction: From social perception to social reality. *Journal of Experimental Social Psychology,* 1978, *14,* 148–162.

Snyder, M., Tanke, E. D., & Berscheid, E. Social perception and interpersonal behavior: On the self-fulfilling nature of social stereotypes. *Journal of Personality and Social Psychology,* 1977, *35,* 656–666.

Solomon, R. L., & Corbit, J. D. An opponent-process theory of motivation: I. Temporal dynamics of affect. *Psychological Review,* 1974, *81,* 119–145.

Soper, P. H., & L'Abate, L. Paradox as a therapeutic technique: A review. *International Journal of Family Counseling,* 1977, *5,* 10–21.

Soskin, W. F., & John, V. P. The study of spontaneous talk. In R. G. Barker (Ed.), *The stream of behavior.* New York: Appleton-Century-Crofts, 1963.

Spanier, G. B. Measuring dyadic adjustment: New scales for assessing the quality of marriage and similar dyads. *Journal of Marriage and the Family,* 1976, *38,* 15–28.

Spanier, G. B. The changing profile of the American family. *Journal of Family Practice,* 1981, *13*, 61–69.

Spanier, G. B., & Casto, R. F. Adjustment to separation and divorce: A qualitative analysis. In G. Levinger & O. C. Moles (Eds.), *Divorce and separation.* New York: Basic Books, 1979.

Spanier, G. B., Lewis, R. A., & Cole, C. L. Marital adjustment over the family life cycle: The issue of curvilinearity. *Journal of Marriage and the Family,* 1975, *37*, 263–275.

Speck, R. V., & Attneave, C. L. *Family networks.* New York: Pantheon Books, 1973.

Speck, R. V. Psychotherapy of the social network of a schizophrenic family. *Family Process,* 1967, 6, 208–214.

Speer, D. C. Marital dysfunctionality and two-person non-zero sum game behavior: Cumulative monadic measures. *Journal of Personality and Social Psychology,* 1972, *21*, 18–24.

Spence, J. T., & Helmreich, R. L. *Masculinity and femininity: Their psychological dimensions, correlates and antecedents.* Austin: University of Texas Press, 1978.

Stagner, R. (Ed.). *The dimensions of human conflict.* Detroit: Wayne State University Press, 1967.

Stambul, H. B. Stages of courtships: The development of premarital relationships. Unpublished doctoral dissertation, University of California, Los Angeles, 1975.

Stapleton, J., & Bright, R. *Equal marriage.* New York: Harper & Row, 1976.

Steck, L., Levitan, D., McLane, D., & Kelley, H. H. Care, need, and conceptions of love. *Journal of Personality and Social Psychology,* 1982, *43*, 481–491.

Steers, R. M. Antecedents and outcomes of organizational commitment. *Administrative Science Quarterly,* 1977, *22*, 45–56.

Steiner, I. D. Whatever happened to the group in social psychology? *Journal of Experimental Social Psychology,* 1974, *10*, 94–108.

Steinglass, P. The home observation assessment method (HOAM): Real-time naturalistic observation of families in their homes. *Family Process,* 1979, *18*, 337–354.

Steinmetz, S. K. Violence between family members. *Marriage and Family Review,* 1978, *1*(3), 1–16.

Steinmetz, S. K. Disciplinary techniques and their relationship to aggressiveness, dependency, and conscience. In W. R. Burr, R. Hill, F. I. Nye, & I. L. Reiss (Eds.), *Contemporary theories about the family: Research-based theories.* New York: Free Press, 1979.

Stone, L. J., Murphy, L. B., & Smith, H. T. *Competent infant: Research and commentary.* New York: Basic Books, 1973.

Straus, M. A., Gelles, R. J., & Steinmetz, S. K. *Behind closed doors: Violence in the American family.* Garden City, N.Y.: Anchor Books, 1980.

Straus, M. A., & Hotaling, G. T. (Eds.). *The social causes of husband–wife violence.* Minneapolis: University of Minnesota Press, 1980.

Straus, M. A., & Tallman, I. SIMFAM: A technique for observational measurement and experimental study of families. In J. Aldous, T. B. Condon, R. Hill, M. Straus, & I. Tallman (Eds.), *Family problem solving.* Hinsdale, Ill.: Dryden Press, 1971.

Strodtbeck, F. L. Husband–wife interaction over revealed differences. *American Sociological Review,* 1951, *16*, 468–473.

Strodtbeck, F. L. The family as a three-person group. *American Sociological Review,* 1954, *19,* 23–29.

Stroebe, W. Ähnlichkeit und Komplementarität der Bedürfnisse als Kriterium der Partnerwahl: Zwei spezialle Hypothesen. In G. Mikula & W. Stroebe (Eds.), *Sympathie, Freundschaft, und Ehe.* Bern, Switzerland: Huber, 1977.

Strongman, K. T. *The psychology of emotion* (2nd ed.). New York: Wiley, 1978.

Stryker, S. *Symbolic interactionism.* Menlo Park, Calif.: Benjamin/Cummings, 1980.

Stuart, R. B. An operant interpersonal program for couples. In D. H. L. Olson (Ed.), *Treating relationships.* Lake Mills, Iowa: Graphic, 1976.

Stuart, R. B. *Helping couples change: A social learning approach to marital therapy.* New York: Guilford, 1980.

Sudman, S., Bradburn, N. M., Blair, E., & Stocking, C. Modest expectations: The effects of interviewer's prior expectations on responses. *Sociological Methods and Research,* 1977, 6, 171–182.

Sullaway, M., & Christensen, A. Couples and families as participant observers of their interaction. In J. P. Vincent (Ed.), *Advances in family intervention, assessment, and theory* (Vol. 3). Greenwich, Conn.: JAI Press, in press.

Swensen, C. H., Jr. The behavior of love. In H. A. Otto (Ed.), Love today: A new exploration. New York: Association Press, 1972.

Swensen, C. H., Jr. *Introduction to interpersonal relations.* Glenview, Ill.: Scott, Foresman, 1973.

Symons, D. *The evolution of human sexuality.* New York: Oxford University Press, 1979.

Szinovacz, M. E. Relationship among marital power measures: A critical review and an empirical test. *Journal of Comparative Family Studies,* 1981, *12,* 151–169.

Szinovacz, M. E. Marital dynamics and family power. In M. Sussman & S. Steinmetz (Eds.), *Handbook of marriage and the family* (2nd ed.). In preparation, 1982.

Tarpy, R. M. The nervous system and emotion. In D. K. Candland, J. P. Fell, E. Keen, A. I. Leshner, R. Plutchik, & R. M. Tarpy, *Emotion.* Monterey, Calif.: Brooks/Cole, 1977.

Taylor, S. E., & Fiske, S. T. Point of view and perceptions of causality. *Journal of Personality and Social Psychology,* 1975, *32,* 439–445.

Tedeschi, J. T., Schlenker, B. R., & Lindskold, S. The exercise of power and influence: The source of influence. In J. T. Tedeschi (Ed.), *The social influence processes.* Chicago: Aldine, 1972.

Terman, L. M., & Wallin, P. The validity of marriage prediction and marital adjustment tests. *American Sociological Review,* 1949, *14,* 503–504.

Tesser, A., & Paulhus, D. Toward a causal model of love. *Journal of Personality and Social Psychology,* 1976, *34,* 1095–1105.

Theodorson, G. A., & Theodorson, A. G. *A modern dictionary of sociology.* New York: Crowell, 1969.

Thibaut, J. W., & Kelley, H. H. *The social psychology of groups.* New York: Wiley, 1959.

Thomas, E. A. C., & Malone, T. W. On the dynamics of two-person interactions. *Psychological Review,* 1979, *36,* 331–360.

Thomas, E. A. C., & Martin, J. A. Analysis of parent–infant interaction. *Psychological Review,* 1976, *83,* 141–156.

Thompson, S. C., & Kelley, H. H. Judgments of responsibility for activities in close relationships. *Journal of Personality and Social Psychology,* 1981, *41,* 469–477.

Thorndike, E. L., & Lorge, I. *The teacher's word book of 30,000 words.* New York: Columbia University Teachers College, 1944.

Toomey, D. M. Conjugal roles and social networks in an urban working class sample. *Human Relations,* 1971, *24,* 417–431.

Traupmann, J., & Hatfield, E. Love and its effect on mental and physical health. In R. W. Fogel, E. Hatfield, S. B. Kiesler, & E. Shanas (Eds.), *Aging: Stability and change in the family.* New York: Academic Press, 1981.

Triandis, H. C. *The analysis of subjective culture.* New York: Wiley-Interscience, 1972.

Triandis, H. C., Vassiliou, V., & Nassiakou, M. Three cross-cultural studies of subjective culture. *Journal of Personality and Social Psychology Monograph Supplement,* 1968, *8,* 1–42.

Tronick, E., Als, H., & Adamson, L. Structure of early face-to-face communicative interactions. In A. Bullowa (Ed.), *Before speech: The beginnings of human communications.* Cambridge: Cambridge University Press, 1979.

Tronick, E., Als, H., Adamson, L., Wise, S., & Brazelton, T. B. The infant's response to entrapment between contradictory messages in face-to-face interaction. *Journal of the American Academy of Child Psychiatry,* 1978, Spring, 1–13.

Truax, C. B., & Carkhuff, R. R. *Toward effective counseling and psychotherapy: Training and practice.* Chicago: Aldine, 1967.

Turk, J. L. Power as the achievement of ends: A problematic approach in family and small group research. *Family Process,* 1974, *13,* 39–52.

Turk, J. L., & Bell, N. W. Measuring power in families. *Journal of Marriage and the Family,* 1972, *34,* 215–222.

Turner, J. H. *The structure of sociological theory* (rev. ed.). Homewood, Ill.: Dorsey, 1978.

Turner, R. H. The normative coherence of folk concepts. *Research Studies of the State College of Washington,* 1957, *25,* 127–136.

Turner, R. H. Role-taking: Process versus conformity. In A. R. Rose (Ed.), *Human behavior and social processes: An interactionist approach.* Boston: Houghton Mifflin, 1962.

Turner, R. H. Role: Sociological aspects. In D. Sills (Ed.), *The international encyclopedia of the social sciences* (Vol. 13). New York: Macmillan & Free Press, 1968, pp. 552–557.

Turner, R. H. *Family interaction.* New York: Wiley, 1970.

Turner, R. H. Strategy for developing an integrated role theory. *Humboldt Journal of Social Relations,* 1979/80, *7*(1), 123–139.

Turner, R. H. Role taking as process. Unpublished manuscript, University of California, Los Angeles, undated.

Tyler, L. E. *The work of the counselor* (3rd ed.). New York: Appleton-Century-Crofts, 1969.

United Nations Demographic Yearbook 1981. New York: United Nations Department of International Economic and Social Affairs Statistical Office, 1982.

U.S. Bureau of the Census. *Statistical Abstract of the United States: 1979.* Washington, D.C.: U.S. Government Printing Office, 1979.

Vincent, J. P., Friedman, L. C., Nugent, J., & Messerly, L. Demand characteristics in observations of marital interaction. *Journal of Consulting and Clinical Psychology,* 1979, *47,* 557–566.

Vogel, E. F., & Bell, N. W. The emotionally disturbed child as the family scapegoat. In N. W. Bell & E. F. Vogel (Eds.), *A modern introduction to the family.* Glencoe, Ill.: Free Press, 1960.

Wachtel, P. L. Interaction cycles, unconscious processes, and the person–situation issue. In D. Magnusson & N. S. Endler (Eds.), *Personality at the crossroads: Current issues in interactional psychology.* Hillsdale, N.J.: Erlbaum, 1977. (a)

Wachtel, P. L. Psychoanalysis and behavior therapy: Toward an integration. New York: Basic Books, 1977. (b)

Walker, K. Time spent by husbands in household work. *Family Economics Review,* 1970, *3,* 8–11.

Walker, K., & Woods, M. *Time use: A measure of household production of family goods and services.* Washington: Home Economics Association, 1976.

Waller, W. *The family: A dynamic interpretation.* New York: Cordon, 1938.

Waller, W. *The old love and the new: Divorce and readjustment.* Carbondale, Ill.: Southern Illinois University Press, 1967. (Originally published, 1930.)

Waller, W., & Hill, R. *The family: A dynamic interpretation* (rev. ed.). New York: Dryden Press, 1951.

Walster, E., Aronson, V., Abrahams, D., & Rottman, L. Importance of physical attractiveness in dating behavior. *Journal of Personality and Social Psychology,* 1966, *4,* 508–516.

Walster, E., & Walster, G. W. Equity and social justice. *Journal of Social Issues,* 1975, *31*(3), 21–43.

Walster, E., Walster, G. W., & Berscheid, E. *Equity: Theory and research.* Boston: Allyn & Bacon, 1978.

Walster, E., Walster, G., Piliavin, J., & Schmidt, L. "Playing hard-to-get": Understanding an elusive phenomenon. *Journal of Personality and Social Psychology,* 1973, *26,* 113–121.

Walster, E., Walster, G. W., & Traupmann, J. Equity and premarital sex. *Journal of Personality and Social Psychology,* 1978, *36,* 82–92.

Watson, J., & Potter, R. J. An analytic unit for the study of interaction. *Human Relations,* 1962, *15,* 245–263.

Watson, J. B. *Behaviorism.* Chicago: University of Chicago Press, 1924.

Watzlawick, P. The psychotherapeutic technique of "reframing." In J. L. Claghorn (Ed.), *Successful psychotherapy.* New York: Brunner/Mazel, 1976.

Watzlawick, P., Beavin, J. H., & Jackson, D. D. *The pragmatics of human communication.* New York: Norton, 1967.

Waxler, N. E. Parent and child effects on cognitive performance: An experimental approach to the etiological and responsive theories of schizophrenia. *Family Process,* 1974, *13,* 1–22.

Weakland, J. Communication theory and clinical change. In P. Guerin (Ed.), *Family Therapy.* New York: Gardner, 1976.

Weary-Bradley, G. Self-serving biases in the attribution process: A re-examination of the fact or fiction question. *Journal of Personality and Social Psychology,* 1978, *36,* 56–71.

Weber, M. *The theory of social and economic organization.* New York: Free Press, 1947.

Weick, K. E. Group processes, family processes, and problem solving. In J. Aldous, T. Condon, R. Hill, M. Straus, & I. Tallman (Eds.), *Family problem solving.* Hinsdale, Ill.: Dryden, 1971.

Weiner, B., & Peter, N. A cognitive–developmental analysis of achievement and moral judgments. *Developmental Psychology,* 1973, *9,* 290–309.

Weinstein, E. A., & Deutschberger, P. Some dimensions of altercasting. *Sociometry,* 1963, *26,* 455–466.

Weiss, R. L. Strategic behavioral marital therapy: Toward a model for assessment and intervention. In J. P. Vincent (Ed.), *Advances in family intervention, assessment, and theory* (Vol. 1.). Greenwich, Conn.: JAI Press, 1980.

Weiss, R. L., Hops, H., & Patterson, G. R. A framework for conceptualizing marital conflict: A technology for altering it, some data for evaluating it. In L. A. Hamerlynck, L. C. Handy, & E. J. Mash (Eds.), *Behavior change: Methodology, concepts, and practice.* Champaign, Ill.: Research Press, 1973.

Weiss, R. L., & Margolin, G. Marital conflict and accord. In A. R. Ciminero, K. S. Calhoun, & H. E. Adams (Eds.), *Handbook for Behavioral Assessment.* New York: Wiley, 1977.

Weiss, R. S. *Marital separation.* New York: Basic Books, 1975.

Weiss, R. S. The emotional impact of marital separation. *Journal of Social Issues,* 1976, *32,* 135–146.

Welch, G. J., & Granvold, D. K. Seminars for separated/divorced: An educational approach to postdivorce adjustment. *Journal of Sex and Marital Therapy,* 1977, *3,* 31–39.

Weller, R. H. The employment of wives, dominance, and fertility. *Journal of Marriage and the Family,* 1968, *30,* 437–443.

Wells, R. A., & Dezen, A. E. The results of family therapy revisited: The non-behavioral methods. *Family Process,* 1978, *17,* 251–274.

White House Conference on Families. *Listening to America's families: Action for the 80's.* Washington, D.C.: U.S. Government Printing Office, 1980.

Wilkes, R. E. Husband–wife influence in purchase decisions: A confirmation and extension. *Journal of Marketing Research,* 1975, *7,* 224–227.

Williams Moore, J., & McClintock, E. Coding manual for mother–child interactions. Unpublished manuscript, Family Care Center, Santa Barbara, 1979.

Wills, T. A., Weiss, R. L., & Patterson, G. R. A behavioral analysis of the determinants of marital satisfaction. *Journal of Consulting and Clinical Psychology,* 1974, *42,* 802–811.

Wilson, E. D. *Sociobiology: The new synthesis.* Cambridge, Mass.: Belknap Press of Harvard University Press, 1975.

Wimberly, H. Conjugal-role organization and social networks in Japan and England. *Journal of Marriage and the Family,* 1973, *35,* 125–130.

Winch, R. F. *Mate selection: A study of complementary needs.* New York: Harper & Bros., 1958.

Winch, R. F., Ktsanes, T., & Ktsanes, V. The theory of complementary needs in mate selection: An analytic and descriptive study. *American Sociological Review,* 1954, *19,* 241–249.

Wish, M., Deutsch, M., & Kaplan, S. J. Perceived dimensions of interpersonal relations. *Journal of Personality and Social Psychology,* 1976, *33,* 409–420.

Wolfe, D. M. Power and authority in the family. In D. Cartwright (Ed.), *Studies in social power.* Ann Arbor, Mich.: Institute for Social Research, 1959.

Yankelovich, D. New rules in American life: Searching for self-fulfillment in a world turned upside down. *Psychology Today,* April 1981, pp. 36–91.

Yarrow, M. R., & Waxler, C. Z. Observing interaction: A confrontation with methodology. In Robert D. Cairns (Ed.), *The analysis of social interactions: Methods, issues and illustrations.* Hillsdale, N.J.: Erlbaum, 1979.

Yogev, S. Do professional women have egalitarian marital relationships? *Journal of Marriage and the Family,* 1981, *43,* 865–871.

Young, M., & Willmott, P. *The symmetrical family.* New York: Random House, 1973.

Young, P. T. Studies in affective psychology. *American Journal of Psychology,* 1927, *38,* 157–193.

Zegiob, L. E., Arnold, S., & Forehand, R. An examination of observer effects in parent–child interactions. *Child Development,* 1975, *46,* 509–512.

Zegiob, L. E., & Forehand, R. Parent–child interactions: Observer effects and social class differences. *Behavior Therapy,* 1978, *9,* 118–123.

Zelditch, M. Role differentiation in the nuclear family. In T. Parsons & R. F. Bales (Eds.), *Family socialization and interaction process.* Glencoe, Ill.: Free Press, 1955.

Zinner, J., & Shapiro, R. Projective identifications as a mode of perception and behavior in families of adolescents. *International Journal of Psychoanalysis,* 1972, *53,* 523–530.

Index of Names

Index of Topics